Advances in VIRUS RESEARCH

VOLUME **83**

Bacteriophages, Part B

Advances in VIRUS RESEARCH

VOLUME **83**

Bacteriophages, Part B

Edited by

MAŁGORZATA ŁOBOCKA
Autonomous Department of Microbial Biology
Faculty of Agriculture and Biology
Warsaw University of Life Sciences
Nowoursynowska 159, Warsaw, Poland
Department of Microbial Biochemistry
Institute of Biochemistry and Biophysics
Polish Academy of Sciences
Pawińskiego 5A, Warsaw, Poland

WACŁAW T. SZYBALSKI
Professor Emeritus of Oncology
McArdle Laboratory for Cancer Research
University of Wisconsin Medical School
Madison, Wisconsin, USA

ELSEVIER

AMSTERDAM • BOSTON • HEIDELBERG • LONDON
NEW YORK • OXFORD • PARIS • SAN DIEGO
SAN FRANCISCO • SINGAPORE • SYDNEY • TOKYO
Academic Press is an imprint of Elsevier

Academic Press is an imprint of Elsevier

32 Jamestown Road, London, NW1 7BY, UK
The Boulevard, Langford Lane, Kidlington, Oxford, OX51GB, UK
Radarweg 29, PO Box 211, 1000 AE Amsterdam, The Netherlands
225 Wyman Street, Waltham, MA 02451, USA
525 B Street, Suite 1900, San Diego, CA 92101-4495, USA

First edition 2012

Library of Congress Cataloging-in-Publication Data

A catalog record for this book is available from the Library of Congress

British Library Cataloguing-in-Publication Data

A catalogue record for this book is available from the British Library

ISBN: 978-0-12-394438-2
ISSN: 0065-3527

For information on all Academic Press publications
visit our website at store.elsevier.com

Printed and bound by CPI Group (UK) Ltd, Croydon, CR0 4YY

Transferred to digital print 2012

CONTENTS

Section 2: Interactions and Applications of Phage Proteins and Protein Complexes

CONTRIBUTORS

Sonja Blasche
Deutsches Krebsforschungszentrum, Heidelberg, Germany
Email: s.blasche@dkfz-heidelberg.de

Jan Borysowski
Department of Clinical Immunology, Transplantation Institute, Medical University of Warsaw, Warsaw, Poland
Email: jborysowski@interia.pl

Sherwood Casjens
Division of Microbiology and Immunology, Pathology Department, University of Utah School of Medicine, Salt Lake City, Utah
Email: sherwood.casjens@path.utah.edu

Nina Chanishvili
Laboratory for Genetics of Microorganisms and Bacteriophages, Eliava Institute of Bacteriophage, Microbiology and Virology, Tbilisi, Georgia, USA
Email: n_chanish.ibmv@caucasus.net

Krystyna Dąbrowska
Bacteriophage Laboratory, Ludwik Hirszfeld Institute of Immunology and Experimental Therapy, Polish Academy of Sciences, Wrocław, Poland
Email: dabrok@iitd.pan.wroc.pl

Kamil Dąbrowski
Department of Microbial Biochemistry, Institute of Biochemistry and Biophysics, Polish Academy of Sciences; and Autonomous Department of Microbial Biology, Faculty of Agriculture and Biology, Warsaw University of Life Sciences, Warsaw, Poland
Email: kamild@ibb.waw.pl

Gian Marco De Donatis
Nanobiotechnology Center, Department of Pharmaceutical Sciences, and Markey Cancer Center, University of Kentucky, Lexington, Kentucky, USA
Email: gianmarco.dedonatis@gmail.com

Laurent Debarbieux
Institut Pasteur, Molecular Biology of the Gene in Extremophiles Unit, Department of Microbiology, Paris, France
Email: laurent.debarbieux@pasteur.fr

Terje Dokland
Department of Microbiology, University of Alabama at Birmingham, Birmingham, Alabama, USA
Email: dokland@uab.edu

Shengli Dong
Department of Biochemistry and Molecular Genetics, University of Alabama at Birmingham, Birmingham, Alabama, USA
Email: sdong@lsuhsc.edu

David M. Donovan
Animal Biosciences and Biotechnology Laboratory, ANRI, ARS, USDA, Beltsville, Maryland, USA
Email: david.donovan@ars.usda.gov

Wojciech Fortuna
Bacteriophage Laboratory and Phage Therapy Unit, Ludwik Hirszfeld Institute of Immunology and Experimental Therapy, Polish Academy of Sciences, Wrocław, Poland
Email: fortuna@iitd.pan.wroc.pl

Aleksandra Głowacka
Department of Microbial Biochemistry, Institute of Biochemistry and Biophysics, Polish Academy of Sciences, and Autonomous Department of Microbial Biology, Faculty of Agriculture and Biology, Warsaw University of Life Sciences, Warsaw, Poland
Email: aleksandra_sggw@o2.pl

Andrzej Górski
Bacteriophage Laboratory and Phage Therapy Unit, Ludwik Hirszfeld Institute of Immunology and Experimental Therapy, Polish Academy of Sciences, Wrocław; and Department of Clinical Immunology, Transplantation Institute, Medical University of Warsaw, Warsaw, Poland
Email: agorski@ikp.pl

Jan Gawor
Laboratory of DNA Sequencing and Oligonucleotide Synthesis, Institute of Biochemistry and Biophysics, Polish Academy of Sciences, Warsaw, Poland
Email: gaworj@ibb.waw.pl

Aneta Gołaś
Department of Clinical Immunology, Transplantation Institute, Medical University of Warsaw, Warsaw, Poland
Email: aneg@poczta.fm

Agnieszka Gozdek
Department of Microbial Biochemistry, Institute of Biochemistry and Biophysics, Polish Academy of Sciences, Warsaw, Poland
Email: agnes@ibb.waw.pl

Peixuan Guo
Nanobiotechnology Center, Department of Pharmaceutical Sciences, and Markey Cancer Center, University of Kentucky, Lexington, Kentucky, USA
Email: peixuan.guo@uky.edu

Roman Häuser
Institute of Toxicology and Genetics, Karlsruhe Institute of Technology, Karlsruhe, Germany; and Deutsches Krebsforschungszentrum, Heidelberg, Germany
Email: r.haeuser@Dkfz-Heidelberg.de

Elisabeth Haggård-Ljungquist
Department of Genetics, Microbiology and Toxicology, Stockholm University, Stockholm, Sweden
Email: Elisabeth.Haggard@gmt.su.se

Monika S. Hejnowicz
Department of Microbial Biochemistry, Institute of Biochemistry and Biophysics, Polish Academy of Sciences, Warsaw, Poland
Email: hejmonika@ibb.waw.pl

Ewa Jończyk
Bacteriophage Laboratory, Ludwik Hirszfeld Institute of Immunology and Experimental Therapy, Polish Academy of Sciences, Wrocław, Poland
Email: ewa.jonczyk@iitd.pan.wroc.pl

Marlena Kłak
Bacteriophage Laboratory, Ludwik Hirszfeld Institute of Immunology and Experimental Therapy, Polish Academy of Sciences, Wrocław, Poland
Email: mklak@iitd.pan.wroc.pl

Danuta Kłosowska
Department of Clinical Immunology, Transplantation Institute, Medical University of Warsaw, Warsaw, Poland
Email: rovanemi1@wp.pl

Ewelina Kaniuga
Department of Clinical Immunology, Transplantation Institute, Medical University of Warsaw, Warsaw, Poland
Email: ewelkaniuga@wp.pl

Jochen Klumpp
Institute of Food, Nutrition and Health, ETH Zurich, Zurich, Switzerland
Email: jochen.klumpp@hest.ethz.ch

Grażyna Korczak-Kowalska
Department of Clinical Immunology, Transplantation Institute, Medical University of Warsaw; and Department of Immunology, Warsaw University, Warsaw, Poland
Email: korczak-kowalska@wp.pl

Jarosław Kosakowski
Department of Microbial Biochemistry, Institute of Biochemistry and Biophysics, Polish Academy of Sciences, Warsaw, Poland
Email: jarkos@ibb.waw.pl

Helena Kosowska
Laboratory of DNA Sequencing and Oligonucleotide Synthesis, Institute of Biochemistry and Biophysics, Polish Academy of Sciences, Warsaw, Poland
Email: hela@ibb.waw.pl

Magdalena Kwiatek
Military Institute of Hygiene and Epidemiology, Puławy, Poland
Email: magda.kwiatek09@gmail.com

Sławomir Letkiewicz
Phage Therapy Unit, Ludwik Hirszfeld Institute of Immunology and Experimental Therapy, Polish Academy of Sciences, Wrocław; and Katowice School of Economics, Katowice, Poland
Email: letkiewicz1@o2.pl

Małgorzata Łobocka
Department of Microbial Biochemistry, Institute of Biochemistry and Biophysics, Polish Academy of Sciences; and Autonomous Department of Microbial Biology, Faculty of Agriculture and Biology, Warsaw University of Life Sciences, Warsaw, Poland
Email: lobocka@ibb.waw.pl; malgorzata_lobocka@sggw.pl

Marzena Łusiak-Szelachowska
Bacteriophage Laboratory, Ludwik Hirszfeld Institute of Immunology and Experimental Therapy, Polish Academy of Sciences, Wrocław, Poland
Email: marzena@iitd.pan.wroc.pl

Ryszard Międzybrodzki
Bacteriophage Laboratory and Phage Therapy Unit, Ludwik Hirszfeld Institute of Immunology and Experimental Therapy, Polish Academy of Sciences, Wrocław, Poland
Email: mbrodzki@iitd.pan.wroc.pl

Ian Molineux
Molecular Genetics and Microbiology, Institute for Cell and Molecular Biology, University of Texas–Austin, Austin, Texas, USA
Email: molineux@mail.utexas.edu

Kenan C. Murphy
Department of Microbiology and Physiological Systems, University of Massachusetts Medical School, Worcester, Massachusetts, USA
Email: kenan.murphy@umassmed.edu

Daniel C. Nelson
Institute for Bioscience and Biotechnology Research, University of Maryland, Rockville; and Department of Veterinary Medicine, Virginia-Maryland Regional College of Veterinary Medicine, College Park, Maryland, USA
Email: nelsond@umd.edu

Monika Ohams
Department of Clinical Immunology, Transplantation Institute, Medical University of Warsaw, Warsaw, Poland
Email: mkniotek@wp.pl

Natasza Olszowska-Zaremba
Department of Clinical Immunology, Transplantation Institute, Medical University of Warsaw, Warsaw, Poland
Email: nataszaolszowska@wp.pl

Sylwia Parasion
Military Institute of Hygiene and Epidemiology, Puławy, Poland
Email: sparasion@gmail.com

Zdzisław Pawełczyk
Phage Therapy Unit, Ludwik Hirszfeld Institute of Immunology and Experimental Therapy, Polish Academy of Sciences, Wrocław, Poland
Email: pawelczykzd@wp.pl

David G. Pritchard
Department of Biochemistry and Molecular Genetics, University of Alabama at Birmingham, Birmingham, Alabama, USA
Email: pritchard_d@bellsouth.net

Sylwia Purchla
Department of Clinical Immunology, Transplantation Institute, Medical University of Warsaw, Warsaw, Poland
Email: sylwia.purchla@gmail.com

Lorena Rodriguez-Rubio
Instituto de Productos Lácteos de Asturias (IPLA-CSIC), Villaviciosa, Asturias, Spain
Email: lorenarguez@ipla.csic.es

Paweł Rogóż
Phage Therapy Unit, Ludwik Hirszfeld Institute of Immunology and Experimental Therapy, Polish Academy of Sciences; and Department of Orthopedics, Medical Academy in Wrocław, Medical University, Wrocław, Poland
Email: pawel.rogoz1@wp.pl

Margarita Salas
Centro de Biología Molecular "Severo Ochoa" (CSIC-UAM), Cantoblanco, Madrid, Spain
Email: msalas@cbm.uam.es

Emilie Saussereau
Institut Pasteur, Molecular Biology of the Gene in Extremophiles Unit, Department of Microbiology, Paris, France
Email: emilie.saussereau@pasteur.fr

Mathias Schmelcher
Animal Biosciences and Biotechnology Laboratory, ANRI, ARS, USDA, Beltsville, Maryland, USA
Email: mathias.schmelcher@hest.ethz.ch

Chad Schwartz
Nanobiotechnology Center, Department of Pharmaceutical Sciences, and Markey Cancer Center, University of Kentucky, Lexington, Kentucky, USA
Email: chad.schwartz@uky.edu

Krzysztof Szufnarowski
Phage Therapy Unit, Ludwik Hirszfeld Institute of Immunology and Experimental Therapy, Polish Academy of Sciences Wrocław; and Regional County Hospital and Research Center, Wrocław, Poland
Email: szufnarowski@wssk.wroc.pl

Peter Uetz
Center for the Study of Biological Complexity, Virginia Commonwealth University, Richmond, Virginia, USA
Email: peter@uetz.us

Magdalena Ulatowska
Department of Microbial Biochemistry, Institute of Biochemistry and Biophysics, Polish Academy of Sciences, Warsaw; and Laboratory of Bacteriophages, Hirszfeld Institute of Immunology and Experimental Therapy, Polish Academy of Sciences, Wrocław, Poland
Email: lmi1@poczta.onet.pl

Albrecht von Brunn
Max-von-Pettenkofer-Institut, Lehrstuhl Virologie, Ludwig-Maximilians-Universität, München, Germany
Email: vonbrunn@mvp.uni-muenchen.de

Beata Weber-Dąbrowska
Bacteriophage Laboratory and Phage Therapy Unit, Ludwik Hirszfeld Institute of Immunology and Experimental Therapy, Polish Academy of Sciences, Wrocław, Poland
Email: weber@iitd.pan.wroc.pl

Piotr Wierzbicki
Department of Clinical Immunology, Transplantation Institute, Medical University of Warsaw, Warsaw, Poland
Email: selket-anat@wp.pl

Magdalena Witkowska
Department of Microbial Biochemistry, Institute of Biochemistry and Biophysics, Polish Academy of Sciences, Warsaw, Poland
Email: magda_witkowska@ibb.waw.pl

Elżbieta Wojtasik
Department of Pharmacognosy and Molecular Basis of Phytotherapy, Faculty of Pharmacy, Medical University of Warsaw, Warsaw, Poland
Email: wojtasik.elzbieta@gmail.com

Hui Zhang
Nanobiotechnology Center, Department of Pharmaceutical Sciences, and Markey Cancer Center, University of Kentucky, Lexington, Kentucky, USA
Email: hui.zhang@uky.edu

PREFACE

When writing about phages, one must remember to consider their practical applications, as even results of early phage studies opened a plethora of phage usage possibilities. Phage therapy, phage typing, phage display of antibodies, transductional mapping of bacterial chromosomes and phage-mediated bacterial mutagenesis are only a few of the examples. As we can now see, the application of phages in the past was significantly curtailed by limited possibilities of exploring the phage world, as well as by technical difficulties in getting an insight into phage structure and biology. These limitations have been decreasing with the advent of new research possibilities initiated, e.g., by the development of omics and mass spectrometry, as well as different kinds of fluorescent, atomic force and electron microscopy technologies.

Ironically, despite the above mentioned possibilities, the most complex obligatorily virulent phages revealed their genomic structure only within the recent decade, after the publication of the first draft of the human genome sequence. Numerous genes of these phages encoding proteins toxic to bacteria prevented the determination of their complete genomic sequences until the development of cloning-independent sequencing technologies and their common use. Thus, only since the beginning of the XXI century, which has been called by some "the age of phage", have we been able to discover the full richness of the phage world, its remarkable diversity and the huge impact of phages on bacterial populations. We can use modern research tools to get a deeper insight into those phage properties and structures that were detected or predicted by researchers a long time ago. Finally, having nowadays the possibility to study phages at a molecular level, we can widen our understanding of their nature, interactions with each other and with organisms of the surrounding world to the extent that enables a more and more deliberate use of phage-encoded proteins and complex macromolecular structures as well as complete phages, single or in mixtures. This volume concerns the above-mentioned issues.

An early interest in phages was partially motivated by their potential bactericidal activity and hence, the possibility of their therapeutic use. Although the therapeutic application of phages has been nearly abandoned for years, especially in Western countries, it once again became the focus of attention with the emergence and spread of antibiotic-resistant bacterial pathogens. Thus, the first part of this volume is devoted entirely to phage therapy. In the context of a possible return to these clinical approaches, results of early studies on phage therapy that have remained unknown for most of the readers for decades, but are outlined here in a chapter by Nina Chanishvili, should become of special value. Although their original interpretation by former researchers may seem currently too speculative and not always justified, and although proper controls were often lacking, now, at the time of costly and restricted studies on humans, they are a valuable source of information. The molecular basis of some of the methods that were used at that time and may seem "suspicious" at first glance can be understood today as a reader may learn from Chapter 5. Also, the efficacy of phage therapy that was often estimated in the past just on the basis of a "patient's feeling of getting better", can be verified now in well designed scientific or medical experiments, as one can learn from chapters by Andrzej Górski and co-authors, by Ryszard Międzybrodzki and co-authors, and by Emilie Saussereau and Laurent Debarbieux. It is becoming clear that one can consider various phages as specific medicines. To be effective, they have to be applied in a certain way and at a certain time, similarly to traditional medicines that work more or less efficiently depending on the illness, patient and certain other factors. The molecular structure of many of these potential phage medicines is no longer a mystery, which opens a way for future genetic engineering to obtain phages of better therapeutic potential and wider specificities. Phage engineering is relatively fast and easy in the present era of synthetic biology.

Interactions and selected applications of phage proteins and macromolecular complexes are the subject of the second part of this volume. A chapter by Roman Häuser and co-authors reveals how phage protein interactomes can be reconstructed based on the results of high-throughput studies and presents a detailed overview of protein interactomes of several model phages. Hui Zhang and co-authors describe the structure, action and potential application of the most powerful biological nanomotors – complexes of phi29 phage portal proteins with other macromolecular components of phage DNA packaging machinery. A chapter by Kenan Murphy is focused on phage recombinases, which became a primary tool in various types of DNA manipulations *in vivo* and *in vitro*, even supplanting restriction enzymes in cloning protocols. Other phage enzymes, endolysins, are the subject of a chapter by Daniel Nelson and co-workers. As bacterial peptidoglycan-hydrolysing proteins, they are an attractive alternative to analogously acting antibiotics, and, in addition to phages that encode them, may become useful in the fight with antibiotic-resistant bacterial pathogens.

Małgorzata Łobocka
Warsaw, Poland
e-mail: malgorzata_lobocka@sggw.pl, lobocka@ibb.waw.pl
Wacław T. Szybalski
Wisconsin, USA
e-mail: szybalski@oncology.wisc.edu

Section 1
Phages from Medical Perspective

CHAPTER 1

Phage Therapy—History from Twort and d'Herelle Through Soviet Experience to Current Approaches

Nina Chanishvili[1]

Contents

Laboratory for Genetics of Microorganisms and Bacteriophages, Eliava Institute of Bacteriophage, Microbiology & Virology, Tbilisi, Georgia, USA
[1] Corresponding author, E-mail: n_chanish.ibmv@caucasus.net

Advances in Virus Research, Volume 83
ISSN 0065-3527, DOI: 10.1016/B978-0-12-394438-2.00001-3

Abstract Felix d'Herelle proposed the use of bacteriophages for the therapy of human and animal bacterial infections at the beginning of the 20th century. This approach, however, was not widely accepted in the West. After the emergence of antibiotics in 1940s, phage research was diverted to a more fundamental level. At the same time, phage therapy was widely practiced in the Soviet Union due to collaboration of Felix d'Herelle with his Georgian colleagues. The majority of the articles dedicated to this subject are from the 1930s and 1940s. The old Soviet literature indicates that phage therapy was used extensively to treat a wide range of bacterial infections in the areas of dermatology (Beridze, 1938), ophthalmology (Rodigina, 1938), urology (Tsulukidze, 1938), stomatology (Ruchko and Tretyak, 1936), pediatrics (Alexandrova *et al.*, 1935; Lurie, 1938), otolaryngology (Ermolieva, 1939), and surgery (Tsulukidze, 1940, 1941). These articles were published in Russian and thus were not readily available to Western scientists. The Western skepticism toward phage therapy itself was again followed by renewed interest and reappraisal, mainly due to the emergence of drug-resistant bacteria. Often the experiments described in the old Soviet articles were not designed properly: the use of placebos and the coding of preparations were absent from most of the studies, number of patients in the experimental and control groups was unequal or missing, sometimes no control groups were used at all, or patients treated previously unsuccessfully with antibiotics were employed as an experimental group and as control. The results obtained and the efficiency of phage prophylaxis were estimated by comparing with results obtained in previous years. In most publications, phage titers and descriptions of methods used for evaluation of the results are not specified. Nevertheless, past experience indicates some effectiveness of phage therapy and prophylaxis. Therefore, these clinical results should not be neglected when designing any future studies.

I. GENERAL BACKGROUND

A. Discovery of bacteriophages

During World War I, rather sensational news was spread—the viruses "eaters of micro-organisms" have been discovered by Felix d'Herelle, who developed a phage preparation to treat World War I soldiers affected by dysentery (Hausler, 2008). Doctors all over the world were excited by this news, especially because reports published in medical journals that viruses–bacteriophages are harmless for humans and animals and can be applied successfully as therapeutic means (Alessandrini and Doria, 1924; Bruynoghe and Maisin, 1921; Compton, 1929; Costa Cruz, 1924; Parfitt, 2005; Rice, 1930; Spence and McKinley, 1924; Sulakvelidze *et al.*, 2001).

Actually, the history of bacteriophages started long before this event. In 1896, the English bacteriologist Ernest Hanbery Hankin tried to enumerate the number of *Vibrio cholera* in a cubic millimeter of water in the river Ganges. Samples were taken in places where the river enters and exits the city Agra (Adhya and Merril, 2006; Kazhal and Iftimovich, 1968). Surprisingly, Hankin (1896) counted 100,000 infective units in the cubic millimeter of water in the entrance to the city, while their number decreased to only 90 where the river exited the city. This strange Ganghes self-purification could not find its explanation by that time and was referred to as "Hankin's phenomenon."

In 1906–1909, Felix d'Herelle traveled across Mexico, where he met an entomologist who attracted his attention to the violent epidemics among locust. d'Herelle isolated bacterium *Coccobaccilus aeridiorum* (presently known as *Enterobacter aerogenes*) (Summers, 1999, 2001). Motivated by an ambitious desire to use these bacteria deliberately against plagues of locust, he conducted a number of experiments in Tunis and Guyana. During these experiments, d'Herelle observed the appearance of transparent holes ("*taches vierges*") in the bacterial lawn. He presumed that they were caused by a virus that was mistakenly supposed to be a cause of the locust infection. d'Herrele isolated this virus and used it against locust but without any success. He started to pursue an idea to use the host bacteria against locust again and did not take into consideration the strange "*taches vierges*" (clear plaques).

In 1915, the well-known British journal *The Lancet* published an article written by Frederick Twort about "the transmissible bacterial lyses" (Twort, 1915), in which Twort described his observation of "the eaten edges of the colonies of *Staphylococcus*." He managed to filter the appropriate cultures of *Staphylococcus* and spotted the filtrate on the lawn of different *Staphylococcus* strains. Thus, he received a clear zone of lysis again and again. However, Twort could not explain the observed event and provided only its description. This was the very first publication on bacteriophages. d'Herelle read this article, which reminded him of his own observations in Mexico and Tunis (Summers, 2001). He suspected that the filtered agent was a bacterial virus—an invisible invader that destroys bacteria.

The discovery or rediscovery of bacteriophages by Felix d'Herelle after Frederick Twort is frequently associated with an outbreak of severe hemorrhagic dysentery among French troops stationed at Maisons-Laffitte (on the outskirts of Paris) in July–August 1915 (Sulakvelidze *et al.*, 2001; Summers, 2001). Several soldiers were hospitalized, and d'Herelle was assigned to conduct an investigation of the outbreak. During these studies, he made bacterium-free filtrates of the patients' fecal samples and mixed and incubated them with *Shigella* strains isolated from the patients. A portion of the mixtures was inoculated into experimental animals (as part of d'Herelle's studies on developing a vaccine against bacterial dysentery), and a portion was spread on agar medium in order to observe the growth of the bacteria. On these agar cultures d'Herelle observed again the appearance of small, clear areas, which he initially called *taches*, then *taches vierges*, and, later, *plaques* (Summers, 1999). D'Herelle's findings were presented during the September 1917 meeting of the Academy of Sciences and were subsequently published in the meeting's proceedings (d'Herelle, 1917). In contrast to Hankin and Twort, d'Herelle had little doubt about the nature of the phenomenon, and he proposed that it was caused by a virus capable of parasitizing bacteria. The name "bacteriophage" (from "bacteria" and Greek φᾰγεῖν *phagein* "to eat") was also proposed by d'Herelle (Summers, 1999).

The discovery of bacteriophages was inevitable. Similar phenomena have been observed in remote regions of the world by different scientists. At the end of the 19th century, N.F. Gamaleya (later a Honorary Member of the Academy of the USSR) published an article in the *Russian Archives of the Pathological and Clinical Medicine* (Gamaleya, 1898). In this article he described the lysis of *Bacillus antracis* in distilled water, after which the water obtained an ability to lyse other strains of *B. antracis*. In 1917, a young Georgian scientist George Eliava had observed mysterious disappearance of *V. cholera* cells (d'Herrelle, 1935; Georgadze and Makashvili, 1979).

The greatest merit of Felix d'Herelle is that he advanced the idea of using bacteriophages for the treatment of human and animal bacterial diseases. For this idea he deserved the Noble Prize, to which he was nominated eight times, every year since 1925, although he was never awarded one [cited by Hausler (2008) according to Nobel Archives].

B. Early clinical trials

Not long after his discovery, d'Herelle used phages to treat dysentery, which was probably the first attempt to use bacteriophages therapeutically. The studies were conducted at the Hôpital des Enfants-Malades in Paris in 1919 under the clinical supervision of Professor Victor-Henri Hutinel, the hospital's chief of pediatrics (Summers, 1999). The phage preparation was ingested by d'Herelle, Hutinel, and several hospital interns in order to confirm its safety before administering it the next day to a 12-year-old boy with severe dysentery. The patient's symptoms ceased after a single administration of d'Herelle's antidysentery phages, and the boy fully recovered within a few days. The efficacy of the phage preparation was "confirmed" shortly afterward when three additional patients having bacterial dysentery and treated with one dose of the preparation started to recover within 24 hr of treatment. However, results of these studies were not published immediately and, therefore, the first reported application of phages to treat infectious diseases of humans came later from Richard Bruynoghe and Joseph Maisin (1921), who used bacteriophages to treat staphylococcal skin disease. The bacteriophages were injected into and around surgically opened lesions, and the authors reported regression of the infections within 24 to 48 hr. Several similarly promising studies followed (Rice, 1930; Schelss, 1932; Stout, 1933). Encouraged by these early results, d'Herelle and others continued studies of the therapeutic use of phages (e.g., d'Herelle used various phage preparations to treat thousands of people having cholera and/or bubonic plague in India (Kazhal and Iftimovich, 1968; Summers, 1999; 2001).

In 1916–1930, d'Herelle and collaborators undertook numerous expeditions to China, Laos, India, Vietnam, and Africa to combat epidemics caused by cholera and plague with bacteriophages against these pathogens. According to Romanian medical historians Kazhal and Iftimovich (1968), the first attempts to use cholera bacteriophages for treatment and prophylaxis were performed by Felix d'Herelle and George Eliava in 1931. According to these authors, the Institute of Vaccine and Sera in Tbilisi produced the first commercial anticholera phage preparation, which was reported to be used successfully for the control of epidemics threatening the southeast territories of the USSR (Kazhal and Iftimovich, 1968). According to the estimations published at that time, due to the application of bacteriophages it became possible to reduce the mortality of cholera in India to 10% (Kazhal and Iftimovich, 1968). This fact is described by d'Herelle's himself in the book "Bacteriophage and the Phenomenon of Recovery" published in the Russian language in 1935 in Tbilisi, Georgia.[1] According to d'Herelle (1935), cholera epidemics occurred in Punjab region in 1927. Patients were treated orally with 2 ml of cholera phages diluted in 20 ml of water but the titer of phage in these preparations is unknown. If the patient vomited, a repeated dose of phages (5 ml diluted in 100 ml of water) was administered slowly with a teaspoon. The control group of patients treated themselves using a folk medicine (plant extracts). Of 14,450 people who lived in nine villages of the Punjab region, only 73 have been treated with the phages. d'Herelle explained it by mentioning that people in India opposed any new medical measures and rarely permitted him and his colleagues to use phages for treatment. Thus, only desperately ill patients were subjected to phage therapy. Altogether 118 persons were included in the control group in which 74 lethal outcomes (62.7%) were registered, whereas in the experimental group the mortality rate was almost one-tenth, with 5 cases out of 73 (6.8%) (d'Herelle, 1935).

[1] After the execution of George Eliava, the book was impounded and access to it was limited to "for professional use only."

C. Early commercial production of bacteriophages

In his book, d'Herelle mentioned the establishment of two industrial centers for the production of bacteriophages against cholera in 1931 in India (d'Herelle, 1935; Kazhal and Iftimovich, 1968). D'Herelle's commercial laboratory in Paris produced at least five phage preparations against various bacterial infections. The preparations were called *Bacté-coli-phage*, *Bacté-rhino-phage*, *Bacté-intesti-phage*, *Bacté-pyo-phage*, and *Bacté-staphy-phage* and were marketed by a company that later became the large French company L'Oréal (Sulakvelidze *et al.*, 2001; Summers, 1999, 2001).

The Oswaldo Cruz Institute in Rio de Janeiro, Brazil, started production of the antidysentery bacteriophages in 1924 to combat dysentery in Latin American countries (d'Herelle, 1935; Dublanchet and Bourne, 2007). Within a year the institute produced 10,000 vials of phages, which were sent to hospitals around Brazil (Hausler, 2008). Therapeutic phages were also produced in the United States. In the 1940s, the Eli Lilly Company (Indianapolis, IN) produced seven phage products for human use, including preparations targeted against staphylococci, streptococci, *Escherichia coli*, and other bacterial pathogens (Sulakvelidze *et al.*, 2001). These preparations consisted of phage-lysed, bacteriologically sterile broth cultures of the targeted bacteria (e.g., *Colo-lysate, Ento-lysate, Neiso-lysate*, and *Staphylo-lysate*) or the same preparations in a water-soluble jelly base (e.g., *Colo-jel, Ento-jel*, and *Staphylo-jel*). They were used to treat various infections, including abscesses, suppurating wounds, vaginitis, acute and chronic infections of the upper respiratory tract, and mastoid infections. However, the efficacy of phage preparations was controversial (Eaton and Bayne-Jones, 1934; Krueger and Scribner, 1941). This could be caused by the absence of viable phages, low phage titer, or narrow strain range of phage in these preparations as, for example, discovered in the case of some commercial antistaphylococcal phage preparations (Rakieten, 1932). As a result, with the advent of antibiotics, commercial production of therapeutic phages ceased in most Western countries.

Despite apparently promising results of phage therapy in certain cases, it had been abandoned in Western countries for more than 80 years, but not in the former Soviet Union where d'Herelle, together with his close friend and associate George Eliava, founded the Bacteriophage Institute in Tbilisi (Georgia).

D. George Eliava and the Eliava Institute of Bacteriophage

In the history of medicine little is known about George Eliava, who was a colorful central figure in phage history. Without the support of Eliava provided to Felix d'Herelle, much of the knowledge on phage therapy would not be achieved. Because of his progressive thinking, tireless activities, and close collaboration with many foreign scientists, including d'Herelle, George Eliava became a victim of Stalin's regime in 1937. He was pronounced to be a "people's enemy" and was executed. All photographs and documents belonging to George Eliava were destroyed by the KGB. His memory was restored only after the reassessment of the outcomes of the Red Terror and Stalin's regime by Gorbachev in 1989 and rehabilitation of its victims (Smith, 1996).

George Eliava was born on January 13, 1892, in the village of Sachkhere, West Georgia. In 1909 he entered the Novorosyisk University in Odessa (Georgadze and Makashvili, 1979). He wished to study literature. During his first years at the university he joined the student revolutionary movement and was expelled from the university with no rights to enter any other university in Russia. However, his rich and powerful relatives managed to send him to Geneva (Switzerland) where in 1912–1914 Eliava took a course in bacteriology. During the summer of 1914, Eliava came back to Georgia for vacations, but after the start of the World War I he could not return to the university in Geneva for his studies. Due to the efforts of his relatives he received permission to continue his studies at the Faculty of Medicine of the Moscow University in Russia, which he finished in 1916. In the same year he was appointed as head of the Caucasian front-line bacteriology laboratory in Trabzon. Beginning in 1917, he took lead of the Tbilisi Bacteriology Laboratory, which belonged to the Caucasian Cities' Union. In 1919–1921, 1925–1927, and 1931, Eliava worked at the Pasteur Institute in Paris, together with famous bacteriologists: Émile Roux, Charles Nicolle (Nobel Prize winner in Medicine in 1903), Albert Calmette, and Gaston Ramon, among others. In the beginning of the 1920's he met Felix d'Herelle there. In 1923, Eliava initiated foundation of the Institute of Bacteriology (the present Eliava Institute of Bacteriophage, Microbiology & Virology; EIBMV). He became its first director. Simultaneously with his administrative activities in 1927–1937, Eliava was teaching at Tbilisi State University and leading the department of hygiene and later the department of microbiology. In 1934, he initiated establishment of the Anti Plague Station in Tbilisi (the present National Center of Disease Control)(Georgadze and Makashvili, 1979). Although Eliava did not publish many articles, all of them touch significant topics, sounding rather contemporarily even now (d'Herelle and Eliava, 1921a,b; Eliava, 1930; Eliava and Legraoux, 1921; Eliava and Pozersky, 1921a,b; Eliava and Suarez 1927a,b; Nattan-Larrier *et al.*, 1931). Together with d'Herelle, Eliava discovered bacteriophage lysins (d'Herelle and Eliava, 1921a) and, in collaboration with E. Pozersky (1921a), found that the substance quinine, which was known already in ancient China as a treatment against malaria, specifically affected bacteriophages as well. Eliava was one of the first who drew out an assumption that bacteriophages may change the nature of the host bacteria (Eliava and

Pozersky,1921b).[2] He studied the immune response to phage therapy already in 1921 (d'Herelle and Eliava, 1921b), adsorption of bacteriophages on leukocytes (Eliava, 1930), and permeability through placenta (Nattan-Larrier *et al.*, 1931).

In November 1921, Eliava returned to Georgia with scientific equipment worth approximately 100,000 FF, which was a gift of the Pasteur Institute to Georgian colleagues (Georgadze, 1974).

George Eliava met Felix d'Herelle in Paris during one of his early stays at the Pasteur Institute and was fascinated by d'Herelle's ideas of using bacteriophages for therapy. Eliava invited d'Herelle to Georgia where they spent altogether 18 months in 1933 and 1934 collaborating with other Georgian colleagues (Georgadze, 1974). D'Herelle intended to move to Tbilisi permanently (a cottage built for his use still stands on the institute's grounds); however, his intentions could not be realized. In 1937, Eliava, together with his wife Amelia Vol-Levitskaya (Polish opera singer), was arrested and executed as a "people's enemy" for being in intellectual opposition with Laurenti Beria, the chief of the secret police to Joseph Stalin. According to another version, Beria considered Eliava as his competitor with him for a woman, Tinatin Jikia (librarian at the Bacteriophage Institute).

Frustrated and disillusioned, d'Herelle never returned to Georgia. Nonetheless, the institute survived and later became one of the largest facilities in the world engaged in the development of therapeutic phage preparations. During the best times the scientific staff of the institute enumerated about 100–120 people, including technicians, while the industrial part employed approximately 500–600 people, including specialists and support personnel. At that time the institute produced phage preparations (often several tons a day) against a dozen bacterial pathogens, including staphylococci, *Pseudomonas*, *Proteus*, and many enteric pathogens (Georgadze, 1974). Most of the Soviet studies reviewed in this article involved phages developed and produced at the EIBMV (Chanishvili, 2009).

II. MASS APPLICATION OF PHAGES

A. Wound treatment

From a review of historic literature, it is apparent that phage therapy trials were active in the 1930s and 1940s throughout Georgia, Russia, Ukraine, Belarus, and Azerbaijan in the Soviet Union. Observations of cases

[2] This is known now as lysogenic conversion—a change in the properties of the bacterium as a result of phage infection and lysogeny.

associated with road accidents and septic infection carried out at the Ostroumovkaya hospital in Moscow by Kokin and colleagues (cited by Krestnikova, 1947) led to the development of methods and instructions for intramuscular and even intravenous use of phages, which was crucial in cases of generalized infections. Results of these observations were reported at conferences held in March, June, and December of 1940. These methods and instructions were approved by the Soviet Supreme Red Army Military-Sanitarian Office and were applied to the treatment of soldiers in the Red Army during World War II and continued after it.

The application of phage therapy to surgical and wound treatment began during the Finnish Campaign in 1939–1940, with the first review of this work published by Kokin (1941, 1946). Kokin described the application of mixtures of bacteriophages infecting anaerobes, *Staphylococcus* and *Streptococcus*, and produced by the EIBMV, Tbilisi, Georgia, for the treatment of gas gangrene. The mixture was applied to 767 infected soldiers with a lethal outcome (death) observed in 18.8% of cases compared with 42.2% in the control group of soldiers treated with other methods. Using the same mixture of phages, other authors observed a 19.2% lethal outcome in a group of soldiers compared with 54.2% in a group treated with other medications (Lvov and Pasternak, 1947—cited by Krestovnikova, 1947). In addition to its therapeutic use, this phage preparation was used by the mobile sanitary brigades as an emergency treatment for wounds (prophylaxis of gas gangrene). Krestovnikova (1947) summarizes the observations of three mobile sanitary brigades carried out over periods of 2–6 weeks following evacuation to front-line hospitals.

The first brigade treated 2500 soldiers with phages. According to reports from that time, only 35 soldiers (1.4%) in this group revealed symptoms of gas gangrene, whereas in the control group of 7918 wounded soldiers, 342 (4.3%) were infected. The second brigade applied phage therapy to 941 soldiers, of whom only 14 (1.4%) suffered from gas gangrene, in contrast to 6.8% of soldiers in the control group treated by other methods. The third brigade treated 2584 soldiers, 18 (0.7%) of whom developed symptoms of gas gangrene, whereas 2.3% in the control group developed these symptoms. Comparison of data described by these three independent brigades showed an average 30% decrease in the number of gas gangrene as a consequence of the prophylactic treatment of wounds through application of the phage mixtures (Kokin, 1946; Krestnikova, 1947).

One of the pioneers in the application of phages in surgery was A.P. Tsulukidze, a professor of medicine, who began using such preparations in 1931 for the treatment of various diseases. In 1938 a prominent Soviet surgeon, Burdenko, recommended the use of phages against purulent infections. However, the same Burdenko, on Stalin's orders, falsified the results of Soviet studies on Katyń murders of Polish officers, committed by Soviet NKWD in 1941, and thus his scientific credibility may remain, at least in part, in question (Cienciala *et al.*, 2007).

An experimental series of phage preparations containing components against bacteria related to *Staphylococcus, Streptococcus, Escherichia coli,* and *Proteus* species was tested at the end of the 1930s in the surgical and gynecology clinics in Moscow (cited by Krestnikova, 1947).

According to Tsulukidze (1940), the wounds were analyzed bacteriologically by blood analysis carried out before the initiation of phage therapy and during surgical manipulations (bandages, puncture, application of phages, etc.). Prior to obtaining the results of bacteriological analysis, a Pio-bacteriophage preparation or a mixture of Strepto- and Staphylo-bacteriophages was applied topically or directly to the accessible part of the wound. The condition of the wound was described thoroughly, and temperature, pulse, patient's rate of breathing, and so on were all recorded. These examinations were also performed after each phage application. The majority of patients arriving from the front line with wounds were bedridden and in serious condition: 38.3% had soft tissue injuries and 61.7% had bone injuries. In the majority of cases, phages were administered during the first 6 days after initial infection.

Initially the application of phage therapy was used only for the most severe cases where a lethal outcome was expected. Later a wider group of patients was involved in the study. In the majority of cases, bacteriological analysis indicated the presence of mixed bacterial infections (Tsulukidze, 1940). Only in rare cases were patients arriving directly from the battlefield characterized with a monoinfection. Subcutaneous injection of phages was performed three to four times every second day to avoid the development of antiphage antibodies. Additionally, phages were sprayed onto the top of the wound each time bandages were changed.

All patients with injuries of the soft tissues (38.3%) underwent "ordinary therapy," which in this instance implied treatment with chloramines, rivanol, and Vishnevsky ointment.[3] These patients often had major tissue damage with penetrating or perforating wounds. The wounds were characterized by the accumulation of pus, infections, and surrounding inflammation, sometimes with necrotic foci. A number of cases with an abscess/phlegmon around a bullet or mine fragment underwent cuts performed during first aid in a field hospital prior to the start of phage therapy. Many of these patients suffered from serious intoxication,

[3] Vishnevky ointment was developed by a prominent Russian surgeon, Alexander Vasilyevich Vishnevsky. It consists of birch tar (3 g), xeroform (3 g), and castor oil (94 g). The ointment has both predrying and antiseptic effects and facilitates tissue regeneration. It is used for the treatment of wounds, burns, bed sores, chilblain, fistulas, skin ulcers, psoriasis, trophic sores, and capillary diseases of lower extremities such as trombophlebitis, obliterating endarteritis.

fever, and gangrenous inflammation and required the removal of wood or bullet particles. Treatment of such patients was performed by removal of tampons (used to fill the space of missing tissue), purification of wounds by treatment with iodine and alcohol, and then washing with 2% of a sodium chloride solution followed by spraying with phages. Simultaneously, 5–10 ml of phages (titer unknown) was injected remote from the wound into the stomach wall, shoulder, or hip. The wound was bandaged with gauze soaked in phages. Tampons and drainages were not applied. According to reports from that time, no cases treated with this method required additional surgical intervention. After the first one to two phage applications the body temperature used to be normalized. Only three to four such treatments were normally required to achieve a complete cure and blood test results were also improved. Because the recovery from traumatic injuries and numerous lesions required an extended period, wounds were stitched after phage therapy on the 6th–8th day of treatment so that further infection was unlikely. In general, treatment with phage therapy took a number of days, whereas "ordinary" or standard therapy took several weeks.

Patients with bone injuries or open fractures arrived following first aid using splints. The wounds were purified, tampons with chloramines, rivanol and Vishnevsky ointment were removed, and wounds were treated with 2% sodium chloride solution and then sprayed with phages. Simultaneously, phages were injected intramuscularly or subcutaneously at a site remote from the wound. Following this a blind plaster cast was applied. A second approach involved phage therapy, stitching of the wound, and then the application of plaster. A third approach involved application of a plaster cast directly after a single treatment with phages. According to direct reports from that time, treatment with phages resulted in a faster release of pain, improvement in patients' general conditions, and clear symptoms of wound healing after 2 or 3 days. Tsulukidze and colleagues reported that they were able to avoid moisturizing of the plaster, necessity of its frequent change, and the development of foul odor (usually associated with secondary infections) by using phage therapy prior to plastering. The plaster could remain unchanged for up to 60 days, allowing for the earlier start of physical therapy (Tsulukidze, 1940, 1941). They also claimed that in the cases of severe hips, shins, forearms, and shoulders injuries, which normally require amputation, no amputation was necessary if wounds were left for 10–30 days in blind plasters. Moreover, phage therapy allowed the application of stitches and/or plaster casts as early as the 10th–11th day after the start of treatment.

The war and the need for therapeutic preparations inspired Soviet doctors to perform new trials with phages and to develop novel methods for phage administration. This period was one of the most fruitful in the development of phage therapy in the former Soviet Union (Kokin, 1941, 1946; Krestnikova, 1947). A book written by Alexander Tsulukidze "Experience of the Use of Bacteriophages in Conditions of War Trauma" (1941), which summarizes the results obtained after Finnish Campaign, is especially interesting for military surgeons.

Tsulukidze (1941) described 20 hospital cases where bacteriophages against anaerobic bacteria were used in combination with Strepto- and Staphylo-bacteriophages. Seventeen out of 20 wounded soldiers received a mixture of phages directly on the battlefield, while 3 were treated with phages on arrival at the hospital. All 20 patients were in a severe condition when they arrived at the hospital, and bacteriological studies showed that all were infected with *Clostridium perfringens*. In each case the wounds were serious, including 19 cases of injuries of the lower extremities. According to Tsulukidze (1941) there were significant differences in the development of infection between soldiers treated with phages soon after injury and those that received phages later. In soldiers treated earlier, the wounds were clear of infection sooner and granulation appeared rapidly, temperature was normalized in a shorter period of time, and foul odors did not develop or were insignificant. Tsulukidze (1941) reported that soldiers who had received phages on the battlefield did not develop any complications. However, 2 patients who were not treated with phages prior to arrival at the hospital developed generalized sepsis and died. In these severe cases, other therapies were applied in addition to phage therapy, such as ultraviolet irradiation, alcohol, oxygen treatment of wounds, and blood transfusion, but they did not help.

On the basis of aforementioned observations, Tsulukidze *et al.* (1941) assumed that infection with *C. perfringens* was usually accompanied with streptococcal and staphylococcal infections, as the latter produced conditions favorable for the growth of *Clostridia* species. Use of a phage mixture targeting all three bacteria was therefore considered beneficial. Tsulukidze (1941) proposed that a strategy for the treatment of anaerobic infections should be based on the combined use of phage therapy and antigangrenous serum. He presumed that phages would lyse the bacteria causing the infection, while the serum would neutralize the toxins.

B. Persistence of phages in animals and humans

In order to work out the proper strategy for phage therapy and prophylaxis it was necessary to answer the question: How long will the various forms of phages persist *in vivo*?

Kaplan (1941) studied the persistence of dysenteric phages in 178 people. Each was given 10 ml of phage suspension orally three times at

5-day intervals. Prior to prophylactic "phaging"[4] these people were studied for phage carriage. The dysenteric phages in a very low titer[5] were found in the stools of only two patients. According to Kaplan (1941), phages given prophylactically were released with the stool samples for a maximum of 10 days, with titers decreasing gradually and always lower than the initial titer of orally received phages. No phages were found in the urine even 48 hr after phage administration (Kaplan, 1941).

Karpov and Yavorskya (1949) performed a number of animal studies. Based on the results obtained, they concluded that parenteral administration of phages was followed by its rapid spread in the organism. Depending on the dose of phages administered, it was found in blood for 2 to 9 days, as was estimated by a standard plaque assay with a specific bacterial host. The phages emerged in the spleen between the 2nd and 10th day, in the liver between the 3 rd and 8th day, and typically remained in the intestines from 4 to 10 days. In a number of cases the phages were identified in the intestines within 24–48 hr in a titer 10^2–10^3 (by Gratia, 1939). Later Karpov (1950) used a murine model including 134 animals (67 in the experimental group and 67 in the control group) to demonstrate that mice fed with phages against typhoid and/or paratyphoid bacteria and then injected with appropriate bacteria intravenously or parenterally (5×10^7 CFU/ml) were prevented from the development of infection, opposite to mice of the control group. In addition, Karpov noted an activation of the body's immune response (phagocytosis[6]) in the experimental group of mice. The phages were given to mice in a dose of 1 ml (titer unknown) on a piece of white bread soaked in soda solution. Bacterial injections were carried out between the 1st and 11th days after phage administration (Karpov, 1950)

To demonstrate the survival of phages in the human body, Karpov and Yavorskaya (1949) undertook human studies. They focused their interests on oral administration of phages and the doses administered. The study was carried out on 112 patients 7–19 years of age. Seventy-three people received 10 ml of phage suspension, and 39 people received 30 ml of similar suspension orally (titer 10^{-7} by Appelmans, 1921). During the following 5–7 days, the stool samples of phage-treated people were examined for the presence of phages. For this purpose, a small portion of the stool sample was inoculated into a meat–peptone medium and the mixture was incubated at 37 °C for 18–20 hr. After incubation, chloroform was added to each mixture and the samples were filtered. The filtrate was

[4] Application of bacteriophage preparations for prophylaxis of bacterial infections, mainly oral.

[5] Presumably the phages titers of prophylactic preparation were in the range of 10^7–10^8 plaque-forming unit (PFU)/ml, whereas the titer of the phage derived from the patients' body (feces, urine) prior to "phaging" was about 10^1–10^2 PFU/ml by Gratia (1939).

[6] The method of phagocytosis assay is not described in the original source.

spot tested for the presence of specific phages on the lawn of the phage-sensitive typhoid culture. Phages that were specific for the lawn bacteria were detected in several stool samples. The proportion of people whose stools contained specific phages within 24–48 hr was much higher among those people who were given a 30-ml dose (56%) than among those who received 10 ml (15.1%). The authors also observed a sharp decrease in the number of specific phages in stool samples after 24 and 48 hr in the first group who only received 10 ml as opposed to the other group where this number increased and then decreased gradually (Karpov and Yavorskaya, 1949). The increase of number of phages followed by their decrease in the stool most likely resulted from the propagation of phages in the intestines occurring soon after their administration and their persistence in intestines as long as their host bacteria were present there.

C. Treatment of intestinal *Salmonella* and *Shigella* infections

In the 1920s and 1940s the intestinal infections caused by *Salmonella* and *Shigella* species were a huge problem all over the world (Alessandrini and Doria, 1924; Chanishvili *et al.*, 2001; Compton, 1929; Costa-Cruz, 1924; Eaton and Bayne-Jones, 1934; Karamov, 1938; Krestnikova, 1947). Karamov (1938) provided epidemiological data on mortality rates at different times and at various geographic locations. The mortality rate in case of typhoid fever varied between 7 and 10%. In Baku (Azerbaijan) in 1932 the mortality index was equal to 5.8%. Similarly, in one of the main hospitals in Leningrad (Russia) the mortality rate in 1931 also attained 5.8%. During the water outbreak in Rostov (Russia) in 1926, this index was 8.24% and in 1926 in Hannover, Germany, was 11.4%. These figures indicated the urgent need of introducing novel therapeutic means to combat infections. According to Karamov (1938), clinical studies on phage therapy against *Salmonella typhi* and *S. paratyphi* infections performed on 60 patients in Baku (Azerbaijan) were unsuccessful. The phages were administered orally once a day in 10-ml doses for 10 days. Karimov (1938) mentioned that phage therapy did not decrease the mortality rate, which remained at the level of 12% in 1936 and was comparable to the mortality rate in the case of other sorts of treatment used.

Despite this, studies on the development of therapeutic phages in the former Soviet Union were continued. Zabrezhinsky and Gorstkina-Shevandrova (1946) published an article that focused on the potential of phage therapy for the treatment of typhoid fever. The authors cited previous investigations carried out by Braude and Koshkina (cited by Zabrezhinsky and Gorstkina-Shevandrova, 1946), who tried to treat 35 patients suffering from typhoid fever. Bacteriophages were administered orally in 10-ml doses (titer unknown) three times every 12 hr. Braude and Koshkina reported a slight improvement in patients' conditions, in particular, shortening of the shivering period. However, a cure was not achieved. According to Zabrezhinsky and Gorstkina-Shevandrova (1946), Alexandrov *et al.* (1940) applied phages to treat 57 cases of typhoid fever. In the first group of 20 patients they applied intramuscular phage injections (2–3 ml) two to three times and observed an improvement in 12 cases. In the second group of 20 patients, phages were administered intramuscularly and orally; an improvement was reported in 14 cases. In the third group, where the phages were applied only orally, an improvement was reported in only 5 cases. Thus, (Alexandrov *et al.*, 1940) concluded that the best result was achieved using a combined intramuscular and oral application of phages (cited by Zabrezhinsky and Gorstkina-Shevandrova, 1946). Unfortunately, the authors did not provide any data concerning the titer of phages in their preparations.

Zabrezhinsky and Gorstkina-Shevandrova (1946) referred to contemporary studies and concluded that the anti-infectious effect of the bacteriophages was not limited to its ability to lyse bacteria, but that the phages also induced bacterial mutations. Perhaps they could have had in mind lysogenic conversion. In particular, according to these authors, the phages deprived the bacteria of Duran–Reynolds[7] and, thus, made them unable to colonize the epithelial tissues. They supposed that the phages are characterized by invasive activity, that is, can enter the tissues even when administered orally, and do not neutralize toxins. Due to the latter, they could not be assured of a 100% success rate of phage therapy. In their studies on patients suffering from typhoid fever, experimental and control groups included 50 patients each. The diagnosis of all patients was approved bacteriologically or serologically by methods that were used commonly at that time. A daily dose of 30 ml of the typhoid phages (titer 10^{-8}–10^{-9} by Appelmans, 1921)[8] that, according to previously performed *in vitro* tests, lysed the infecting bacteria, was given to patients over 3 consecutive days. The authors reported a positive effect in 32 cases (64%). The persistence of fever was shorter than 20 days and was observed in 56% of patients in this group in comparison with the control group in which such a short period of fever was observed only in 22% of patients. The authors concluded that early start of phage treatment leads to a higher efficacy of therapy. Start of phage treatment after the

[7] Diffusion factors, agents of an enzymatic nature capable of increasing the permeability of connective tissue (chiefly by depolymerizing its basic substance, hyaluronic acid) and thus of speeding the diffusion and passage into the lymphatic capillaries of the water, salts, metabolic products of tissues, and infecting bacteria. These factors were discovered by the Spanish scientist F. Duran-Reynals in 1928 (Wikipedia).

[8] It should be noted that the Appelmans test does not give a possibility of precise enumeration of the phage particles per milliliter. Usually, the titer by Appelmans (1921) differs significantly from the titers by Gratia (1939). According to observations, titers equal to 10^{-9} by Appelmans correspond to 10^{4} or 10^{5} by Gratia. However, according to the Appelmans test it is possible to determine the time in which phage-resistant bacterial mutants may emerge. According to regulations in use in Russia and Georgia, lysis of bacterial cells treated with preparations of therapeutic phages should continue for 24–48 hr (Appelmans, 1921)

15th day of infection was not advised (Zabrezhinsky and Gorstkina-Shevandrova, 1946).

Astashkevich (1950) reported on the treatment of 52 cases of typhoid and paratyphoid diseases and compared the conditions of phage-treated patients with the conditions of 40 patients in the control group. Data concerning diagnosis, age, severity of infection, and rapidity of hospitalization of phage-treated and untreated patients were comparable. A polyvalent bacteriophage against *S. typhi* and *S. paratyphi* produced by the Sverdlovsk Institute of Epidemiology and Microbiology was used in these studies. Based on the estimation that the phages remain in the body no longer than 6–7 days, the authors applied the following regimen of phage therapy. Patients were given the phages for 3 successive days each week throughout their stay in the clinic. A single dose of phages was 10 ml (titer unknown). The phages were administered orally before meals together with an equal amount of 5% soda solution. This method of administration was assumed to ensure the maintenance of a sufficient concentration of bacteriophages in the body throughout the duration of infection, that is, for approximately 14 days after the decrease of temperature. In the majority of cases (35), phage therapy was started on the 6th day of infection and no later than the 20th day. A patient with severe infection received a total of 225 ml of bacteriophage lysates,[9] a patient with middle severity received 145 ml, and a patient with a mild form of infection received 105 ml. The authors reported that fever persisted for a shorter period in the experimental group of patients (38 days) as compared to the control group (52 days). At the same time the authors observed a higher rate of postinfection complications in the experimental group (37.7%) as compared to the control group (27.5%). The authors explained these results by the fact that the experimental group included more patients with severe and middle severity infections (80.8%) than the control group (70%). The frequency of relapse of infection in both groups was the same (13.5%).

An interesting study was performed by Manolov *et al.* (1948), who reported on intravenous applications of bacteriophages for the treatment of typhoid fever. The authors applied 20–25 ml of appropriate bacteriophage suspension (titer unknown) intravenously. The phages were suspended in a saline solution so that it contained a minimal amount of organic ballast.[10] The safety of these bacteriophages was proven in experiments on rabbits and mice. Intravenous administration of the typhoid bacteriophages in animal models was reported by the authors to protect the mice from the development of infection when challenged by a double

[9] Lysates contained live bacteriophages,solid cellular debris and soluble cell components, including bacterial antigens. The latter may have an impact on the final outcome of therapy independently of bacteriophages due to stimulation of the body immune system, as is observed in the case of vaccination.

[10] Mainly proteins and polysaccharides, that is, substances that may cause allergic reactions.

dose of the typhoid culture (a single dose was not defined in the original source). Fifteen to 20 min after intravenous administration of the phages to patients, the patients claimed to be shivering. After 2–3 hr, a rise in temperature was reported, which in some cases was followed with nausea and vomiting. At 12–14 hr after phage injection the temperature was normalized; however, after 24 hr it rose again to the same level. This led the authors to apply several phage injections every day or every other day. Unfortunately, the number of patients was not determined in this study. However, comparison of the results of aforementioned studies, with results of phage therapy using orally administered phages, led the authors to the following conclusions:

1. Low doses (2–10 ml) (titer unknown) of typhoid bacteriophages administered orally are inefficient. Additionally, oral administration of bacteriophages is not reasonable because of the specificity of pathogenesis of typhoid fever.

2. Intravenous administration of 20–25 ml of typhoid bacteriophage prepared in a saline solution every day over 3 days was not followed by any serious side effects. However, this treatment led to a decrease of temperature and shortening of the fever period, improvement of the general patient's condition, and a complete cure (Manolov, 1948).

How often the cases of complete cure were observed is unknown.

Mikaelyan (1949) reported on the results of phage therapy in carriers of *S. typhi* or *S. paratyphi* who were food-processing industry workers. The study was performed in 1941–1945. Monovalent phages against these pathogens produced by the Tbilisi Institute of Vaccine and Sera were used in this experiment (titer 10^{-7} by Appelmans, 1921). The phages were given only to those patients that remained carriers of the pathogenic bacteria a long time after initial conventional treatment had ceased (the time period between cessation of conventional treatment and initiation of phage treatment was not precised). These so-called convalescents received a dose of phages (15 ml) (titer 10^{-7} by Appelmans, 1921) mixed with an equal volume of 10% sodium bicarbonate ($NaHCO_3$) solution. Doses were administered three times once per 5 days. Bacteriological study of the stool samples was performed three to five times, 7–10 days after the start of phage treatment. In 1941, 65 food industry workers were tested for being bacterial carriers. Twelve of them stopped producing typhoid bacteria within 14 days, while the rest (53) continued to be carriers for 15–20 days, as determined based on analysis of stool samples. These patients were divided into two groups: one consisting of 29 persons who underwent specific treatment with appropriate phages and the other consisting of 24 persons left as a control. All patients from the phage-treated group were examined bacteriologically four to six times at intervals of 7–10 days. Results of the examination showed that 26 patients out of 29

(89.6%) were cured from a bacterial carrier state of *S. typhi* and *S. paratyphi* within 14 days, which was determined based on the results of stool analysis. Three remaining patients continued the production of pathogenic bacteria in stool for 20–22 days, after which these cases were cured as well. Repeated examination of stool samples after 6–8 months (21 patients) and 1 year (8 patients) following the phage treatment did not reveal the presence of bacterial pathogens in any case (the author did not specify which identification method was used in this case). Moreover, a study of the health status of family members of the phage-treated patients and relative bacteriological analysis did not reveal any cases of the disease. The study of stool samples of 24 patients in the control group demonstrated that in 22 cases the bacterial carrier state continued for 15–84 days; in 2 cases this period was as long as 104–120 days after initial identification of the pathogens. In 1942–1945, a group of 64 food workers was included into the experiment. Altogether in 1941–1945 129 patients underwent phage treatment and a positive outcome was reported in 101 cases (89.4%) (Mikaelyan, 1949).

A similar study was published by Gnutenko (1951), who also reported the results of bacteriophage treatment of workers in the food processing industry—carriers of *S. typhi* or *S. paratyphi*. Twenty-one carriers, mainly over 40 years of age (18 women and 3 men), underwent phage treatment during 1940–1941 and in 1944. Previous treatments using other methods were reported to be unsuccessful and it was decided to apply a combined phage therapy by intramuscular phage injections, as well as rectal and duodenal administrations of phages. In those cases where duodenal treatment was not possible, oral administration was used. Duodenal administration was considered to have two important advantages: (a) phages were transported directly into the area colonized by the pathogenic bacteria and (b) the phage preparation was not affected by gastric acids and therefore was supposed to retain its initial activity. The rectal treatment was believed to provide a supplementary treatment, as bacteria surviving in the gallbladder were supposed to be killed by phages in the rectum. To evaluate the potential for the migration of phages from the rectum to the gallbladder, samples of bile were checked for the presence of both administered therapeutic phages and naturally occurring phages. Differences in plaque tests and host range differences between phages served to distinguish therapeutic phages from naturally occurring phages. This method of phage differentiation was commonly used by researchers at that time. Similarly, stool samples of the patients were studied for the presence of specific bacteriophages prior to administration of the therapeutic phage preparation. Thirty milliliters of phage preparation was administered as an enema after washing the intestines with a cleansing enema (titer of phage preparation is unknown). After 12–14 hr, a probe was introduced directly into the gallbladder to take a bile sample, followed by administration of a 20-ml phage preparation. After 5–6 hr, patients were given an intramuscular injection of phages (5–10 ml) (titer unknown). The aforementioned procedures were repeated three to five times as a single course of treatment. This complete course was repeated two to three times, presumably with a week interval. After each course of treatment, feces were examined bacteriologically for the presence of pathogenic bacteria (the original source does not give a precise description of the method applied). After this series of treatments, 6 of the 10 carriers of *S. typhi* were reported to be clear of the tested bacteria, whereas 4 showed no change, presumably meaning that the titer of pathogenic bacteria remained the same before and after treatment. Eight of the 11 *S. paratyphi* carriers were reported to be cleared, whereas 3 did not respond to the treatment. The authors recommended carrying out such treatments under hospital conditions, performed with at least three courses of treatment over 10–15 days (Gnutenko, 1951).

The broadest clinical study on the therapeutic effect of dysenteric phages was reported by Sapir (1939). The author described altogether 1064 cases of dysentery treated with bacteriophages in two different clinics in Moscow. The patients' group included 767 men and 297 women ranging in age from newborns to 79 years old. Dysentery was diagnosed using bacteriological tests (362 patients), clinical observations (512 cases), and clinical colitis (190). Bacteriological analysis of 362 patients indicated that dysentery in 289 cases was caused by *Shigella Shiga-Kruse*[11] (22 resulting in a lethal outcome), 69 cases by *S. Hiss-Flexneri*,[12] and 4 cases by *S. Shmitz-Shtitzeri*.[13] A standard phage therapy, using the dysenteric bacteriophage preparation developed by the Mechnikov Institute in Moscow, was applied to every age group as described below. The exact content of this preparation has not been specified.

A daily dose of phage for an adult was 20 ml and for a child was 10 ml (titer 10^{-9-11} by Appelmans, 1921). The dose was divided into two portions and given to patients at midnight on the day of arrival and at 4:00 a.m. to minimize the inactivation of phage by any meal residues. The comparable time of phage administration to all patients facilitated the evaluation of the results of phage therapy. Patients were given a magnesium–soda solution (magnesium 10 g/liter + soda 20 g/liter) initially 6 hr prior to phage therapy and then every 2 hr repeatedly six times following 12 hr. Adults were given 100 ml of the solution at each administration, and children were given 10–50 ml depending on their age. The solution was given with the aim of providing optimal conditions for phage propagation and also to

[11] *Shigella dysenteriae* type 1.
[12] *Shigella flexneri*.
[13] *Shigella dysenteriae* type 2.

help clear the intestines. Patients were kept on a strict diet during the first 48 hr. The author concluded that the application of this dose of phage divided into two portions and administered over the course of a single day was sufficient and did not need to be repeated. Some of the patients were subjected either to serotherapy or to nonspecific (i.e., symptomatic) therapies. Sometimes even a combination of phage- and serotherapies has been applied. In cases of severe intoxication, intravenous transfusion of a saline solution with adrenaline or glucose was applied. In the 15 most severe cases, antidysenteric serum was applied in addition to the phage preparation; however, a lethal outcome was not avoided. The author concluded that in those cases where serotherapy was unsuccessful, phage therapy also appeared inefficient (Sapir, 1939). Patients were dived into groups based on the severity of infection (with severe infection, cases with middle severity, mild severity, and clinical colitis) (Table I). Although the result of therapy is provided for each group, what fraction of patients in each group was treated only with phages is unclear. The average duration of disease was 43.3 hospital days.

According to the author, the application of phage therapy decreased the duration of the hospital stay significantly in comparison with other methods based only on symptomatic treatment or even specific (serological) treatments (Table II). The author underlined a key role of the early start of phage therapy for the shortening of hospitalization time. Additionally, he reported that usually, after 1–2 days of phage therapy, a dramatic improvement in the patient's condition was observed, as reflected by less frequent and less watery stools that did not contain blood and/or mucus. According to the author, the most visible effects were observed in cases where pathological changes did not develop deeper than catarrh and inflammation. The author underlined that sometimes phages from the preparation appeared to be unable to lyse the particular strain, or the toxic syndrome was too strong or secondary infections developed, and perhaps hence the apparent lack of the effectiveness of phage therapy in these cases (Sapir, 1939).

TABLE I Fractions of patients with different severity of dysentery included in Sapir's studies (according to Sapir, 1939)

Disease forms and outcomes	Number of patients	%
Severe forms, with lethal outcome	47	4.4
Severe forms, cured	155	14.6
Middle severity form, cured	491	46.1
Mild severity form, cured	181	17.0
Clinical colitis, cured	190	17.9
Total	1064	100

TABLE II Duration of hospitalization after application of different treatment methodologies (according to Sapir, 1939)

Type of disease	Applied therapy	Duration of hospital stay (days)
Severe form	Serotherapy	21.9
Severe form	Phage therapy	20.8
Middle severity form	Nonspecific therapy	16.5
Middle severity form	Serotherapy	14.9
Middle severity form	Phage therapy	11.5

Sapir (1939) also reported that after 1 day of phage treatment the number of patients with bloody stools decreased from 100 to 74; on the 5th day of treatment, only 4 patients remained suffering from this symptom. After 1 week of phage therapy, 95% of patients did not show pathological symptoms and could be released from the hospital. A lethal outcome was registered in 47 cases (4.4%). However, the author emphasized that these patients suffered from dysenterial pancolitis and severe degenerative changes of the parenchyma of various organs, ulcers of thick gut, and so on, which are typical for long-term and severe intoxication, as verified by postmortem pathoanatomical studies. The author concluded that:

1. Phages should be given to every patient that shows symptoms of dysentery, independently of whether the patient is arriving to the hospital, visiting ambulance, or asking for medical help at home. This measure would have both therapeutic and prophylactic effects.

2. Combined phage- and serotherapy should be used in special cases such as hypertoxicosis among adults and for the treatment of toxic and subtoxic syndrome among young children. In all the aforementioned cases, the author also recommended the application of proper doses of serum and its injection at least 6 hr prior to the start of phage therapy (Sapir, 1939). What is the exact basis for the latter recommendation is unclear.

Lipkin and Nikolskaya (1940) performed phage therapy on 100 patients suffering from dysentery. A control group of 50 patients received ordinary medication, such as purgative salts, which were used in most cases. In 21 cases the patients underwent sero-therapy. In 5 severe cases, a combined phage- and serotherapy was used. All patients were maintained under the same conditions, in terms of care, diet, and so on. Phages produced by the Tbilisi Institute of Vaccine and Sera and Kuibishev Institute of Epidemiology and Microbiology were used in these studies. The titers of these phage preparations were 10^{-9-11} (by Appelmans, 1921). Five milliliters of phages were given to patients orally together with 2% soda solution three times per day. After receiving the phages, patients fasted all day. In almost every case the phage treatment was continued for 1 day. Only in 6 cases was the phage treatment at the same dose performed over 2 days. Sixty-six percent of patients received the phages within the first 5 days of the start of infection. Development of the disease was evaluated through observations of stool frequency, presence of mucus, blood, tensions, and so on.

Lipkin and Nikolskaya (1940) reported a significant effect of phage therapy even in cases where treatment was started rather late. Twenty-five percent of patients (out of 100) stopped reporting painful symptoms by the 2nd day of treatment; 79% did not show pathological symptoms by the 4th day and 100% by the 6th day after which the stool was normalized. These data are in contrast to results obtained with "ordinary therapy," where only 2% (1 case out of 50 patients) showed an improvement on the 2nd day of treatment, 14% on the 4th day, and 46% on the 6th day. It is noteworthy that patients with relatively easy cases were included in the control group, whereas those with relatively severe illnesses were included in the experimental group. According to the authors, the fact that these patients showed an improvement as soon as they got phage treatment illustrated the effectiveness of this method (Lipkin and Nikolskaya, 1940). The relief of symptoms in patients treated with serotherapy was recorded in 33% of cases (7 persons out of 21) on the 4th day of treatment and in 67% on the 6th day, indicating that serotherapy resulted in a slower relief of symptoms than phage therapy. Five patients subjected to serotherapy remained sick over 10 days. These patients later underwent phage therapy as well (without further serotherapy).

Phage preparations were considered to be particularly efficient for the treatment of intestinal infections. Vlasov and Artemenko (1946) described the results of treating with phages of 30 persons with chronic dysentery. Many of these chronic patients were exhausted by infection and were bedridden. A dry tablet preparation known as "phage-vaccine"—a combined preparation comprising 10^6 killed cells/ml and 10^{-7} of bacteriophages—was used (by Appelmans, 1921). The patients had suffered with infections for 1–2 years, and in 70% of cases, rectoscopic examination indicated the presence of bleeding ulcers. Prior to combined "phage-vaccine" therapy, all the patients underwent unsuccessful multiple courses (one to eight times) of therapy with antibiotics and sulfonamide preparations. After "phage-vaccine" therapy, the authors reported a cure in the case of 26 patients (86.7%) within 10–20 days. Assessment of the results was based on improvements in the general condition of patients, including normalization of comprogram test[14] results formation of stools and recovery of the mucous layer of the sigmoid colon and rectum (Vlasov and Artemenko, 1946).

[14] A type of stool analysis —one of the methods allowing evaluation of the function of the digestive system. Fecal analysis helps to reveal gastric, small and large intestines, and pancreas, liver, and gallbladder problems. Coprogram A coprogram gives an information about the stool quality, according to which a doctor can judge on the status of the digestive process. In addition, analysis of feces is useful for the detection of intestinal parasitic diseases (helminthes, giardiasis, etc.)

D. Combined use of phage- and sero- therapies

Authors of early publications suggested that an effective use of combined bacteriophage therapy with sero-therapy required consideration of the existing "competition" between phages and anti-bacterial serum as illustrated in 1934 by Lavrik (cited by Krestnikova, 1947). This author showed that the bacterial cells treated with anti-bacterial serum were unable to be infected with phages. This phenomenon was confirmed in an *in vitro* and *in vivo* experiments by Belkina (1939) (cited by Krestnikova, 1947), who studied the interaction of phages with cells of *Proteus* sp. and *Shigella shigae* strains. Phages added to a mixture of bacteria and antibacterial serum that was specific to these bacteria did not prevent agglutination and did not lyse bacteria, indicating that the specific serum binds to the bacterial cell surface and thus prevents phage infection. In animal studies with mice infected with *Proteus* sp. and *S. shigae*, the best results were achieved when the phages were administered 3 hours prior to application of the specific serum, which resulted in the survival of 70--80% of mice. When the serum was administered prior to the phages, 100% of mice in both experimental and control groups died (cited by Krestnikova, 1947). However, these findings were not always taken into consideration and, according to other authors, specific antibacterial sera were often given to the patients prior to phage therapy. Articles describing the combined use of specific antibacterial sera and bacteriophage preparations were discussed in Section II.C (Karamov, 1938; Sapir, 1939; Zabrezhinsky and Gorstkina-Shevandrova, 1946). In most of the articles devoted to the combined use of phage- and serotherapy, prophylactic antidysenteric or anti-*Salmonella* sera were given to the patients immediately on arrival at the hospital and preferably 6 hr prior to the start of phage therapy (Sapir, 1939).

Both antidysenteric serum and appropriate bacteriophage preparations were reported to be used to treat dysentery. Ionov *et al.* (1939) published their observations on the specific action of antidysenteric serum and bacteriophages. This study was carried out in 1935–1936 and involved 502 patients (175 children and 327 adults). Bacteriological analysis of stool samples showed the presence of *S. shigae* and *S. flexneri*. Of 327 adults, 175 were treated only with phages, 90 with antidysenteric serum, and 62 received a combined phage and serotherapy. A polyvalent antidysenteric phage preparation and monovalent sera against *S. shigae*, *S.flexneri*, and *S. Hiss-Russel*[15] were used in these studies. In the group of patients treated with phages only, the preparations were initially administered orally as a single 10- to 15-ml dose (for adults) and then the same dose of phages (10–15 ml) was given simultaneously both orally and per

[15] *S. Hiss-Russel*=*S. flexneri*

rectum as an enema (titer of phage in the preparations was not provided by the authors). A control group consisted of patients receiving standard medications ("ordinary therapy"). The disappearance of blood from the stools was observed in 50% of patients treated with phages as compared to 20–30% in the control group (Ionov *et al.*, 1939). The combination of phage- and serotherapy was used in the treatment of 62 adults who did not demonstrate any symptoms of improvement after phage therapy. The serum was administered as a subcutaneous injection. In 27 cases the serum was administered during the first 4 days after phage therapy with a positive effect being observed in 14 cases during the first week and in 7 cases after 3 weeks. Six patients died. Of the 35 patients that received serotherapy later than the 5th day following phage therapy, only 8 recovered within a week, 17 recovered within 3 weeks, and 10 patients died. It was therefore recommended to develop a combined phage- and serotherapy. The authors emphasized that to optimize the effectiveness of phage therapy it is important to start it at the early stages of infection. They noted that the stool samples of 43% of the patients undergoing phage therapy were free of pathogenic bacteria—the cause of infection—after 2–3 days. Bacterial carriage was reported to be much less frequent in the phage-treated patients compared with the control group treated by "ordinary therapy" (i.e., with standard medication). Of patients treated with ordinary therapy, 47.9% continued to have pathogenic bacteria in their stools after 15 days, whereas only 20.2% of the patients undergoing phage therapy retained pathogenic bacteria in their stools. Of patients treated with phages, 1.7% died compared with 3.5% of those treated with serum and 5.8% of those receiving ordinary treatment, the details of which are not specified in the original source. The authors indicated that ordinary therapy was usually applied in cases of mild severity, whereas combined phage- and serotherapy was applied in the most severe cases (Ionov *et al.*, 1939).

Golubtsov (1940) reported a comparative study carried out in 1939 of the effect of sero- and phage therapies among children suffering from dysentery. The children were divided into two groups: group I consisted of 22 children treated with serotherapy and group II comprised 18 children treated with specific bacteriophages. The majority of children in each group were aged between 5 and 18 months, with 7 children between 1.5 and 2 years old.[16] Bacteriological analysis was only successful in 59% of cases but indicated that *S. Shiga-Kruse* and *S. Hiss-Russel* predominated.

Golubtsiov (1940) subgrouped the dysenteric diseases in children according to their severity into three types: nontoxic (mild), toxic, and

[16] The author did not mention how these seven were distributed between the two groups.

chronic. Evaluation of the effectiveness of therapy was based on body temperature, presence of toxic syndrome and cramps, and stool characteristics (frequency, presence of blood, mucus, pus, etc.).

Children of group I received antidysenteric serum during the first days after infection. The serum was administered intramuscularly in doses of 10,000–20,000 antitoxic units per injection within a 3- to 4-day period. In 47% of cases, toxic syndrome and vomiting were reported to be relieved and an improvement in the general condition of patients was observed, as evidenced by better reactions of the infants to surrounding people and events, better sleep patterns, and reemergence of appetite. Serotherapy did not have a significant effect on body temperature and stool quality and frequency. A decrease in temperature was observed in only 27% of cases with changes of stool frequency in 37% of cases.

Children of group II received the specific antidysenteric phages, which were administered orally in the morning before meals, in a dose of 1.0 ml for infants below 1 year, 1.5 ml for children of 1.5 years, and 2.0 ml for 2 year olds (titer unknown). The therapeutic effect was indicated by a decrease in temperature in 77% of patients with an improvement in stool characteristics and a decrease in stool frequency observed in 60%, and relief of cramps in 53.3%. A relief of toxic syndrome[17] was observed only in 17% of cases (Golubtsov, 1940).

Summarizing these data, Golubtsov (1940) concluded that the effects of sero- and phage therapies are different. Serotherapy decreased the intensity and frequency of toxic syndrome significantly, whereas phage therapy had no such effect but gave better results for the improvement of pathological changes in the colon (such as damage of the mucous layer). The author recommended the application of specific phages in the early stages of the disease and in combination with serotherapy (Golubtsov, 1940).

E. Use of bacteriophages for prophylaxis

Phages have also been used extensively in the former Soviet Union for prophylaxis in regions with a high incidence of infections and also in communities where the rapid spread of infections may occur, such as kindergartens, schools, and military accommodation (Agafonov *et al.*, 1984; Anpilov and Proskudin, 1985; Belikova, 1941; Blankov, 1941; Blankov and Zherebtsov, 1941; Florova and Cherkass, 1965;Kagan *et al.*, 1964; Karpov, 1946).

[17] Toxic syndrome (TS) is a toxic condition that occurs in response to the impact of toxic substances, e.g., produced inside the body by bacteria. TS is characterized by severe metabolic disorders and functions of various organs and systems, primarily the central nervous and cardiovascular systems.

The application of phages for prophylaxis was carried out in 1929–1930 against the bacterial diseases that were the most serious problems at that time, such as dysentery, typhoid fever, and staphylococcal infections. The first mass application of dysenteric bacteriophages in the USSR was performed in Alchevsk (Donbas region) in Ukraine in 1930 (Ruchevski, "*Vrachebnaya gazeta*," 1931, 21: 1586, cited by Krestnikova, 1947). An experiment on the prophylactic use of phages was carried out successfully later in 1935 on thousands of people in regions with a high incidence of dysentery (Melnik *et al.*, 1935). Results were reported at scientific conferences in 1934 and 1936 in Kiev and in 1939 in Moscow after which the dysenterial phage preparation was finally approved as a preventive measure for mass application (Krestnikova, 1947). According to Krestnikova (1947), it was recommended that repeated seasonal prophylactic "phaging" be carried out in areas where the dysentery was endemic. Later modifications included the supply of dysenterial phages in dry tablet forms, which also began to be included in clinical studies.

One of the later studies by Babalova *et al.* (1968) in 1963–1964 describes the results of preventive treatment carried out with phage tablets having an acid-resistant coating. This polyvalent antidysenteric preparation contained different phages against the following bacteria: *S. flexneri* (10^{-7}–10^{-9} by Appelmans, 1921), *S. sonnei* (10^{-7}–10^{-8}), *S. grigorieva-shiga* (10^{-7}–10^{-9}), *S. stutzeri*[18] (10^{-6}–10^{-7}), and *S. newcastle*[19] (10^{-6}). A study of the persistence of bacteriophages in children was also carried out and showed that phages could be detected in the body for 5–7 days. The authors mentioned that phages active against *S. flexneri* serotypes 1a and 2a and *S. grigorieva-shiga*[20] were especially long lasting and could be isolated from stool samples longer than other phages included in this preparation, such as that active against *S. newcastle*. Perhaps the latter was dependent rather on the individual properties of these phages by themselves and could have nothing to do with their specificity.

According to Babalova *et al.* (1968), a prophylactic experiment was carried out by specially trained staff of the Tbilisi Institute of Vaccine and Sera[21] in collaboration with the regional sanitary-epidemic stations in two Tbilisi districts characterized by distinct epidemic situations, one of which was known to have above average infrastructure and hence a lower susceptibility to infection outbreaks. Prophylactic "phaging" was carried out during the period from May 8 to October 25, 1963 and from May 15 to October 31, 1964. Special attention was given to the method of selecting the experimental and control groups of children. Children living in each

[18] *Shigella dysenteriae* type 1 and 2, respectively.
[19] *Shigella flexneri* type 6.
[20] *S. dysenteriae* type 1.
[21] Former name of the Eliava Institute of Bacteriophage, Microbiology & Virology in Tbilisi, Georgia.

district were registered and divided into two equal groups according to sanitary-epidemic conditions. Prophylactic "phaging" was carried out on children living on one side of the street, whereas those living on the other side did not get the phage treatment and thus formed a control group. Doctors visited each child from both experimental and control groups at least once a week, challenging the experimental group with a new dose of phages and performing observations, registering any cases of disease, and providing this information to the center.

Phages were given to children aged between 6 months and 7 years. Tablets were administered before meals or 2 hr after meals. Children aged 6 months–5 years received one tablet (equal to 20 ml of phage suspension; the titers of each component of this polyvalent dysenteric preparation were given earlier), and children over 5 years received two tablets. The average number of observation days in experimental groups and control groups was comparable, 108.6 and 109.5, respectively. The effect of phage prophylaxis was evaluated on the basis of clinical symptoms rather than on bacteriological analysis, as only single suspected cases of intestinal disease were studied bacteriologically. The incidence of acute dysentery in the control group was 3.8 times higher than in the experimental group. The observed difference was significant, as verified by the statistical analysis of results. The effects of phage prophylaxis differed between different age groups. For example, in the age group 6 months–1 year, the epidemic index (EI)[22] was 5.7; in the age group 1–3 years the EI was 3.7, and in the age group 3–5 years the EI was 2.4. In the 5- to 7-year-old age group, no cases were registered. The authors concluded that a 3.8-fold decrease of disease incidence in children of 6 months–7 years of age due to prophylactic "phaging" was a promising result. They suggested using the tablet form of dysenteric bacteriophages for mass application (Babalova *et al.*, 1968).

Later, studies on mass prophylaxis of intestinal diseases by the application of phages were performed in the Red Army units by military doctors as well. For prevention of dysentery and typhoid epidemics, specific phages were also used with two tablets administered once every 5–7 days during the outbreak season. The authors reported that the prophylactic use of phages resulted in a 6- to 8-fold decrease in the number of intestinal infection cases in the tested groups in comparison with the control groups (Agafonov *et al.*, 1984; Anpilov and Proskudin, 1984; Kurochka *et al.*, 1987).

An interesting experiment is described by Sayamov (1963) and Plankina *et al.* (1961), who present the results of therapeutic and prophylactic

[22] EI is the ratio between the expected and the observed number of infected persons calculated per 1000 persons. EI is used here as a tool to measure the effectiveness of prophylactic measures.

trials using anticholera bacteriophage preparation, performed by Soviet doctors in East Pakistan in 1958 and in Afghanistan in 1960. The bacteriophage preparation was developed by an *in vitro* cultivation of bacteriophages on *V. cholera* strains isolated directly in epidemic site on media reconstructing *in vivo* conditions, as suggested by Nikonov in 1959 (cited according to Sayamov, 1963). In particular, these media contained bile and fragments of small intestine in Tyrode's solution.[23] The ability of selected phage clones to propagate *in vivo* was increased via a procedure known as adaptation, which is based on sequential, alternate passages of bacteriophages through the small intestine of guinea pigs[24] and through bile.

During a cholera outbreak in Dacca (East Pakistan) in 1958, Soviet doctors used "adapted" phages to treat 22 patients, all in serious conditions and with signs of anuria[25] (Sayamov, 1963). Typically, each patient was given 5–10 ml of phage suspension in a saline solution intravenously once and 30 ml of phage suspension orally each day for 3 consecutive days (phage titer in these suspensions was not provided). Only two, the most seriously sick and dehydrated, patients died, whereas the fatality rate among Kandahar hospital patients treated symptomatically and/or with antibiotics (mainly oxytetracycline) was officially estimated to be about 50%. At the same time, to prevent the spread of cholera, Soviet doctors were reported to treat about 30,000 people with phages prophylactically in seven highly populated centers of East Pakistan (Sayamov, 1963). Details of this treatment are not specified. No cholera cases were reported in any of these centers following such a prophylactic action.

According to Sayamov (1963), the epidemic in Afganistan started on August 6, 1960. Only persons suffering from severe forms of cholera were sent to the hospital, where the usual symptomatic treatment, together with antibiotics (mainly oxytetracycline), was given. On the recommendation of the Soviet doctors, all patients were transferred to an isolation block. On October 7 and 8, phage prophylaxis started, which included the patients as well as the healthy hospital staff and students being on training at that time (altogether over 1600 persons). A single dose of the phages in saline solution (5–10 ml) was administered intravenously[26] or intramuscularly,

[23] A solution containing salts and glucose.
[24] A segment of small intestine of an anesthetized, living guinea pig was isolated carefully using two ligatures so that the branches of the blood vessels supplying the isolated loop were left intact. This technique was suggested by De and Chatterjee in 1953 (cited by Sayamov, 1963). The isolated loop was infected with *V. cholera* using 100–1000 *Vibrio* cells. After 24 hr the majority of infected animals died. If anticholera bacteriophage was injected into the same loop simultaneously or after a certain period of time, animals survived. However, the aim of this experiment was to obtain bacteriophage clones that would be able to propagate efficiently in *in vivo* conditions. Therefore, phages isolated from one animal were used to cure another animal, and so forth (Sayamov, 1963).
[25] Complete suppression of urine formation and excretion.
[26] Purification of phages for injections by CsCl or $CsSO_4$ gradient centrifugation had never been applied.

followed by an oral daily dose (30 ml, children received 20 ml) for 3 days. As a result of these measures, a lethal outcome was observed only in 4 out of 119 cholera patients (3.5%). What is more, no new cases of cholera were detected in the hospital after the start of therapeutic and prophylactic phage treatment between October 7 and 10. In 80% of patients treated with bacteriophages in the Kabul hospitals, the initial diagnosis was confirmed bacteriologically as well. No complications were observed after intramuscular injection of phages, whereas chills and headaches were sometimes observed after intravenous administration.

As described by Sayamov (1963), in October 1960 cholera appeared in the Kataghan province in the north of Afghanistan. A number of patients were accommodated in hospitals set up in the towns of Pul-i-Khumri and Kunduz, but because of a shortage of hospital beds and transportation problems, many sick people remained at home. To prevent the spread of cholera, Soviet doctors, in cooperation with Afghan colleagues, arranged a massive therapeutic and prophylactic action among the cholera-stricken rural population. Most of the patients treated at home (about 90%) received a dose of 20–30 ml (or, in particularly severe cases, as much as 50 ml) of phage suspension in one or two intramuscular injections (intravenous administration was not used). A single dose of cholera vaccine was administered simultaneously as well. The healthy part of each village population was treated the same way. No new cases of cholera occurred in rural populations as a result of this combined therapeutic and prophylactic action but, as underlined by Sayamov (1963), only when the whole population in a given village was subjected to the phage treatment. Altogether approximately 270,000 persons underwent phage treatment. Whether any control groups were included in these therapeutic/prophylactic actions, as well as what was the contribution of phages and what was the contribution of vaccine in the observed prophylactic effect, is unknown. However, Sayamov (1963) concluded that the suspension of anticholera bacteriophages is a valuable preparation, which could be used as an important supplement to the standard measures for cholera control.

III. CURRENT STATUS OF GEORGIAN RESEARCH AND THEIR IMPACT ON WORLDWIDE PHAGE THERAPY STUDIES

Ongoing research and clinical trials

Production and use of phages for therapy and prophylaxis never stopped at the Eliava Institute of Bacteriophage; however, the scale is much smaller than before the break off of the Soviet Union. Therapeutic and prophylactic phages are currently produced by the small spin-off company "Eliava BioPreparations, Ltd." established by the Eliava IBMV. In the past, the scale of phages production covered the needs of the greatest part of the Soviet countries, including the Caucasian and Middle Asian republics. Today, production of the "Eliava BioPreparations, Ltd." fully satisfies the Georgian market. In 2010 the company started to export phages outside Georgia into Azerbaijan. Another spin-off company, "Eliava Diagnostics," receives, on average, 50–60 patients daily seeking the phage treatment, often after unsuccessful antibiotic therapy. In 2010, a day clinic, "The Eliava Phage Therapy Center," was opened, which renders services to local and foreign patients.

These achievements would be impossible without intensive research held at the Eliava Institute of Bacteriophage, most of which are conducted in close collaboration with the international scientific community. On the one hand, this helps Georgian scientists (1) to develop and introduce modern techniques into their research and (2) to design proper experiments corresponding to Western standards. On the other hand, collaborative projects help Western researchers to avoid the "discovery of bicycles," which may occur easily without having background knowledge in the development of phage preparations and their application in practice. Over 50 collaborative international projects have been accomplished during the period of 1995–2010, which resulted in a number of significant publications (Ceyssens *et al.*, 2011; Chanishvili *et al.*, 2001, 2009; Glonti *et al.*, 2010; Khawaldeh *et al.*, 2011; Kutateladze and Adamia, 2010; Kutter *et al.*, 2010; Markoishvili *et al.*, 2002; Merabishvili *et al.*, 2009; Pirnay *et al.*, 2011; Sulakvelidze *et al.*, 2001; Verbeken *et al.*, 2007).

Among other achievements in phage therapy, it is necessary to mention the treatment of cases of secondary infections associated with cystic fibrosis. Since 2008, together with the National Center of Cystic Fibrosis in Tbilisi, phage preparations against secondary infections have been used in eight patients with cystic fibrosis. Phages were applied in infants and adults via a nebulizer several times for 6–10 days. Simultaneously, patients were treated by conventional antibiotics, antimucus medications, and vitamins. Phage application caused a substantial decrease of bacterial counts in the patients' sputum samples and an improvement of the general health condition. Due to bacteriophage therapy, long-term remissions of infections were achieved (Kutateladze and Adamia, 2010).

Despite economic difficulties during 1990–2000, the institute's staff managed to elaborate new phage-based products, such as PhageBioDerm, which is a novel wound-healing preparation consisting of a biodegradable polymer impregnated with an antibiotic and lytic bacteriophages (Markoishvili *et al.*, 2002). It has been licensed for sale in the Republic of Georgia (one of the former Soviet Union republics). In 1999–2000, in Tbilisi, Georgia, 107 patients who had ulcers that had failed to respond to conventional therapy were treated with PhagoBioDerm alone or in combination with other interventions. The wounds/ulcers healed completely in 67 (70%) of 96 patients for whom follow-up data were available. In 22 cases in which microbiological data were available, healing was associated with a concomitant elimination of, or a reduction in, specific pathogenic bacteria in the ulcers. Bacteriophages in PhagoBioDerm helped clear the wounds from multidrug-resistant *S. aureus*. Results suggest that this slow-release biopolymer is safe and of possible benefit in the management of refractory wounds and support the apparent utility of bacteriophages in this setting. Further studies, including carefully designed clinical trials, will be required to evaluate rigorously the efficacy of this novel wound-dressing preparation (Markoishvili *et al.*, 2002; Sulakvelidze *et al.*, 2001).

While the phage preparations are in everyday use in Georgia and in a number of ex-Soviet countries, there are many obstacles for the clinical application of bacteriophages in Western countries, such as the perception of viruses as "enemies of life" (Villareal, 2005), the lack of a specific frame for phage therapy in current medicinal product regulation (Pirney *et al.*, 2011), and the absence of well-defined and safe bacteriophage preparations. To evaluate the safety and efficacy of bacteriophages in the treatment of burn wound infections in a controlled clinical trial, a highly purified and fully defined bacteriophage cocktail (BFC-1) was prepared, which is active against the *P. aeruginosa* and the *S. aureus* strains actually circulating in the Burn Centre of the Queen Astrid Military Hospital (Merabishvili *et al.*, 2009). Based on successive selection rounds, three bacteriophages were retained from an initial pool of 82 *P. aeruginosa* and 8 *S. aureus* bacteriophages, specific for prevalent *P. aeruginosa* and *S. aureus* strains in the Burn Centre of the Queen Astrid Military Hospital in Brussels, Belgium. This cocktail, consisting of *P. aeruginosa* phages 14/1 (Myoviridae) and PNM (Podoviridae) and *S. aureus* phage ISP (Myoviridae), was produced and purified of endotoxin. Quality control included stability (shelf life); determination of pyrogenicity, sterility, and cytotoxicity; confirmation of the absence of temperate bacteriophages; and transmission electron microscopy-based confirmation of the presence of the expected virion morphologic particles, as well as of their specific interaction with the target bacteria. Bacteriophage genome and proteome analysis confirmed the lytic nature of the bacteriophages, the absence of toxin-encoding genes, and showed that the selected phages 14/1, PNM, and ISP are close relatives of, respectively, F8, φKMV, and phage G1. The bacteriophage cocktail is currently being evaluated in a pilot clinical study cleared by a leading medical ethical committee. No adverse reactions were observed (Merabishvili *et al.*, 2009). A detailed description of the quality-controlled, small-scale phage preparation was considered as a first step to promote the concept of phage treatment in Western medicine. In addition, it supported the creation of a discussion platform for regulatory framework for approval of phage therapy (Pirney *et al.*, 2011).

Antibiotics are becoming ineffective as important bacterial pathogens evolve to outsmart them. However, the antibiotic pipeline is running dry with only a few new antibacterial drugs expected to make it to the market in the foreseeable future. Bacteria resistant to all available antibacterial drugs, so-called superbugs, are emerging worldwide. Evolutionary ecology might inform practical attempts to bring these pathogens under stronger human control (Williams, 2010).

In this context, various laboratories worldwide and a handful of small pharmaceutical companies are turning to phages (Thiel, 2004). There is hope that commercial interest shown by these companies will finally lead to recognition of phage therapy by the Western world.

Despite its long (eastern European) history, phage therapy is not currently authorized for routine use on humans in the West. Today, it is only approved in some former Soviet republics such as Russia and Georgia, where commercial phage preparations are sold in pharmacies (Chanishvili, 2009). In Poland, a recent member of the European Union, phage therapy is considered an "experimental treatment" covered by the Physician Practice Act (*Polish Law Gazette* N° 28 of 1997) and the Declaration of Helsinki, administered only when other therapeutic options do not exist (Górski *et al.*, 2009; Pirney *et al.*, 2011). In France, therapeutic made-to-order phage preparations from the Institute Pasteur (Paris and Lyon) were used until the beginning of the nineties. Today, a French practitioner, Alain Dublanchet, still uses commercial phage preparations (purchased in Russia and Georgia) to treat severe infections (Pirney *et al.*, 2011). Despite the absence of a specific framework for phage therapy (Górski *et al.*, 2009; Pirney *et al.*, 2011), a pilot clinical trial in burn wounds was approved by an ethical committee in Belgium (Kutter *et al.*, 2010). In the United States, a Food and Drug Administration (FDA)-approved phase I clinical trial was conducted. No safety concerns were found (Kutter *et al.*, 2010). A British phage therapy company conducted a phase I/II clinical trial in chronic otitis. This study was approved through the UK Medicines and Healthcare products Regulatory Agency and the Central Office for Research Ethics Committees ethical review process (Kutter *et al.*, 2010).

To avoid drug-licensing difficulties, some U.S.-based phage companies decided to first develop phage products for the decontamination of food, plants, fields, and livestock (Thiel, 2005). They hope to create revenue to fund research into human therapeutics and to familiarize the authorities and the general public about phages. Phages for decontaminating food plants, ready-to-eat meat, poultry products, cheese, and live animals that will be slaughtered for human consumption were approved by the FDA and are now in use. An increasing number of publications dedicated to this subject appears every year (Abuladze *et al.*, 2008; Leverentz *et al.*, 2003; Sabour and Griffiths, 2010). However, many scientists are concerned about the impact of such massive and widespread applications of phages on natural microbial communities because of their role in the horizontal spread of drug resistance and virulence genes (Sabour and Griffiths, 2010). It is presumed that phage-derived products (e.g., cell wall-degrading enzymes such as endolysins) can and probably will be licensed and marketed within a few years. Of course, these phage products lack the capacity of self-replication and adaptation in the infectious site, which is a disadvantage in comparison with live bacteriophages. Because of their chemical nature and resemblance to antibiotics, they may also select resistance, but presumably at a lower rate than antibiotics (Fenton *et al*, 2010; Hoopes *et al.*, 2009; Sabour and Griffiths, 2010).

Phage therapy has great potential in clinical practice, especially in chronic cases. As with antibiotic treatment, there are likely to be important evolutionary consequences related to phage therapy (Levin and Bull, 2004) if it is implemented widely and without sufficient oversight. In the past, phages have been used massively for the prophylaxis of bacterial infections in humans, as is shown in this chapter. In numerous cases, the descriptions of observed effects indicated positive results of phage therapy, although the parallel control studies either were not performed or were documented incompletely in view of current requirements. Also, the specificity of phages allows applying them for diagnostic purposes, sometimes even leading to the identification of nonculturable bacteria (Loessner *et al.*, 1996; Rees and Dodd, 2006). This practice should not be underestimated in planning future phage research.

REFERENCES

Abuladze, T., Li, M., Menetrez, M. Y., Dean, T., Senecal, A., and Sulakvelidze, A. (2008). Bacteriophages Reduce reduce experimental contamination of hard surfaces, tomato, spinach, broccoli, and ground beef by *Escherichia coli* O157:H7. *Appl. Environ. Microbiol.* **74**:6230–6238.

Adhya, S., and Merril, C. (2006). The road to phage therapy. *Nature* **443**:754–755.

Agafonov, B. I., Khokhlov, D. T., and Zolochevsky, M. A. (1984). Epidemiology of typhoid-paratyphoid infections and their prophylactics. *Military-Med. J.* **6**:36–40.

Alessandrini, A., and Doria, R. (1924). Il batteriofago nella terapia del tifo addominale. *Policlinico, sez prat.* **31**:109.

Alexandrov, M. B., Dyakova, A. A., Raikhelson, I. A., and Melnik, E. G. (1940). Bacteriophage therapy of typhoid fever. *In* "Abstracts of the Inter-institutional Conference on Bacteriophages Research, December 21–25, 1940, Moscow, 30".

Alexandrova, M. B., Zhivago, N. L., Alekseeva, L. N., and Zeitlenok, N. A. (1935). Phage therapy against dysentery. *J. Microbiol. Epidemiol. Immunol.* **11**:860–868.

Anpilov, L. I., and Prokudin, A. A. (1984). Prophylactic effectiveness of the dry polyvalent dysentery bacteriophage in organized communities. *Military-Med. J.* **5**:39–40.

Appelmans, R. (1921). Le dosage du bactériophage. *C.R. Soc. Biol.* **85**:1098–1099.

Astashkevich, H. M. (1950). Experience of application of phage therapy for treatment of typoid and paratyphoid diseases. *J. Microb. Epidemiol. Immun.* **7**:42.

Babalova, E. G., Katsitadze, K. T., Sakvarelidze, L. A., Imnaishvili, N. Sh., Sharashidze, T. G., Badashvili, V. A., Kiknadze, G. P., Meipariani, A. N., Gendzekhadze, N. D., Machavariani, E. V., Gogoberidze, K. L., Gozalov, E. I., *et al.* (1968). On the issue of prophylactic importance of the dry dysenteric bacteriophage. *J. Microbiol. Epidemiol. Immunol.* **2**:143–145.

Belikova, M. A. (1941). Experience of phage prophilaxis of summer dysentery among the young children performed in the city Stalingrad. *J. Microbiol. Epidemiol. Immunol.* **5–6**:142.

Beridze, M. A. (1938). Role of Bacteriophage Therapy in Combating of Purulent Skin Infections. Medgiz, Tbilisi, Georgia.

Blankov, B. I. (1941). Analysis of the results of phage prophylaxis of dysentery among the contacting people. Report # 1. *J. Microbiol. Epidemiol. Immunol.* **5–6**:125–131.

Blankov, B. I., and Zherebtsov, I. D. (1941). Experience on the multiple phaging of the contacting population in the fight against dysentery. Report # 2. *J. Microbiol. Epidemiol. Immunol.* **5–6**:131–136.

Bruynoghe, R., and Maisin, J. (1921). Essais de thérapeutique au moyen du bacteriophage. *C.R. Soc. Biol.* **85**:1120–1121.

Ceyssens, P.-J., Glonti, T., Kropinski, A. M., Lavigne, R., Chanishvili, N., Kulakov, L., Lashkhi, N., Tediashvili, M., and Merabishvili, M. (2011). Phenotypic and genotypic variations within a single bacteriophage species. *Virol. J.* **8**:134.

Chanishvili, N. (2009). A Literature Review of the Practical Application of Bacteriophage Research. EIBMV, Tbilisi, Georgia.

Chanishvili, N., Chanishvili, T., Tediashvili, M., and Barrow, P. A. (2001). Phages and their application against drug-resistant bacteria. *J. Chem. Technol. Biotechnol.* **76**:689–699.

Cienciala, A. M., Lebiedieva, N. S., and Materski, W. (2007). *In* "Katyń: A crime without punishment", pp. 226–229. Yale University Press.

Compton, A. (1929). Antidysentery bacteriophage in treatment of dysentery: A record of 66 cases treated, with inferences. *Lancet* **2**:273.

Costa Cruz, J. (1924). Le traitement des dysenteries bacillaires par le bacteriophage. *C.R. Soc Biol.* **91**:845.

De, S. N., and Chatterjee, D. N. (1953). Experimental study on mechanism of action of *Vibrio cholerae* on intestinal mucus membrane. *J. Pathol. Bacteriol.* **66**:559–562.

d'Herelle, F. (1935). Bacteriophage and Phenomenon of Recovery. TSU Press, Tbilisi, Georgia.

d'Herelle, F., and Eliava, G. (1921a). Unicite du bacteriae sur la lysine du bacteriophage. *C.R. Soc. Biol.* **85**:701–702.

d'Herelle, F., and Eliava, G. (1921b). Sur le serum antibacteriophage. *C.R. Soc. Biol.* **84**:719–721.

d'Herelle, F. (1917). Sur un microbe invisible antagoniste des bacilles dysentériques. *C.R. Acad. Sci. (Paris)* **165**:373–375.

Dublanchet, A., and Bourne, S. (2007). The epic of phage therapy. *Can. J. Infect. Dis. Med. Microbiol.* **18**:15–18.

Eaton, M. D., and Bayne-Jones, S. (1934). Bacteriophage therapy: Review of the principles and results of the use of bacteriophage in the treatment of infections. *JAMA* **23**:1769–1939.

Eliava, G. (1930). Au sujet de l 'adsorption du bacteriophage par les leucocytes. *C.R. Soc. Biol.* **105**:829–831.

Eliava, G., and Legraoux, R. (1921). Sur un liquide eu se maintiant invariable les nombre des bacteries des cultures. *Ann. Institute Pasteur* **35**:713–717.

Eliava, G., and Pozersky, E. (1921a). D'laction destructive des sels de quinine sur le bacteriophage de D'Herelle. *C.R. Soc. Biol.* **85**:139–141.

Eliava, G., and Pozersky, E. (1921b). Sur les caracter nouveaux presentes par le Bacille de Shiga avant resiste a l'action du bacteriophage de d'Herelle. *C.R. Soc. Biol.* **84:**708–710.

Eliava, G., and Suarez, E. (1927a). Au sujet de l 'ultrafiltration du corpuscule bacteriophage. *C.R. Soc. Biol.* **96:**462–464.

Eliava, G., and Suarez, E. (1927b). Dimensions du carpuscule bacteriophage. *C.R. Soc. Biol.* **96:**460–462.

Ermolieva, V. V. (1939). About bacteriophage and its application. *J. Microbiol. Epidemiol. Immunol.* **9:**9–17.

Fenton, M., Ross, P., McAuliffe, O., O'Mahony, J., and Coffey, A. (2010). Recombinant bacteriophage lysins as antibacterials. *Bioeng. Bugs* **1:**9–16.

Florova, N. N., and Cherkass, F. K. (1965). Results of mass application of polyvalent dysenteric bacteriophage. *J. Microbiol. Epidemiol. Immunol.* **3:**118–125.

Gamaleya, N. F. (1898). Bacterial lyzins: The enzymes destroying bacteria. *Russ. Arch. Pathol. Clin. Med. Bacteriol.* **6:**607–613.

Georgadze, I. A. (1974). Fifty years of the Tbilisi Scientific-Research Institute of Vaccine and Sera of the Ministry of Health of the USSR. *In* "Selected Articles of the Jubilee Dedicated to 50th Anniversary of the Tbilisi Institute of Vaccine and Sera". TIVS, Tbilisi, Georgia.

Georgadze, I. A., and Makashvili, E. G. (1979). George Eliava. *In* "The Georgian Soviet Encyclopedia," Vol. 4, p. 125.

Glonti, T., Chanishvili, N., and Taylor, P. W. (2010). Bacteriophage-derived enzyme that depolymerizes the alginic acid capsule associated with cystic fibrosis isolates of *Pseudomonas aeruginosa. J. Appl. Environ. Microbiol.* **108:**695–702.

Gnutenko, M. P. (1951). Treatment of *S. typhi* and *paratyphi* bacterial carriers with bacteriophages. *J. Microbiol. Epidemiol. Immunol.* **5:**56–60.

Golubtsov, G. V. (1940). Sero- and phage- therapies of dysentery among babies and infants. *In* "Selected Articles of the 1st Scientific Conference of the Bashkir Medical University, March 23–25, 1939", pp. 13–16. Ufa, Russia.

Górski, A., Międzybrodzki, R., Borysowski, J., Weber-Dąbrowska, B., Łobocka, M., Fortuna, W., Letkiewicz, S., Zimecki, M., and Filby, G. (2009). Bacteriophage therapy for the treatment of infections. *Curr. Opin. Invest. Drugs* **10:**766–774.

Hankin, E. H. (1896). L'action bactericide des eaux de la Jumna et du Gange sur le vibrion du cholera. *Ann. Inst. Pasteur* **10:**511.

Hausler, T. (2008). "Viruses vs. Superbugs". MacMillan, New York.

Hoopes, T., Stark, C. J., Kim, H. A., Sussman, D. J., Donovan, D. M., and Nelson, D. C. (2009). Use of a bacteriophage lysine PlyC, as an enzyme disinfectant against *Streptococcus equi. Appl. Environ. Microbiol.* **75:**1388–1394.

Ionov, I. D., Erez, S. L., and Goldenberg, E. Y. (1939). Specific therapy of dysentery. *In* "Variability of Microorganisms and Bacteriophage Research" (Proceedings of the Scientific Conference, 1936), pp. 337–346. Kiev, Ukraine.

Kagan, M. I., Kuznetsova, E. V., and Teleshevskaya, E. A. (1964). To the issue of epidemic effectiveness of the planned phaging in the day nurseries. *J. Microbiol. Epidemiol. Immunol.* **7:**89–102.

Kaplan, A. S. (1941). Duration of maintenance of the polyvalent dysenteric bacteriophage in the human organism in case of its prophylactic use. *J. Microbiol. Epidemiol. Immunol.* **5–6:**163.

Karamov, S. (1938). Experience of phage therapy for treatment of typhoid fever. *In* "Selected Articles of Azerbaijani Institute of Epidemiology and Microbiology", Vol. 6, pp. 101–105. Baku, Azerbaijan SSR.

Karpov, S. P. (1946). The specific bacteriophage in relation to the issue of combating typhoid and paratyphoid diseases. *J. Microbiol. Epidemiol. Immunol.* **1–2:**40–44.

Karpov, S. P. (1950). Mechanism of prophylactic action of typhoid-paratyphoid bacteriophage (dynamics of the phage release from the organism of the parenterally phaged white mice). *J. Microbiol. Epidemiol. Immunol.* **7:**43–45.

Karpov, S. P., and Yavorskaya, B. M. (1949). Mechanism of prophylactic action of typhoid paratyphoid bacteriophage (dynamics of the phage release from the organism of the phaged people). *J. Microbiol. Epidemiol. Immunol.* **7:**40–42.

Kazhal, N., and Iftimovich, R. (1968). From the History of Fight Against Bacteria and Viruses. Nauchnoe Izdatelstvo, Buchares.

Khawaldeh, A., Morales, S., Dillon, B., Alavidze, Z., Ginn, A. N., Thomas, L., Chapman, S. J., Dublanchet, A., Smithyman, A., and Iredell, J. R. (2011). Bacteriophage therapy for refractory *Pseudomonas aeruginosa* urinary tract infection. *J. Med. Microbiol.* **60:**1697–1700.

Kokin, G. A. (1941). Use of bacteriophages in surgery. Soviet Medicine (''Sovietskaya Meditsina'') **9**, 15–18.

Kokin, G. A. (1946). Phage therapy and phage prophylaxis of gas gangrene. *In* ''Experience of the Soviet Military Medicine during the Great Patriotic War 1941–1945'', Vol. 3, pp. 56–63. Medgiz, Moscow.

Krestovnikova, V. A. (1947). Phage treatment and phage prophylactics and their approval in the works of the Soviet researchers. *J. Microb. Epidemiol. Immunol.* **3:**56–65.

Krueger, A. P., and Scribner, E. J. (1941). Bacteriophage therapy. II. The bacteriophage: Its nature and its therapeutic use. *JAMA* **19:**2160–2277.

Kurochka, V. K., Karniz, A. F., and Khodyrev, A. P. (1987). Experiences of implementation of preventive anti-epidemic measures in the center of intestinal infections with water transmission mechanism of morbidity. *Military-Med. J.* **7:**36–37.

Kutateladze, M., and Adamia, R. (2010). Bacteriophages as potential new therapeutics to replace or supplement antibiotics. *Trends Biotechnol.* **28:**591–595.

Kutter, E., De Vos, D., Gvasalia, G., Alavidze, Z., Gogokhia, L., Kuhl, S., and Abedon, S. T. (2010). Phage therapy in clinical practice: Treatment of human infections. *Curr. Pharm. Biotechnol.* **11:**69–86.

Leverentz, B., Convay, W. C., Camp, M. J., Janisiewicz, W. J., Abuladze, T., Yang, M., Saftner, R., and Sulakvelidze, A. (2003). Biocontrol of *Listeria monocytogenes* on fresh-cut produce by treatment with lytic bacteriophages and a bacteriocin. *Appl. Environ. Microbiol.* **69:**4519–4526.

Levin, B. R., and Bull, J. J. (2004). Population and evolutionary dynamics of phage therapy. *Nat Rev. Microbiol.* **2:**166–173.

Lipkin, and Nikolskaya (1940). Experience of phage therapy of dysentery. *In* ''Selected Articles of the Kuibishev Red Army Military-Medical Academy'', Vol. 4, pp. 193–198. KRAMMA, Kuibishev, Russian SSR.

Loessner, M. J., Rees, C. E., Stewart, G. S., and Scherer, S. (1996). Construction of luciferase reporter bacteriophage A511::luxAB for rapid and sensitive detection of viable *Listeria* cells. *Appl. Environ. Microbiol.* **62:**1133–1140.

Lurie, M. N. (1938). Treatment of dysentery and haemolytic intestinal diseases among children and adults. *In* ''Selected articles of Azerbaijani Institute of Epidemiology and Microbiology, '' Vol. 6, pp. 31–34.

Manolov, D. G., Sekunova, V. N., and Somova, E. E. (1948). Experience of therapy of typhoid fever by intravenous administration of the phage. *J. Microbiol. Epidemiol. Immunol.* **4:**33.

Markoishvili, K., Tsitlanadze, G., Katsarava, R., Morris, J. G., and Sulakvelidze, A. (2002). A novel sustained-release matrix based on biodegradable poly(ester amide)s and impregnated with bacteriophages and an antibiotic shows promise in management of infected venous stasis ulcers and other poorly healing wounds. *Intl. J. Dermatol.* **41:**453–458.

Melnik, M. I., Nikhinson, I. M., and Khastovich, R. I. (1935). Phage prophylaxis of dysentery. *In* "Proceedings of the Mechnikov Institute in Kharkov", Vol. 1(1):MIKH, Kharkov, Ukrainian SSR89.

Merabishvili, M., Pirnay, J.-P., Verbeken, G., Chanishvili, N., Tediashvili, M., Lashkhi, N., Glonti, T., Krylov, V., Mast, J., Van Parys, L., Lavigne, R., Volckaert, G., *et al.* (2009). Quality-controlled small-scale production of a well-defined bacteriophage cocktail for use in human clinical trials. *PLoS One* **4:**e4944.

Mikaelyan, V. G. (1949). Phage therapy of the bacterial carriers of typhoid bacteria. *In* "Proceedings of the Yerevan Medical Institute", Vol. 6, pp. 54–59. YMI, Yerevan, Armenian SSR.

Nattan-Larrier, L., Eliava, G., and Riegard, L. (1931). Bacteriophage et permeabilite placentaire. *C.R. Soc. Biol.* **106:**794–797.

Parfitt, T. (2005). Georgia: An unlikely stronghold for bacteriophage therapy. *Lancet* **365:**2166–2167.

Pirnay, J.-P., De Vos, D., Verbeken, G., Merabishvili, M., Chanishvili, N., Vaneechoutte, M., Zizi, M., Laire, G., Lavigne, R., Huys, I., Van den Mooter, G., Buckling, A., *et al.* (2011). The phage therapy paradigm: Prêt-à-porter or sur-mesure? *Pharm. Res.* **28:**934–937.

Plankina, Z. A., Nikonov, A. G., Sayamov, R. M., and Kotliarova, R. I. (1961). Control of cholera in Afghanistan. *J. Microbiol. Epidemiol. Immunol.* **32:**202–204.

Rakieten, M. L. (1932). Studies with *Staphylococcus* bacteriophage. I. The preparation of polyvalent Staphylococcus bacteriophage. *Yale J. Biol. Med.* **4:**807–818.

Rees, C., and Dodd, C. (2006). Phage for rapid detection and control of bacterial pathogens in food. *Adv. Appl. Microbiol.* **59:**159–186.

Rice, T. B. (1930). Use of bacteriophage filtrates in treatment of suppurative conditions: Report of 300 cases. *Am. J. Med. Sci.* **179:**345–360.

Rodigina, A. M. (1938). Pneumococcal bacteriophage: Its application for treatment of the ulcerous corneal serpens. Perm, Russia.

Ruchko, I., and Tretyak, K. (1936). Therapeutic effect of *Staphylococcus* phage for oral and dental infections. *Soviet Stomatology ("Sovetskaya Stomatologia")* **1:**11–20.

Sabour, P. M., and Griffiths, M. W. (2010). Bacteriophage in the Control of Food and Waterborne Pathogens. ASM Press, Washington, DC.

Sapir, I. B. (1939). Observations and recommendations related to phage therapy of dysentery. *In* "Proceedings of the Moscow Institute of Infectious Diseases after I.I. Mechnikov", pp. 135–151.

Sayamov, R. M. (1963). Treatment and prophylaxis of cholera with bacteriophage. *Bull. WHO* **28:**361–367.

Schless, R. A. (1932). *Staphylococcus aureus* meningitis: Treatment with specific bacteriophage. *Am. J. Dis. Child.* **44:**813–822.

Smith, K. E. (1996). Remembering Stalin's Victims: Popular Memory and the End of the USSR. Cornell University Press, NY.

Stout, B. F. (1933). Bacteriophage therapy. *Texas State J. Med.* **29:**205–209.

Sulakvelidze, A., Alavidze, Z., and Morris, J. G., Jr. (2001). Bacteriophage Therapy. *Antimicrob. Agents Chemother.* **45:**649–659.

Summers, W. C. (1999). Felix d'Herelle and the Origins of Molecular Biology. Yale University Press, New Haven, CT.

Summers, W. C. (2001). Bacteriophage therapy. *Annu. Rev. Microbiol.* **55:**437–451.

Thiel, K. (2004). Old dogma, new tricks: 21st century phage therapy. *Nat. Biotechnol.* **22:**31–36.

Tsulukidze, A. P. (1938). Application of phages in urology. *Urology ("Urologia")* **15:**10–13.

Tsulukidze, A. P. (1940). Phage treatment in surgery. *Surgery ("Khirurgia")* **12:**132–133.

Tsulukidze, A. P. (1941). Experience of Use of Bacteriophages in the Conditions of War Traumatism. Gruzmedgiz, Tbilisi, Georgia.

Twort, F. (1915). An investigation on the nature of ultramicroscopic viruses. *Lancet* **11:**1241.

Verbeken, G., De Vos, D., Veneechoutte, M., Merabishvili, M., Zizi, M., and Pirnay, J. P. (2007). European regulatory conundrum of phage therapy. *Future Microbiol.* **2:**485–491.

Villareal, L. P. (2005). Overall issues of viral and host evolution. *In* "Virus and Evolution of Life" (I. P. Villareal, ed.), pp. 1–28. ASM Press, Washington, DC.

Vlasov, K. F., and Artemenko, E. A. (1946). Treatment of chronic dysentery. *Soviet Med. ("Sovetskaya Medicina")* **10:**22–28.

Williams, P. D. (2010). Darwinian interventions: Taming pathogens through evolutionary ecology. *Trends Parasitol.* **26:**83–92.

Zabrezhinsky, L. M., and Gorstkina-Shevendrova, L. A. (1946). Phage therapy of typhoid fever. *J. Microbiol. Epidemiol. Immunol.* **4,** 25–27. (1940). Experience of phage therapy of dysentery. *In* "Selected Articles of the Kuibishev Red Army Military-Medical Academy", **4,** pp. 193–198. KRAMMA, Kuibishev, Russian SSR.

CHAPTER 2

Phage as a Modulator of Immune Responses: Practical Implications for Phage Therapy

Andrzej Górski,[*,†,#,1] Ryszard Międzybrodzki,[*,#] Jan Borysowski,[†] Krystyna Dąbrowska,[*] Piotr Wierzbicki,[†] Monika Ohams,[†] Grażyna Korczak-Kowalska,[†,‡] Natasza Olszowska-Zaremba,[†] Marzena Łusiak-Szelachowska,[*] Marlena Kłak,[*] Ewa Jończyk,[*] Ewelina Kaniuga,[†] Aneta Gołaś,[†] Sylwia Purchla,[†] Beata Weber-Dąbrowska,[*,#] Sławomir Letkiewicz,[#,§] Wojciech Fortuna,[*,#] Krzysztof Szufnarowski,[#,||] Zdzisław Pawełczyk,[#] Paweł Rogóż,[#,¶] and **Danuta Kłosowska[†]**

Contents

[*] Bacteriophage Laboratory, Ludwik Hirszfeld Institute of Immunology and Experimental Therapy, Polish Academy of Sciences, Wrocław, Poland
[#] Phage Therapy Unit, Ludwik Hirszfeld Institute of Immunology and Experimental Therapy, Polish Academy of Sciences, Wrocław, Poland
[†] Department of Clinical Immunology, Transplantation Institute, Medical University of Warsaw, Warsaw, Poland
[‡] Department of Immunology, Warsaw University, Warsaw, Poland
[§] Katowice School of Economics, Katowice, Poland
[||] Regional County Hospital and Research Center, Wrocław, Poland
[¶] Department of Orthopedics, Wrocław Medical University, Wrocław, Poland
[1] Corresponding author, E-mail: agorski@ikp.pl

Advances in Virus Research, Volume 83
ISSN 0065-3527, DOI: 10.1016/B978-0-12-394438-2.00002-5

Abstract

Although the natural hosts for bacteriophages are bacteria, a growing body of data shows that phages can also interact with some populations of mammalian cells, especially with cells of the immune system. In general, these interactions include two main aspects. The first is the phage immunogenicity, that is, the capacity of phages to induce specific immune responses, in particular the generation of specific antibodies against phage antigens. The other aspect includes the immunomodulatory activity of phages, that is, the nonspecific effects of phages on different functions of major populations of immune cells involved in both innate and adaptive immune responses. These functions include, among others, phagocytosis and the respiratory burst of phagocytic cells, the production of cytokines, and the generation of antibodies against nonphage antigens. The aim of this chapter is to discuss the interactions between phages and cells of the immune system, along with their implications for phage therapy. These topics are presented based on the results of experimental studies and unique data on immunomodulatory effects found in patients with bacterial infections treated with phage preparations.

ABBREVIATIONS:

PPhs	purified phages
PhLs	phage lysates
Ig	Immunoglobulin
IFN	interferon
Il	interleukin
PT	phage therapy
RB	respiratory burst
ROS	reactive oxygen species
TLRs	Toll-like receptors
TCR	T-cell receptor
mAbs	monoclonal antibodies
SPL	Staphage Lysate

I. BACKGROUND

The vast majority of studies on phage biology have traditionally focused on the interactions of bacteriophages with bacterial cells. However, bacteriophages can also interact with some populations of mammalian cells, especially with immune cells. In fact, the first studies on the interactions between bacteriophages and immune cells were conducted by Felix d'Herelle shortly after the discovery of phages (d'Herelle, 1922). Since then, a considerable body of experimental data has accumulated to show that phages can substantially affect cells of the immune system both *in vitro* and *in vivo*.

Generally, the interactions between phages and immune cells include two main aspects. The first is phage immunogenicity, that is, the natural capacity of phages to induce specific immune responses, especially the production of antibodies against phage antigens (largely capsid proteins). The other aspect includes the immunomodulatory activity of phage preparations, that is, the nonspecific effects of phages on different functions of major populations of immune cells involved in both innate and adaptive immune responses. In fact, phages can affect immune functions as diverse as phagocytosis or the respiratory burst of phagocytic cells, the production of antibodies, and T-cell proliferation.

Studies on the effects of phages on immune cells are important for different medical applications of phages. For example, phage phi X174 has been used for over 30 years as an antigen to evaluate humoral immunity in patients with immunodeficiencies (Bearden *et al.*, 2005). Furthermore, phage particles can be used as vehicles for vaccine antigens; promising results of experimental studies suggest that such vaccines can also be used in humans (Clark and March, 2004). Obviously, knowledge about phage immunogenicity is essential for the rational design of such vaccines. Moreover, studies on phage interactions with immune cells are very important for the rational use of phage therapy. For instance, specific antibodies are considered to be one of the most important factors limiting the therapeutic efficacy of phages *in vivo* (Sulakvelidze *et al.*, 2001). Therefore, of paramount importance are data showing that the intensity of the anti-phage humoral response can vary depending on the route of administration of phage preparations. Research into the effects of phages on phagocytic cells may allow for better understanding of mechanisms involved in the elimination of bacteria by phages *in vivo*.

The main objective of this chapter is to present the current state of research on the interactions between bacteriophages and immune cells. It also discusses implications of those interactions for biomedical applications of bacteriophages, especially for phage therapy. This chapter presents data regarding not only purified phages (PPhs), but also phage lysates (unpurified phage preparations; PhLs), because in fact it is Phls rather than PPhs that have been used at major centers of phage therapy, such as the Hirszfeld Institute of Immunology and Experimental Therapy, Wrocław, Poland (Ślopek *et al.*, 1987; Weber-Dąbrowska *et al.*, 2000a).

II. PHAGE IMMUNOGENICITY

A. Anti-phage humoral responses

Antiviral antibodies are one of the main components of antiviral immune responses. In the case of pathogenic viruses, these antibodies can exhibit four main activities: virus neutralization, antibody-dependent cellular cytotoxicity, antibody-dependent cell-mediated virus inhibition, and phagocytosis (Forthal and Moog, 2009).

Antibodies that were examined in the vast majority of studies on phage immunogenicity are neutralizing antibodies. Essentially, these are defined as antibodies that bind epitopes within those parts of the virion essential for infecting the host cells (Forthal and Moog, 2009). In the case of phages, neutralizing antibodies inhibit the infection of the host bacterial cells by binding to the phage tails (Jerne and Avegno, 1956). However, one must be aware that phage–antibody interactions do not necessarily mean phage inactivation. Usually, binding of antibodies to proteins engaged in infection may impede it and result in visible loss of phage antibacterial activity. Other proteins may be bound by antibodies without any visible effect on phage viability (Jerne and Avegno, 1956).

Anti-phage-neutralizing antibodies are considered one of the most important factors potentially limiting the efficacy of phage therapy (Sulakvelidze *et al.*, 2001). Indeed, a considerable body of experimental data indicates that these antibodies may reduce the therapeutic effectiveness of phages. First, many studies have shown that sera of nonimmunized humans and animals (or, more specifically, animals not immunized with phage during the experiment) in fact have a low level of phage-neutralizing antibodies (so-called "natural antibodies") (Jerne 1956; Kamme 1973; Kucharewicz-Krukowska and Ślopek, 1987). Thus, in some individuals, phage-neutralizing antibodies can be present in the serum even before phage administration. It is believed that the presence of such antibodies can be explained by the well-documented omnipresence of phages in different environments, as well as in food and within normal microflora, where phages are present in very large numbers along with bacteria (Górski and Weber-Dąbrowska 2005). This implies constant natural "immunization" of humans (as well as animals) with phage antigens. Second, neutralizing antibodies are produced in high titers following the systemic administration of phages to animals (Jerne 1956).

Third, the clearance of T7 phage from the blood of mice is slower in B-cell-deficient animals compared with wild-type mice (Srivastava *et al.*, 2004). Taken together, these data show that neutralizing antibodies indeed may decrease the efficacy of phage therapy substantially. However, the intensity of anti-phage humoral response may vary depending on the phage type; some phages are very weak immunogens and require repeated injection coupled with administration of adjuvant to induce detectable antibody titers (Sulakvelidze and Barrow 2005).

Our group has also examined anti-phage-neutralizing antibodies in sera of patients with bacterial infections treated with PhLs at the Phage Therapy Unit at Hirszfeld Institute of Immunology and Experimental Therapy, Wrocław, Poland. Weak activities in sera were detected in patients prior to phage administration up to a serum dilution of 1:100. Antibody activity increased in some patients during the treatment up to a serum dilution of 1:1500. Interestingly, no significant increase in antibody activity was noted in patients who received PhLs orally (Figs. 1 and 2).

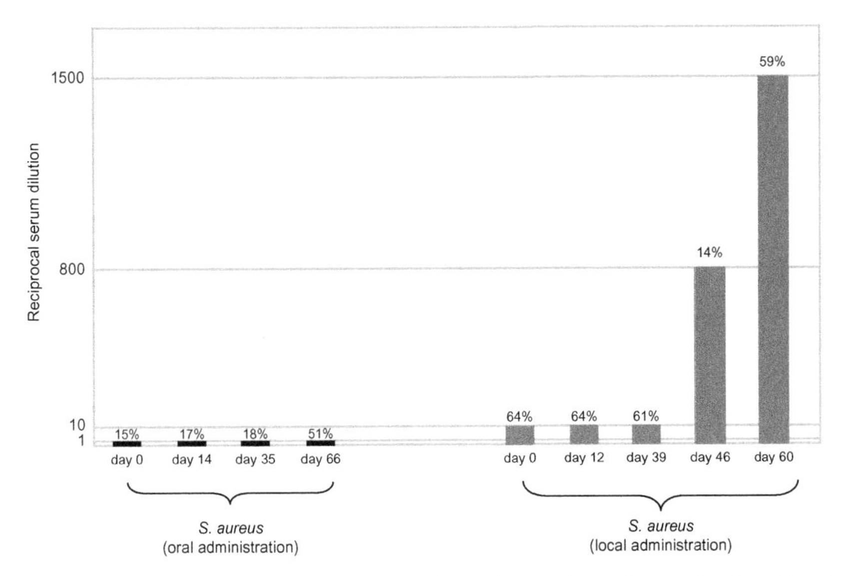

FIGURE 1 Inactivation of phage by patients' sera during phage therapy (oral and local administration). Anti-phage activity of patients' sera was assayed as described in detail earlier (Kucharewicz-Krukowska and Ślopek, 1987). Briefly, 50 μl of phage (10^6 PFU/ml) was mixed with 450 μl of serum and incubated at 37 °C for 30 min. A sample was subsequently added to a bacterial strain on agar plates and incubated at 37 °C for 8 hr. The degree of phage neutralization by sera was determined in reference to the control (bacterial broth). Results obtained for two patients are shown: one patient treated orally (left) and another locally (right) with *S. aureus* phage. The inactivation of phage is presented as the highest reciprocal serum dilution that inactivated at least 10% of phages, and the exact percentage of neutralization of phages by patients' sera during phage therapy is shown over the bars.

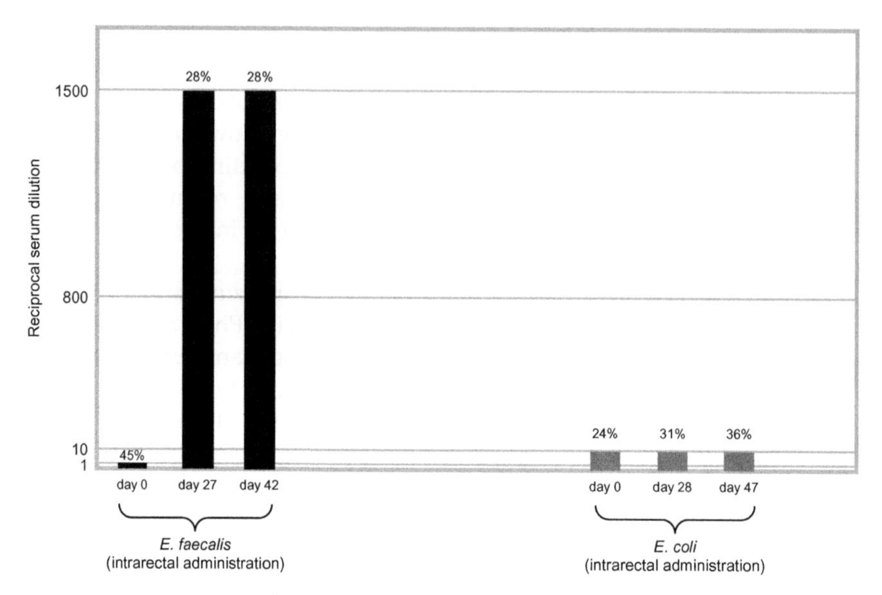

FIGURE 2 Inactivation of phage by patients' sera (intrarectal administration). Anti-phage-neutralizing antibody responses in two patients are shown: one patient treated intrarectally with *Enterococcus faecalis* phage (left) and another intrarectally with *E. coli* phage (right). Data presented as in Figure 1.

Antibody responses to phages appear to be also dependent on the initial status of patients, who could be divided into ''responder'' and ''nonresponder'' groups (Łusiak-Szelachowska *et al.*, manuscript in preparation).

Immunogenicity of phages, coupled with ease of their quantification and lack of toxicity, has allowed for the use of phi X174 phage for the evaluation of humoral immunity in diagnosing and monitoring of patients with primary and secondary immunodeficiencies (Ochs *et al.*, 1971; Wedgwood *et al.*, 1975). This phage is a potent T-cell-dependent neoantigen that triggers typical primary, secondary, and tertiary humoral responses after intravenous injection to patients (Wedgwood *et al.*, 1975). It is currently considered one of the standard antigens used to evaluate humoral immunity in clinical medicine (Bearden *et al.*, 2005). However, in fact the generation of specific antibodies to phi X174 requires cooperation among antigen-presenting cells, B cells, and T cells. Therefore, its abnormalities can result not only from the dysfunction of B cells, but of T cells, complement, and adhesive molecules as well (Andrews *et al.*, 1997). So far, phi X174 has been employed for diagnosing and monitoring of patients with a wide range of both primary and secondary immunodeficiencies (Bearden *et al.*, 2005).

When interpreting the results of studies on anti-phage antibodies, one should take into account that their production can be stimulated not only by phage particles themselves, but also by some components of bacterial cells present in phage preparations, especially lipopolysaccharides (LPS). This was shown in a study in which the production of anti-phage antibodies was induced by the systemic administration of T2 phage to mice (Michael and Kuwatch, 1969). The study revealed that systemic administration of 50 μg of LPS of *Escherichia coli* B (without preimmunizing animals with the phage) increased the level of anti-T2-neutralizing antibodies in the blood of the mice; a similar effect was found following the administration of Shigella LPS. However, the level of anti-phage antibodies was substantially lower than that found following the administration of T2 phage itself at the dose of 10^9 plaque-forming units (PFU)/mouse. Moreover, the phage-neutralizing activity of the sera of mice was elevated for a shorter time following the administration of LPS compared with the T2 phage itself. Unfortunately, the phenomenon of LPS-induced generation of anti-phage antibodies was not examined in more detail. In particular, the authors did not attempt to determine the minimal dose of LPS capable of inducing anti-phage antibodies. Nevertheless, the results of this study are important because LPS is one of the components of bacterial cells known to be present in small concentrations, even in PPhs. Thus, its presence in phage preparations needs to be taken into account when designing (and interpreting the results of) studies on anti-phage antibodies, as well as in phage therapy. However, immunization with intact bacterial cells has no effect against phages and vice versa, as bacteriophages are distinct from bacteria in their antigenic characteristics (Kańtoch 1956).

B. Anti-phage cellular responses

It is well known that T cells play an important role in the response to viral infections, with both virus-specific $CD8^+$ T cells and $CD4^+$ T cells being engaged in the cellular arm of adaptive immunity triggered during such infections (Amanna and Slifka 2011). Therefore, it could be expected that administered phage should also induce some form of cellular response. However, data on anti-phage cellular responses are very scanty compared with the knowledge about anti-phage humoral responses. In fact, to our knowledge, there is only one report in the literature clearly indicating that administration of phage can elicit cellular responses both *in vitro* and *in vivo*. In that study, Langbeheim *et al.* (1978) investigated the response of guinea pigs to MS-2 phage and a synthetic conjugate corresponding to a defined region of phage coat protein. The animals were sensitized by the antigens, and the sensitization *in vivo* was evaluated by intradermal injection of the test antigens, resulting in local erythema and induration. *In vitro* cellular sensitization was determined by measuring proliferative responses of lymph node cells to the test antigens. Strong *in vivo* reactions to injected phage were noted in all animals. Interestingly, whole phage particles induced stronger sensitization than the conjugate. Lymphocytes from animals sensitized with phage gave brisk responses to the virus but no reaction was found to the conjugate. Accordingly, animals sensitized with the conjugate had lymphocytes that reacted only to the specific antigen. However, Srivastava *et al.* (2004) demonstrated that the kinetics of phages in the blood is similar in immunocompetent and T-cell-deficient mice. This result shows that T cells do not have any significant role in the inactivation of phages *in vivo*.

C. Bacteriophage proteins' effects on mammalian immunity

Immunological effects of bacteriophages are fairly complex, as most bacteriophages are very complex structures, with a number of different capsid proteins exposed for external interactions. Therefore, the general effect of a phage is a resultant of multicomponent activity of capsid proteins. By breaking this effect down for particular proteins, we can gain a sensitive tool for dealing with anti-phage immunity.

The most spectacular effects of phages are antibodies induced by them (see Section II.A). At the early stage of phage investigations, phage proteins' antigenicity and antibody induction served as tools for characterization of evolutionary relatedness. The idea referred to serological cross-reactions, as the intensity of these reactions reflects the relatedness between phages, generally decreasing with growing evolutionary distance between species (Stent 1963). Permissible interactions of phage capsids with antisera reflect homologies and similarities among phages. Nowadays we know that the homologies of phage genomes usually accumulate in specific conserved regions (or even parts of a gene), whereas other regions are highly variable (Krisch and Comeau, 2008). Similarly, the arrangement of antigens on a capsid is not even, that is, potential similarities between phages may occur in selected parts only. Some capsid components are antigenically specific for a phage, whereas others can be shared by different phage species (Tikhonenko *et al.*, 1976).

Antibodies against selected phage gene products were also one of the main tools for recognition of gene function in phages. Structural proteins can be localized on capsids by means of specific antibodies, giving accurate and exact information on virus particle arrangement (Ishii and Yanagida, 1977; Yanagida, 1972).

Some interesting conclusions on phage proteins' immunogenicity can be drawn from investigations on engineered phages. They cannot always be simply transferred to wild-type phages, but these studies give important information about molecular details that probably determine phage effects on immunity. Studies of interactions of phage-displayed short peptides (T7 phage) made by Sokoloff *et al.* (2000) revealed that a dependency exists between the properties of phage surface peptides and the innate immune system response. Natural antibodies, that is, those preexisting in mammalian serum, recognize phages and start complement activation. However, providing capsids with carboxy-terminal lysine or arginine residues protects the phage against complement-mediated inactivation by binding C-reactive protein in rats (Sokoloff *et al.*, 2000).

Another instance of molecular manipulations that change the effect of phage on innate immunity was described by Merril *et al.* (1996), who investigated phage λ derivatives in germ-free mice. The overall mechanism of phage clearance involves reticuloendothelial system filtration, which is usually proposed as the main path of phage removal from nonimmunized mammals. A single mutation in the major capsid protein gene (G → A transition) caused a substitution of the amino acid lysine for glutaminic acid, which determined a long-circulating phenotype. These long-circulating mutants were able to persist in the mice circulation much longer and also had better antibacterial activity in bacteriemic mice than the corresponding wild-type strain (Merrill *et al.*, 1996). Further studies have confirmed that indeed this single specific substitution confers the "long circulating phenotype" (Vitiello *et al.*, 2005), which emphasizes the importance of such subtle differences in phage antigenic structure for their potential use in therapy.

In the phage T4, an opposite phenomenon, that is, "short circulating phage," was demonstrated by Dąbrowska *et al.* (2007). After T4 lost its nonessential decorating protein Hoc, its clearance from the mouse circulation occurred faster. This was not a change of a protein but its deletion, which strongly suggests that some proteins are able to mediate or moderate phage interactions with the immune system. The phage defective in the *hoc* gene was also described as more adhesive to mammalian cells than the wild-type phage (Dąbrowska *et al.*, 2004). It is not very likely that adhesion is really specific (see Section III.B); nevertheless, its intensity might influence phage susceptibility to filtration and removal from a higher organism. Elevated adhesion properties of Hoc-deprived phage are in line with the observations of Sathaliyawala *et al.* (2010), who reported that *hoc*$^-$*soc*$^-$ phages aggregate frequently, as visualized in cryoelectron microscopic images, apparently through interactions between capsids. Hoc is a prokaryotic protein that contains immunoglobulin-like domains in its structure (Bateman *et al.*, 1997). Possibly this makes it a modulator of T4 phage effects in mammals (Dąbrowska *et al.*, 2006, 2007), while it is still questionable what the exact role of Hoc on T4 capsid is. It was proposed that Hoc might prevent the aggregation of phage particles in the infected cell, where the concentration of newly assembled phage can be quite high or, oppositely, the exposed Ig-like domains may interact weakly and nonspecifically to surface carbohydrates and other molecules of bacteria. Adhesion to bacteria or biofilms may provide the phage with a survival advantage (Sathaliyawala *et al.*, 2010). Bioinformatics studies revealed that Ig-like domains are quite common in phage genomes (Fraser *et al.*, 2006), which may suggest that a similar adhesion-modulating effect of some capsid proteins can be expected in other phage strains.

Receptor-targeted, that is, chemically modified, phage M13 was also investigated for its time of clearance. Molenaar *et al.* (2002) showed that conjugation of either galactose or succinic acid groups to phage coat proteins resulted in a substantially reduced plasma half-life of the phage. However, a phage display of random peptides (C-X7-C library) on the T7 phage coat was shown not to influence the clearance of phages in murine blood (Srivastava *et al.*, 2004). All these studies may play a crucial role in vaccine design and other molecular medicine approaches that use phages as a platform or delivery vector, but they obviously can also shed some light on phage effectiveness and side activities in phage therapy.

Of note, chemical modification of phages by conjugation of monomethoxy-polyethylene glycol (mPEG) to its proteins renders phages less immunogenic: PEGylated phages elicit diminished levels of cytokine [interferon (IFN)-γ and interleukin (IL)-6] and their circulation half-life is increased markedly, which suggests that this approach may be of interest for increasing efficacy of PT (Kim *et al.*, 2008).

III. IMMUNOMODULATORY ACTIVITY OF PHAGE PREPARATIONS

A. Effects of phages on phagocytic cells

The activity of phagocytic cells constitutes one of the essential functions of antibacterial immune responses (Silva 2010). Investigating phage interactions with major populations of phagocytic cells may thus verify whether phages could eliminate bacteria *in vivo* not only by direct bactericidal activity, but also by activating phagocytic cells. Furthermore, interactions between endogenous phages (known to be present in very large numbers within normal microflora; Górski and Weber-Dąbrowska 2005) and phagocytic cells may be relevant for understanding the complexity of the immune defenses against bacterial infections.

1. Effects of phages on phagocytosis

The first description of the influence of phage on phagocytosis was a report by Felix d'Herelle, who studied the effects of a *Shigella* phage on phagocytosis of *Shigella* by guinea pig "leukocytes" (the author did not specify whether he was dealing with peritoneal cells, of which the majority are macrophages, or peripheral leukocytes, most of which are granulocytes) (d'Herelle 1922). After the coincubation of bacteria, phage, and leukocytes for 10 min, the phagocytic index of the cells increased dramatically compared with the control (cells cultured with bacteria in the absence of phage). Another interesting finding was that the development of phage resistance by bacteria was associated with the development of their resistance to phagocytosis. d'Herelle suggests that phages act as specific opsonins markedly facilitating bacterial phagocytosis. This effect is mediated by a soluble factor(s) present in phage preparations.

Kańtoch *et al.* (1958) demonstrated that the T5 phage does not affect the phagocytosis of *E. coli* by guinea pig granulocytes. Furthermore, phages adsorbed to bacteria may remain biologically active upon their uptake by these cells. In contrast, after the completion of bacterial phagocytosis, phages are no longer able to lyse intracellular bacteria (Kańtoch and Szalaty, 1960). However, work by the same authors suggests that the T2 phage may diminish the ability of horse leukocytes to phagocytose bacteria from different species, including *Staphylococcus aureus*, *E. coli*, and *Mycobacterium tuberculosis* (Kańtoch *et al.*, 1958). This effect was found to be dose dependent—phage in a concentration of 10^{10}/ml caused almost complete inhibition of phagocytosis, whereas lower concentrations of phage caused only a slight decrease of granulocyte activity. The process of inhibition was also time dependent (increasing incubation time and ingestion of phage particles decreased phagocyte ability to ingest bacteria) and was brought about by living as well as by heat- and serum-inactivated phage. Interestingly, the most efficient inhibition occurred when phages were inactivated by antibodies, which suggests that immune complexes consisting of phages and anti-phage antibodies may be especially active in diminishing granulocyte phagocytosis of bacteria.

Another study focused on the effects of two PPhs, T4 and F8 (a *Pseudomonas aeruginosa* phage), on the phagocytosis of *E. coli* (Przerwa *et al.*, 2006). *In vitro* experiments showed that both T4 and F8 inhibited the phagocytosis of bacteria in a dose-dependent manner when preincubated with phagocytic cells. However, preincubation of *E. coli* with T4 resulted in a small increase in the efficiency of phagocytosis. Coincubation of phages, bacteria, and phagocytic cells led to a reduction in the intensity of phagocytosis of *E. coli*. Similar results were obtained with neutrophils and monocytes. In the same study in experiments performed on mice inoculated with bacterial cells, T4 was found to stimulate the intensity of phagocytosis by neutrophils. However, in uninfected mice, T4 weakly reduced the phagocytosis by monocytes while having no effect on the intensity of phagocytosis by neutrophils.

In vitro monitoring of phagocytosis of *S. aureus* by neutrophils isolated from patients subjected to phage therapy has revealed that such therapy may further decrease this function in patients in whom phagocytic activity was lowered prior to the beginning of treatment. However, no correlation was found between alterations in phagocytosis and the outcome of therapy. Three months after completion of therapy, neutrophil functions returned to normal. Moreover, phage therapy accelerated the turnover of neutrophils as manifested by an increase in the number of their immature forms and a concomitant decrease in the number of mature cells (Weber-Dąbrowska *et al.*, 2002).

2. Effects of phages on the respiratory burst

Respiratory burst (RB) is a rapid increase in the production of reactive oxygen species (ROS) during the phagocytosis of microbes. On the one hand, RB is an essential component of innate immunity enabling phagocytic cells to eliminate microbes. However, on the other hand, excessive production of ROS may result in the induction of oxidative stress in cells. Both pathogenic bacteria and viruses can induce oxidative stress in host cells during infection (Schwarz, 1996; Victor *et al.*, 2004). Therefore, examining the effects of phages on the production of ROS is essential in verifying the safety of phage therapy.

The first study to evaluate the influence of phages on ROS generation showed that purified T4 phage preparation induced a very weak RB compared with bacterial cells *in vitro* both in monocytes and in neutrophils (Przerwa *et al.*, 2006). It is noteworthy that the titer of phage used in these experiments (10^9 PFU/ml) was one order of magnitude higher than the titer of bacteria (10^8 colony-forming unit/ml). Furthermore, phages reduced, in a dose-dependent manner, *E. coli*-induced RB when preincubated with bacteria, the effect being significant with higher phage titers (10^9 and 10^{10} PFU/ml).

Another study showed that a purified T4 phage preparation can inhibit the RB induced by both LPS and bacterial cells (Międzybrodzki *et al.*, 2008). It was also shown that neither purified preparations nor lysates of staphylococcal A3/R phage induced a significant RB in monocytes or neutrophils *in vitro* (Borysowski *et al.*, 2010).

3. Effects of phages on other functions of phagocytic cells

Our group also examined the effects of phages on the migration of phagocytic cells, intracellular killing of bacteria, and expression of Toll-like receptors (TLRs). The majority of PPhs and PLs did not affect human granulocyte and mononuclear cell migration *in vitro* [the only exception was some inhibitory activity of T4 phage known to contain in its capsid Hoc protein, which, according to our hypothesis (Dąbrowska *et al.*, 2007), may have some immunomodulatory activities]. Furthermore, phage preparations did not influence intracellular killing of bacteria by these cells (Kurzepa, 2011).

Toll-like receptors are one of the most important classes of receptors of phagocytic cells involved in innate immune responses, including inflammatory reactions (Kumar *et al.*, 2009). In particular, the TLR-mediated activation of monocytes or macrophages by pathogenic bacteria and viruses results in an increase in the production of pro-inflammatory cytokines, ROS, and nitric oxide. The natural ligand for TLR4 and TLR2 receptors is LPS. Our study showed that T4 phage preparations have no effect on the percentage of $TLR2^{+}$ and $TLR4^{+}$ cells in either unstimulated or LPS-activated monocytes *in vitro*. Furthermore, a purified preparation of T4 had no effect on the expression of either TLR2 or TLR4, but the T4 lysate slightly increased the expression of TLR4 on $CD14^{+}/CD16^{-}$ cells (G. Korczak-Kowalska *et al.*, unpublished data). These findings are important for phage therapy, as they indicate that phage preparations are not likely to exert pro-inflammatory activity in a TLR-dependent mechanism.

B. Phage interactions with T and B cells

Data show that phage preparations may also modulate immune functions by direct interactions with T and B cells. In general, PPhs seem to cause immunosuppressive effects; lysates of anti-staphylococcal phage may be stimulatory, while other preparations may mediate immunosuppression.

Mankiewicz *et al.* (1974) reported that tuberculous guinea pigs inoculated intraperitoneally with mycobacteriophages showed depression of their skin reactions to tuberculin. Furthermore, these authors demonstrated that mycobacteriophages could inhibit, in a dose-dependent manner, the phytohemagglutinin (PHA)-induced activation of lymphocytes when added to lymphocyte cultures. That phage may indeed exert immunosuppressive activity *in vitro* has been confirmed by studies showing that a purified T4 phage preparation inhibited human T-cell proliferation induced via the CD3–TCR complex (Górski *et al.*, 2006). However, Zimecki *et al.* (2003) showed that purified preparations of an *S. aureus* phage exert costimulatory effects on splenocytes activated by a suboptimal concentration of the mitogen concanavalin A (Con A).

Another line of research is the adhesion of phages to immune cells. Kniotek *et al.* (2004a) showed that human T cells exhibit adhesive interactions with immobilized T4 and HAP1 (a T4 mutant lacking the Hoc protein). Experiments using purified recombinant phage proteins have revealed that T cells adhere to gp24 (a T4 capsid protein) but not Hoc (Ohams *et al.*, unpublished data). Monoclonal antibodies (mAbs) blocking beta1 and beta3 common chains of integrins diminish these interactions significantly, suggesting that this reactivity depends—at least in part—on the engagement of receptors belonging to both families of integrins (Fig. 3). This seems to be confirmed by the results of preliminary experiments with blocking mAbs to specific integrins, which revealed that VLA-5 appears to be the major integrin responsible for human T-cell adhesion to gp24 (Fig. 4), while the involvement of the beta3 family was confirmed in experiments showing that eptifibatide (a beta3 function inhibitor) was also functional in this system (Ohams *et al.*, unpublished data). Moreover, phages could inhibit the interactions of platelets with their major ligand,

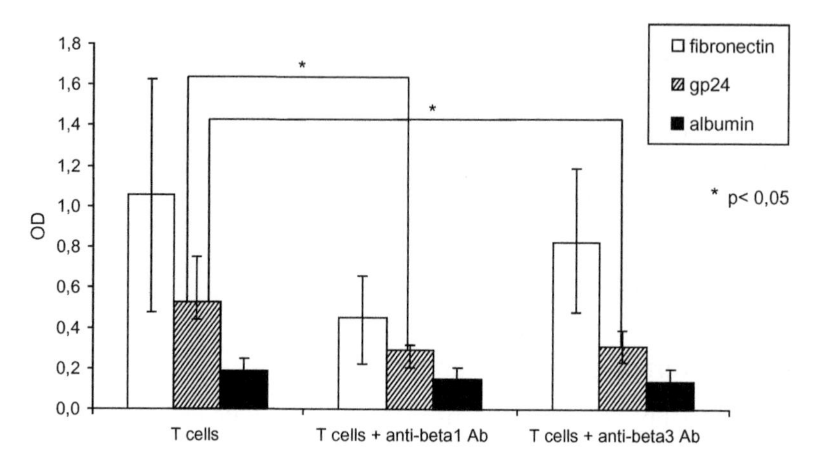

FIGURE 3 Influence of anti-β1 and anti-β3 mAbs on T-cell adhesion to gp24 protein. The effects of anti-β1 and anti-β3 blocking monoclonal antibodies (mAbs) on human T-cell adhesion to gp24 (a T4 phage capsid protein) and fibronectin (FN) are shown. Experiments were conducted as described elsewhere (Jerzak *et al*., 2002). Purified T cells were added (2×10^5 per well) to microtiter plates coated with FN, gp24, or control protein (albumin, all 10 μg/ml) and incubated for 1 hr at 37 °C in a humidified atmosphere in the presence of phytohemagglutinin (50 μg/ml), as well as blocking mAbs to common chains of β1 and β3 integrins (Millipore). Controls received irrelevant antibody (mouse antihuman Ig, Daco). Following incubation, nonadherent cells were washed out with phosphate-buffered saline, the remaining cells stained with 0.1% crystal violet in alcohol, and color yields measured with an ELISA reader at 600 n*M*. Mean values of 10 experiments ± SD are shown. Both integrins appear to be involved in mediating the reactivity to gp24, while only β1 is functional in the case of FN.

fibrinogen, while their effects on T-cell adhesion to that ligand were weak (Kniotek *et al*., 2004a).

Phages may also diminish alloantigen-induced immunoglobulin production *in vitro* as well as specific antibody responses in mice (Kniotek *et al*., 2004b). Moreover, phages inhibit activation of NF-κB, a key transcription factor regulating the expression of many genes, including those encoding pro-inflammatory cytokines (Górski *et al*., 2006). These *in vitro* immunosuppressive effects of phage were confirmed by *in vivo* experiments, which have revealed an ability of phages to significantly extend the survival of allogeneic skin transplants in both normal and sensitized mice, as well as to diminish the development of an inflammatory infiltrate at the allograft site (Górski *et al*., 2006).

C. Effects of phages on the production of cytokines

Several studies have shown that phages can substantially affect the production of various cytokines. One of the mechanisms mediating these effects could be inhibition of the activity of NF-κB (Górski *et al*., 2006).

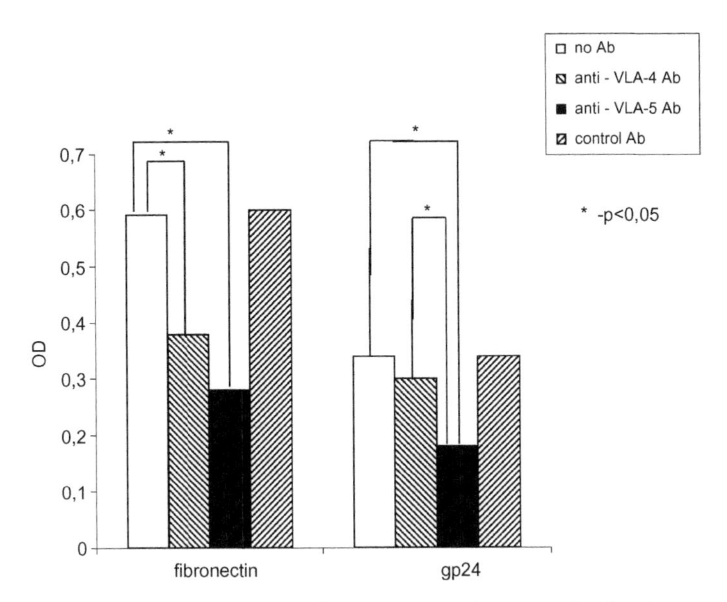

FIGURE 4 Influence of anti-VLA-4 and anti-VLA-5 mAbs on T-cell adhesion to gp24 protein. To identify which specific integrins mediate T-cell adhesion to gp24, experiments identical to those shown in Figure 3 were performed using blocking mAbs to VLA-4 and VLA-5 integrin (Millipore). As an isotypic control, murine antihuman IgG3 mAb (BioLegend) was used. Results of one typical experiment out of four performed that gave similar results are shown. T-cell adhesion to gp24 is only inhibited by anti-VLA5 mAb, while the reactivity to fibronectin is inhibited by both VLA-4 and VLA-5.

Kleinschmidt *et al.* (1970) showed that a purified T4 phage preparation administered intravenously to mice increased the level of interferon in animal sera, which had strong inhibitory activity to vesicular stomatitis virus (this effect was unrelated to LPS). Zimecki *et al.* (2003) found that purified preparations of an *S. aureus* phage can induce IL-6 production in splenocyte cultures and enhance Con A-induced production of this cytokine. However, studies suggest that purified T4 does not induce significant intracellular interleukin (IL)-6 and tumor necrosis factor (TNF)-α synthesis in human monocytes, whereas T4 lysate causes a considerable increase in the production of IL-6 (P.Wierzbicki *et al.*, unpublished data). Furthermore, purified T4 inhibited mitogen-induced IL-2 production by human mononuclear cells (Przerwa *et al.*, 2005).

It was also shown that purified staphylococcal A3 phage does not induce the production of IL-6, whereas both purified preparations and lysates of this phage stimulate the synthesis of TNF-α (P. Wierzbicki *et al.*, unpublished data). In accord with the authors' data, Krishnan and Ganfield (1994) demonstrated that a PhL of *S. aureus*, called Staphage Lysate (SPL; see Section III.D), induces TNF-α, IL-1β, IFN-γ, and IL-10 synthesis by human mononuclear cells and the human monocytic cell line

THP-1. Data from experimental phage therapy of animals indicate that successful treatment corrects increased levels of pro-inflammatory cytokines associated with bacterial infections. Zimecki *et al.* (2009) described such effects for TNF-α and IL-6 in sera of mice infected with *S. aureus* and treated with phage. Similar results were obtained by Hung *et al.* (2011) in mice with *Klebsiella pneumoniae*-induced liver abscesses and bacteremia treated with phages. Furthermore, Kumari *et al.* (2010) showed that successful treatment of *K. pneumoniae*-induced burn wound infection complicated by bacteremia and organ injury was associated with lowering of IL-1β, TNF-α, and IL-10 in sera and lungs of phage-treated mice. Data from experimental phage therapy were confirmed by a clinical study by Weber-Dąbrowska *et al.* (2000b), who demonstrated that phage therapy may influence cytokine production in treated patients. Interestingly, these effects varied depending on the initial responsiveness of the patients found prior to treatment: in those with a low or moderate serum TNF-α level, phage therapy upregulated this cytokine production, lowering it in patients in whom TNF-α levels were initially high. Phage acted in a similar way on the LPS-induced cytokine production by patients' mononuclear cells *in vitro*.

D. Immunomodulatory effects of Staphage Lysate

Staphage Lysate, also called SPL, is a PhL of *S. aureus* produced by Delmont Laboratories (www.delmont.com). Its immunomodulatory activity was studied in much detail and found some applications in experimental medicine. The majority of normal human subjects demonstrate a positive delayed-type reactivity to SPL (Dean *et al.*, 1975), which suggests that this reagent could be used for *in vitro* assessment of cell-mediated reactivity. These authors also performed more detailed studies to evaluate the effects of SPL on immune cells. SPL used contained approximately 2×10^9 *S. aureus* cells and 10^{10} phage particles per milliliter (7 mg/ml of protein). The authors showed that the optimal SPL concentration causing maximal lymphocyte proliferation ranged between 30 and 50 μg of protein/well and the proliferative response peaked on day 6 of culture. The mean stimulation index was 127 (approximately 50% of that caused by PHA, a well-known T-cell mitogen). Further studies have revealed that, in contrast to PHA, the pure T-cell mitogen SPL activates and induces the proliferation of both T and B cells, as well as causes inhibition of leukocyte migration in two out of three donors (another measure of cell-mediated immunity dependent on the production of migration inhibitory factor by T cells) (Dean *et al.*, 1975). These data suggest that some components of SPL can be recognized as specific antigens by lymphocytes of sensitized individuals but probably can also act as a nonspecific inducer of immunity. This assumption has been strengthened by data indicating that SPL is a polyclonal activator of immunoglobulin production by mouse and human B cells (Lee *et al.*, 1985). SPL can also increase antibody titers up to 10-fold as well as susceptibility to phagocytosis (Mills, 1962). In addition, SPL was found to induce IL-1β, TNF-α, and IFN-γ production by human lymphocytes *in vitro* (Krishnan and Ganfield, 1994).

Cell reactivity to SPL measured as lymphocyte proliferation has been used as an *in vitro* measure of immunocompetence in patients with immunodeficiencies (Dean *et al.*, 1977). Studies in mice have revealed that SPL can enhance specific and nonspecific resistance to bacterial infections acting probably on T cells and macrophages (Esber *et al.*, 1981). The product has been licensed for the treatment of pyoderma caused by staphylococci in dogs with up to 77% efficacy (www.delmont.com).

Bogden and Esber (1978) advanced a model of immunotherapy referred to as "induction and elicitation" in which natural immunity to *S. aureus* or that caused by administration of these bacteria is associated with subsequent administration of SPL in order to boost the immune system. Using a metastasizing animal tumor model, the authors demonstrated that this form of immunotherapy was more effective than surgery. That the concept of induction and elicitation may be applied effectively as an immunotherapeutic modality in experimental cancer was confirmed using a similar experimental model by (Mathur *et al.*, 1988). The authors showed that rats receiving SPL treatment had significantly smaller lymph node involvement compared with rats from the control group. Moreover, the best results were achieved using induction (administration of dead staphylococcal cells with subsequent SPL treatment). However, SPL vaccination was not found to have a beneficial effect in a rabbit model of staphylococcal blepharitis and catarrhal infiltrates (Giese *et al.*, 1996).

Staphage Lysate has also been used in the treatment of patients with staphylococcal infections; published evidence dates back to the late 1970s. Mills (1956) used SPL as aerosol therapy of sinusitis, emphasizing the anti-allergic effects of the lysate; this author suggested that polysaccharides and nucleoproteins present in the preparation may bring about a gradual hyposensitization to bacterial allergy present in the majority of sinusitis patients. In his subsequent work, the author extended these data, stressing SPL stimulatory effects at the level of local and systemic immunity. He also observed nonspecific positive effects related to SPL administration: infections caused by other bacteria and viruses seemed to heal quicker than formerly (Mills, 1962). In a study done by Baker *et al.* (1963), SPL was used to stimulate active immunity in patients exposed to or having chronic or recurrent staphylococcal infections. Notably, SPL was found to be more effective than any other staphylococcal vaccine. Salmon and Symonds *et al.* (1963) reported excellent results of SPL treatment of more than 600 patients with chronic staphylococcal infections. The authors believed that those beneficial effects were mediated by its vaccine-like action (SPL was used as an antigen to stimulate production of antibodies, although no data on their actual levels were provided). Although some patients received intensive therapy (one cubic centimeter subcutaneously daily), no alarming or untoward effects were observed.

The concept of induction and elicitation has also been applied in the treatment of chronic osteomyelitis in children. It was demonstrated that the combination of anti-staphylococcal vaccine and SPL influenced the course of the disease positively and reduced the number of relapses (Pillich *et al.*, 1978). What is more, SPL is registered in the Czech Republic and Slovakia for the treatment of staphylococcal infections.

E. Immunomodulatory effects of phage therapy

The authors' group evaluated selected immune parameters in 70 patients with antibiotic-resistant infections (44 males, 26 females, median age 46.5) treated with PhLs at the Phage Therapy Unit at the Hirszfeld Institute of Immunology and Experimental Therapy, Wrocław, Poland, between 2007 and 2011 (Table I). Blood samples were collected from patients no earlier than 14 days before beginning phage therapy and no later than 14 days after suspending or ending the phage application (in some cases, one short break up to 14 days in the application of phages between days 21 and 84 was allowed). The following parameters were studied: T- and B-cell proliferation induced by the mitogens OKT3 (Muromonab CD3), PHA, and *Staphylococcus aureus* Cowan 1 (SAC), natural killer (NK) cell activity, T and NK cell percentage, phagocytosis of zymosan particles, and phorbol myristate acetate (PMA)-induced ROS production by phagocytes. Their changes after 7–20 days, 21–48 days, and 49–84 days of phage therapy compared to values before therapy were analyzed. The overall results of immunomonitoring of the patients are shown in Figure 5. Up to 50% of patients had signs of immunodeficiency prior to phage therapy, and on initial analysis (with the use of McNemar's test) of the number of patients below or above the lower limit of normal range, there were no clear effects of therapy except for a significant double increase (from 34 to 69%) in the number of patients with normal and heightened B-cell responses to SAC on days 49–84 of treatment. There were no significant changes when the numbers of patients below or above the upper limit of normal range were analyzed, which means that no significant overstimulation of the immune system by the phage lysates were observed.

Further analysis was performed to determine whether the clinical outcome of phage therapy could be linked to alterations in patients' immune reactivity (changes in the values of the particular immunological parameter before and during therapy were compared using sign test or Wilcoxon matched-pairs test). A significant increase in phagocytosis

TABLE I Patient characteristics (total n = 70)

Gender	Males: n = 44 Females: n = 26
Median age	Males: 49.5 (23–79) years Females: 35.0 (20–82) years Total: 46.5 (20–82) years
Diagnosis	Urogenital infections: 31 cases (including 16 men with chronic bacterial prostatitis) Prosthetic joint infection: 6 cases Osteomyelitis: 8 cases Joint infection: 4 cases Skin and soft tissue infection: 13 cases Upper respiratory tract infection 6 cases (including 2 cases of chronic sinusitis) External ear infection: 2 cases
Pathogens causing infection	Gram-positive bacteria (n = 47): *S. aureus* (n = 31, including three MRSA cases) *E. faecalis* (n = 16) Gram-negative bacteria (n = 12): *E. coli* (n = 5) *P. aeruginosa* (n = 5) *P. putida* (n = 1) *K. pneumoniae* (n = 1) Mixed Gram-negative/Gram-positive infections (n = 11) *E. coli/E. faecalis* (n = 6) *E. faecalis/K. pneumoniae* (n = 1) *E. faecalis/K. oxytoca* (n = 1) *E. faecalis/P. aeruginosa* (n = 1) *S. aureus/E. cloace* (n = 1) *E. faecalis/E. coli/P. vulgaris* (n = 1)
Concomitant antibacterial treatment	Antibiotics/chemotherapeutics: n = 13 Disinfectants: n = 2
Response to phage treatment[a]	Pathogen eradication and/or recovery: 9 cases Good clinical result: 11 cases Clinical improvement: 13 cases Questionable clinical improvement: 7 cases Transient clinical improvement observed only during phage therapy: 12 cases No response to treatment: 17 cases Clinical deterioration: 1 case

[a] The response to phage treatment was established based on the results of control microbiological tests (bacterial cultures), assessment of the intensity of disease symptoms by the physician during patient's clinical examination, and evaluation of the results of other control relevant diagnostic tests, as well as the opinion of consulting medical specialists. Pathogen eradication was confirmed by bacterial cultures. Recovery means wound healing or complete subsidence of the symptoms of infection. A good clinical result means almost complete subsidence of some symptoms of infection (in some cases confirmed by the results of laboratory tests, such as normalization of inflammatory parameters) and significant improvement of general patient's condition. Clinical improvement means discernible decrease of the intensity of some symptoms of infection not observed during periods without treatment. Questionable clinical improvement means that the degree of reduction in the intensity of some symptoms of infection could have been observed in patient also during periods without treatment.

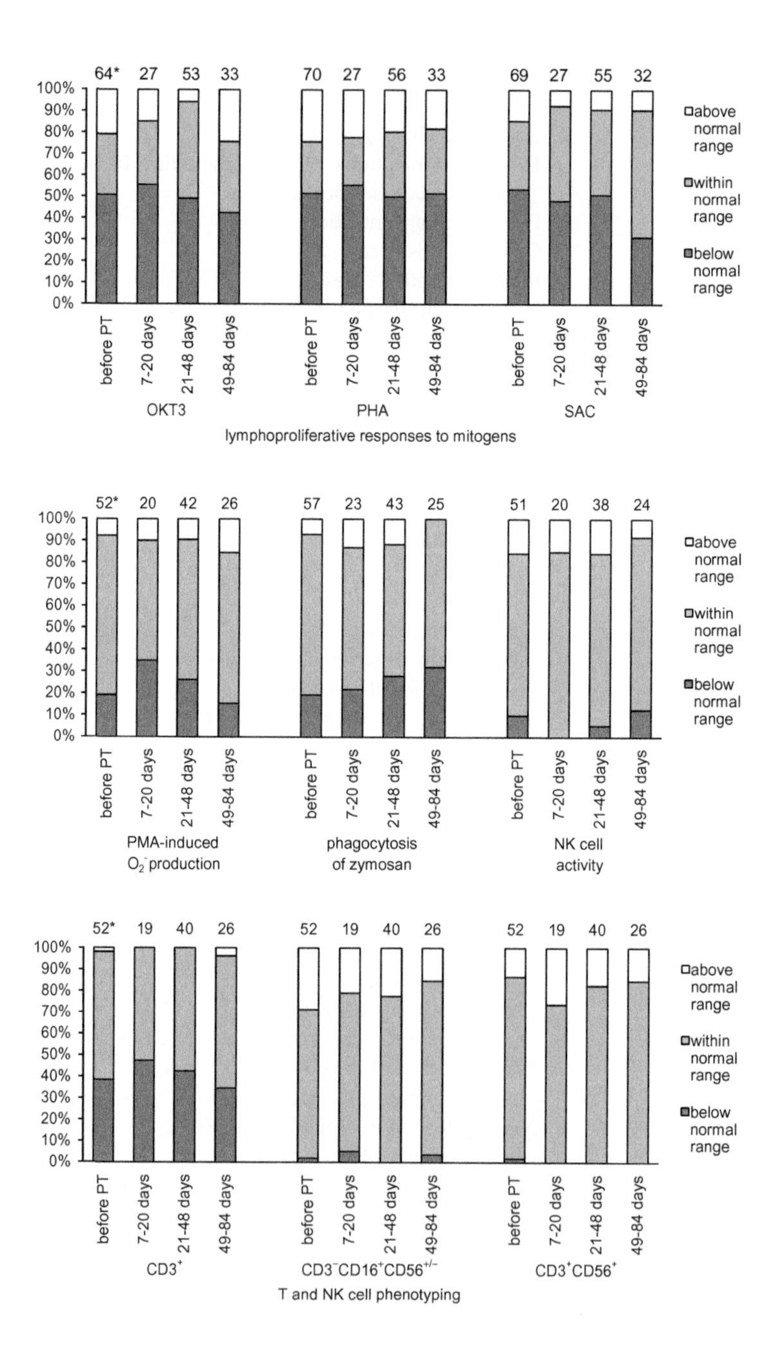
64*
27
53
33
70
27
56
33
69
27
55
32
100%
90%
80%
70%
60%
50%
40%
30%
20%
10%
0%
before PT
7-20 days
21-48 days
49-84 days
OKT3
PHA
SAC
lymphoproliferative responses to mitogens
above normal range
within normal range
below normal range
52*
20
42
26
57
23
43
25
51
20
38
24
PMA-induced O2- production
phagocytosis of zymosan
NK cell activity
52*
19
40
26
CD3+
CD3-CD16+CD56+/-
CD3+CD56+
T and NK cell phenotyping

evaluated on days 7–20 of phage therapy was associated with a good prognosis of the treatment (pathogen eradication and/or recovery, good clinical result, or clinical improvement according to Table I, was achieved in almost 50% of patients). Analysis of the relationship between phage type and immune reactivity revealed that patients treated with enterococcal phages or phages specific to Gram-negative bacterial species had diminished T-cell responses to OKT3. Patients who received only phages of Gram-negative bacteria presented a significant decrease in the number of $CD3^+CD56^+$ cells at days 21–48 of treatment. A group of patients receiving only enterococcal phages had a significantly increased median value of phagocytosis after 21–48 days of phage therapy (however, only six patients were available for this comparison). Interestingly, in patients receiving only staphylococcal phages, no significant changes were observed for any immunological parameter analyzed. Moreover, analysis of the influence of the routes of phage administration (phages were given by oral, intrarectal, or topical route) on immune parameters suggests that intrarectal administration (alone or combined with topical application) may reduce the number of circulating NK cells after 49–84 days of phage therapy. Finally, patients treated with phages may develop phage-neutralizing antibodies, although this effect is rather weak or absent following oral phage administration and probably depends on the initial responder or nonresponder status and phage type used (see Section II.A).

A separate study performed on 37 patients treated with PhLs at the Phage Therapy Unit at the Hirszfeld Institute (Miedzybrodzki *et al.*, 2009)

FIGURE 5 Changes in different immunological parameters during phage treatment. Mitogen-induced proliferation of peripheral blood mononuclear cells (PBMCs) was measured by an [^{3}H]thymidine incorporation assay using the OKT3 monoclonal antibody (normal value: 22,832–51,111 cpm), phytohemagglutinin (normal value: 34,247–61,405 cpm), and *Staphylococcus aureus* Cowan 1 (normal value: 775–5692 cpm)(Górski *et al.*, 2006). Phorbol myristate acetate-induced ROS production by neutrophils was determined by the cytochrome *c* reduction assay as described elsewhere (Sustiel *et al.*, 1989) (normal value: 7.06–12.04 nmol O_2^-/2.5 min $\times 10^6$ cells). Neutrophil phagocytosis of zymosan particles (Invitrogen) was evaluated by flow cytometry according to the manufacturer's instructions (normal value: 75–95%). The percentages of $CD3^+$ (normal value: 65–81%), $CD3^-CD16^+CD56^{+/-}$ (normal value: 6–18%), and $CD3^+CD56^+$ cells (normal value: 2–10%) in PBMCs was determined by flow cytometry using anti-CD3, anti-CD16, and anti-CD56 mAbs from Becton-Dickinson according to the manufacturer's instructions. NK cell activity (normal value: 3–17% of dead target cells) was measured by flow cytometry as described elsewhere (Chang *et al.*, 1993). Bars show the cumulative percentage of patients below, above, or in a range of normal values before beginning phage therapy and in periods of 7–20 days, 21–48 days, and 49–84 days of the phage preparation application. *Values above bars show number of patients whose data were analyzed in individual subgroups.

showed that phage therapy results in a decrease in the C-reactive protein serum level and white blood cell count; however, administration of PhLs to patients had no significant effect on the erythrocyte sedimentation rate. Thus therapeutic effects of phage preparations may result not only from the elimination of bacteria, but also from normalization of some inflammatory markers associated with bacterial infections.

IV. PHAGE VIRION INACTIVATION BY IMMUNE CELLS

Inchley (1969) showed that the majority of T4 phage particles are cleared rapidly from the circulation of mice by cells of the reticuloendothelial system of the liver and, to a lesser extent, of the spleen. However, experiments performed on nonimmune germ-free mice demonstrated that it is the spleen rather than the liver that clears most of the phage particles from the blood regardless of the route of phage administration (Geier *et al.*, 1973). Interestingly, both studies consistently showed that phage particles are inactivated by the spleen more slowly than by other organs, especially the liver. Although these classic studies clearly demonstrated that the majority of phage virions are eliminated *in vivo* by cells of the reticuloendothelial system of the spleen and the liver, data show that neutrophils and macrophages may also be involved to some extent in the degradation of phage virions.

Detailed studies on *in vivo* interactions between guinea pig peritoneal macrophages and T2 phage were carried out by Kańtoch (1961). Essentially, those interactions occur 30 min after phage administration. However, there was a great variation among animals in the amount of phage-positive cells (from 1:20 to 1:1630), which means that the majority of peritoneal leukocytes did not contain biologically active phage. *In vitro* studies revealed that phage uptake depended on virus concentration and did not occur below a phage concentration of 10^7/ml. The process also depended on the presence of serum and was optimal at 2 °C. Studies of the uptake of phage by macrophages were also performed by Aronow *et al.* (1964) with the use of electron microscopy. These authors showed that the uptake of phage by rabbit peritoneal macrophages occurred after approximately 7 min of incubation and reached maximal intensity between 15 and 30 min. Intracellular destruction of ingested phages occurred some 2 hr after their uptake as no virus disintegration could be observed before that time. However, in neutrophils, disrupted phage particles were seen just after 15 min of incubation.

Ferrini *et al.* (1989) described the inactivation of λ phage vector particles *in vitro* when the phage was exposed to neutrophils stimulated with PMA. An important observation was that the presence of the phage did not stimulate neutrophils, and incubation of active phages with unstimulated neutrophils did not cause any significant decrease in phage titer. However, incubation with stimulated neutrophils caused a decrease in phage titer, probably as a result of phage inactivation by hypochloric acid produced by the stimulated cells. Interestingly, rabbit macrophages previously "immune" (being able to make contact with a phage) remove the T1 phage faster than "nonimmune" cells *in vitro* (Nelstrop *et al.*, 1967).

Although the aforementioned studies showed that both neutrophils and macrophages can degrade phage virions (at least in some experimental settings), two studies demonstrated that neither of those populations is substantially involved in the clearance of phages from blood in mice following systemic administration of phages (Srivastava *et al.*, 2004, Uchiyama *et al.*, 2009).

It was also suggested that phages may retain their antibacterial activity after internalization to mammalian cells (Ivanekov *et al.*, 1999). Preliminary *in vitro* studies (E. Jończyk *et al.*, unpublished data) showed that T4 phage particles released from lysed human phagocytes after their incubation with phages *in vitro* retain their lytic activity. Moreover, when neutrophils or monocytes with phagocytosed *E. coli* B (cells initially prepared for the intracellular killing test) were incubated with the purified T4 phage, almost complete lysis (99%) of bacteria was observed compared to 63–82% in the control without phage. This may suggest that some phages possibly lyse intracellular bacteria and help phagocytic cells in the elimination of bacteria.

V. CONCLUDING REMARKS

The aim of this chapter was to update and summarize existing knowledge on phage effects on the immune system. Evidently, such a report should be helpful in fully understanding the possible mechanisms of action of phage therapy and for appropriate planning of clinical trials, which would take into account different aspects of possible interactions of phages with the human organism. In this context, two questions are of paramount importance, namely: How does the immune system react to the administration of phages and how could these reactions, especially the production of antibodies, neutralize phage activities?

In general, when PPhs are studied *in vitro*, the prevailing effect is the suppression of lymphocyte activation coupled with anti-inflammatory effects (e.g., the reduction of ROS production and the inhibition of NF-κB activation). Furthermore, similar effects were observed *in vivo* (extension of skin allograft survival in mice). In contrast, SPL is a polyclonal lymphocyte activator and an inducer of immunoglobulin production, while its purified analogue does not have such effects (A. Górski *et al.*, unpublished data). Thus, SPL appears to be exceptional among phage preparations used in phage therapy as it can cause both immune enhancement and pathogen eradication.

Current efforts to obtain PPhs that could meet the stringent requirements for parenteral administration in the clinic have been underway among centers involved in experimental and human phage therapy. However, it is believed that their eventual development in sufficient quantities should not entirely eliminate the need for PhLs currently applied. Patients with immunodeficiency syndrome need eradication of an offending pathogen, as well as enhancement of their immunity. Thus, when dealing with patients with difficult *S. aureus* infections and concurrent immunodeficiency, the treatment of choice would be SPL rather than its purified analogue.

Patients with chronic infections frequently have immune deficits that may predispose them to life-threatening infections. Furthermore, prolonged treatment with antibiotics (known to diminish immune responses) constitutes an additional risk factor perpetuating and aggravating their condition in a vicious circle. A significant proportion of the authors' patients had signs of immunodeficiency prior to therapy (e.g., approximately 50% for T and B lymphoproliferative responses). However, this condition does not constitute a contraindication for phage treatment, as pointed out in an earlier review (Borysowski and Górski, 2008).

There are no data in the literature to provide a definite answer to the important scientific and practical question of whether phage therapy significantly modifies immune functions in human and, if so, how such effects could contribute to the success or failure of treatment. Because current therapy involves PhLs rather than PPhs (at least in the majority of patients being treated), it could also be argued that phage preparations act in a similar manner to bacterial vaccines enhancing immunity and thereby contributing to the clearance of infection and improvement in patients' condition. However, data do not appear to support the notion that the major effect of phage therapy is to enhance immunity. In fact, the percentage of patients showing immune deficiencies at different stages of the treatment, as well as upon its completion, did not differ significantly from initial values (except for B-cell activation, which normalized in some patients, so their number was significantly higher upon completion of phage therapy). Because this does not exclude observed fluctuations in some immune parameters evaluated (e.g., T and B lymphoproliferative responses), it cannot be excluded that a more in-depth, detailed analysis performed at the level of individual patients would reveal more significant alterations of immunity, at least in some of them. This will be the subject of the authors' next report, including an enlarged cohort of treated patient. In fact, the authors previously described an increase in immunoglobulin production by B cells of patients treated with SPL (Kniotek *et al.*, 2004b), which corresponds to the trend toward enhanced B-cell proliferative responses currently reported.

It appears that the most interesting immunomodulatory phenomenon observed during the first weeks of therapy was the enhancement of granulocyte phagocytosis, which was indicative of concurrent patients' improvement. This observation is important for at least two reasons: (a) it has a prognostic value pointing to a potentially positive outcome of therapy, a factor of utmost clinical importance, and (b) it suggests that phage preparations may enhance granulocyte functions supporting and extending the assumptions suggesting that the mechanism of phage therapy is associated with an ability of an administered phage to reduce the number of offending bacteria to the threshold at which the immune system can successfully combat infection, a hypothesis originally put forward by d'Herelle (1922) and subsequently further developed by others (Górski and Weber-Dąbrowska, 2005).

When analyzing the effects of phage, one should keep in mind that *in vitro* data were obtained using PPhs that exerted suppressive effects on some immune functions (e.g., T-cell activation). However, similar effects could also be observed using these preparations *in vivo* in mice (an extension of allograft survival) and in patients treated with lysates of enterococcal phage or lysates of phage specific to Gram-negative bacteria (diminished T-cell responses to OKT3 at days 7–20 of phage therapy), as well as in some patients receiving PhLs intrarectally (a diminished level of NK-cell activity). Therefore, the final effect of phage preparations on the immune system may be regulated by a number of factors, especially initial immune status, the phage type, its form (PPhs vs PLs), route of phage administration, the length of treatment, and neutralizing antibody production.

Because patients received preparations that contained active phage particles and remnants of bacteria lysed by phage, the immunomodulatory effects may be mediated by both components of the preparations. As pointed out, the authors noted an increase in phagocytosis on days 7–20 of therapy in patients with a good prognosis. *In vitro* data obtained with purified phage suggest that phages inhibit human neutrophil phagocytosis in a dose-dependent manner, while they were without such effect when given in mice. Thus, the observed effect in patients was rather unrelated to the direct action of the phage on neutrophils and could have been caused by an improvement of phagocytic functions due to decreased bacterial load resulting from efficient phage therapy. Furthermore, the increased B-cell proliferative response observed on days 49–84 of the treatment could not be linked to the activity of purified phage proteins, as the authors' and other authors data cited in this chapter suggest that such an enhancement of immunity is caused by bacterial products present in phage preparations, especially in SPL (e.g., SPL but not its purified phage component causes human B-cell activation; see earlier discussion). At any rate, the reported immunomodulatory effects of phage therapy appear to be so subtle that the issue of which component of the phage preparation is actually responsible for the phenomena observed is of a rather minor practical importance.

As far as the issue of antibody production is concerned, preliminary data contained in this chapter indicate that the problem of an antibody response to phage administered during therapy is more complex than has usually been presented in that phage therapy and neutralizing antibody formation do not necessarily parallel each other. Experimental animals injected with phage intravenously or intraperitoneally may indeed produce such antibodies, but it has been determined that using such efficient routes does not always produce positive results and some 10% of animals may not respond (Delmastro *et al.*, 1997). Some studies also demonstrated significant systemic and local (mucosal) humoral responses against phage coat proteins following oral administration of phage to rodents (Clark and March, 2004). Data suggest that neutralizing antibody responses may indeed occur in patients receiving phage locally and intrarectally, but oral administration may be less likely to induce such antibodies. This confirms the report of Kucharzewicz-Krukowska and Ślopek (1987) suggesting that the immunogenic effect of phage administered by the oral route in patients is very weak. Likewise, Bruttin and Brussow (2005) showed that oral administration of T4 phage to healthy volunteers does not result in the production of antibodies. In general, the anti-phage humoral response seems to be dependent on a number of factors, especially the phage type, the route of phage administration, and the responder/nonresponder status.

ACKNOWLEDGMENTS

This work was supported by the European Regional Development Fund within the Operational Program Innovative Economy, 2007-2013, Priority axis 1. Research and Development of ModernTechnologies, Measure 1.3 Support for R&D projects for entrepreneurs carried out by scientific entities, Submeasure 1.3.1 Development projects as project No. POIG 01.03.01-02-003/08 entitled "Optimization of the production and characterization of bacteriophage preparations for therapeutic use", by statutory funds from the Medical University of Warsaw—1MG/N/2011, and by a Ministry of Science and Higher Education grant N N402 268036 entitled "Involvement of bacteriophages T4 and A5 in the activation of human dendritic cells and macrophages" (2009-2012).

REFERENCES

Amanna, I. J., and Slifka, M. K. (2011). Contributions of humoral and cellular immunity to vaccine-induced protection in humans. *Virology* **15:**206–215.

Andrews, R. G., Winkler, A., Potter, J., Bryant, E., Knitter, G. H., Bernstein, J. D., and Och, H. D. (1997). Normal immunologic response to a neoantigen, bacteriophage phi X174 in baboons with long-term lymphohematopoietic reconstitution from highly purified CD34+ lin- allogeneic marrow cells. *Blood* **90:**1701–1708.

Aronow, R., Danon, D., Shahar, A., and Aronson, M. (1964). Electron microscopy of in vitro endocytosis of T2 phage by cells from rabbit peritoneal exudate. *J. Exp. Med.* **120:**943–954.

Baker, A. G. (1963). Staphylococcus bacteriophage lysate: Topical and parenteral use in allergic patients. *Pa Med. J.* **66:**25–28.

Bateman, A., Eddy, S. R., and Mesyanzhinov, V. V. (1997). A member of the immunoglobulin superfamily in bacteriophage T4. *Virus Genes* **14:**163–165.

Bearden, C. M., Agarwal, A., Book, B. K., Vieira, C. A., Sidner, R. A., Ochs, H. D., Young, M., and Pescovitz, M. D. (2005). Rituximab inhibits the in vivo primary and secondary antibody responses to a neoantigen, bacteriophage phi X174. *Am. J. Tranplant* **5:**50–57.

Bogden, A. E., and Esber, H. J. (1978). The concept of induction and elicitation as an immunotherapeutic approach. *Natl. Cancer Inst. Monogr.* **49:**365–367.

Borysowski, J., and Gorski, A. (2008). Is phage therapy acceptable in the immunocompromised host? *Int. J. Infect. Dis.* **12:**466–471.

Borysowski, J., Wierzbicki, P., Kłosowska, D., Korczak-Kowalska, G., Weber-Dabrowska, B., and Górski, A. (2010). The effects of T4 and A3/R phage preparations on whole-blood monocyte and neutrophil respiratory burst. *Viral Immunol.* **23:**541–544.

Bruttin, A., and Brussow, H. (2005). Human volunteers receiving *Escherichia coli* phage T4 orally: A safety test of phage therapy. *Antimicrob. Agents Chemother.* **49:**2874–2878.

Chang, L., Gusewitch, G. A., Chritton, D. B., Folz, J. C., Lebeck, L. K., and Nehlsen-Cannarellas, S. L. (1993). Rapid flow cytometric assay for the assessment of natural killer cell activity. *J. Immunol. Methods* **166:**45–54.

Clark, J. R., and March, J. B. (2004). Bacterial viruses as human vaccines? *Expert Rev Vaccines* **3:**463–476.

Dabrowska, K., Opolski, A., Wietrzyk, J., Switala-Jelen, K., Boratynski, J., Nasulewicz, A., Lipinska, L., Chybicka, A., Kujawa, M., Zabel, M., Dolinska-Krajewska, B., Piasecki, E., *et al.* (2004). Antitumor activity of bacteriophages in murine experimental cancer models caused possibly by inhibition of beta3 integrin signaling pathway. *Acta Virol.* **48:**241–248.

Dabrowska, K., Switała-Jeleń, K., Opolski, A., and Górski, A. (2006). Possible association between phages, Hoc protein, and the immune system. *Arch. Virol.* **15:**209–215.

Dabrowska, K., Zembala, M., Boratynski, J., Switala-Jelen, K., Wietrzyk, J., Opolski, A., Szczaurska, K., Kujawa, M., Godlewska, J., and Gorski, A. (2007). Hoc protein regulates the biological effects of T4 phage in mammals. *Arch. Microbiol.* **187:**489–498.

Dean, J. H., Connor, R., Herberman, R. B., Silva, J., McCoy, J. L., and Oldham, R. K. (1977). The relative proliferation index as a more sensitive parameter for evaluating lymphoproliferative responses of cancer patients to mitogens and alloantigens. *Int. J. Cancer* **20:**359–370.

Dean, J. H., Silva, J. S., McCoy, J. L., Chan, S. P., Baker, J. J., Leonard, C., and Herberman, R. B. (1975). *In vitro* human reactivity to staphylococcal phage lysate. *J. Immunol.* **115:**1060–1064.

Delmastro, P., Meola, A., Monaci, P., Cortese, R., and Galfre, G. (1997). Immunogenicity of filamentous phage displaying peptide mimotopes after oral administration. *Vaccine* **15:**1276–1285.

d'Herelle, F. (1992). Opsonic power of the lysins. *In* "Bacteriophage: Its Role in Immunity", p. 125. Williams & Wilkins, Baltimore.

Esber, H. J., DeCourcy, S. J., Jr., and Bogden, A. E. (1981). Specific and nonspecific immune resistance enhancing activity of staphage lysate. *J. Immunopharmacol.* **3:**79–92.

Ferrini, U., Mileo, M. M., Nista, A., Mattei, E., and Orofino, A. (1989). Polymorphonuclear leucocyte stimulation measured by phage inactivation. *Int. Arch. Allergy Appl. Immunol.* **90:**207–212.

Forthal, D. N., and Moog, C. (2009). Fc receptor-mediated antiviral antibodies. *Curr. Opin. HIV AIDS* **4:**388–393.

Fraser, J. S., Yu, Z., Maxwell, K. L., and Davidson, A. R. (2006). Ig-like domains on bacteriophages: A tale of promiscuity and deceit. *J. Mol. Biol.* **359:**496–507.

Geier, M. R., Trigg, M. E., and Merril, C. R. (1973). Fate of bacteriophage lambda in non-immune germ-free mice. *Nature* **246:**221–223.

Giese, M. J., Adamu, S. A., Pitchekian-Halabi, H., Ravindranath, R. M., and Mondino, B. J. (1996). The effect of *Staphylococcus aureus* phage lysate vaccine on a rabbit model of staphylococcal blepharitis, phlyctenulosis, and catarrhal infiltrates. *Am. J. Ophthalmol.* **122:**245–254.

Górski, A., and Weber-Dąbrowska, B. (2005). The potential role of endogenous bacteriophages in controlling invading pathogens. *Cell. Mol. Life Sci.* **62:**511–519.

Górski, A., Kniotek, M., Perkowska-Ptasińska, A., Mróz, A., Przerwa, A., Gorczyca, W., Dąbrowska, K., Weber-Dąbrowska, B., and Nowaczyk, M. (2006). Bacteriophages and transplantation tolerance. *Transplant Proc.* **38:**331–333.

Hung, C. H., Kuo, C. F., Wang, C. H., Wu, C. M., and Tsao, N. (2011). Experimental phage therapy in treating *Klebsiella pneumonia*-mediated liver abscesses and bacteremia in mice. *Antimicrob. Agents Chemother.* **55:**1358–1365.

Inchley, C. J. (1969). The activity of mouse Kupffer cells following intravenous injection of T4 bacteriophage. *Clin. Exp. Immunol.* **5:**173–187.

Ishii, T., and Yanagida, M. (1977). The two dispensable structural proteins (soc and hoc) of the T4 phage capsid; their purification and properties, isolation and characterization of the defective mutants, and their binding with the defective heads in vitro. *J. Mol. Biol.* **109:**487–514.

Ivanekov, V., Felici, F., and Menon, A. G. (1999). Uptake and intracellular fate of phage display vectors in mammalian cells. *Biochim. Biophys.* **1448:**450–462.

Jerne, N. K. (1956). The presence in normal serum of specific antibody against bacteriophage T4 and its increase during the earliest stages of immunization. *J. Immunol.* **76:**209–216.

Jerne, N. K., and Avegno, P. (1956). The development of the phage-inactivating properties of serum during the course of specific immunization of an animal: Reversible and irreversible inactivation. *J. Immunol.* **76:**200–208.

Jerzak, M., Baranowski, W., Rechberger, T., and Gorski, A. (2002). Enhanced T cells interactions with extracellular matrix proteins in infertile women with endometriosis. *Immunol. Lett.* **81:**65–70.

Kamme, C. (1973). Antibodies against staphylococcal bacteriophages in human sera. I. Assay of antibodies in healthy individuals and in patients with staphylococcal infections. *Acta Path. Microbiol. Scand. B* **81:**741–748.

Kańtoch, M. (1961). The role of phagocytes in virus infections. *Arch. Immunol. Ther. Exp.* **9:**261–268.

Kańtoch, M. (1956). Wybrane zagadnienia z nauki o bakteriofagach. *Post Hig Med Dośw* **10:**47–486.

Kańtoch, M., Skurski, A., and Wieczorek, Z. (1958). *In vitro* blockade of bacterial phagocytosis of leukocytes by means of bacterial viruses. *Schweiz Z Pathol. Mikrobiol.* **21:**1106–1119.

Kańtoch, M., and Szalaty, H. (1960). Lytic activity of bacteriophages toward phagocytized bacteria. *Arch. Immunol. Ther. Exp.* **8:**407–416.

Kim, K. P., Cha, J. D., Jang, E. H., Klumpp, J., Hagens, S., Hardt, W. D., Lee, K. Y., and Loessner, M. J. (2008). PEGylation of bacteriophages increases blood circulation time and reduces T-helper type 1 immune response. *Microb. Biotechnol.* **1:**247–257.

Kleinschmidt, W. J., Douthart, R. J., and Murphy, E. B. (1970). Interferon production by T4 coliphage. *Nature* **228:**27–30.

Kniotek, M., Ahmed, A. M. A., Dabrowska, K., Świtała-Jeleń, K., Weber-Dąbrowska, B., Boratyński, J., Nowaczyk, M., Opolski, A., and Górski, A. (2004a). Bacteriophage interactions with T cells and platelets. *In* "Genomic Issues, Immune System Activation and Allergy" (Immunology 2004a), Medimond International Proceedings (Bologna: Monduzzi Editors), pp. 189–193.

Kniotek, M., Weber-Dąbrowska, B., Dąbrowska, K., Świtała-Jeleń, K., Boratyński, J., Wiszniewski, W., Glinkowski, I., Babiak, I., Gorecki, M., and Nowaczyk, M. (2004b). Phages as immunomodulators of antibody production. *In* "Genomic Issues, Immune System Activation and Allergy" (Immunology 2004), Medimond International Proceedings (Bologna: Monduzzi Editors), pp. 33–36.

Krisch, H. M., and Comeau, A. M. (2008). The immense journey of bacteriophage T4: From d'Hérelle to Delbrück and then to Darwin and beyond. *Res. Microbiol.* **59:**314–324.

Krishnan, G., and Ganfield, D. J. (1994). "Cytokines Produced by Staphage Lysate". 12th Eur Immun Meeting Barcelona, Abstract Book, p. 395.

Kucharewicz-Krukowska, A., and Slopek, S. (1987). Immunogenic effect of bacteriophages in patients subjected to phage therapy. *Arch. Immunol. Ther. Exp.* **35:**553–561.

Kumar, H., Kawai, T., and Akira, S. (2009). Toll-like receptors and innate immunity. *Biochem. Biophys. Res. Commun.* **388:**621–625.

Kumari, S., Harjai, K., and Chhibber, S. (2010). Evidence to support the therapeutic potential of bacteriophage Kpn5 in burn wound infection caused by *Klebsiella pneumonia* in BALB/c mice. *J. Microbiol. Biotechnol.* **20:**935–941.

Kurzepa, A. (2011). "The Influence of Bacteria on Migration and Intracellular Killing of Bacteria by Human Phagocytes". Ph.D. thesis, Hirszfeld Institute of Immunology and Experimental Therapy, Wrocław, Poland.

Langbeheim, H., Teitelbaum, D., and Arnon, R. (1978). Cellular immune responses toward MS-2 phage and a synthetic fragment of its coat protein. *Cell. Immunol.* **38:**193–197.

Lee, B., Murakami, M., Mizukoshi, M., Shinomiya, N., and Yada, J. (1985). Effects of Staphylococcal Phage Lysate (SPL) on immunoglobulin production in human peripheral lymphocyte cultures. *Yakugaki Zasshi* **105:**574–579.

Mankiewicz, E., Kurti, V., and Adomonis, H. (1974). The effect of mycobacteriophage particles on cell-mediated immune reactions. *Can. J. Microbiol.* **20:**1209–1218.

Mathur, A., Narayanan, K., Zerbe, A., Ganfield, D., Ramasastry, S. S., and Futrell, J. W. (1988). Immunomodulation of intradermal mammary carcinoma using staphage lysate in a rat model. *J. Invest. Surg.* **1:**117–123.

Merril, C. R., Biswas, B., Carlton, R., Jensen, N. C., Creed, G. J., Zullo, S., and Adhya, S. (1996). Long-circulating bacteriophage as antibacterial agents. *Proc. Natl. Acad. Sci. USA* **93:**3188–3192.

Michael, J. G., and Kuwatch, C. J. (1969). Contribution of endotoxin to antibody formation against T2 phage. *Nature* **222:**684–685.

Miedzybrodzki, R., Fortuna, W., Weber-Dabrowska, B., and Gorski, A. (2009). A retrospective analysis of changes in inflammatory markers in patients treated with bacterial viruses. *Clin. Exp. Med.* **9:**303–312.

Międzybrodzki, R., Switala-Jelen, K., Fortuna, W., Weber-Dabrowska, B., Przerwa, A., Lusiak-Szelachowska, M., Dabrowska, K., Kurzepa, A., Boratynski, J., Syper, D., Pozniak, G., Lugowski, C., and Gorski, A. (2008). Bacteriophage preparation inhibition of reactive oxygen species generation by endotoxin-stimulated polymorphonuclear leukocytes. *Virus Res.* **131:**233–242.

Mills, A. E. (1956). Staphylococcus bacteriophage lysate aerosol therapy of sinusitis. *Laryngoscope* **66:**846–858.

Mills, A. E. (1962). Staphylococcus phage lysates: An immune-biological therapy for the prevention and control of staphylococcal disease. *Laryngoscope* **72:**367–383.

Molenaar, T. J., Michon, I., de Haas, S. A., van Berkel, T. J., Kuiper, J., and Biessen, E. A. (2002). Uptake and processing of modified bacteriophage M13 in mice: Implications for phage display. *Virology* **293:**182–191.

Nelstrop, A., Taylor, G., and Collard, P. (1967). Studies on phagocytosis. *II. In vitro phagocytosis by macrophages. Immunology* **14:**339–346.

Ochs, H. D., Davis, S. D., and Wedgwood, R. J. (1971). Immunologic responses to bacteriophage phi X174 in immunodeficiency diseases. *J. Clin. Invest.* **50:**2559–2568.

Pillich, J., Tovarek, J., and Fait, M. (1978). Contribution to the treatment of acute haematogenous and chronic secondary osteomyelitis in children. *Z. Orthop. Ihre Grenzgeb.* **116:**40–46.

Przerwa, A., Kniotek, M., Nowaczyk, M., Weber-Dabrowska, B., Switala-Jelen, K., Dabrowska, K., and Gorski, A. (2005). Bacteriophages inhibit interleukin-2 production by human T lymphocytes. *In* "12th Congress of the European Society for Organ Transplantation," Geneva, Switzerland.

Przerwa, A., Zimecki, M., Świtała-Jeleń, K., Dąbrowska, K., Krawczyk, E., Łuczak, M., Weber-Dąbrowska, B., Syper, D., Międzybrodzki, R., and Górski, A. (2006). Effects of bacteriophages on free radical production and phagocytic functions. *Med. Microbiol. Immunol.* **195:**143–150.

Salmon, G. G., Jr., and Symonds, M. (1963). Staphage lysate therapy in chronic staphylococcal infections. *J. Med. Soc. N. J.* **60:**188–193.

Sathaliyawala, T., Islam, M. Z., Li, Q., Fokine, A., Rossmann, M. G., and Rao, V. B. (2010). Functional analysis of the highly antigenic outer capsid protein, Hoc, a virus decoration protein from T4-like bacteriophages. *Mol. Microbiol.* **77:**444–455.

Schwarz, K. B. (1996). Oxidative stress during viral infection: A review. *Free Radic. Biol. Med.* **21:**641–649.

Silva, M. T. (2010). When two is better than one: Macrophages and neutrophils work in concert in innate immunity as complementary and cooperative partners of a myeloid phagocyte system. *J. Leukocyte Biol.* **87:**93–106.

Ślopek, S., Weber-Dąbrowska, B., Dąbrowski, M., and Kucharewicz-Krukowska, A. (1987). Results of bacteriophage treatment of suppurative bacterial infections in the years 1981-1986. *Arch. Immunol. Ther. Exp. (Warsz)* **35:**569–583.

Sokoloff, A. V., Bock, I., Zhang, G., Sebestyén, M. G., and Wolff, J. A. (2000). The interactions of peptides with the innate immune system studied with use of T7 phage peptide display. *Mol. Ther.* **2:**131–139.

Srivastava, A. S., Kaido, T., and Carrier, E. (2004). Immunological factors that affect the in vivo fate of T7 phage in the mouse. *J. Virol Methods* **115:**99–104.

Stent, G. S. (1963). Molecular Biology of Bacterial Viruses. W.H Freeman and Company, San Francisco/London.

Sulakvelidze, A., Alavidze, Z., and Morris, J. G., Jr. (2001). Bacteriophage therapy. *Antimicrob. Agents Chemother.* **45:**649–659.

Sulakvelidze, A., and Barrow P. (2005). Phage therapy in animals and agribusiness. *In* "Bacteriophages. Biology and Applications" (E. Kutter and A. Sulakvelidze, eds.). CRC Press, Boca Raton, FL.

Sustiel, A. M., Joseph, B., Rocklin, R. E., and Borish, L. (1989). Asthmatic patients have neutrophils that exhibit diminished responsiveness to adenosine. *Am. Rev. Respir. Dis.* **140:**1556–1561.

Tikhonenko, A. S., Gachechiladze, K. K., Bespalova, I. A., Kretova, A. F., and Chanishvili, T. G. (1976). Electron-microscopic study of the serological affinity between the antigenic components of phages T4 and DDVI. *Mol. Biol. (Mosk)* **10:**667–673.

Uchiyama, J., Maeda, Y., Takemura, J., Chess-Williams, R., Wakiguchi, H., and Matsuzaki, M. (2009). Blood kinetics of four intraperitoneally administered therapeutic candidate bacteriophages in healthy and neutropenic mice. *Microbiol. Immunol.* **53:**301–304.

Victor, V. M., Rocha, M., and de la Fuente, M. (2004). Immune cells: Free radical and antioxidants in sepsis. *Int. Immunopharmacol.* **4:**327–347.

Vitiello, C. L., Merril, C. R., and Adhya, S. (2005). An amino acid substitution in a capsid protein enhances phage survival in mouse circulatory system more than a 1000-fold. *Virus Res.* **114:**101–103.

Weber-Dąbrowska, B., Mulczyk, M., and Górski, A. (2000a). Bacteriophage therapy of bacterial infections: An update of our institute's experience. *Arch. Immunol. Ther. Exp. (Warsz)* **48:**547–551.

Weber-Dąbrowska, B., Zimecki, M., and Mulczyk, M. (2000b). Effective phage therapy is associated with normalization of cytokine production by blood cell cultures. *Arch Immunol Ther Exp (Warsz)* **48:**31–37.

Weber-Dąbrowska, B., Zimecki, M., Mulczyk, M., and Górski, A. (2002). Effect of phage therapy on the turnover and function of peripheral neutrophils. *FEMS Immunol. Med. Microbiol.* **34:**135–138.

Wedgwood, R. J., Ochs, H. D., and Davis, S. D. (1975). The recognition and classification of immunodeficiency diseases with bacteriophage phi X174. *Birth Defects Orig. Artic Ser.* **11:**331–338.

Yanagida, M. (1972). Identification of some antigenic precursors of bacteriophage T4. *J. Mol. Biol.* **65:**501–517.

Zimecki, M., Artym, J., Kocieba, M., Weber-Dabrowska, B., Borysowski, J., and Górski, A. (2009). Effects of prophylactic administration of bacteriophages to immunosuppressed mice infected with *Staphylococcus aureus*. *BMC Microbiol.* **9:**169.

Zimecki, M., Weber-Dabrowska, B., Łusiak-Szelachowska, M., Mulczyk, M., Boratyński, J., Poźniak, G., Syper, D., and Górski, A. (2003). Bacteriophages provide regulatory signals in mitogen-induced murine splenocyte proliferation. *Cell. Mol. Biol. Lett.* **8:**699–711.

CHAPTER 3

Clinical Aspects of Phage Therapy

Ryszard Międzybrodzki,[*,†,1] **Jan Borysowski,**[‡] **Beata Weber-Dąbrowska,**[*,†] **Wojciech Fortuna,**[*,†] **Sławomir Letkiewicz,**[†,§] **Krzysztof Szufnarowski,**[†,||] **Zdzisław Pawełczyk,**[†] **Paweł Rogóż,**[†,¶] **Marlena Kłak,**[*] **Elżbieta Wojtasik,**[#] and **Andrzej Górski**[*,†,‡]

Contents

* Bacteriophage Laboratory, Ludwik Hirszfeld Institute of Immunology Experimental Therapy, Polish Academy of Sciences, Wrocław, Poland
† Phage Therapy Unit, Ludwik Hirszfeld Institute of Immunology and Experimental Therapy, Polish Academy of Sciences, Wrocław, Poland
‡ Department of Clinical Immunology, Transplantation Institute, Medical University of Warsaw, Warsaw, Poland
§ Katowice School of Economics, Katowice, Poland
|| Regional County Hospital and Research Center, Wrocław, Poland
¶ Department of Orthopedics, Wrocław Medical University, Wrocław, Poland
# Department of Pharmacognosy and Molecular Basis of Phytotherapy, Faculty of Pharmacy, Medical University of Warsaw, Warsaw, Poland
1 Corresponding author, E-mail: mbrodzki@iitd.pan.wroc.pl

Advances in Virus Research, Volume 83
ISSN 0065-3527, DOI: 10.1016/B978-0-12-394438-2.00003-7

Abstract

Phage therapy (PT) is a unique method of treatment of bacterial infections using bacteriophages (phages)—viruses that specifically kill bacteria, including their antibiotic-resistant strains. Over the last decade a marked increase in interest in the therapeutic use of phages has been observed, which has resulted from a substantial rise in the prevalence of antibiotic resistance of bacteria, coupled with an inadequate number of new antibiotics. The first, and so far the only, center of PT in the European Union is the Phage Therapy Unit (PTU) established at the Ludwik Hirszfeld Institute of Immunology and Experimental Therapy, Wrocław, Poland in 2005. This center continues the rich tradition of PT in Poland, which dates from the early 1920s. The main objective of this chapter is to present a detailed retrospective analysis of the results of PT of 153 patients with a wide range of infections resistant to antibiotic therapy admitted for treatment at the PTU between January 2008 and December 2010. Analysis includes the evaluation of both the efficacy and the safety of PT. In general, data suggest that PT can provide good clinical results in a significant cohort of patients with otherwise untreatable chronic bacterial infections and is essentially well tolerated. In addition, the whole complex procedure employed to obtain and characterize therapeutic phage preparations, as well as ethical aspects of PT, is discussed.

I. BACKGROUND

The increasing prevalence of antibiotic-resistant bacteria on a global scale is currently considered one of the greatest therapeutic challenges (Giamarellou, 2010; Köck *et al.*, 2010). This problem is exacerbated by the crisis in the search for new antibiotics (Freire-Moran *et al.*, 2011; Jabes, 2011). An alternative option to eliminate bacteria, including their antibiotic-resistant strains, is phage therapy (PT), which relies on the use of bacteriophages—viruses that specifically kill bacterial cells

(Górski *et al.*, 2009). The main features of phages as antibacterial agents include a novel mode of action (distinct from the mechanisms of action of antibiotics); a narrow antibacterial range, resulting in an ability to selectively eliminate pathogenic bacteria without affecting normal microflora adversely; and a capacity for multiplication at the site of infection (Górski *et al.*, 2009; Hanlon, 2007). A high potential of phages to eliminate antibiotic-resistant bacteria has been shown in a number of preclinical studies (Biswas *et al.*, 2002; Capparelli *et al.*, 2007; Matsuzaki *et al.*, 2003; Wang *et al.*, 2006). Potential fields of application of phages include not only clinical medicine, but also the food and agriculture industry (Balogh *et al.*, 2010; Górski *et al.*, 2009; Mahony *et al.*, 2011). In fact, the first phage products, such as ListShield and Listex P100, which are used for the prevention of food-borne infections with *Listeria monocytogenes*, and AgriPhage for the prevention and control of bacterial spot or bacterial speck on tomato and pepper plants, were registered and marketed a few years ago.

The first attempts to treat bacterial infections in humans with phages were conducted around 1920, just a few years after the discovery of bacteriophages. Since then, a great number of uncontrolled "clinical" studies have been carried out to evaluate the efficacy and safety of phage therapy (Sulakvelidze and Kutter, 2005). However, such studies do not meet the current rigorous standards for clinical trials. Only recently have the results of the first randomized controlled trials (RCTs) of phage therapy been published (see Table I). Currently, the status of phage treatment in humans differs among individual countries. While in the United Kingdom, United States, Switzerland, and India phages have been used in RCTs, in Poland and Belgium PT is conducted as an experimental treatment under the Declaration of Helsinki; in Georgia and Russia it is considered a standard treatment (Pirnay *et al.*, 2011).

The history of PT in Poland encompasses over 80 years; to our knowledge, the first paper on the therapeutic use of bacteriophages was published in Poland in 1923 (Kalinowski and Czyż, 1923). Since then, PT has been performed on a relatively small scale at different Polish centers (Jasieński, 1927; Lityński, 1950; Trelińska *et al.*, 2010; Witoszka and Strumiłło, 1963). The turning point in research on bacteriophages and PT in Poland was establishment of the Institute of Immunology and Experimental Therapy (IIET) in Wrocław by Professor Ludwik Hirszfeld in 1952. The IIET played a crucial role in the development of clinical PT in Poland by producing therapeutic phage preparations and coordinating PT conducted at different Silesian hospitals. In the 1980s and 1990s, about 2000 patients were treated with phage preparations produced by the IIET. The results of PT performed in Poland in this period of time were published in several articles, which are of primary importance to the English-language literature on clinical PT (Ślopek *et al.*, 1987; Weber-Dąbrowska *et al.*, 2000, 2003). In 2005 the first center of PT in the European Union (EU) was

TABLE I Summary of published results of randomized controlled trials of phage therapy

Aim of the trial	Organizer/ sponsor	Time of trial	Comments	Reference
A single-center, randomized, placebo-controlled trial on the safety and bioavailability measure of oral phage	Nestlé Research Center, Nestec Ltd., Lausanne, Switzerland	June 2003	Fifteen healthy adult volunteers received two doses [10^3 or 10^5 plaque-forming units (PFU)/ml] of purified T4 phage or placebo in 150 ml of drinking water. Neither adverse events nor significant change in population of commensal *E. coli* following phage application was observed. Phages were detected in stools in all volunteers administered phage preparations in a dose-dependent manner. However, neither phage particles nor anti-phage antibodies were detected in blood of the subjects.	Bruttin and Brüssow, 2005
A double-blind, placebo-controlled initial phase I/II clinical trial targeting chronic ear infections caused by *P. aeruginosa*	Biocontrol Ltd., London, UK	July 2006–November 2007	Twelve patients with otitis media caused by antibiotic-resistant *P. aeruginosa* were administered a single dose of a preparation containing six phages and another 12 were treated with placebo. A significant reduction of clinical symptoms at day 42 in the	Wright *et al.*, 2009

			bacteriophage-treated group compared with the control group was found (55 and 104% of total clinical score at day 0, respectively). There was also a 76% decrease in the mean count of bacteria in samples taken from the patient's ears 6 weeks after phage application, while in the control group a small increase (9%) in the bacterial count was observed. No adverse events following the administration of phage preparations were reported.	
A prospective, randomized, double-blind controlled study of WPP-201 for the safety and efficacy of treatment of venous leg ulcers	Southwest Regional Wound Care Center in Lubbock, Texas	September 2006–May 2008	Phase I study of WPP-201—a cocktail of eight lytic bacteriophages against *P. aeruginosa*, *S. aureus*, and *E. coli* developed by Intralytix Inc. It contained approximately 1×10^9 PFU/ml of each phage. The primary objective of this study was to evaluate the safety of the topical use of WPP-201 in patients with full thickness venous leg ulcers of greater than 30 days duration. WPP-201 had no significant effect on ulcer healing. No other adverse events were found in subjects following the administration of phages either.	Rhoads *et al.*, 2009

opened at the IIET. This center, the Phage Therapy Unit (PTU), conducts PT in a form of experimental therapy (Międzybrodzki *et al.*, 2007).

The main objective of this chapter is to present a detailed retrospective analysis of the results of PT of patients admitted for treatment at the PTU between January 2008 and December 2010, as well as to discuss some ethical, regulatory, and methodological aspects of PT.

II. ETHICAL ASPECTS OF BACTERIAL DRUG RESISTANCE AND PHAGE THERAPY

As pointed out by Selgelid (2007), drug resistance is not only a clinical problem but also an ethical issue, as it may be linked to overconsumption of antibiotics by the wealthy and their underconsumption by the poor or otherwise marginalized. Antibiotics are overprescribed by half of the physicians in the United States and Canada, and up to 90% of prescriptions in teaching hospitals are inappropriate. The emergence of drug resistance can also be prompted by prematurely terminated treatment, and the poor often cannot afford to complete it. Selgelid (2007) considers drug resistance a matter of ethical urgency and appeals for the development of novel means to combat it.

The ethical issues linked to antibiotic resistance were also discussed by Aiello *et al.* (2006). The authors emphasize the existing conflicts between private interests (e.g., corporations) and the public interest of society at large. Therefore, one of the major causes of the present crisis is the fact that antibiotics are developed on the basis of market criteria (profit seeking and ability to pay) rather than on the basis of benefit to the public. For the pharmaceutical industry, the potential risks in the development of new antimicrobials outweigh the incentives (considerable costs, relatively short period of use, risk of resistance, etc.). This contrasts with the clear potential for profit making with lifestyle drugs and drugs for chronic conditions. Furthermore, producers are interested in increasing sales of their products when the expiry of their patents approaches—contrary to public interest, as limited consumption of antibiotics may preserve new drugs, thus allowing their continued efficacy and preventing the appearance of resistance. The authors propose decisive changes in the policies governing the development of antimicrobials: they should be considered as a public good, and the state could take over their development and production. An alternative policy could be based on underwriting the development of particular drugs by industry. This would require funds and decisions about the development of new agents based on sound scientific data and interest of the public. Probably the best option would be partnership allowing the private and public sectors to share risks and benefits—market exclusivity, tax incentives, expedient approval, and so on—in return for the successful development of new drugs. An example to follow might be the Orphan Drug Act (extended tax credit and guaranteed 7 years of market exclusivity for developers of drugs for rare conditions). Finally, the authors also recommend development and use of alternative therapies, believing it may provide "an important avenue for interrupting the cycle of resistance and obsolesce associated with new antimicrobial use." They conclude that "long-term solutions that do not rely on market forces to govern the production and consumption of antimicrobials" are needed immediately.

The economic aspects of phage therapy were discussed in report by Międzybrodzki *et al.* (2007), in which average costs of treatment of methicillin-resistant *Staphylococcus aureus* (MRSA) infections with modern antibiotics (a 10-day course) against phage therapy (mean duration based on patients' records: 6.5 weeks) were compared. The cost of phage therapy was about half the cost of vancomycin and several times less compared with other drugs (linezolid, teicoplanin, quinupristin). The authors are currently assessing the option of shortening the initial course of phage therapy to a period comparable to antibiotic treatment, which could further widen this already spectacular gap between the costs of phage therapy and modern antibiotics. This suggests that wider introduction of phage therapy should lead to substantial savings in healthcare costs, which is presently one of the greatest challenges of modern civilization. Currently phage therapy is not covered by public health insurance, which poses problems to at least some patients who cannot afford to bear the costs or do not have additional private insurance covering it. However, it should be noted that Switzerland has decided to reimburse complementary medicine following a referendum in which two-thirds of voters supported that decision (Stafford, 2011). Among therapies included are homeopathy, herbal medicine, and traditional Chinese medicine. The effectiveness of such therapy must be reviewed by an internationally recognized institution whose assessment is then evaluated and the final decision on insurance coverage is taken by the relevant Swiss authority. Of note, the Swiss Medical Association concluded that this move was a "superb decision."

The public threat of growing antimicrobial resistance is so great that some experts believe it requires special health regulations as it constitutes a public health emergency of international concern (Wernli *et al.*, 2011), and PT deserves its own regulatory framework in Europe (Pirnay *et al.*, 2011; Verbeken *et al.*, 2007). Current European pharmacological regulations/definitions and standards are not adequately adapted to phage preparations used for treatment because they do not directly consider bacterial viruses as medicinal products. This results in many doubts and controversies when interpreting these regulations in the context of the therapeutic use of phages in humans. In 2009, a nonprofit organization called Phages for Human Applications Group Europe (P.H.A.G.E.) was even established by Belgian research groups and members of the Pasteur Institute to develop a specific regulatory framework for the use of bacteriophages (http://www.p-h-a-g-e.org/Home.html).

Of special concern is the recent emergence of carbapenem-resistant *Enterobacteriaceae* for which no reliable therapy currently exists, and the proposed combination protocols have high toxicity. An outbreak of *Escherichia coli* 0104:H4 (enterohemorrhagic *E. coli*; EHEC) infections in Europe (considered to be the worst ever) with its many fatalities constitutes a spectacular example of our civilization's continued vulnerability to the danger of bacterial infections and the inability of antibiotics to control this challenge; in fact, it is believed that antibiotics could have contributed to the spread of the bacterium and are contraindicated in the treatment of this syndrome, where therapy is rather symptomatic (Turner, 2011a). What is more, data on the efficacy of eculizumab in the treatment of hemolytic uremic syndrome prompted attempts to use the drug in this infection, which may cause devastating consequences. However, a year's treatment costs more than $400,000 (Turner, 2011b). Thus, this outbreak constitutes an important additional burden to the economy of healthcare, which should give food for thought. According to some authors, "a global mechanism incorporating both systemic surveillance and effective public health response is urgently needed" (Wernli *et al.*, 2011). Evidently, a wider application of phage therapy could be part of such a response.

The authors strongly believe that the time has come for state authorities to consider inclusion of PT in governmental insurance plans, which is prompted by the recent fatalities resulting from the outbreak of EHEC. The fact that the tragedy took place in Germany where social care and healthcare are so well developed and organized is an ominous sign of the existing danger. In fact, the president of the city of Wroclaw has established a program covering the costs of PT for the residents of the city from public funds, a clear example to follow. Finally, sound ethical and social reasons exist supporting necessary changes in the existing state coverage plans.

Phage therapy has been used extensively in Poland over the last three decades. However, after accession of Poland to the EU in 2004, the rules for conducting phage therapy had to be adapted to the new regulations. To ensure the continuation of phage treatment in Poland, the IIET opened the PTU in Wrocław (which operates as a nonprofit unit of IIET) and introduced new standards to enable the use of phage preparations prepared by the Bacteriophage Laboratory of the IIET. These standards require, among other things, approval of the bioethics committee for PT to be conducted by an individual physician or a health center that intends to use phages from the IIET collection for treatment. In Poland, phage preparations are used for PT under the rules of therapeutic experiments, which are included in the Act on the Medical Profession and comply to a significant degree with the norms set by the World Medical Association in the Declaration of Helsinki (Międzybrodzki *et al.*, 2007). The primary purpose of a therapeutic experiment (in contrast to clinical research) is the introduction by a physician of new or only partially tested diagnostic, therapeutic, or prophylactic methods for the direct benefit of the person being treated. Thus far, 12 regional bioethical commissions in Poland have approved the use of PT to meet local demands for treatment. Furthermore, a regional bioethical commission for Lower Silesia has granted such approval for all regional hospitals intending to use phage preparations from the IIET; these hospitals are only obliged to notify the commission of the initiation of such treatment conducted under the special protocol supervised by the IIET.

The therapeutic protocol employed at the PTU (its formal title is "Experimental phage therapy of drug-resistant bacterial infections, including MRSA infections") is still being improved and is modified periodically with approval of the bioethics committee. At first, phage preparations were employed according to the procedures developed by Ślopek *et al.* (1983). However, in 2008 some major changes were made to these methods. In particular, the administration of phages by the intrarectal route and as inhalations of aerosol was started, and the therapeutic protocols were modified. Moreover, new specialists joined the PTU team. Therefore, the analysis presented in this chapter includes patients admitted for PT since 2008.

III. THERAPEUTIC BACTERIOPHAGES FROM THE COLLECTION OF THE INSTITUTE OF IMMUNOLOGY AND EXPERIMENTAL THERAPY IN WROCŁAW

The IIET bacteriophage collection includes over 500 virulent phages of bacteria from many different genera (see Table II). The collection was initiated in 1948 by Professor Ludwik Hirszfeld. It was enriched with phages isolated by the Bacteriophage Laboratory of IIET, as well as phages obtained as a result of cooperation with Polish (Department of Microbiology at the Medical University in Gdańsk, Institute of Microbiology at the University of Łódź, National Institute of Hygiene in Warsaw, Department of Microbiology at the Medical University of Wrocław) and foreign laboratories (Research Institute of Epidemiology and Microbiology in Sofia, Bulgaria; National Bacteriological Laboratory in Stockholm, Sweden; Institute of Microbiology, Parasitology, and Epidemiology in Bucharest, Romania; National Institute of Hygiene in Budapest, Hungary; Institute of Experimental Epidemiology in Wernigerode, Germany;

TABLE II Status of bacteriophage collection of the Institute of Immunology and Experimental Therapy (May 2011)

Bacteriophage host	Number of phages
Escherichia coli	121
Klebsiella pneumoniae or *Klebsiella oxytoca*	95
Enterococcus faecalis or *Enterococcus faecium*	73
Enterobacter cloacae	48
Shigella flexneri or Shigella sonnei	39
Citrobacter freundii	38
Pseudomonas aeruginosa or *Pseudomonas fluorescens*	37
Salmonella enteritidis or *Salmonella typhimurium*	32
Stenotrophomonas maltophilia	18
Serratia marcescens or *Serratia liquefaciens*	17
Proteus mirabilis	17
Morganella morganii	14
Staphylococcus aureus	7
Acinetobacter baumannii	5
Burkholderia cepacia	2

Department of Microbiology at the University of Adelaide, Australia; Institute of Hygiene at the University of Cologne, Germany). Over the last decade it has been significantly enriched with phages against different bacterial strains (including multidrug-resistant ones) cultured from patients. The present composition of the institute's therapeutic phage collection is shown in Table II. The collection has been enriched with new phages (including phages specific to *Streptococcus*) as part of the project entitled "Optimization of the production and characterization of bacteriophage preparations for therapeutic use" covered by EU funds within the Operational Program Innovative Economy 2007-2013. They have been isolated from environmental samples (crude and purified city sewage), inland and sea waters, or filtrates of biological samples. The majority of phages characterized by electron microscopy were classified as members of the *Myoviridae* and *Siphoviridae* families and, in a few cases, the *Podoviridae* family (Ślopek and Krzywy, 1985; unpublished data).

Since 2006, a few thousand bacterial strains have been tested for their sensitivity to phages from the IIET collection. It was found that bacteria sensitive to phages isolated long ago are less frequent among newly isolated strains than among strains isolated in the past. Such differences in bacterial sensitivity to phages have been especially observed for *E. coli*, *Klebsiella pneumonia*, *Pseudomonas aeruginosa*, and, to a lesser degree, *Staphylococcus aureus* (unpublished data). Thus, it is necessary to control the lytic spectrum of phages in the collection and to isolate new phages with a

TABLE III Characteristics of the 157 patients admitted for phage therapy at the Phage Therapy Unit in 2008–2010

Gender	Males: $n = 89$
	Females: $n = 68$
Median age	Males: 44.5 (21–79) years
	Females: 47 (20–82) years
	Total: 46 (20–82) years
Diagnosis	Genital and urinary tract infection in men: chronic bacterial prostatitis ($n = 14$); urinary infection ($n = 10$); chronic bacterial prostatitis/urinary infection ($n = 6$)
	Genital and urinary tract infection in women: urinary infection ($n = 14$); vaginal infection ($n = 3$); urinary/vaginal infection ($n = 5$)
	Soft tissue infection: postoperative wound infection ($n = 6$); leg ulcer ($n = 8$); abscess, phlegmon, or empyema penetrating to the body cavities ($n = 5$); deep tissue infection ($n = 11$)
	Skin infection: external ear infection ($n = 4$); unspecified local infection of skin ($n = 3$); atopic dermatitis complicated by staphylococcal infection ($n = 1$); acne ($n = 1$); eczema ($n = 1$); furunculosis ($n = 1$)
	Orthopedic infections: prosthetic joint infection ($n = 8$); osteomyelitis ($n = 22$); joint infection ($n = 5$); osteomyelitis/joint infection ($n = 2$); discitis ($n = 1$)
	Respiratory tract infections: upper respiratory tract infections ($n = 17$); lower respiratory tract infections ($n = 4$); upper and lower respiratory tract infections ($n = 3$)
	Other: internal ear infection ($n = 1$), recurrent bacteremia ($n = 1$)

(*continued*)

TABLE III *(continued)*

Pathogens causing the infection	Monoinfections: *S. aureus* ($n = 76$, including seven MRSA cases); *S. haemolyticus* ($n = 1$); *E. faecalis* ($n = 17$); *E. coli* ($n = 15$); *P. aeruginosa* ($n = 13$); *P. putida* ($n = 1$); *E. cloacae* ($n = 1$); *K. pneumoniae* ($n = 1$); *Salmonella* group C ($n = 1$) Polyinfections: *S. aureus* (MRSA)/*S. mitis* ($n = 1$); *S. aureus*/*A. junii* ($n = 1$); *S. aureus*/*E. cloacae* ($n = 1$); *S. aureus*/*M. morganii* ($n = 1$); *S. aureus*/*P. aeruginosa* ($n = 1$); *E. faecalis*/*K. oxytoca* ($n = 1$); *E. faecalis*/*C. freundii* ($n = 1$); *E. faecalis*/*E. coli*/*K. oxytoca* ($n = 1$); *E. faecalis* /*P. aeruginosa* ($n = 1$); *E. faecalis*/*P. vulgaris* ($n = 1$); *E. faecalis*/*S. aureus* ($n = 1$); *E. faecalis*/coagulase negative staphylococcus ($n = 1$); *E. coli*/*E. faecalis* ($n = 9$); *E. coli* – two different strains ($n = 1$); *E. coli*/*K. pneumoniae* ($n = 1$); *P. aeruginosa*/*S. aureus* ($n = 1$); *P. aeruginosa*/*S. aureus*/*P. vulgaris* ($n = 1$); *P. aeruginosa*/*S. maltophilia* ($n = 1$); *P. aeruginosa* /group F β-hemolytic *Streptococcus* ($n = 1$); *P. aeruginosa*/*A. baumannii* ($n = 1$); *P. vulgaris*/*E. coli*/*E. faecalis* ($n = 1$)
Concomitant antibacterial treatment	Antibiotics/chemotherapeutics: $n = 31$ Antibiotics/chemotherapeutics and disinfectants: $n = 6$ Antibiotics/chemotherapeutics and herbs or supplements for treatment of urinary infections: $n = 4$ Disinfectants: $n = 3$
Number of patients included in analysis	Patients for whom information on the treatment effect was available: $n = 153$ Patients for whom at least one laboratory result during PT was available: $n = 154$ Patients for whom complete information on cumulative treatment duration was available: $n = 49$

broader host range. Both polyvalent phages with a broad host range and monovalent phages (specific to one bacterial species or even to a particular strain) are used in PT. Analysis of the sensitivity of bacteria isolated from patients treated in the PTU showed that for 90% of *S. aureus*, 72% of *P. aeruginosa*, and 77% of *Enterococcus faecalis* strains it was possible to find an active specific phage in the collection (unpublished data). Importantly, for all tested *E. faecalis* high-level aminoglycoside resistance strains, this percentage was 100%.

Phage lysates for therapeutic purposes are prepared by the Bacteriophage Laboratory of IIET in broth culture or peptone water by the standard method of Ślopek *et al.* (1983) with some modifications (Letkiewicz *et al.*, 2009; Zimecki *et al.*, 2003). The phage titer in a preparation used for PT should be between 10^6 and 10^9 plaque-forming units/ml. The procedure of preparation of phage formulations against *Staphylococcus* and *Pseudomonas* has been under patent protection since 2002 (US Patent 7232564 B2). Since 2006 all therapeutic staphylococcal phages (marked as A5/80, 676/F, 676/Ż, P4/6409, A3/R, liz/80, and fi 200/6409), as well as some enterococcal (marked as 1 K/N, 1 C/K, and 15/P) and *Pseudomonas* (marked as F-8, 119x, and col21) phages used for experimental PT, have been produced by BIOMED SA, Kraków, Poland under good manufacturing practice conditions according to the techniques described earlier. No human immunodeficiency or hepatitis C viruses have been found in these preparations.

IV. CLINICAL RESULTS OF PHAGE THERAPY CONDUCTED IN THE PHAGE THERAPY UNIT IN WROCŁAW

A. Patients and treatment protocol

One hundred fifty-seven patients (68 women, 89 men) were admitted for treatment at the PTU between January 2008 and December 2010 because of various infections resistant to antibiotic treatment defined as (i) infection caused by multidrug-resistant bacteria; (ii) infection that persisted despite treatment with targeted antibiotics, when, in the opinion of the qualified specialist, the previous antibiotic treatment was ineffective and/or its application did not allow a significant improvement to be expected but there was a need for noninvasive treatment); or (iii) cases where targeted antibiotic treatment was impossible, for example, because of medical contraindications. Only patients at least 18 years old who signed informed consent were eligible for PT. A prerequisite for receiving PT was the sensitivity of bacteria isolated from the site of infection to at least one phage from the bacteriophage collection of the IIET. Patients disqualified from PT included individuals with allergy to components of phage preparations and pregnant or breastfeeding women, as well as some cases of malabsorption syndrome, allergy to food or animal protein, and advanced hepatic insufficiency. Generally, 447 patients were disqualified from PT during the analyzed period because they did not fulfill the eligibility criteria or resigned from the treatment.

Patients with the following types of infections were eligible for PT: genital and urinary tract infections, soft tissue infections, orthopedic infections, and respiratory tract infections. Etiological agents of these infections are listed in Table III. A polyinfection was defined as the presence of at least two different bacterial pathogens before the beginning of PT or the appearance of new pathogens during the break between PT cycles.

Phage preparations against *Staphylococcus, Enterococcus, E. coli, Pseudomonas, Klebsiella, Enterobacter, Proteus, Citrobacter, Salmonella,* and/or *Stenotrophomonas* (obtained according to the methods and in standards already mentioned in the part III of this chapter) were administered to patients topically, orally, intrarectally, intravaginally, or as inhalations of aerosol. In some patients, a combination of oral and topical or intrarectal and topical routes of administration was applied. The topical route of administration of phage preparations (twice daily) included gargling, fistular irrigation, irrigation of the abscess cavity, and sitz baths, as well as the use of wet compresses, nose drops, ear drops, vaginal irrigation, or inhalations. By the oral route, phage preparations were administered in the dose of 10–20 ml three times daily, at least 30 min before meals. Ten milliliters of oral suspension of dihydroxyaluminium sodium carbonate (68 mg/ml) was administered to these patients up to 20 min prior to a phage preparation to protect phage virions against inactivation by gastric juice. By the intrarectal route, phage preparations were administered in the dose of 10–20 ml twice daily. The maximal cumulative duration of the use of phage preparation during one course of PT was 12 weeks, which could be prolonged up to another 12 weeks in cases when the infection persisted, despite a good response to PT (categories A–C as further explained). Interruptions during the course of PT were possible for justified reasons (e.g., if there was a need for preparation of new phages for treatment). If these breaks lasted longer than 4 weeks they divided PT into cycles. If the infection recurred, a new course of phage treatment could be started only after an obligatory break lasting at least 4 months. The treatment had to be stopped when the patient withdrew his or her consent or any serious adverse events related to PT occurred. It could also be stopped when, in the physician's and patient's opinion, the results of PT were not satisfactory, a satisfactory effect of PT was achieved in the physician's opinion, or the physician acknowledged that continuation of PT would threaten the patient's health.

Determination of the sensitivity of isolated bacterial strains to phages from the IIET collection (called the phage-typing procedure) was conducted according to the standard technique of Adams (1959). For each patient, only preparations of specific lytic phage that was confirmed to be active against the pathogenic bacterial strain were used for treatment, as was verified by testing the ability of such phage to form clear plaques on a lawn of the tested bacterial cells. Preparations containing only one pure virus line were used. In the case of a mixed infection with two bacterial strains at the same time, if it was possible to prepare phages against both bacteria, the patient was administered individual doses of both phage preparations alternately. If the use of phage against only one bacterial strain (called the targeted pathogen) was possible, use of a targeted antibiotic as concomitant treatment was allowed. The use of other medications for concomitant diseases was also allowed.

B. Methods of evaluation of the efficacy of phage therapy

The effectiveness of PT was evaluated by the physician conducting PT based on the results of control microbiological tests, especially bacterial cultures, assessment of the intensity of infection symptoms, results of control laboratory assays, and opinions of consulting medical specialists. The results of PT fell into one of seven major categories: A—pathogen eradication and/or recovery (pathogen eradication was confirmed by the results of bacterial cultures; recovery refers to wound healing or complete subsidence of the infection symptoms); B—good clinical result (almost complete subsidence of some infection symptoms in some cases confirmed by the results of laboratory assays, together with a significant improvement of the patient's general condition after completion of PT); C—clinical improvement (discernible reduction in the intensity of some infection symptoms after completion of PT to a degree not observed before PT, when no treatment was used, or during breaks between successive cycles of PT); D—questionable clinical improvement (reduction in the intensity of some infection symptoms to a degree that could also be observed before PT or during breaks between successive cycles of PT); E—transient clinical improvement (reduction in the intensity of some infection symptoms observed only during application of phage preparations and not after termination of PT); F—no response to the treatment; and G—clinical deterioration (exacerbation of symptoms of infection at the end of PT). In addition, categories A–C were considered good responses to PT and categories D–G were considered inadequate responses to PT. In the case of PT conducted in cycles, the assessment of clinical effects of the treatment was based on the analysis of its course during all completed cycles. Samples for bacteriological testing were taken at least once before each PT course in all patients. Control samples for bacterial culture and phage typing were taken obligatorily when there were signs of superinfection or when the results of PT were not satisfactory in the opinion of the physician conducting the treatment—usually during breaks in PT cycles and after treatment.

Cumulative treatment duration was calculated as the sum of the number of days during which the patient used the phage preparation irrespective of its dose and mode of application. To evaluate the influence of the route of administration of phage preparations on the efficacy of PT, vaginal application and inhalations were analyzed separately from other topical routes (which included gargling, fistular irrigation, irrigation of the abscess cavity, and sitz baths, as well as the use of wet compresses, nose drops, and ear drops). Changes in patients' C-reactive protein (CRP) serum concentration, erythrocyte sedimentation rate (ESR), and white blood cell count (WBC) were evaluated for two periods of PT: earlier, during days 5–8; and later, during days 9–32 after starting treatment. These intervals were chosen on the basis of the methodology used in previous such analysis (Międzybrodzki *et al.*, 2009). Changes in selected hematological and biochemical serum parameters were evaluated for four periods of cumulative phage use: 3–6, 7–20, 21–48, and 49–84 days. If two or more results of laboratory tests were obtained during the analyzed period, their average was calculated for general analysis. Results of the most recent laboratory tests done before PT were used as baseline.

Because variables in the analyzed groups were not distributed normally, nonparametric statistical tests were used for analysis. Kruskal–Wallis one-way analysis of variance was applied for testing differences in the cumulative treatment duration in different kinds of infections. Differences in values of the laboratory test before starting and during PT were analyzed using the Wilcoxon matched pairs test. The relation between the duration of PT and its effectiveness was assessed using Spearman's rank correlation coefficient. The statistical significance of differences between observed and expected frequencies was determined using χ^2 statistics (Pearson's χ^2 test, Fisher's exact test, or Yates' χ^2 test when appropriate). Values for p less than 0.05 were considered significant.

C. General evaluation of the results of phage therapy

Evaluation of the efficacy of PT is based on data from 153 patients (in 4 out of the total of 157 patients who started PT it was not possible to evaluate the effectiveness of the treatment because they dropped out of the clinical observation). Their detailed characteristics are presented in Table III. Overall, a good response to PT was observed in 61 patients (39.9%); in particular, pathogen eradication and/or recovery was achieved in 28 subjects (18.3%) (Table IV). However, inadequate response to the treatment was noted in 92 patients (60.1%). No significant difference in the efficacy of PT between women ($n=68$) and men ($n=85$) was found (data

TABLE IV General evaluation of results of phage therapy (type of infection and use of antibacterials during treatment were also taken into consideration)

			Type of infection				Use of antibacterials			
	General evaluation[b] ($n = 153$)		Monoinfection ($n = 123$)		Polyinfection[c] ($n = 30$)		No antibacterials ($n = 109$)		Antibacterial[d] used during PT ($n = 44$)	
Category of response to treatment[a]	*n*	%	*n*	%	*n*	%	*n*	%	*n*	%
A - pathogen eradication and/or recovery	28	18.3	22	17.9	6	20.0	22	20.2	6	13.6
B - good clinical result	13	8.5	11	8.9	2	6.7	7	6.4	6	13.6
C - clinical improvement	20	13.1	14	11.4	6	20.0	15	13.8	5	11.4
D - questionable clinical improvement	10	6.5	8	6.5	2	6.7	8	7.3	2	4.6
E - transient clinical improvement	33	21.6	27	22.0	6	20.0	22	20.2	11	25.0
F - no response to treatment	39	25.5	32	26.0	7	23.3	28	25.7	11	25.0
G - clinical deterioration	10	6.5	9	7.3	1	3.3	7	6.4	3	6.8
Good response (total A–C):	**61**	**39.9**	**47**	**38.2**	**14**	**46.7**	**44**	**40.4**	**17**	**38.6**
Inadequate response (total D–G):	**92**	**60.1**	**76**	**61.8**	**16**	**53.3**	**65**	**59.6**	**27**	**61.4**

[a] Pathogen eradication was confirmed by the results of bacterial cultures. Recovery refers to wound healing or complete subsidence of the infection symptoms. Good clinical result means almost complete subsidence of some infection symptoms, in some cases confirmed by the results of laboratory assays, together with a significant improvement of the patient's general condition. Clinical improvement means a discernible reduction in the intensity of some infection symptoms after completion of PT to a degree not observed before PT when no treatment was used or during breaks between successive cycles of PT. Questionable clinical improvement means a discernible reduction in the intensity of some infection symptoms to a degree not observed before PT or during breaks between successive cycles of PT. Transient clinical improvement means improvement of the patient's condition observed only during PT.

[b] Relevant data were available for 153 subjects because 4 out of the total 157 patients who started PT in the years 2008–2010 dropped out of the clinical observation (one case with internal ear infection, one case with chronic bacterial prostatitis, one case with osteomyelitis, and one case with external ear infection).

[c] Defined as the presence of at least two different bacterial pathogens before the beginning of PT or the appearance of new pathogens during the break between PT cycles.

[d] Including antibiotics, chemotherapeutics, disinfectants, and other potentially antibacterial agents.

not shown). A comparison of results obtained in patients with monoinfections (123 cases) and those with polyinfections (30 cases) revealed no significant differences in the efficacy of PT in terms of either the percentage of good responses to the treatment or the rate of pathogen eradications and/or recoveries (Table IV).

A separate analysis was performed to compare the efficacy of PT in diabetic and nondiabetic patients (12 and 141 cases, respectively). However, no significant differences were found between these two groups either in the percentages of good responses to PT (50 and 39.1% for diabetic and nondiabetic patients, respectively; data not shown) or in the rate of pathogen eradications and/or recoveries (16.7 and 18.4% for diabetic and nondiabetic patients, respectively; data not shown).

Phage preparations that were applied most frequently included bacteriophages specific to staphylococci (51.6%), enterococci (11.1%), *E. coli* (11.1%), and *Pseudomonas* (9.8%). The effectiveness of PT varied considerably depending on the specificity of the phage preparations (Table V). The highest percentage of good responses to the treatment (64.7%), as well as the highest percentage of subjects in whom PT resulted in bacterial eradication and/or recovery (47.1%), was found in the case of enterococcal phages. Their therapeutic effectiveness was significantly higher compared with staphylococcal phages (36.7% of good responses to PT, $p=0.035$) and *Pseudomonas* phages (20% of good responses to PT, $p=0.029$).

An important question was whether the efficacy of PT is dependent on the route of administration of phage preparations. Relevant analysis is presented for the individual routes or their combinations (phage administration by different routes at the same time or in sequences during treatment), which were applied in at least three patients (Table VI). The highest percentage of good responses to PT was found in patients receiving phages by the oral route (72.2%), whereas the highest percentage of subjects in whom PT resulted in pathogen eradication and/or recovery was noted in the group that received phages intrarectally (44%). The difference between groups of patients administered phage preparations orally and topically (28.6% of good responses to PT) was statistically significant ($p=0.012$). Also, rectal phage application and combined oral/topical application was significantly more effective than topical only ($p=0.026$ and $p=0.032$, respectively).

While in 109 patients (71.2%) phage preparations were applied as the only form of therapy, 44 subjects (28.8%) used other antibacterial agents over the course of PT or during breaks between therapeutic cycles. These agents included antibiotics/chemotherapeutics (41 cases), disinfectants (9 cases), and/or herbs and supplements prescribed for some kinds of infections (4 cases); they were used either separately or in combination. In the majority of patients, they were administered as a continuation of antibiotic therapy started prior to PT or started during PT because of

TABLE V Detailed evaluation of results of phage therapy depending on the used phage preparation[a]

	Staphylococcal phage preparations ($n = 79$)		Enterococcal phage preparations ($n = 17$)		*E. coli* phage preparations ($n = 17$)		*Pseudomonas* phage preparations ($n = 15$)	
Category of response to treatment	*n*	%	*n*	%	*n*	%	*n*	%
A - pathogen eradication and/or recovery	12	15.2	8	47.1	2	11.8	1	6.7
B - good clinical result	7	8.9	2	11.8	1	5.9	1	6.7
C - clinical improvement	10	12.7	1	5.9	3	17.6	1	6.7
D - questionable clinical improvement	6	7.6	0	0.0	0	0.0	1	6.7
E - transient clinical improvement	18	22.8	0	0.0	6	35.3	6	40.0
F - no response to treatment	20	25.3	6	35.3	5	29.4	2	13.3
G - clinical deterioration	6	7.6	0	0.0	0	0.0	3	20.0
Good response (total A–C):	**29**	**36.7**	**11**	**64.7**	**6**	**35.3**	**3**	**20.0**
Inadequate response (total D–G):	**50**	**63.3**	**6**	**35.3**	**11**	**64.7**	**12**	**80.0**

[a] The most frequently applied phage preparations (specific to staphylococci, enterococci, *E. coli*, and *Pseudomonas*) were included.

TABLE VI Detailed evaluation of results of phage therapy depending on the route of administration of the phage preparation[a]

Category of response to treatment	Topical ($n = 70$)		Oral ($n = 11$)		Oral/ topical[b] ($n = 22$)		Rectal ($n = 18$)		Rectal/ topical[b] ($n = 14$)		Vaginal ($n = 6$)		Inhalations of aerosol ($n = 3$)	
	n	%	*n*	%	*n*	%	*n*	%	*n*	%	*n*	%	*n*	%
A - pathogen eradication and/or recovery	9	12.9	3	27.3	4	18.2	8	44.4	3	21.4	1	16.7	0	0.0
B - good clinical result	5	7.1	4	36.4	1	4.5	1	5.6	1	7.1	0	0.0	1	33.3
C - clinical improvement	6	8.6	1	9.1	7	31.8	1	5.6	1	7.1	2	33.3	0	0.0
D - questionable clinical improvement	5	7.1	0	0.0	2	9.1	0	0.0	2	14.3	0	0.0	1	33.3
E - transient clinical improvement	19	27.1	1	9.1	4	18.2	2	11.1	2	14.3	1	16.7	0	0.0
F - no response to treatment	18	25.7	2	18.2	4	18.2	6	33.3	5	35.7	2	33.3	0	0.0
G - clinical deterioration	8	11.4	0	0.0	0	0.0	0	0.0	0	0.0	0	0.0	1	33.3
Good response (total A–C):	**20**	**28.6**	**8**	**72.7**	**12**	**54.5**	**10**	**55.6**	**5**	**35.7**	**3**	**50.0**	**1**	**33.3**
Inadequate response (total D–G):	**50**	**71.4**	**3**	**27.3**	**10**	**45.5**	**8**	**44.4**	**9**	**64.3**	**3**	**50.0**	**2**	**66.7**

[a] Analysis is presented for individual routes or their combinations, which were applied in at least three patients.
[b] Routes of phage administration were applied concomitantly or consecutively.

aggravation of the infection symptoms (25 cases); in other subjects, they were applied due to the occurrence of another infection not subjected to PT (7 cases), lack of an active phage against one or more pathogens in patients with polyinfections (5 cases), superinfection that developed during PT (5 cases), or aggravation of infection symptoms that occurred during breaks between the therapeutic cycles of PT only (2 cases). The authors compared the efficacy of PT in patients who received phage preparations only and those who had incidents of the use of other antibacterial agents during PT (Table IV). However, no significant difference was found between these two groups—the percentages of good responses to PT were 40.9% in the group of patients treated with phage preparations only and 38.6% in those in whom other antibacterial agents were also applied. The rate of pathogen eradication was even insignificantly higher (20.2%) in the group that received no antibacterials than in the group in which other antibacterials were used (13.6%). Apparently, the use of antibacterials by the patients during PT did not influence the final result of therapy.

Likewise, the rate of pathogen eradications and/or recoveries was slightly higher in patients administered phage preparations only than in those who also used other antibacterial agents (20 and 13.6%, respectively); however, the difference between these two groups fell short of statistical significance.

Another important question was whether any correlation exists between the duration of PT and its effectiveness. A relevant analysis was performed based on data from 149 patients for whom the date of final use of the phage preparation was available (Table VII). The

TABLE VII Cumulative duration of phage treatment

Category of the response to treatment	No. of patients	Median (days)	Minimum (days)	Maximum (days)
A - pathogen eradication and/or recovery	27	43.0	6	165
B - good clinical result	13	71.0	21	165
C - clinical improvement	19	87.0	12	209
D - questionable clinical improvement	10	68.5	14	161
E - transient clinical improvement	33	63.0	16	328
F - no response to treatment	37	46.0	4	144
G - clinical deterioration	10	34.5	3	89
Total:	149	55.0	3	328

cumulative treatment duration (i.e., the sum of days on which phage preparations were administered) varied considerably depending on the results of PT, the occurrence of side effects, and the route of administration of phage preparations. The median cumulative treatment duration was 55 days (the minimum treatment duration was 3 days, the maximum 328 days). No significant correlation was found between the duration of PT and its effectiveness, with the Spearman's rank correlation coefficient being 0.15 ($p=0.076$). In addition, the Kruskal–Wallis test showed no significant differences in the cumulative treatment duration calculated for each of the six different kinds of infections listed in Table VIII (its medians were in a range of 43.5–66.0 days).

D. Detailed evaluation of the results of phage therapy

A detailed comparative analysis of results obtained in patients with six different kinds of infections is shown in Table VIII (it does not include one subject suffering from recurrent bacteremia caused by *E. faecalis* treated with oral enterococcal phage and antibiotics and in whom a good clinical result was obtained; this single case does not fall into any kind of the diseases shown in Table VIII). Good responses to PT were achieved in 48.3% of men with genital or urinary tract infections, 45.9% of patients with orthopedic infections, 36.7% of subjects with soft tissue infections, 36.4% of women with genital or urinary tract infections, 30% of patients with skin infections, and 29.2% of patients with respiratory tract infections. However, differences among the individual groups were not statistically significant. The highest rate of pathogen eradication and/or recovery was found in men with genital or urinary tract infections (37.9%). This value was significantly higher than corresponding values obtained in groups of patients with respiratory tract infections and skin infections (respectively: 8.3%, $p=0.01$ and 0%, $p=0.021$).

Separate analyses were conducted for results obtained within each of the individual groups of patients with different kinds of infections (Tables IX–XIV). The group of patients with orthopedic infections included 37 cases (Table IX). In this group, the frequencies of good responses to PT (categories A–C) were significantly higher in patients administered phages only orally (100%) as well as by a combination of the oral and the topical route (75%) when compared with patients who received the preparations topically (28%; $p=0.026$ and $p=0.014$, respectively). There were no significant changes in the results of pathogen eradication and/or recovery (category A of the response to PT) between these groups (Table IX). The majority of subjects in this group (34 cases) received staphylococcal phages; good responses were achieved in 16 of these patients (47.1%). Interesting results were obtained by comparison of PT outcome between two groups of patients with orthopedic infections.

TABLE VIII Detailed evaluation of results of phage therapy in patients with different disorders (does not include one patient with recurrent bacteremia included in general analysis)

Category of response to treatment	Genital and urinary tract infections in men[a] ($n = 29$)		Genital and urinary tract infections in women[b] ($n = 22$)		Soft tissue infections[c] ($n = 30$)		Skin infections[d] ($n = 10$)		Orthopedic infections[e] ($n = 37$)		Respiratory tract infections[f] ($n = 24$)	
	n	%	*n*	%	*n*	%	*n*	%	*n*	%	*n*	%
A - pathogen eradication and/or recovery	11	37.9	3	13.6	5	16.7	0	0.0	7	18.9	2	8.3
B - good clinical result	2	6.9	0	0.0	2	6.7	2	20.0	3	8.1	3	12.5
C - clinical improvement	1	3.4	5	22.7	4	13.3	1	10.0	7	18.9	2	8.3
D - questionable clinical improvement	2	6.9	0	0.0	2	6.7	0	0.0	3	8.1	3	12.5
E - transient clinical improvement	5	17.2	4	18.2	8	26.7	5	50.0	8	21.6	3	12.5
F - no response to treatment	8	27.6	10	45.5	6	20.0	1	10.0	7	18.9	7	29.2
G - clinical deterioration	0	0.0	0	0.0	3	10.0	1	10.0	2	5.4	4	16.7
Good response (total A–C):	**14**	**48.3**	**8**	**36.4**	**11**	**36.7**	**3**	**30.0**	**17**	**45.9**	**7**	**29.2**
Inadequate response (total D–G):	**15**	**51.7**	**14**	**63.6**	**19**	**63.3**	**7**	**70.0**	**20**	**54.1**	**17**	**70.8**

[a] Including chronic bacterial prostatitis ($n = 13$), urinary infection ($n = 10$), and chronic bacterial prostatitis/urinary infection ($n = 6$).
[b] Including urinary infection ($n = 14$), vaginal infection ($n = 3$), and urinary/vaginal infection ($n = 5$).
[c] Including postoperative wound infection ($n = 6$); leg ulcer ($n = 8$); abscess, phlegmon, or empyema penetrating to body cavities ($n = 5$); and deep tissue infection ($n = 11$).
[d] Including external ear infection ($n = 3$), unspecified local infection of skin ($n = 3$), atopic dermatitis complicated by staphylococcal infection ($n = 1$), acne ($n = 1$), eczema ($n = 1$), and furunculosis ($n = 1$).
[e] Including prosthetic joint infection ($n = 8$), osteomyelitis ($n = 21$), joint infection ($n = 5$), osteomyelitis/joint infection ($n = 2$), and discitis ($n = 1$).
[f] Including upper respiratory tract infections ($n = 17$), lower respiratory tract infections ($n = 4$), and upper and lower respiratory tract infections ($n = 3$).

TABLE IX Detailed evaluation of results of phage therapy in patients with orthopedic infections

	Way of administration of the phage preparation						Type of phage preparations applied			
	Topical[a] ($n = 25$)		Oral/topical[a] ($n = 8$)		Oral ($n = 4$)		Staphylococcal ($n = 34$)		Other[b] ($n = 3$)	
Category of response to treatment	*n*	%	*n*	%	*n*	%	*n*	%	*n*	%
A - pathogen eradication and/or recovery	3	12.0	2	25.0	2	50.0	7	20.6	0	0.0
B - good clinical result	1	4.0	0	0.0	2	50.0	3	8.8	0	0.0
C - clinical improvement	3	12.0	4	50.0	0	0.0	6	17.6	1	33.3
D - questionable clinical improvement	3	12.0	0	0.0	0	0.0	3	8.8	0	0.0
E - transient clinical improvement	6	24.0	2	25.0	0	0.0	6	17.6	2	66.7
F - no response to treatment	7	28.0	0	0.0	0	0.0	7	20.6	0	0.0
G - clinical deterioration	2	8.0	0	0.0	0	0.0	2	5.9	0	0.0
Good response (total A–C):	**7**	**28.0**	**6**	**75.0**	**4**	**100.0**	**16**	**47.1**	**1**	**33.3**
Inadequate response (total D–G):	**18**	**72.0**	**2**	**25.0**	**0**	**0.0**	**18**	**52.9**	**2**	**66.7**

[a] Topical application included fistular irrigation and/or wet compresses on the external orifice of the fistula.
[b] Including *Pseudomonas* ($n = 2$) and staphylococcal/*Enterobacter* ($n = 1$).

TABLE X Detailed evaluation of results of phage therapy in patients with soft tissue infections

	Way of administration of the phage preparation						Type of phage preparations applied					
	Topical[a] ($n = 26$)		Oral/topical[a] ($n = 3$)		Rectal/topical[a] ($n = 1$)		Staphylococcal ($n = 14$)		*Pseudomonas* ($n = 8$)		Other[b] ($n = 8$)	
Category of response to treatment	*n*	%	*n*	%	*n*	%	*n*	%	*n*	%	*n*	%
A - pathogen eradication and/or recovery	4	15.4	1	33.3	0	0.0	3	21.4	0	0.0	2	25.0
B - good clinical result	2	7.7	0	0.0	0	0.0	1	7.1	0	0.0	1	12.5
C - clinical improvement	3	11.5	0	0.0	1	100.0	1	7.1	0	0.0	3	37.5
D - questionable clinical improvement	1	3.8	1	33.3	0	0.0	0	0.0	1	12.5	1	12.5
E - transient clinical improvement	7	26.9	1	33.3	0	0.0	4	28.6	4	50.0	0	0.0
F - no response to treatment	6	23.1	0	0.0	0	0.0	4	28.6	2	25.0	0	0.0
G - clinical deterioration	3	11.5	0	0.0	0	0.0	1	7.1	1	12.5	1	12.5
Good response (total A–C):	**9**	**34.6**	**1**	**33.3**	**1**	**100.0**	**5**	**35.7**	**0**	**0.0**	**6**	**75.0**
Inadequate response (total D–G):	**17**	**65.4**	**2**	**66.7**	**0**	**0.0**	**9**	**64.3**	**8**	**100.0**	**2**	**25.0**

[a] Topical application included wet compresses applied on the wound, fistular irrigation, and irrigation of the abscess cavity.
[b] Including Staphylococcal /*Pseudomonas* ($n = 3$), enterococcal ($n = 2$), *E. coli* ($n = 1$), *Proteus* ($n = 1$), and *Salmonella* ($n = 1$).

TABLE XI Detailed evaluation of results of phage therapy in men with genital or urinary tract infections

	Way of administration of the phage preparation						Type of phage preparations applied					
	Rectal ($n = 15$)		Rectal/topical[a] ($n = 12$)		Other[b] ($n = 2$)		Enterococcal ($n = 12$)		*E. coli* ($n = 5$)		Other[c] ($n = 12$)	
Category of response to treatment	*n*	%	*n*	%	*n*	%	*n*	%	*n*	%	*n*	%
A - pathogen eradication and/or recovery	7	46.7	3	25.0	1	50.0	7	58.3	1	20.0	3	25.0
B - good clinical result	1	6.7	1	8.3	0	0.0	0	0.0	1	20.0	1	8.3
C - clinical improvement	1	6.7	0	0.0	0	0.0	0	0.0	0	0.0	1	8.3
D - questionable clinical improvement	0	0.0	2	16.7	0	0.0	0	0.0	0	0.0	2	16.7
E - transient clinical improvement	2	13.3	2	16.7	1	50.0	0	0.0	2	40.0	3	25.0
F - no response to treatment	4	26.7	4	33.3	0	0.0	5	41.7	1	20.0	2	16.7
G - clinical deterioration	0	0.0	0	0.0	0	0.0	0	0.0	0	0.0	0	0.0
Good response (total A–C):	**9**	**60.0**	**4**	**33.3**	**1**	**50.0**	**7**	**58.3**	**2**	**40.0**	**5**	**41.7**
Inadequate response (total D–G):	**6**	**40.0**	**8**	**66.7**	**1**	**50.0**	**5**	**41.7**	**3**	**60.0**	**7**	**58.3**

[a] Topical application included wet compresses applied on the glans of penis (they were applied in some patients with chronic bacterial prostatitis, as well as those with urinary infection when concomitant infection under foreskin of the penis was suspected).

[b] Including oral/topical ($n = 1$) and oral/ rectal ($n = 1$).

[c] Including *Pseudomonas* ($n = 3$), enterococcal/staphylococcal ($n = 2$), staphylococcal ($n = 1$), *Klebsiella* ($n = 1$), *E. coli/Klebsiella* ($n = 1$), Enterococcal//*Klebsiella* ($n = 1$), enterococcal/*E. coli* ($n = 1$), enterococcal/*Citrobacter* ($n = 1$), enterococcal/*Pseudomonas* ($n = 1$).

TABLE XII Detailed evaluation of results of phage therapy in women with urinary or vaginal infections

	Way of administration of the phage preparation								Type of phage preparations applied					
	Vaginal ($n = 6$)		Oral ($n = 4$)		Rectal ($n = 3$)		Other[a] ($n = 9$)		*E. coli* ($n = 10$)		Enterococcal/*E. coli* ($n = 7$)		Other[b] ($n = 5$)	
Category of response to treatment	n	%	n	%	n	%	n	%	n	%	n	%	n	%
A - pathogen eradication and/or recovery	1	16.7	1	25.0	1	33.3	0	0.0	1	10.0	1	14.3	1	20.0
B - good clinical result	0	0.0	0	0.0	0	0.0	0	0.0	0	0.0	0	0.0	0	0.0
C - clinical improvement	2	33.3	1	25.0	0	0.0	2	22.2	2	20.0	1	14.3	2	40.0
D - questionable clinical improvement	0	0.0	0	0.0	0	0.0	0	0.0	0	0.0	0	0.0	0	0.0
E - transient clinical improvement	1	16.7	1	25.0	0	0.0	2	22.2	3	30.0	1	14.3	0	0.0
F - no response to treatment	2	33.3	1	25.0	2	66.7	5	55.6	4	40.0	4	57.1	2	40.0
G - clinical deterioration	0	0.0	0	0.0	0	0.0	0	0.0	0	0.0	0	0.0	0	0.0
Good response (total A–C):	**3**	**50.0**	**2**	**50.0**	**1**	**33.3**	**2**	**22.2**	**3**	**30.0**	**2**	**28.6**	**3**	**60.0**
Inadequate response (total D–G):	**3**	**50.0**	**2**	**50.0**	**2**	**66.7**	**7**	**77.8**	**7**	**70.0**	**5**	**71.4**	**2**	**40.0**

[a] Including sitz bath ($n = 1$); topical ($n = 1$); oral/topical ($n = 1$); oral/rectal/topical ($n = 1$); rectal/topical ($n = 1$); vaginal/rectal ($n = 1$); vaginal/oral ($n = 1$); vaginal/topical ($n = 1$); and vaginal/oral/topical ($n = 1$), where topical refers to compresses soaked with the phage preparation applied to the vulva.

[b] Including enterococcal ($n = 2$), staphylococcal ($n = 1$), *Enterobacter* ($n = 1$), and enterococcal/*Proteus* ($n = 1$).

TABLE XIII Detailed evaluation of results of phage therapy in patients with respiratory tract infections

	Way of administration of the phage preparation								Type of phage preparations applied			
	Topical[a] ($n = 12$)		Oral/topical ($n = 6$)		Inhalation of aerosol ($n = 3$)		Other[b] ($n = 3$)		Staphylococcal ($n = 20$)		Other[c] ($n = 4$)	
Category of response to treatment	*n*	%	*n*	%	*n*	%	*n*	%	*n*	%	*n*	%
A - pathogen eradication and/or recovery	2	16.7	0	0.0	0	0.0	0	0.0	2	10.0	0	0.0
B - good clinical result	1	8.3	0	0.0	1	33.3	1	33.3	2	10.0	1	25.0
C - clinical improvement	0	0.0	2	33.3	0	0.0	0	0.0	2	10.0	0	0.0
D - questionable clinical improvement	1	8.3	1	16.7	1	33.3	0	0.0	3	15.0	0	0.0
E - transient clinical improvement	2	16.7	0	0.0	0	0.0	1	33.3	2	10.0	1	25.0
F - no response to treatment	4	33.3	3	50.0	0	0.0	0	0.0	7	35.0	0	0.0
G - clinical deterioration	2	16.7	0	0.0	1	33.3	1	33.3	2	10.0	2	50.0
Good response (total A–C):	**3**	**25.0**	**2**	**33.3**	**1**	**33.3**	**1**	**33.3**	**6**	**30.0**	**1**	**25.0**
Inadequate response (total D–G):	**9**	**75.0**	**4**	**66.7**	**2**	**66.7**	**2**	**66.7**	**14**	**70.0**	**3**	**75.0**

[a] Topical application included gargling and phage application using nasal drops.
[b] Including oral ($n = 1$) and oral/inhalations of aerosol ($n = 2$).
[c] Including *Pseudomonas* ($n = 2$), *Pseudomonas/Stenotrophomonas* ($n = 1$), and *E. coli* ($n = 1$).

TABLE XIV Detailed evaluation of results of phage therapy in patients with skin infections

	Way of administration of phage preparation						Type of phage preparation applied			
	Topical[a] ($n = 6$)		Oral/topical[a] ($n = 3$)		Oral ($n = 1$)		Staphylococcal ($n = 9$)		*Pseudomonas* ($n = 1$)	
Category of response to treatment	*n*	%	*n*	%	*n*	%	*n*	%	*n*	%
A - pathogen eradication and/or recovery	0	0.0	0	0.0	0	0.0	0	0.0	0	0.0
B - good clinical result	1	16.7	1	33.3	0	0.0	1	11.1	1	100
C - clinical improvement	0	0.0	1	33.3	0	0.0	1	11.1	0	0.0
D - questionable clinical improvement	0	0.0	0	0.0	0	0.0	0	0.0	0	0.0
E - transient clinical improvement	4	66.7	1	33.3	0	0.0	5	55.6	0	0.0
F - no response to treatment	0	0.0	0	0.0	1	100	1	11.1	0	0.0
G - clinical deterioration	1	16.7	0	0.0	0	0.0	1	11.1	0	0.0
Good response (total A–C):	**1**	**16.7**	**2**	**66.7**	**0**	**0.0**	**2**	**22.2**	**1**	**100**
Inadequate response (total D–G):	**5**	**83.3**	**1**	**33.3**	**1**	**100**	**7**	**77.8**	**0**	**0.0**

[a] Topical application included the use of wet compresses, and ear drops.

The first group included patients who received phages orally and/or topically ($n=12$), whereas the second group were those treated only topically ($n=25$). Both groups were similar ($p>0.05$) with regard to gender (percentage of women was 25, and 24% respectively), median age (61 and 53 years, respectively), cumulative time of treatment (53 and 70 days, respectively), use of antimicrobials (25 and 24% of cases, respectively), and the percentage portion of *S. aureus* monoinfections (91.7 and 92.0%, respectively). The median time of disease duration was 18 and 36 months, respectively, but this difference was not statistically significant. The authors observed a significantly higher ($p=0.005$) frequency of good responses to PT in the first group (83.3%) than in the second group (28.0%). This difference was also significant ($p=0.006$) when subjects treated only with phage were considered. In this case, good responses to PT were achieved in 58.3% of patients from the first group ($n=8$), while in the second group they were achieved only in 20.0% of patients ($n=19$).

The group of patients with soft tissue infections comprised 30 individuals (Table X). The topical route of administration of phage preparations, which was the most frequently used one resulted in good responses in 9 out of 26 patients (34.6%); pathogen eradication and/or recovery was found in 4 of these subjects (15.4%). Staphylococcal phages were more effective than *Pseudomonas* phages in this group; their use resulted in good responses in 5 out of 14 subjects (35.7%) and in 0 out of 8 subjects (0%), respectively.

Another group subjected to PT comprised men with genital or urinary tract infections (29 patients including 13 cases with chronic bacterial prostatitis, 6 diagnosed with chronic bacterial prostatitis/urinary infection, and 10 with urinary infection only; Table XI). Good responses to PT in this group were achieved in 9 out of 15 patients (60%) administered phage preparations intrarectally and in 4 out of 12 patients (33%) administered bacteriophages by a combination of the intrarectal and the topical route. Some differences in the effectiveness of PT depending on the specificity of phage preparations were also found in this group. A higher, but not statistically significant, rate of good responses (7 out of 12 patients; 58.3%) was noted for enterococcal phages compared with *E. coli* phages (2 out of 5 patients; 40%). Likewise, a higher rate of pathogen eradication and/or recovery was found in patients who received enterococcal phages (7 out of 12 subjects; 58.3%) compared with those treated with *E. coli* phages (1 out of 5 patients; 20%).

The group of women with genital or urinary tract infections included 22 patients (3 cases with vaginal infection, 5 diagnosed with urinary/vaginal infection, and 14 with urinary infection; Table XII). The most effective routes of administration of phage preparations in these patients were the intravaginal route and the oral route, both of which resulted in 50% rates of good responses to treatment (3 out of 6 and 2 out of 4 patients, respectively). However, the highest rate of patients with pathogen eradication and/or recovery was found in women who received phages intrarectally (1 out of 3 subjects; 33%) rather than orally (1 out of 4 subjects; 25%) or intravaginally (1 out of 6 subjects; 16.7%).

The group of patients with respiratory tract infections included 24 subjects (Table XIII). For the most frequently used route of administration of phage preparations in this group, that is, the topical route, good responses were achieved in 3 out of 12 subjects (25%); in 2 of these patients (16.7%), pathogen eradication and/or recovery was achieved. Staphylococcal phage preparations used in 20 subjects from this group were effective in 6 patients (30%).

The group of patients with skin infections included 10 subjects (Table XIV). The most frequently used route of administration of phage preparations in this group was the topical route, which resulted in good responses to PT in 1 out of 6 subjects (16.7%). Nine out of 10 patients in this group received staphylococcal phages, which were effective in 2 subjects (22.2%). However, none of these patients achieved pathogen eradication and/or recovery.

E. Effects of phage preparations on selected inflammatory markers

Two separate statistical analyses were performed to evaluate the effects of PT on selected inflammatory markers associated with infection, including WBC count and CRP level in the peripheral blood, as well as the ESR. In each analysis, values of these parameters found after 5–8 and 9–32 days of PT (early and the late phase of the treatment, respectively) were compared with those found in the most recent assays done before the start of PT (baseline). These time periods were chosen to enable comparison of current results with previous analysis of the changes in inflammatory parameters obtained for patients admitted for PT between January 2006 and October 2007 (Międzybrodzki *et al.*, 2009).

The first general analysis included all subjects for whom data on the values of these parameters before, during, and/or after PT were available, irrespective of the kind of infection and the use of antibacterial agents other than phages (Table XV). The comparison performed using the Wilcoxon test did not reveal any significant differences between values of the studied parameters found prior to the administration of phage preparations and during PT. The second analysis was based on data from patients who were selected for analysis based on the same criteria used in previous work (Międzybrodzki *et al.*, 2009): only those with orthopedic infections, skin infections, soft tissue infections, or lower respiratory tract infections who did not receive any other antibacterial agents during PT were included (62 patients, in over 70% of cases infected

TABLE XV Changes of inflammatory markers (WBC count and level of C-reactive protein in peripheral blood, as well as erythrocyte sedimentation rate) during PT[a]

Biomarker		All patients			Selected group[b]		
		n	Mean	± SE	*n*	Mean	± SE
WBC count ($10^3/mm^3$)	Baseline:	32	6.5	± 0.3	15	7.0	± 0.3
	After 5–8 days:	32	6.7	± 0.4	15	7.6	± 0.5
	Baseline:	110	7.0	± 0.2	46	7.1	± 0.3
	After 9–32 days:	110	6.9	± 0.2	46	7.3	± 0.3
Erythrocyte sedimentation rate (mm/hr)	Baseline:	22	30.1	± 6.9	12	42.4	± 10.1
	After 5–8 days:	22	29.0	± 6.5	12	43.3	± 9.6
	Baseline:	77	24.4	± 3.1	31	34.8	± 5.6
	After 9–32 days:	77	23.9	± 3.0	31	36.1	± 5.7
C-reactive protein (mg/ml)	Baseline:	22	8.3	± 2.5	12	13.7	± 4.1
	After 5–8 days:	22	9.0	± 2.7	12	14.7	± 4.4
	Baseline:	79	10.3	± 1.6	36	16.2	± 2.9
	After 9–32 days:	79	11.1	± 1.6	36	16.7	± 3.0

[a] Results of all patients are shown, as well as a subgroup of patients selected for analysis according to the same criteria used in a previous (Międzybrodzki *et al.*, 2009) analysis [*n*, number of patients for whom relevant (to the analyzed time period) results were available].

[b] This group included only patients with osteomyelitis, prosthetic joint infection, joint infection, skin and soft tissue infection, or lower respiratory tract infection who did not receive any antibacterial during the analyzed period of PT and for whom relevant laboratory data were available.

with *S. aureus*). This analysis also did not show any significant changes in the mean values of analyzed inflammatory markers. However, in a subgroup of patients who had baseline CRP above 10 mg/liter ($n=26$), a statistically significant reduction in the mean CRP concentration by 22.5% (from 26.2 to 20.3 mg/liter; $p=0.006$) during days 9–32 of treatment was observed.

F. Changes in phage typing profile of bacterial strains isolated during treatment

The authors also analyzed changes in the phage typing profile of bacterial strains isolated from patients and the development of bacterial resistance to the therapeutic phages (Table XVI). This analysis included subjects with infections caused by *S. aureus* ($n=53$), *E. faecalis* ($n=14$), *E. coli* ($n=14$), and *P. aeruginosa* ($n=11$). Some of these patients used other antibacterial agents during PT (detailed data are presented in Table XVI). A high frequency of changes in the phage typing profile of the analyzed bacterial strains was found. It occurred in all cases of *E. faecalis*- and *E. coli*-infected patients, in 10 out of 11 *P. aeruginosa*-infected patients (90.9%), and in 37 out of 53 *S. aureus*-infected patients (69.8%). Resistance of the targeted

TABLE XVI Changes in phage typing profile and acquisition of phage resistance in strains of most frequent bacterial species during PT[a]

Pathogen	No. of patients included in analysis[b]	No. of patients in whom a change in phage typing profile of the pathogen was observed	No. of patients in whom the pathogen acquired resistance to the phage used during PT	No. of patients in whom the pathogen acquired resistance to any species-specific phage from the IIET collection
Staphylococcus aureus	53	37 (69.8%)	9 (17.0%)	4 (7.5%)
Enterococcus faecalis	14	14 (100.0%)	6 (42.9%)	3 (21.4%)
Escherichia coli	14	14 (100.0%)	12 (85.7%)	4 (28.6%)
Pseudomonas aeruginosa	11	10 (90.9%)	4 (36.4%)	3 (27.3%)

[a] Phage typing of the bacterial pathogen was done using phages from the IIET collection. Analysis included only cases in whom results of phage typing of bacteria isolated before, during, and/or after PT were available.
[b] Numbers of patients who used antibacterials during PT were nine for *S. aureus*, six for *E. faecalis*, six for *E. coli*, and four for *P. aeruginosa* infection.

pathogen to a therapeutic phage developed in 85.7% of *E. coli* cases (this was a cause of the frequent change of phages used in the course of PT in this group of patients), 42.9% of *E. faecalis* cases, 36.4% of *P. aeruginosa* cases, and 17% of *S. aureus* cases. Complete resistance to any phage from the IIET collection was observed in 28.6% of *E. coli* cases, 27.3% of *P. aeruginosa* cases, 21.4% of *E. faecalis* cases, and 7.5% of *S. aureus* cases.

G. Safety of phage therapy

The effects of PT on organ and system function in patients by analyzing selected hematological, urinary, and biochemical serum parameters were also evaluated. Their mean values $\pm$SE were determined in four periods of cumulative phage use (3–6, 7–20, 21–48, and 49–84 days of treatment) and were compared with results of tests done before PT (Tables XVII and XVIII). When analyzing parameters of bone marrow function, no significant changes were observed for hemoglobin, hematocrit, erythrocytes, leukocytes, lymphocytes, monocytes, neutrophils, eosinophils, or basophils (Table XVII). The only significant change was a slight decrease in platelet (PLT) count found after 21–48 days of PT from 263.9$\pm$10.2 to 252.1$\pm$10.4 cells $\times 10^3/\text{mm}^3$($n=79$; $p<0.01$) (Table XVII). However, the decrease in PLT count was minimal (one cannot exclude that it could reflect an anti-inflammatory effect of PT) and rather transient because no difference between the mean PLT count in this period and its corresponding baseline value was observed (257.7$\pm$11.7 and 256.2 $\pm$11.9 cells $\times 10^3/\text{mm}^3$, respectively; $n=51$) during 49–84 days of PT. In a group of 129 patients in whom the PLT count at baseline was normal (150–400 cells $\times 10^3/\text{mm}^3$), the authors observed its slight decrease to the range 109–147 cells $\times 10^3/\text{mm}^3$ in three cases (in one case it was confirmed to return to normal). In another three cases a slight increase in the PLT count up to 410–486 cells $\times 10^3/\text{mm}^3$ was noted, and in one case its value reached 610 cells $\times 10^3/\text{mm}^3$. Although no significant changes were recorded for total protein, total bilirubin, alanine transaminase, alkaline phosphatase, and amylase levels, statistical analysis suggested a small but significant increase in aspartate transaminase (AST) activity after 21–48 days of PT from 23.2$\pm$1.3 to 24.2$\pm$1.0 UI ($n=71$, $p<0.05$) (Table XVIII). However, among 139 patients in whom baseline AST was in the normal range (0–40 IU), a transient increase in the activity of this enzyme (up to 41–138 UI) during PT was noted in 4 patients and was elevated up to 43–58 IU in another four cases. Furthermore, a slight but significant decrease in γ-glutamyl transpeptidase activity was observed (Table XVIII).

Moreover, analysis revealed a significant increase in fasting blood sugar (FBS) after 49–84 days of PT (from 81.1$\pm$5.1 to 96.2$\pm$2.9 mg/dl; $n=46$; $p<0.05$). However, the mean FBS did not exceed normal values

TABLE XVII Mean changes in selected patients' hematological parameters during PT

Parameter (unit)	Cumulative phage use	Number of analyzed cases	Baseline value		Value during PT		p^a
			Mean	± SE	Mean	± SE	
Erythrocytes	3–6 days	17	4.416	± 0.129	4.453	± 0.147	ns[b]
(cells × $10^6/mm^3$)	7–20 days	103	4.697	± 0.051	4.676	± 0.052	ns
	21–48 days	80	4.723	± 0.062	4.715	± 0.065	ns
	49–84 days	51	4.625	± 0.061	4.598	± 0.064	ns
Hemoglobin	3–6 days	17	13.10	± 0.48	13.17	± 0.48	ns
(g/dl)	7–20 days	103	13.77	± 0.19	13.74	± 0.19	ns
	21–48 days	80	13.78	± 0.21	13.69	± 0.20	ns
	49–84 days	51	13.62	± 0.23	13.56	± 0.23	ns
Hematocrit	3–6 days	17	39.3	± 1.1	39.2	± 1.2	ns
(%)	7–20 days	103	41.3	± 0.5	41.1	± 0.5	ns
	21–48 days	80	41.5	± 0.5	41.2	± 0.5	ns
	49–84 days	51	40.8	± 0.6	40.6	± 0.6	ns
Platelets	3–6 days	16	243.9	± 18.7	250.1	± 21.3	ns
(cells × $10^3/mm^3$)	7–20 days	99	256.9	± 8.5	254.7	± 8.0	ns
	21–48 days	79	263.9	± 10.2	252.1	± 10.4	<0.01
	49–84 days	51	257.7	± 11.7	256.2	± 11.9	ns
Leukocytes	3–6 days	17	6.535	± 0.392	6.888	± 0.477	ns
(cells × $10^3/mm^3$)	7–20 days	103	6.941	± 0.185	6.838	± 0.185	ns
	21–48 days	80	6.977	± 0.217	7.056	± 0.260	ns
	49–84 days	51	6.772	± 0.239	6.705	± 0.261	ns

(continued)

TABLE XVII *(continued)*

Parameter (unit)	Cumulative phage use	Number of analyzed cases	Baseline value		Value during PT		p^a
Lymphocytes (%)	3–6 days	15	32.5	± 2.2	30.8	± 2.5	ns
	7–20 days	95	32.2	± 1.0	32.2	± 1.0	ns
	21–48 days	74	33.7	± 1.1	33.4	± 1.1	ns
	49–84 days	49	32.7	± 1.3	31.7	± 1.3	ns
Monocytes (%)	3–6 days	8	6.2	± 1.2	6.9	± 1.2	ns
	7–20 days	61	4.9	± 0.4	5.1	± 0.5	ns
	21–48 days	47	5.3	± 0.5	5.4	± 0.6	ns
	49–84 days	30	5.3	± 0.6	5.0	± 0.6	ns
Neutrophils (%)	3–6 days	15	59.0	± 2.3	61.5	± 3.0	ns
	7–20 days	95	60.1	± 1.0	60.0	± 1.0	ns
	21–48 days	73	58.2	± 1.1	58.7	± 1.2	ns
	49–84 days	49	59.1	± 1.3	60.5	± 1.5	ns
Eosinophils (%)	3–6 days	8	2.3	± 0.7	2.4	± 0.6	ns
	7–20 days	56	2.5	± 0.3	3.0	± 0.3	ns
	21–48 days	45	2.8	± 0.2	2.8	± 0.2	ns
	49–84 days	30	2.2	± 0.2	2.5	± 0.2	ns
Basophils (%)	3–6 days	7	0.2	± 0.1	0.3	± 0.2	ns
	7–20 days	52	0.3	± 0.1	0.4	± 0.1	ns
	21–48 days	41	0.4	± 0.1	0.2	± 0.0	ns
	49–84 days	28	0.3	± 0.1	0.5	± 0.3	ns

[a] Statistical significance of differences was determined using the Wilcoxon matched pairs test.
[b] Not significant.

TABLE XVIII Mean changes in selected patients' biochemical serum parameters during PT

Parameter (unit)	Cumulative phage use	Number of analyzed cases	Baseline value		Value during PT		p^a
			Mean	± SE	Mean	± SE	
Fasting blood sugar (mg/dl)	3–6 days	14	84.6	± 7.6	98.1	± 9.0	<0.05
	7–20 days	92	84.4	± 3.1	93.3	± 1.7	<0.05
	21–48 days	71	82.9	± 3.9	94.5	± 1.5	ns[b]
	49–84 days	46	82.1	± 5.1	96.2	± 2.9	<0.05
Creatinine (mg/dl)	3–6 days	16	0.92	± 0.14	1.01	± 0.14	ns
	7–20 days	91	0.93	± 0.03	0.93	± 0.03	ns
	21–48 days	69	0.90	± 0.02	0.91	± 0.02	ns
	49–84 days	47	0.91	± 0.03	0.93	± 0.03	ns
Total protein (g/dl)	3–6 days	5	7.50	± 0.32	7.31	± 0.33	ns
	7–20 days	10	7.20	± 0.16	7.08	± 0.13	ns
	21–48 days	23	7.24	± 0.11	7.17	± 0.11	ns
	49–84 days	22	7.38	± 0.11	7.39	± 0.12	ns
Total bilirubin (mg/dl)	3–6 days	17	0.57	± 0.05	0.58	± 0.06	ns
	7–20 days	92	0.64	± 0.03	0.66	± 0.03	ns
	21–48 days	72	0.62	± 0.03	0.65	± 0.04	ns
	49–84 days	46	0.56	± 0.03	0.56	± 0.04	ns
Aspartate transaminase (UI)	3–6 days	17	34.0	± 12.9	31.5	± 10.5	ns
	7–20 days	97	24.7	± 2.4	26.1	± 3.0	ns
	21–48 days	71	23.2	± 1.3	24.2	± 1.0	<0.05
	49–84 days	47	21.4	± 0.9	21.8	± 1.2	ns

(*continued*)

TABLE XVIII *(continued)*

Parameter (unit)	Cumulative phage use	Number of analyzed cases	Baseline value		Value during PT		p^a
Alanine transaminase (UI)	3–6 days	17	29.1	± 7.6	29.8	± 7.5	ns
	7–20 days	94	24.7	± 1.8	24.9	± 2.2	ns
	21–48 days	68	24.9	± 1.8	25.7	± 1.8	ns
	49–84 days	46	21.9	± 1.6	27.0	± 3.8	ns
γ-Glutamyl transpeptidase (UI)	3–6 days	9	49.3	± 13.2	40.1	± 7.8	ns
	7–20 days	22	46.8	± 8.3	41.1	± 7.2	<0.01
	21–48 days	28	30.9	± 4.1	28.2	± 4.2	<0.05
	49–84 days	23	34.8	± 7.0	32.4	± 5.8	ns
Alkaline phosphatase (UI)	3–6 days	5	74.8	± 17.5	90.2	± 18.3	ns
	7–20 days	13	104.4	± 19.4	97.3	± 16.9	ns
	21–48 days	27	85.1	± 7.8	78.5	± 6.9	ns
	49–84 days	17	70.1	± 6.9	75.0	± 8.7	ns
Amylase (UI)	3–6 days	4	57.0	± 6.0	55.8	± 7.3	ns
	7–20 days	10	61.8	± 7.9	68.3	± 9.7	ns
	21–48 days	20	59.1	± 5.1	57.4	± 5.5	ns
	49–84 days	15	63.3	± 7.3	63.0	± 7.2	ns

[a] Statistical significance of differences was determined using the Wilcoxon matched pairs test.
[b] Not significant.

(60–100 mg/dl). A detailed analysis of nondiabetic patients was based on baseline data from 133 patients and results of FBS measures taken during PT (127 subjects had at least one FBS measure during treatment). This analysis showed that (i) before PT only 3 patients had FBS $\geq$126 mg/dl; (ii) of those patients who had FBS $<$126 mg/dl at baseline, only 2 had an incident of FBS $\geq$126 mg/dl but in both cases the result of the next test was $<$126 mg/dl; and (iii) of those who had FBS $<$126 mg/dl at baseline, 22 patients had an incident of FBS in the 100- to 126-mg/dl range (which might suggest impaired fasting glycemia) and in 12 of these cases the result of the next test was normal (60–99 mg/dl).

No significant changes were observed in patients' renal function: the mean creatinine serum concentration did not change significantly (Table XVIII) nor were any worrying results in patients without symptoms of any renal function disorder, urinary tract infection, or diabetes before PT observed. In particular, a protein urine test was positive in 12 of 136 patients in whom it was done at least once during PT (including those with diabetes, urinary tract infection, and/or renal failure). In eight of those cases, maximal protein concentration (PC) did not exceed 0.5 g/liter (in six cases it decreased below the detection level subsequently, and in two other cases it was elevated before PT; similar values were maintained during therapy). The authors observed PC above 0.5 g/liter in urine of four cases during PT: one patient with diabetes and renal insufficiency treated with phages due to urinary tract infection (in this case, PC dropped from 2.8 g/liter before PT to 1.1 g/liter), one patient with chronic urinary tract infection subjected to PT due to purulent infection of deep tissue in whom PC raised to 0.5 g/liter after beginning PT and decreased to 0.2 g/liter at its termination, one patient with periprosthetic infection and diabetes in whom PC raised up to 0.94 g/liter and decreased to 0.26 after PT, and another diabetic man with osteomyelitis in whom PC was 0.42 g/ml before PT and fluctuated between 0.38 and 0.64 g/liter during PT. There were only a few cases where small, mainly transient, and clinically insignificant other abnormal values of the general urine test were found; these included the presence of small amounts of ketones, as well as of red and white cells revealed by microscopic analysis (data not shown). These data thus suggest that PT does not affect significantly the function of the renal glomerular capillary wall.

The most frequent adverse reactions to phage preparations found in patients subjected to PT were symptoms from the digestive tract, local reactions at the site of administration of a phage preparation, superinfections, and a rise in body temperature.

Symptoms from the digestive tract included nausea ($n=3$; 7.9% of patients who received phages orally), abdominal pain ($n=1$), and a transient decrease in appetite ($n=1$). Moreover, five patients reported an unpleasant taste of phage preparations (this was, however, not considered an adverse reaction).

Local reactions at the site of administration of a phage preparation were reported by 26 out of 141 patients (18.4%) administered these preparations in the form of nose drops, wet compresses, a sitz bath, gargling, fistular irrigation, vaginal irrigation, and by intrarectal route. They were usually minor and included redness, smarting, stinging, itching, local pain and irritation of skin, a feeling of local discomfort following intravaginal administration of a phage preparation, drying and irritation of upper respiratory tract mucous membranes, aggravation of atopic dermatitis ($n=1$), urticarial blisters ($n=1$), and purulent blisters ($n=1$). In particular, a feeling of smarting and local pain was reported by 8 patients. In the majority of cases, these reactions were short-lived (up to 30 min) or ceased after several days of using a given phage preparation or its replacement with another preparation. Only in two subjects (1.4%) did PT have to be discontinued due to strong local reactions (in one case because of local pain and in the other because of aggravation of atopic dermatitis in the region of the skin where the phage preparation was applied). Which compounds of phage preparations were responsible for the observed reactions cannot be determined without further studies.

A superinfection that required modification of treatment (discontinuation of PT or use of an antibiotic or an antifungal agent) occurred in 7 out of 153 analyzed patients (4.6%). In 1 patient it required discontinuation of PT after 9 days due to significant exacerbation of the inflammatory signs, such as edema, and pain in the site of infection. It was observed most often in patients who were administered phage preparations topically (five out of seven cases). In two cases, candidosis was diagnosed during PT.

A rise in body temperature was noted in 10 out of 153 patients (6.5%). Five of these patients had subfebrile temperature and 5 had fever. These incidents were observed in patients administered phage preparations topically ($n=5$), by a combination of oral and topical routes ($n=2$), orally ($n=1$), intrarectally ($n=1$), and by inhalations of aerosol ($n=1$) usually a few hours following administration of the first doses of phage preparations. In the majority of these patients the temperature fell to the normal value spontaneously or after the administration of antipyretic agents or replacement of a phage preparation. However, in one case the rise in body temperature required discontinuation of PT, which also resulted in normalization of the temperature.

In addition, the following adverse reactions were observed in or reported by patients: a transient feeling of discomfort in the nasal cavity ($n=1$) and dizziness ($n=1$) following use of a phage preparation in the form of nose drops; small flat ulceration outside the site of topical administration of a phage preparation ($n=1$); itching of the skin in the axillary region outside the site of topical administration of a phage preparation

($n=1$); small ecchymoses on the skin of the chest and leg, together with gum bleeding preceded by an episode of nose and gum bleeding (these symptoms occurred in this patient also prior to PT) without signs of hemorrhagic diathesis following administration of a phage preparation by the intrarectal route and on the glans mucosa ($n=1$; however, these symptoms ceased gradually over the course of the first cycle of PT and did not occur during the second cycle); aching in the renal region and a feeling of general weakness following gargling and use of a phage preparation in the form of nose drops ($n=1$); a transient shortening of the menstrual cycle following gargling and use of a phage preparation in the form of nose drops ($n=1$); an increase in the frequency of defecation and flatulence following intrarectal administration of a phage preparation ($n=1$); a cold/flu with fever ($n=2$); an increase in the amount of purulent discharge in the respiratory tract following inhalations of aerosol ($n=1$); aching of the leg muscles or the knee joint and the sacroiliac joint following administration of a phage preparation by the intrarectal route ($n=2$); constipation following intrarectal administration of phages ($n=1$); limb edema following topical administration of a phage preparation ($n=1$; in this patient, PT was discontinued after edema occurred); pain associated with closure of a fistula and impediment to pus outflow ($n=1$); acute groin and leg swelling in a man following oral administration of a phage preparation ($n=1$; a possible reason for this effect was venous thrombosis; PT was discontinued in this patient); transient itching of the skin of the head following administration of a phage preparation by a combination of intrarectal and topical route ($n=1$); abdominal discomfort following intrarectal administration of a phage preparation ($n=1$); and headache after topical administration of phages ($n=1$). However, whether these symptoms are a direct consequence of the administration of phage preparations is uncertain.

V. DISCUSSION

Treatment results seem to be quite promising, suggesting that it is possible to obtain a satisfactory response to PT in a substantial number of patients (some of them very sick) and achieve a cure in one of five patients who had been treated unsuccessfully with antibiotics previously, sometimes for prolonged periods, with concurrent complications and immunodeficiency demonstrated in some half of them. In fact, in the analyzed patients, a good response to therapy was achieved in 40% of cases, including pathogen eradication and/or recovery in almost 20% of them. Of note, in a clinical trial, a reduction of mean count of bacteria in samples taken from infection sites was considered proof of PT efficacy rather than a complete eradication of infection (Wright *et al.*, 2009); this adds weight to the presented results, which would probably have been even better if such a criterion had been used. Stabilization of the patient's status (this is included in category F of the efficacy of PT) could also be considered a positive effect of PT; this should be particularly relevant, for example, in patients with infected prostheses in whom PT may prolong their functioning. Notably, worsening occurred only in 6.5% of patients, even though more than 70% of them were treated only with phage.

Data presented here summarize results of a therapeutic experiment whose primary purpose was not a research study. Thus, no data are available for a formal control group of patients. However, there are some prerequisites that allow one to suspect that obtained data are not results of the natural course of at least some of the analyzed kinds of infections. In particular, this was demonstrated clearly on a very homogeneous subgroup of patients with orthopedic infections treated only with phage preparations. In this case the percentage of subjects with a good response to PT (categories A–C; 58.3%) in the group of those who received phages orally and/or topically was significantly higher as compared to the group that received phage only topically (20.0%). The significance of this difference was verified by statistical analysis.

The PTU criteria of the evaluation of phage therapy efficacy (categories B and C) were defined to exclude from evaluation any changes of the symptoms of infection that could probably result from spontaneous remission. Additionally, it should be taken into consideration that analyzed cases were especially difficult-to-treat ones (such as osteomyelitis, chronic wound, chronic bacterial prostatitis) (Pronk *et al.*, 2006; Siddiqui and Bernstein, 2010; Spelberg and Lipsky, 2011; Wagenlehner *et al.*, 2009). Indeed, the duration of infection in these patients prior to the start of their treatment with phages was very long [median for all patients was 43 months (with minimum 4 months and maximum 600 months)]. Despite that, overall good responses to PT observed in over one-third of subjects were obtained after a relatively short period of treatment (median disease duration in this subgroup of patients: 30 months; median treatment: 66 days). Pathogen eradication/recovery (category A of the response to PT) in over one-third of men with genital and urinary tract infections after median 66 days of the phage administration or especially 50% rate of pathogen eradication in patients with chronic bacterial prostatitis (confirmed by two consecutive negative bacterial cultures of prostatic fluid and accompanied with significant decrease in prostate volume or number of inflammatory cells in prostatic fluid) as was presented elsewhere (Letkiewicz *et al.*, 2009) is unlikely to be spontaneous.

The efficacy of PT described in this chapter is nevertheless much lower than previously reported results from the IIET. In fact, the historical success rates usually exceeded 90%, varying from 61% (infections of varicose ulcers of lower extremities) to 100% (furunculosis, purulent meningitis) (Weber *et al.*, 2000). The major variable potentially responsible for these discrepancies could be the method of patients' monitoring and evaluation. Prior to establishment of a PTU at the IIET, locally produced phage preparations were distributed among regional hospitals where the actual treatment was carried out, and its results were reported to the institute. With the establishment of the PTU, patients remained under the direct care and observation of physicians, which enabled a more detailed and insightful examination and a longer follow-up, thus allowing for a more critical and reliable assessment of results of therapy.

The authors realize that the scientific value of reports on phage therapy originating from the IIET in the period 1970–2005 is limited; however, they are still meaningful. Regardless of the actual efficacy of treatment, they confirm the lack of significant side effects and strongly suggest high safety of PT. Notably, even though no formal prolonged follow-up of those patients is available, the IIET never received information or any claim related to possible side effects or other harm that could be directly related to prior phage treatment. The same holds true for 6 years of operation of PTU, while the authors treated 293 patients during that period. Therefore, they believe that the major input of those historical years is the confirmation of safety of PT. In this regard, current data from the IIET fully confirm and extend this remarkable safety of PT based on both clinical and laboratory data. PT did not cause any significant alterations in the function of vital organs (kidney, liver, pancreas). What is more, therapy did not affect bone marrow function, nor did it change leukocyte circulation as reflected by normal levels of neutrophils and mononuclear cells (interestingly, the eosinophil level also remained normal, which confirms that patients on PT generally do not develop allergic reactions). The lack of alterations in ESR or CRP, along with these data, strongly suggests that PT is not associated with induction of a clinically significant inflammatory response.

Interestingly, the present analysis did not reveal any significant changes in inflammatory markers, although earlier analysis, which included 37 patients treated with phages between January 2006 and October 2007, showed a significant decrease in CRP serum concentration and WBC during PT (Międzybrodzki *et al.*, 2009). In fact, in a previous study, CRP and WBC decreased significantly between 9 and 32 days of treatment (CRP from 23.3 mg/liter at baseline to 16.1 mg/liter, $n=26$, and WBC from 7.8×10^3 cells/mm^3 at baseline to 7.1×10^3 cells/mm^3). However, for the currently analyzed population, the baseline values of CRP and WBC were much lower (for 9–32 days, mean CRP was 16.2 mg/liter, and WBC was 6.9 cells/mm^3), corresponding rather to their levels during/after treatment in the first study. The apparent discrepancy between the results of these two analyses may result from a substantial difference between the baseline values of the inflammatory markers found in patients prior to PT in the previous and current study. This is likely caused by the lower frequency of subjects with exacerbation of the inflammatory process at the beginning of PT in current analysis. This is probably why no decreases in the general levels of the analyzed markers in the present analysis could be observed, although in a subgroup of patients with active inflammation (baseline CRP above 10 mg/liter), a significant reduction in the serum concentration of CRP was observed, like in the previous study.

A unique aspect of PT performed at the PTU is monitoring of selected immune parameters of treated patients, including the mitogen-induced proliferation of T and B cells, natural killer (NK) cell cytotoxic activity, phagocytosis, and the respiratory burst of granulocytes, as well as T and NK cell percentage in blood (Górski *et al.*, 2012). Relevant assays are performed on immune cells isolated from the peripheral blood collected from patients before, during, and after PT. So far this analysis has included 70 subjects. An interesting finding was that up to 50% of the patients had immunodeficiency prior to PT (manifested as decreased T and B lymphoproliferative responses to mitogens, reactive oxygen species production by phagocytes stimulated with phorbol myristate acetate, and phagocytosis of zymosan particles); however, there were no statistically significant differences in the percentages of subjects with immunodeficiency prior to, during, and after PT, which strongly suggests that the therapeutic effects of phage preparations do not result from their immunostimulatory activity. It was also found that a tendency toward a patient's improvement may correlate with an increase in the intensity of the granulocyte phagocytosis as measured on days 7–20 of PT.

The authors observed significant changes in phage lytic patterns of some bacterial strains isolated from patients during PT. These may result from various reasons, such as genetic instability and/or heterogeneity of bacterial cells in an infecting population and changes in bacterial prophage load, as well as the acquisition of phage resistance (Clokie and Kropinski, 2009; Goerke *et al.*, 2007; Labrie *et al.*, 2010; Sulakvelidze and Kutter, 2005). This may also reflect the problem of infection with a new bacterial strain during phage treatment, which could decrease the efficacy of PT. The low percentage of bacterial resistance to the staphylococcal phages used for treatment most likely results from the polyvalence of these phages (each of them is able to lyse over 50% of *S. aureus* strains isolated from patients and stored in our collection; unpublished data).

The Georgian center, which has treated the largest number of patients, reported high success rates varying from 67% in patients with "lung infections" to 100% in osteomyelitis (Kutateladze and Adamia, 2010). However, the authors have not provided clinical criteria according to which their patients were evaluated. Furthermore, no laboratory data were presented to confirm the efficacy of treatment or to substantiate the claim of lack of side effects of therapy. A summary of available data on the efficacy of phage therapy during its history (until early 2000) was presented by Sulakvelidze and Kutter (2005). The majority of reports analyzed by the authors highlighted rather high rates of success. Heterogeneity of patient cohorts, concomitant use of antibiotics, and different treatment protocols make direct comparison of the historical results with data presented in this chapter difficult.

Interestingly, in the analyzed patients, oral administration of phage was associated with the highest rates of therapy efficacy, which was somewhat unexpected as there are no convincing reports confirming that phage can migrate from the gastrointestinal tract to blood and tissues of healthy people. On the one hand, such a phenomenon should occur in patients, similar to bacterial translocation (Górski *et al.*, 2006). On the other hand, observations indicate that such a route of phage administration may cause in some patients a strong anti-inflammatory effect reflected by a rapid and marked diminution in the CRP serum level, and usually does not result in the production of phage-neutralizing antibodies (Górski *et al.*, 2012; Międzybrodzki *et al.*, 2009; unpublished observations). Therefore, relatively good effects observed using the oral route of phage administration could be associated, at least in part, with these phage activities. However, it should be stressed that the best results of treatment manifested by pathogen eradication and recovery were achieved in almost half the cases (44.4%) using intrarectal administration, which probably leads to satisfactory phage uptake (penetration), as observed in animal experiments (R. Międzybrodzki, M. Kłak, B. Weber-Dąbrowska, and A. Górski, unpublished data) and migration of phages to the site of infection.

As a principle, at the PTU, only patients who had been treated unsuccessfully with antibiotics previously were eligible for PT, and their concomitant use was limited to situations where a patient's condition also required antibiotic treatment (e.g., simultaneous infection with a clinically relevant pathogen for which no phage was available). Although some authors believe that such a combined use of both antibacterials may lower the efficacy of PT, some reports suggest that such a combination may improve the outcome of PT (Sulakvelidze and Kutter, 2005). A report from Georgia appears to confirm this trend (Kutateladze and Adamia, 2010). However, data presented in this chapter do not suggest any benefit of antibiotics added to PT protocols. This important issue requires more studies, including relevant clinical trials.

In the presented material, the mean cumulative duration of PT was 55 days; no association was noted between actual duration of therapy and its efficacy. The duration of PT depends on the individual clinical course of a given patient; so far, no clear data exist on its optimal duration in the human clinic. It is difficult to extrapolate results obtained in animals, which suggest high efficacy of short courses of PT (including one high dose of phage), as those experiments were carried out using models of acute bacterial infections, while all the patients have been victims of chronic infections, sometimes persisting for many years (median duration of the infection was 43 months). Obviously, this issue also requires more in-depth studies to determine the minimal time sufficient to achieve satisfactory results.

Evidently, individual groups/subgroups of patients in this analysis contained relatively small numbers of individuals receiving experimental PT and no formal control group was included. Nevertheless, novel experimental modalities whose preliminary results are subsequently published may involve quite small patient groups. For example, *Lancet* published preliminary results of an experimental treatment of two patients (one with age-related macular degeneration and another with Stargard's macular dystrophy) with human embryonic stem cells; estimated enrollment for this treatment are 12 patients (3 cohorts, each consisting of 3 patients, without any control group) (Schwartz *et al.*, 2012).

The authors are well aware that the final formal proof of efficacy of PT could only be provided by controlled clinical trials. Therefore, it is optimistic that their initial results already published and presented at phage symposia are encouraging (see Table I). However, one should keep in mind that progress in medicine has also been achieved by observational studies (Rosenbaum 2002). They continue to play an important role in clinical research, with a record of successful contributions to medicine proving essential for our knowledge about causes and pathogenesis, for example, infectious causes of disease (Vandenbroucke, 2004). The need and even moral requirement to publish these data has its roots in the Declaration of Helsinki, which sets ethical standards for treatment and research involving human subjects:

> 35. In the treatment of a patient, where proven interventions do not exist or have been ineffective, the physician, after seeking expert advice, with informed consent from the patient or a legally authorized representative, may use an unproven intervention if in the physician's judgement it offers hope of saving life, re-establishing health or alleviating suffering. Where possible, this intervention should be made the object of research, designed to evaluate its safety and efficacy. In all cases, new information should be recorded and, where appropriate, made publicly available.

VI. CONCLUSIONS

In conclusion, the authors' report confirms full safety of PT, thereby confirming that it complies with one of the most important principles of medical practice: *primum non nocere* (first do not harm). Therefore, a physician facing an untreatable bacterial infection may resort to PT with a fair chance of success and practically no risk of side effects. It is hoped that this chapter will prompt appropriate responses from granting agencies, industry, and governments to make further progress to pass regulatory hurdles and fund clinical trials necessary for wider introduction of PT to current therapy, which is clearly in the public interest.

ACKNOWLEDGMENTS

This work was supported by the European Regional Development Fund within the Operational Program Innovative Economy, 2007-2013, Priority axis 1. Research and Development of Modern Technologies, Measure 1.3 Support for R&D projects for entrepreneurs carried out by scientific entities, Submeasure 1.3.1 Development projects as project No. POIG 01.03.01-02-003/08 entitled "Optimization of the production and characterization of bacteriophage preparations for therapeutic use."

REFERENCES

Adams, M. H. (1959). Bacteriophages. Interscience Publishers, Inc., New York.

Aiello, A. E., King, N. B., and Foxman, B. (2006). Antimicrobial resistance and the ethics of drug development. *Am. J. Publ. Health* **96:**1–5.

Balogh, B., Jones, J. B., Iriarte, F. B., and Momol, M. T. (2010). Phage therapy for plant disease control. *Curr. Pharm. Biotechnol.* **11:**48–57.

Biswas, B., Adhya, S., Washart, P., Paul, B., Trostel, A. N., Powell, B., Carlton, R., and Merril, C. R. (2002). Bacteriophage therapy rescues mice bacteremic from a clinical isolate of vancomycin-resistant *Enterococcus faecium*. *Infect. Immun.* **70:**204–210.

Bruttin, A., and Brüssow, H. (2005). Human volunteers receiving *Escherichia coli* phage T4 orally: A safety test of phage therapy. *Antimicrob. Agents Chemother.* **49:**2874–2878.

Capparelli, R., Parlato, M., Borriello, G., Salvatore, P., and Iannelli, D. (2007). Experimental phage therapy against *Staphylococcus aureus* in mice. *Antimicrob. Agents Chemother.* **51:**2765–2773.

Clokie, M. R. J., and Kropinski, A. M. (2009). Bacteriophages: Methods and Protocols, Vol. 1: Humana Press, New York.

Freire-Moran, L., Aronsson, B., Manz, C., Gyssens, I. C., So, A. D., Monnet, D. L., and Cars, O. (2011). ECDC-EMA Working Group: Critical shortage of new antibiotics in development against multidrug-resistant bacteria—Time to react is now. *Drug Resist Updat.* **14:**118–124.

Giamarellou, H. (2010). Multidrug-resistant Gram-negative bacteria: How to treat and for how long. *Int. J. Antimicrob. Agents* **36**(Suppl. 2)**:**S50–S54.

Goerke, C., Gressinger, M., Endler, K., Breitkopf, C., Wardecki, K., Stern, M., Wolz, C., and Kahl, B. C. (2007). High phenotypic diversity in infecting but not in colonizing Staphylococcus aureus populations. *Environ. Microbiol.* **9:**3134–3142.

Górski, A., Międzybrodzki, R., Borysowski, J., Dąbrowska, K., Wierzbicki, P., Ohams, M., Korczak-Kowalska, G., Olszowska-Zaremba, N., Łusiak-Szelachowska, M., Kłak, M., Jończyk, E., Kaniuga, E., *et al.* (2012). Phage as modulator of immune responses: Practical implications for phage therapy. *In* "Advances in Virus Research" (M. Łobocka and W. Szybalski, eds.), Vol. Part B, Chapter 2, pp. 41–72. Elsevier, UK.

Górski, A., Międzybrodzki, R., Borysowski, J., Weber-Dąbrowska, B., Łobocka, M., Fortuna, W., Letkiewicz, S., Zimecki, M., and Filby, G. (2009). Bacteriophage therapy for the treatment of infections. *Curr. Opin. Invest. Drugs* **10:**766–774.

Górski, A., Ważna, E., Weber-Dąbrowska, B., Dąbrowska, K., Świtała-Jeleń, K., and Międzybrodzki, R. (2006). Bacteriophage translocation. *FEMS Immunol. Med. Microbiol.* **46:**313–319.

Hanlon, G. W. (2007). Bacteriophages: An appraisal of their role in the treatment of bacterial infections. *Int. J. Antimicrob. Agents* **30:**118–128.

Jabes, D. (2011). The antibiotic R&D pipeline: An update. *Curr. Opin. Microbiol.* **14:**564–569.

Jasieński, J. (1927). Próby zastosowania bakteriofagji w chirurgji. *Pol Gaz Lek* **4:**67–73 [In Polish].

Kalinowski, W., and Czyż, J. (1923). Sprawozdanie z epidemji czerwonki w roku 1922. *Lekarz Wojsk.* **4:**286–293 [In Polish].

Köck, R., Becker, K., Cookson, B., van Gemert-Pijnen, J. E., Harbarth, S., Kluytmans, J., Mielke, M., Peters, G., Skov, R. L., Struelens, M. J., Tacconelli, E., Navarro Torné, A., *et al.* (2010). Methicillin-resistant *Staphylococcus aureus* (MRSA): Burden of disease and control challenges in Europe. *Euro Surveill.* **15:**19688.

Kutateladze, M., and Adamia, R. (2010). Bacteriophages as potential new therapeutics to replace or supplement antibiotics. *Trends Biotechnol.* **28:**591–595.

Labrie, S. J., Samson, J. E., and Moineau, S. (2010). Bacteriophage resistance mechanisms. *Nat. Rev. Microbiol.* **8:**317–327.

Letkiewicz, S., Międzybrodzki, R., Fortuna, W., Weber-Dąbrowska, B., and Górski, A. (2009). Eradication of *Enterococcus faecalis* by phage therapy in chronic prostatitis: Case report. *Folia Microbiol.* **54:**457–461.

Lityński, M. (1950). Treatment of infection of *E. coli* using specific bacteriophages. *Przegl Lek* **6:**13–19.

Mahony, J., McAuliffe, O., Ross, R. P., and van Sinderen, D. (2011). Bacteriophages as biocontrol agents of food pathogens. *Curr. Opin. Biotechnol.* **22:**157–163.

Matsuzaki, S., Yasuda, M., Nishikawa, H., Kuroda, M., Ujihara, T., Shuin, T., Shen, Y., Jin, Z., Fujimoto, S., Nasimuzzaman, M. D., Wakiguchi, H., Sugihara, S., Sugiura, T., Koda, S., Muraoka, A., and Imai, S. (2003). Experimental protection of mice against lethal *Staphylococcus aureus* infection by novel bacteriophage phi MR11. *J. Infect. Dis.* **187:**613–624.

Międzybrodzki, R., Fortuna, W., Weber-Dąbrowska, B., and Górski, A. (2007). Phage therapy of staphylococcal infections (including MRSA) may be less expensive than antibiotic treatment. *Post Med. Hig Dośw.* **61:**461–465.

Międzybrodzki, R., Fortuna, W., Weber-Dąbrowska, B., and Górski, A. (2009). A retrospective analysis of changes in inflammatory markers in patients treated with bacterial viruses. *Clin. Exp. Med.* **9:**303–312.

Pirnay, J. P., De Vos, D., Verbeken, G., Merabishvili, M., Chanishvili, N., Vaneechoutte, M., Zizi, M., Laire, G., Lavigne, R., Huys, I., Van den Mooter, G., Buckling, A., *et al.* (2011). The phage therapy paradigm: Prêt-à-porter or sur-mesure? *Pharm. Res.* **28:**934–937.

Pronk, M. J., Pelger, R. C., Baranski, A. G., van Dam, A., and Arend, S. M. (2006). Cure of chronic prostatitis presumably due to *Enterococcus spp* and gram-negative bacteria. *Eur. J. Clin. Microbiol. Infect. Dis.* **25:**270–271.

Rhoads, D. D., Wolcott, R. D., Kuskowski, M. A., Wolcott, B. M., Ward, L. S., and Sulakvelidze, A. (2009). Bacteriophage therapy of venous leg ulcers in humans: Results of a phase I safety trial. *J. Wound Care* **18**(237–238)**:**240–243.

Rosenbaum, P. R. (2002). Observational studies. Springer Series in Statistics, 2nd Ed.

Schwartz, S. D., Hubschman, J. P., Heilwell, G., Franco-Cardenas, V., Pan, C. K., Ostrick, R. M., Mickunas, E., Gay, R., Klimanskaya, I., and Lanza, R. (2012). Embryonic stem cell trials for macular degeneration: A preliminary report. *Lancet* **379**(9817)**:**713–720.

Selgelid, M. J. (2007). Ethics and drug resistance. *Bioethics.* **21:**218–229.

Siddiqui, A. R., and Bernstein, J. M. (2010). Chronic wound infection: Facts and controversies. *Clin. Dermatol.* **28:**519–526.

Ślopek, S., Durlakowa, I., Weber-Dabrowska, B., Kucharewicz-Krukowska, A., Dabrowski, M., and Bisikiewicz, R. (1983). Results of bacteriophage treatment of suppurative bacterial infections. I. General evaluation of the results. *Arch. Immunol. Ther. Exp.* **31**:267–291.

Ślopek, S., and Krzywy, T. (1985). Morphology and ultrastructure of bacteriophages: An electron microscopic study. *Arch. Immunol. Ther. Exp.* **33**:1–217.

Ślopek, S., Weber-Dąbrowska, B., Dąbrowski, M., and Kucharewicz-Krukowska, A. (1987). Results of bacteriophage treatment of suppurative bacterial infections in the years 1981-1986. *Arch. Immunol. Ther. Exp.* **35**:569–583.

Stafford, N. (2011). Switzerland is to fund complementary therapies for six years while effectiveness is evaluated. *BMJ* **342**:d819.

Sulakvelidze, A., and Kutter, E. (2005). Phage therapy in humans. *In* "Bacteriophages: Biology and Applications" (E. Kutter and A. Sulakvelidze, eds.), pp. 381–436. CRC Press, Boca Raton, FL.

Trelińska, J., Stolarska, M., Zalewska-Szewczyk, B., Kuzański, W., and Młynarski, W. (2010). The use of bacteriophage therapy in *Pseudomonas aeruginosa* infection in patients with acute leukemias: Two cases report. *Onkol. Pol.* **13**:50–53 [In Polish].

Turner, M. (2011a). Phage on the rampage: Antibiotics use may have driven the development of Europe's deadly. *E. coli. Nature* doi:10.1038/news.2011.360.

Turner, M. (2011b). German *E. coli* outbreak leads to drug trial. *Nature* doi: 10.1038/news.2011.332.

Vandenbroucke, J. P. (2004). When are observational studies as credible as randomized trials? *Lancet* **363**:1728–1731.

Verbeken, G., De Vos, D., Vaneechoutte, M., Merabishvili, M., Zizi, M., and Pirnay, J. P. (2007). European regulatory conundrum of phage therapy. *Future Microbiol.* **2**:485–491.

Wang, J., Hu, B., Xu, M., Yan, Q., Liu, S., Zhu, X., Sun, Z., Reed, E., Ding, L., Gong, J., Li, Q. Q., and Hu, J. (2006). Use of bacteriophage in the treatment of experimental animal bacteremia from imipenem-resistant *Pseudomonas aeruginosa*. *Int. J. Mol. Med.* **17**:309–317.

Weber-Dąbrowska, B., Mulczyk, M., and Górski, A. (2000). Bacteriophage therapy of bacterial infections: An update of our institute's experience. *Arch. Immunol. Ther. Exp.* **48**:547–551.

Weber-Dąbrowska, B., Mulczyk, M., and Górski, A. (2003). Bacteriophages as an efficient therapy for antibiotic-resistant septicemia in man. *Transplant. Proc.* **35**:1385–1386.

Wernli, D., Haustein, T., Conly, J., Carmeli, Y., Kichbusch, I., and Harbuth, S. (2011). A call for action: The application of the international health regulations in the global health threat of antimicrobial resistance. *PLoS Med.* **8**:e1001022.

Witoszka, M., and Strumillo, B. (1963). Bacteriophage therapy of infected surgical wounds. *Pol Tyg Lek.* **18**:430–432.

Wright, A., Hawkins, C. H., Anggård, E. E., and Harper, D. R. (2009). A controlled clinical trial of a therapeutic bacteriophage preparation in chronic otitis due to antibiotic-resistant *Pseudomonas aeruginosa*; a preliminary report of efficacy. *Clin. Otolaryngol.* **34**:349–357.

Zimecki, M., Weber-Dąbrowska, B., Łusiak-Szelachowska, M., Mulczyk, M., Boratyński, J., Syper, D., and Górski, A. (2003). Bacteriophages provide regulatory signals in mitogen-induced murine splenocyte proliferation. *Cell. Mol. Biol. Lett.* **8**:700–711.

CHAPTER 4

Bacteriophages in the Experimental Treatment of *Pseudomonas aeruginosa* Infections in Mice

Emilie Saussereau and **Laurent Debarbieux**[1]

Contents

Institut Pasteur, Molecular Biology of the Gene in Extremophiles Unit, Department of Microbiology, Paris, France
[1] Corresponding author, E-mail: laurent.debarbieux@pasteur.fr

Advances in Virus Research, Volume 83
ISSN 0065-3527, DOI: 10.1016/B978-0-12-394438-2.00004-9

Abstract

The regular increase of drug-resistant pathogens has been a major force in the renewed interest in the use of bacteriophages as therapeutics. In addition to experience acquired in eastern Europe where bacteriophages have been used to treat bacterial infections in humans, in Western countries only experimental models have been developed until recently. The Gram-negative bacterium *Pseudomonas aeruginosa* is an opportunistic pathogen causing particularly severe infections in cystic fibrosis patients. Several experimental models in mice have yielded encouraging results for the use of bacteriophages to treat or prevent septicemia, skin and lungs infections caused by *P. aeruginosa*. Now, a phase II clinical trial conducted in the United Kingdom provides evidence for the efficacy of bacteriophage treatments in chronic otitis due to antibiotic-resistant *P. aeruginosa* strains. Together with experimental models, these results provide an incentive to develop more research and clinical studies to fully appreciate the benefits of the use of bacteriophages in medicine.

I. GENERAL BACKGROUND

A. *Pseudomonas aeruginosa*, an opportunistic pathogen

Pseudomonas is one of the most diverse genera known, with members in diverse environmental niches, such as soil, water, animals, and plants. *Pseudomonas aeruginosa* differs from other species of this genus in being pathogenic to animals (Pier and Ramphal, 2009). This Gram-negative rod-shaped bacterium is an opportunistic pathogen, causing various types of infection (e.g., skin, eyes, ears, respiratory tract, urinary tract, gut-derived sepsis). Immunocompromised patients and patients on immunosuppressive treatments, such as patients suffering from cystic fibrosis, burn wounds, AIDS, and cancer, are the most frequently infected by this organism (Asboe *et al.*, 1998; Lyczak *et al.*, 2000, 2002; Mayhall 2003; Vento *et al.*, 2008).

An ability to adapt to diverse environments is one of the key characteristics of *P. aeruginosa*. This bacterium can grow both aerobically and anaerobically, at environmental temperatures of up to 42 °C, and can make use of many different metabolic compounds, from small to complex molecules. These features have enabled this bacterium to establish itself in many different environments (Pirnay *et al.*, 2009). *P. aeruginosa* is found almost everywhere, from water, to the surface of fruits and vegetables, sinks or shower heads, and, ultimately, medical devices, such as catheters (Kiska *et al.*, 2003; Morrison and Wenzel, 1984).

Several genomic studies have been carried out on *P. aeruginosa*, and nine genomes have been sequenced to date (Table I). The presence of a large number of regulatory genes and quorum-sensing components reflect the comprehensive adaptability of this microbe. Many virulence factors (from secreted proteins to surface structures) have been identified, and *P. aeruginosa* has been shown to adhere strongly to different types of surfaces (biotic and abiotic), resulting in the formation of biofilms that are difficult to eradicate with antibiotics (Bleves *et al.*, 2010; Heiniger *et al.*, 2010; Randall and Irvin, 2008).

B. The drug resistance of *Pseudomonas aeruginosa*

Pseudomonas aeruginosa strains are naturally resistant to several antibacterial drugs (Livermore, 2002). In addition, some clinical isolates have a hyper-mutator phenotype, facilitating the rapid evolution of resistance to drugs to which they were initially sensitive (Ferroni *et al.*, 2009). *P. aeruginosa* strains can also acquire resistance from mobile genetic elements, such as plasmids (Coyne *et al.*, 2010; Poirel *et al.*, 2009; Poole, 2005). Most *P. aeruginosa* strains are still susceptible to colistin, a polymyxin antibiotic, but increasing numbers of reports of pan-resistant strains are emerging (Bonomo and Szabo, 2006; Deplano *et al.*, 2005; Zavascki *et al.*, 2010).

The resistance mechanisms involved are diverse, ranging from enzymes targeting drugs, such as β-lactamases and aminoglycoside-modifying enzymes, to efflux pumps that expel drugs, changes in porin expression or mutations, or modifications to the molecules targeted by the drugs, such as the methylation of 16 S RNA, which confers resistance to aminoglycosides (Bonomo and Szabo, 2006; Falagas *et al.*, 2006; Nordmann *et al.*, 1993; Zavascki *et al.*, 2010). Hopes of discovering new effective drugs have been dashed in the last 10 years by the ability of bacteria to develop resistance mechanisms, in some cases even before the market release of the drug concerned. This has made it difficult to remain one step ahead in the fight against pathogens and has led to a revival of interest in the therapeutic application of bacteriophages, known as phage therapy (Thiel, 2004).

C. *Pseudomonas aeruginosa* infections targeted by bacteriophage treatments

To date, three experimental *P. aeruginosa* infections (septicemia, skin, and lung infections) have been used to study bacteriophage treatments. Bacterial septicemia is a general infection combining a systemic inflammatory response syndrome and bacteremia (presence of viable bacteria in the bloodstream). In the absence of treatment, septicemia may progress to septic shock and, ultimately, death. Most (65%) cases of septicemia are caused

TABLE I Listing of currently sequenced *Pseudomonas aeruginosa* genomes

Strain	Associated disease	Genome size (bp)	Prophages (putative)	Accession number	Reference
P. aeruginosa 39016	Keratitis	6,667,330	4	NZ_AEEX00000000	Stewart *et al.* (2011)
P. aeruginosa PACS2	Not specified	6,492,423	?	NZ_AAQW00000000	NCBI Genome
P. aeruginosa 2192	Cystic fibrosis	6,826,253	≥ 1	NZ_AAKW00000000	Mathee *et al.* (2008)
P. aeruginosa C3719	Cystic fibrosis	6,146,998	≥ 1	NZ_AAKV00000000	Mathee *et al.* (2008)
P. aeruginosa PAb1	Frostbite	6,078,600	?	NZ_ABKZ00000000	Salzberg *et al.* (2008)
P. aeruginosa LESB58	Cystic fibrosis	6,601,757	6	NC_011770.1	Winstanley *et al.* (2009)
P. aeruginosa PA7	Not specified	6,588,339	4	NC_009656.1	Roy *et al.* (2010)
P. aeruginosa UCBPP-PA14	Burn	6,537,648	≥ 1	NC_008463.1	Lee *et al.* (2006)
P. aeruginosa PAO1	Wound	6,264,404	≥ 2	NC_002516.2	Stover *et al.*, (2000)

by Gram-positive bacteria, such as staphylococci or enterococci, but *P. aeruginosa* is the third most frequently isolated Gram-negative bacterium in cases of nosocomial bloodstream infections, after *Escherichia coli* and *Klebsiella* sp. *P. aeruginosa* septicemia is fatal in 38.7 % of cases (Wisplinghoff *et al.*, 2004). Experimental models of *P. aeruginosa* septicemia have proved useful for assessments of the efficacy of phage therapy, as they have shown that bacteriophages can infect bacteria in a wide range of organs.

Pseudomonas aeruginosa skin infections are a serious threat in burn units of care centers. In 1983, 22% of the 887 burns patients treated at Cochin Hospital (Paris, France) had *P. aeruginosa* infections. The mortality rate was 60% for patients with burns covering more than 25% of the body surface, and being positive for blood cultures for *P. aeruginosa*. *P. aeruginosa* and *Staphylococcus aureus* have been identified as the two main antibiotic-resistant pathogens detected in burns patients (Wibbenmeyer *et al.*, 2006). *P. aeruginosa* has also been shown to cause skin infections in immunocompromised patients (Chiller *et al.*, 2001).

Lung infections caused by *P. aeruginosa* are rare in healthy populations. Only a few cases of community-acquired *P. aeruginosa* pneumonia have been identified, mostly in elderly people (Graber Duvernay, 2002). In contrast, immunocompromised patients, such as AIDS patients, are considered to constitute a population at risk (Fichtenbaum *et al.*, 1994). Cystic fibrosis patients comprise the largest population subject to *P. aeruginosa* lung infection, with 80 to 95% of these patients dying from respiratory failure due to chronic status of this infection accompanied by airway inflammation (Lyczak *et al.*, 2002). Indeed, cystic fibrosis, a heritable genetic disease affecting 1 in 2500 people in the Caucasian population, causes a thickening of the airway surface liquid that affects mucociliary clearance, creating a favorable microenvironment for bacterial infections (Corey and Farewell, 1996). The use of bacteriophages to treat lung infections in humans was not reported in eastern European countries before the initiation of such treatments in children with cystic fibrosis in 2007 in Tbilissi (Kutateladze and Adamia, 2010). Moreover, only one report mentioning the use of a murine lung infection model to evaluate the efficacy of bacteriophage treatments was published before 2009 (Meitert *et al.*, 1987).

II. *PSEUDOMONAS AERUGINOSA* BACTERIOPHAGES

A. Prophages (temperate bacteriophages)

Prophages are bacteriophage genomes integrated into the host genome. The sequence of *P. aeruginosa* genomes contains several putative prophages (Table I). Based on their sequences, some prophages may appear to be active, but it is not always possible to reactivate them *in vitro* in laboratory conditions. For example, six putative prophages were identified in the genome sequence of strain LESB58, but only four were found to be active *in vitro* (Winstanley *et al.*, 2009). However, the lack of reactivation *in vitro* of some prophages with apparently complete sequences does not necessarily imply that these prophages are no longer active. We may simply not have discovered the signals driving their release.

Prophages are a major source of new genes and, often, of new functions in bacterial genomes (Brüssow *et al.*, 2004, 2007; Cortez *et al.*, 2009). Functions identified to date include virulence factors and drug resistance mechanisms (Tinsley *et al.*, 2006). For example, the FIZ-15 prophage confers resistance to phagocytosis by macrophages and increases adhesion to human cells, whereas the Pf4 prophage contributes to biofilm formation in the PAO1 strain (Rice *et al.*, 2008; Vaca-Pacheco *et al.*, 1999; Whiteley *et al.*, 2001).

B. Virulent bacteriophages

The large number of virulent bacteriophages infecting *P. aeruginosa* that have been fully sequenced (more than 40 in 2010) bears witness to the high degree of interest in this field (Ceyssens and Lavigne, 2010). Bacteriophage genomes are much less conserved than bacterial genomes. It is therefore not possible to identify a particular subset of sequences present in all bacteriophages infecting a given bacterial species. Given the very high abundance of bacteriophages (10^{32} particles on earth) and the diversity of environments in which *P. aeruginosa* is found, the number of bacteriophages infecting *P. aeruginosa* discovered to date may represent the tip of the iceberg (Fuhrman, 1999).

Phage therapy—the use of bacteriophages to treat bacterial infections—has excited renewed interest in the last 10 years due to the increasing problem of bacterial resistance to multiple antibiotics. Phage therapy is based on the use of a specific bacteriophage infecting a specific strain. As one of the main pathogens displaying resistance to numerous drugs, *P. aeruginosa* has been the focus of several studies using bacteriophages. These studies involved isolation of a large number of bacteriophages, only a few of which were fully characterized. For obvious reasons of safety, the use of temperate bacteriophages for phage therapy is not recommended (Brüssow, 2005). Newly isolated bacteriophages must therefore be characterized to ensure that those not strictly virulent are discarded. It is possible to carry out such characterization by bioinformatics analyses, provided that the full genome sequence has been obtained (Denou *et al.*, 2009; Morello *et al.*, 2011).

III. EXPERIMENTAL PHAGE THERAPY OF *PSEUDOMONAS AERUGINOSA* INFECTIONS IN MICE

A. Treatments for septicemia

1. Experimental models

Most experimental models are based on intraperitoneal injection in mice, resulting in 100% mortality within 24 to 48 hr, depending on the strain used (Hagens *et al.*, 2004; Meitert *et al.*, 1987; Soothill, 1992; Vinodkumar *et al.*, 2008; Wang *et al.*, 2006). This route of infection leads to bacteremia, severe systemic congestion, splenomegaly, acute ascites, and organ hemorrhage (Meitert *et al.*, 1987; Wang *et al.*, 2006). Another model of septicemia resulting from intestinal infection has also been described (Watanabe *et al.*, 2006). In this model, mice receive *P. aeruginosa* in drinking water for 3 days, together with antibiotic treatment to facilitate the intestinal implantation of *P. aeruginosa*. Cyclophosphamide is then administered by intraperitoneal injection to disrupt the intestinal mucosal barrier, allowing septicemia to develop. In this model, which more closely resembles the septicemia observed in human patients, 100% of the mice die within 9 days of the second cyclophosphamide injection.

2. Curative treatments

The renewed interest in phage therapy experiments dates back to the start of the 1980s. During this period, the first advanced study of an experimental bacteriophage-based treatment for *P. aeruginosa* septicemia treatment was carried out (Meitert *et al.*, 1987). This study showed that a single dose of 10^6 plaque-forming unit (PFU) bacteriophages, administered intravenously at the same time as the bacteria, rescued 50% of mice infected with a lethal dose of *P. aeruginosa* [1.5×10^8 colony-forming unit (CFU) intraperitoneally]. More recent studies have shown that the efficacy of bacteriophage treatment increases with the dose given (Soothill, 1992; Wang *et al.*, 2006). If the dose of bacteriophages is increased to a multiplicity of infection (MOI) of 10 to 10^4, then the administration of bacteriophages can be delayed to between 1 and 6 hr after infection, with no effect on rescue rate, which remains at 100% (Hagens *et al.*, 2004; Heo *et al.*, 2009; Vinodkumar *et al.*, 2008; Wang *et al.*, 2006). Phosphate-buffered saline or inactivated bacteriophage solutions were used as a control, with such treatment resulting in 100% mortality.

In addition to the MOI and the timing of bacteriophage administration, the pathway used is also important. The intraperitoneal administration of bacteriophages is much more efficient than intramuscular injection or oral treatment for the treatment of experimental septicemia (Heo *et al*, 2009; Meitert *et al.*, 1987).

The efficacy of bacteriophage treatment for septicemia has also been assessed by counting bacteria in blood samples and in several organs (lungs, spleen, and liver) 1 day after the administration of bacteriophages. Bacteriophage treatment has been reported to decrease bacterial counts by a factor of 200 to 1000 (Heo *et al.*, 2009; Vinodkumar *et al.*, 2008; Watanabe *et al.*, 2006). Bacteriophage treatment also decreases cytokine levels significantly [by factors of five for interleukin (IL)-1β, eight for IL-6, and seven for tumor necrosis factor-α](Watanabe *et al.*, 2006).

3. Preventive treatments

Watanabe *et al.* (2006) did not restrict their studies to the curative treatment of septicemia. They also assessed the efficacy of bacteriophages as a preventive treatment. They showed that the administration of the curative dose *per os* 1 day before infection was ineffective, with all mice dying over the same time interval as untreated animals. This suggests that orally administered bacteriophages do not reside long enough in the body of the animals to protect them from subsequent infections. Consequently, any preventive treatment based on this route of administration is unlikely to be successful unless it is performed only within a few hours before, which reduces its interest considerably.

4. Conclusions

Studies of these experimental models of septicemia have shown that bacteriophages can infect bacteria within the body of an animal. Bacteriophages have been shown to be highly active *in vivo*, rescuing animals from infections with lethal doses of bacteria. However, these results should not be taken as proof that bacteriophages are useful for treating septicemia. As one of the key principles of phage therapy is the use of a bacteriophage able to infect the targeted pathogen, prior identification of the pathogen is required in order to select the most appropriate bacteriophage for treatment. Hospitalized patients with septicemia often require rapid treatment, and the pathogen involved is not always clearly identified. It is therefore difficult to use bacteriophages for treatment of septicemia.

B. Treatment of skin infections

1. Experimental models

In most models, a small area of mouse skin is injured by heating, and bacteria are then injected subcutaneously into the wound or applied directly onto the surface of the skin. When injected, strain PAO1 causes an infection that is fatal within 72 hr, whereas strain 4294, when applied directly to the skin, results in an infection persisting for up to 21 days, with a mortality rate of 10% (Kumari *et al.*, 2009a; McVay *et al.*, 2007; Meitert *et al.*, 1987). Both these models have been used to assess the efficacy of bacteriophages for treating infections of burn wounds.

2. Curative treatments

Direct application of bacteriophages to the burned surface, repeated 10 times every 30 minutes on the first day, results in the healing of 80% of wounds (Meitert *et al.*, 1987). The initial dose of bacteria was 200 times higher than the dose of bacteriophages used. A more recent study by McVay *et al.* (2007) compared the efficacy of bacteriophages administered by three different routes: intraperitoneally, intramuscularly, and subcutaneously. The intraperitoneal route was the most efficient, resulting in 88% survival, whereas survival rates of less than 30% were obtained with the other two routes of administration. This model included a low infectious dose (100 bacteria) and a high treatment dose (3×10^8 PFU). In contrast, Kumari *et al.* (2009a,b) used the same strain (PAO1), but with 10^7 bacteria and treatments with up to 9×10^9 PFU of bacteriophages once daily, without success, with all animals dying within 72 hr. These authors used the same animal model of burn wounds in which they had previously reported the rescue by bacteriophage treatment, within 72 hr, of up to 94% of mice infected with a lethal dose of *Klebsiella pneumoniae* (Kumari *et al.*, 2009a,b). In both studies with the PAO1 strain, the mortality rates of untreated animals were similar, despite a 10^5-fold difference in the bacterial dose used. This suggests that the two groups may have used different clones of this strain, mouse lines with different immune responses, or both.

3. Conclusions

The three articles reporting the use of bacteriophages in animal models of infected burn wounds suggest that direct applications of bacteriophage are the most effective. This led to development of a biopolymer in Georgia. This biodegradable polymer was impregnated with bacteriophages for direct application onto wounds (Markoishvili *et al.*, 2002). A case report in the United Kingdom also showed that bacteriophages applied directly onto burn wounds multiplied actively, helping heal infection of the site and making it possible to carry out skin grafts (Marza *et al.*, 2006). Direct application also ensures that the bacteriophages come into direct contact with the bacteria, increasing the likelihood of efficient treatment. Finally, the work of Kumari and colleagues (Chhibber *et al.*, 2008; Kumari *et al.*, 2010, 2011) in *K. pneumoniae* skin infection models has shown that bacteriophages are at least as efficient as antibiotics.

C. Treatment of lung infections

1. Experimental models

Two different routes of administration have been used to establish bacterial infections of lungs in mice. Bacteria may be introduced directly into the trachea or instilled intranasally. These procedures lead to the development of acute infections, resulting in the death of mice after 24 to 72 hr, depending on the dose and virulence of the strains used (Debarbieux *et al.*, 2010; Meitert *et al.*, 1987; Morello *et al.*, 2011). A few hours after the initiation of infection, a set of inflammation and cell damage markers can be used to demonstrate the rapid progression of lung infection (Morello *et al.*, 2011; Ramphal *et al.*, 2005). Bioluminescent strains of bacteria can be used to monitor infection in a given animal over a period of time. Such strains have been used to determine the kinetics of lung infection (Debarbieux *et al.*, 2010; Ramphal *et al.*, 2008). Furthermore, 20 hr after infection with a strain of *P. aeruginosa* isolated from cystic fibrosis patients, histology and immunohistochemistry analyses of lung sections revealed severe lesions and massive neutrophil recruitment. Large numbers of bacteria were found to be present both within macrophages and outside cells.

2. Curative treatments

Bacteriophage treatments of acute lung infections must be administered rapidly to achieve maximum efficacy, whether an intravenous or intranasal route of administration is used. A single intravenous administration of a bacteriophage dose, which was half the infectious dose of bacteria, immediately after intranasal infection has been shown to rescue 47% of animals (Meitert *et al.*, 1987). Delays to treatment decreased survival greatly, to 27% for a delay of 2 hr and 0% for a delay of 24 hr. In the same model, a single oral administration was tested at several time points, but results obtained were not different from those obtained for intravenous administration. In contrast, a single intranasal administration of bacteriophages 2 hr after the initiation of infection resulted in survival rates of 90 to 100% (Morello *et al.*, 2011). Efficacy was lower for smaller numbers of bacteriophages and for later treatment (Debarbieux *et al.*, 2010). A bioluminescent strain of *P. aeruginosa* has been used for kinetic analyses of treatment. In the first 2 hr after bacteriophage application, no significant differences were found between groups, whereas at 4 hr, bioluminescence levels were lower in the treated group than in the untreated group (Fig. 1) (Debarbieux *et al.*, 2010). This time lag to efficacy probably corresponds to the time required for the bacteriophages to infect and kill enough hosts for a visible effect. Complementary analyses of proinflammatory markers and counts of bacteria and bacteriophages confirmed the lower levels of infection in the treated group than in the control group. Using a clinical strain of *P. aeruginosa* isolated from a cystic fibrosis patient, Morello *et al.* (2011) performed similar experiments with another bacteriophage and obtained similar results. They also carried out histology and immunohistochemistry studies and revealed that lung sections from treated animals were less damaged than those from untreated animals, with no bacterial cells in the intercellular space. These data clearly confirmed that the bacteriophages were able to reach bacteria other than

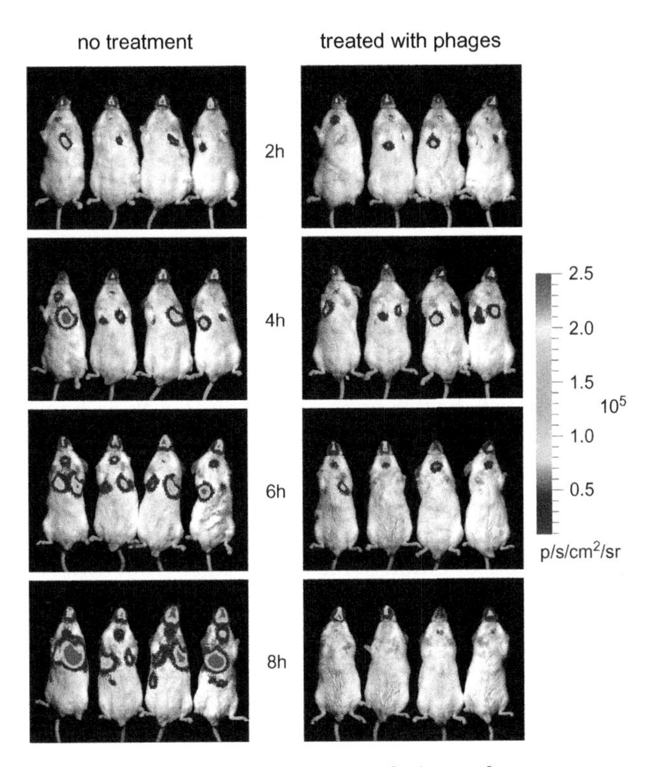

FIGURE 1 Kinetics of bacteriophage treatment of a lung infection in mice. Two groups of mice were infected intranasally with an equal amount of a bioluminescent strain of *Pseudomonas aeruginosa* at $t=0$. Two hours later mice were slightly anesthetized by gas to record images. Right after imaging the group on the left received a phosphate-buffered saline solution while the group on the right received a bacteriophage solution, both by intranasal instillation. Then images were recorded every 2 hr. The color scale varies from low (blue) to high (red) and represents the amount of photons emitted per second per centimeter square per stearadian ($p/s/cm^2/sr$). (See Page 1 in Color Section at the back of the book.)

those already within macrophages, demonstrating an inability of bacteriophages to enter macrophages.

3. Preventive treatments

In their studies of curative bacteriophage-based treatments of lung infection, Debarbieux *et al.* (2010) found that the half-life of bacteriophages in lungs was reasonably high, up to 24 hr. This led to the suggestion that bacteriophages could be used for preventive treatment. The authors showed that prior treatment 24 hr before infection yielded 100% survival in animals subsequently infected with a lethal dose of bacteria (Debarbieux *et al.*, 2010). In a subsequent study, this result was extended by establishing conditions (in relation to endotoxin removal from the bacteriophage solution) in which treatment 4 days before infection yielded survival rates of 90% (Morello *et al.*, 2011).

4. Conclusions

These studies concluded that there is no major inhibitory factor preventing bacteriophages from infecting bacteria within the lungs. Consequently, factors governing bacteriophage efficacy seem to be directly translated from *in vitro* to *in vivo* settings. For example, increasing the dose of bacteriophages results in more rapid bacterial killing, thereby increasing the survival rate. By increasing the efficacy of bacteriophage to kill a particular bacterial strain, it should be possible to improve its therapeutic potential (Morello *et al.*, 2011). Finally, although the lungs may appear to resemble a flask of broth in which bacteriophages encounter their bacterial hosts readily, this is currently an oversimplification, for two main reasons. The first of these reasons is the role of the immune system in controlling treatment efficacy. Bacteriophages may be the principal killer of bacteria in treatment conditions, but they are not the only player. The second reason is the poor representation of bacteriophages from different families used in these studies, precluding the generalization that all potentially therapeutic bacteriophages will be as effective as those described to date, despite their random isolation from the environment.

Similar models of lung infection in mice caused by either *K. pneumonia* or *Burkholderia cenocepacia* have also been studied. In the first case, for nonlethal infections, bacteriophages were administered intraperitoneally and caused a more rapid decrease in the number of bacteria over time than was observed in the control group. However, preventive treatment administered 24 hr before infection was found to be ineffective, whereas treatment 6 hr before infection did have a positive effect (Chhibber *et al.*, 2008). The lack of protection at 24 hr is not surprising because intraperitoneal injection led to the rapid elimination of bacteriophages. In the second case, the efficacy of bacteriophage treatment 24 hr after infection with *B. cenocepacia* was studied, with a comparison of intraperitoneal and intranasal routes of administration. Intraperitoneal administration was found to be slightly less effective than intranasal administration (Carmody *et al.*, 2010). Differences between this model and *P. aeruginosa* infection models may reflect the characteristics of the bacterium, as *B. cenocepacia* does not infect the lung in the same way as *P. aeruginosa* (Ventura *et al.*, 2009).

D. Recent advances in clinical trials on *Pseudomonas aeruginosa* ear infections

1. Chronic otitis in dogs

Two studies focusing on the treatment of chronic otitis in dogs have been published (Hawkins *et al.*, 2010; Marza *et al.*, 2006). Marza *et al.* (2006) published a case report study showing for the first time that topically administered bacteriophages multiply in dog ears, curing otitis, despite the use of a low initial dose of bacteriophages (400 PFU). The second study concerned 10 dogs with chronic otitis caused by antibiotic-resistant *P. aeruginosa* strains. Dogs were treated with a single dose of 10^5 PFU of each of six different bacteriophages applied topically. Clinical signs were scored and both bacteria and bacteriophages were counted. Bacteriophage numbers increased by a factor of 100, whereas bacterial counts decreased slightly, from 3 to 1×10^9 CFU/g. The overall general condition of the dogs improved with treatment, as a 30% decrease in clinical signs was recorded. No adverse event was noted; this single treatment was found to improve the clinical condition of the dogs in the longer term, with no further detection of *P. aeruginosa* over the following 18 months (Hawkins *et al.*, 2010).

2. Chronic otitis in humans

The only advanced clinical trial in accordance with current European regulations was performed in the United Kingdom on 24 patients with unilateral chronic otitis (Wright *et al.*, 2009). In this randomized double-blind study, patients were treated with a single dose of 6×10^5 PFU of a cocktail of six bacteriophages, administered topically. Based on the mean visual analog scale scores obtained, bacteriophage treatment improved symptoms. The decrease in mean *P. aeruginosa* counts by the last day of the trial was small, from 9 to 2×10^9 CFU/g of sample. However, the bacteriophages multiplied during treatment, and some bacteriophage-treated patients displayed improvements of up to 80% at the end of the trial, with no such marked improvement observed in the placebo group. Despite the small number of bacteriophage-treated patients displaying a marked improvement, this study clearly indicates that therapeutic bacteriophages constitute an additional option for combating chronic infections. Furthermore, no serious adverse events were reported. Similarly, a phase I clinical trial on leg ulcers also confirmed an absence of adverse effects of bacteriophage treatment (Rhoads *et al.*, 2009). It should also be noted that pioneering studies of bacteriophage treatments for human *P. aeruginosa* infections have been carried out in eastern Europe since the early 1980s, but these studies either were not detailed enough or were not published in English (Sulakvelidze *et al.*, 2001).

3. Animal models versus human treatments

Human and dog chronic otitis clinical trials can be compared, as treatments were similar, with the same dose of bacteriophages and the same procedure for their local application (Hawkins *et al.*, 2010; Wright *et al.*, 2009). Clinical signs and bacterial counts showed similar patterns of change with treatment in both studies. However, considerable variability between patients was observed in both studies. This aspect is very important in human treatment, but tends not to be picked up in experimental phage therapy studies, which use homogeneous animal populations, such as mouse lines. Experimental phage therapy studies can therefore provide proof of concept for the feasibility of a treatment, but it is not possible to anticipate the results of similar treatments in individual human patients. Consequently, because it is highly unlikely that clinical trials will report 100% efficacy, such trials should therefore focus on demonstrating an improvement in patient status rather than percentage of complete healing.

IV. PERSPECTIVES

A. Future directions in experimental phage therapy

1. Bacteriophage characterization

The bacteriophages used in studies in animal models or in clinical trials were either isolated for that purpose from environmental sources or obtained from collections. They have been characterized to various degrees and it is therefore not possible to determine whether they are all virulent bacteriophages. Further efforts should be made to characterize these bacteriophages. For example, mass spectrometry analyses of major capsid proteins could be carried out, for rapid differentiation between known and unknown bacteriophages, with full genome sequencing then carried out for the unknown bacteriophages. A more detailed molecular approach has also been described (Merabishvili *et al.*, 2007). It should also be noted that cocktails of several bacteriophages have been used to carry out past and current clinical trials in order to prevent the development of bacteriophage-resistant bacteria.

2. Chronic infection in animal models

Animal models have some limitations, but they are an essential tool for investigations in unexplored situations. Animal models were initially established to reproduce deadly infections for the identification of treatments and, subsequently, for studies of virulence factors. Moreover, major differences have been found between closely related animals. For example, fetoplacental *Listeria monocytogenes* infections can develop in gerbils but not in mice due to a single amino acid difference in a cell receptor (Disson *et al.*, 2008). As chronic infections represent an equilibrium between a bacterial infection and its control by the host immune response, it is therefore not a simple task to reproduce this equilibrium in animals with human pathogens. Nevertheless, several models of chronic infection have been developed in recent years, with the aim of more closely mimicking the lung infections occurring in cystic fibrosis patients in particular (Bragonzi, 2010). However, as mortality rates remained high in these experiments, larger numbers of animals need to be included to ensure that a large enough number will survive for at least a few weeks.

B. How far can we translate current results into the development of human phage therapy?

Results obtained in animal models support the development of human phage therapy. However, animal models are invariably designed to study one aspect of infection, whereas human patients often present a clinical history that cannot be reproduced in animals. For example, most, if not all, humans patients likely to benefit from bacteriophage treatment have probably been treated previously with antibiotics. Mimicking such situations with animal models would require long, tedious, and expensive studies. Are such studies really required for the development of human phage therapy? The answer to this question will depend in part on the regulations applied to bacteriophage treatments. Authorization procedures for clinical trials in Europe do not currently appear to be entirely compatible with the real evolving nature of bacteriophages (Verbeken *et al.*, 2007). However, this has not stood in the way of the completion of several trials, which will hopefully supply the regulators with additional data, enabling them to eventually adapt the current regulations (Pirnay *et al.*, 2011; Wright *et al.*, 2009).

ACKNOWLEDGMENT

Work in the author's laboratory is supported by the French cystic fibrosis foundation, Vaincre la Mucoviscidose, Grants IC0704 and IC1011.

REFERENCES

Asboe, D., Gant, V., Aucken, H. M., Moore, D. A., Umasankar, S., Bingham, J. S., Kaufmann, M. E., and Pitt, T. L. (1998). Persistence of *Pseudomonas aeruginosa* strains in respiratory infection in AIDS patients. *AIDS (London, England)* **12:**1771–1775.

Bleves, S., Viarre, V., Salacha, R., Michel, G. P. F., Filloux, A., and Voulhoux, R. (2010). Protein secretion systems in *Pseudomonas aeruginosa*: A wealth of pathogenic weapons. *Int. J. Med. Microbiol.* **300:**534–543.

Bonomo, R. A., and Szabo, D. (2006). Mechanisms of multidrug resistance in *Acinetobacter* species and *Pseudomonas aeruginosa*. *Clin. Infect. Dis* **43**(Suppl 2)**:**S49–S56.

Bragonzi, A. (2010). Murine models of acute and chronic lung infection with cystic fibrosis pathogens. *Int. J. Med. Microbiol.* **300:**584–593.

Brüssow, H. (2005). Phage therapy: The *Escherichia coli* experience. *Microbiol.* **151:**2133–2140.

Brüssow, H. (2007). Bacteria between protists and phages: From antipredation strategies to the evolution of pathogenicity. *Mol. Microbiol.* **65:**583–589.

Brüssow, H., Canchaya, C., and Hardt, W.-D. (2004). Phages and the evolution of bacterial pathogens: From genomic rearrangements to lysogenic conversion. *Microbiol. Mol. Biol. Rev.* **68:**560–602.

Carmody, L. A., Gill, J. J., Summer, E. J., Sajjan, U. S., Gonzalez, C. F., Young, R. F., and LiPuma, J. J. (2010). Efficacy of bacteriophage therapy in a model of *Burkholderia cenocepacia* pulmonary infection. *J. Infect. Dis.* **201:**264–271.

Ceyssens, P.-J., and Lavigne, R. (2010). Bacteriophages of Pseudomonas. *Future Microbiol.* **5:**1041–1055.

Chhibber, S., Kaur, S., and Kumari, S. (2008). Therapeutic potential of bacteriophage in treating *Klebsiella pneumoniae* B5055-mediated lobar pneumonia in mice. *J. Med. Microbiol.* **57:**1508.

Chiller, K., Selkin, B. A., and Murakawa, G. J. (2001). Skin microflora and bacterial infections of the skin. *J. Invest. Dermatol. Symp. Proc.* **6:**170–174.

Corey, M., and Farewell, V. (1996). Determinants of mortality from cystic fibrosis in Canada, 1970–1989. *Am. J. Epidemiol.* **143:**1007–1017.

Cortez, D., Forterre, P., and Gribaldo, S. (2009). A hidden reservoir of integrative elements is the major source of recently acquired foreign genes and ORFans in archaeal and bacterial genomes. *Genome Biol.* **10:**R65.

Coyne, S., Courvalin, P., and Galimand, M. (2010). Acquisition of multidrug resistance transposon Tn6061 and IS6100-mediated large chromosomal inversions in *Pseudomonas aeruginosa* clinical isolates. *Microbiology* **156:**1448–1458.

Debarbieux, L., Leduc, D., Maura, D., Morello, E., Criscuolo, A., Grossi, O., Balloy, V., and Touqui, L. (2010). Bacteriophages can treat and prevent *Pseudomonas aeruginosa* lung infections. *J. Infect. Dis.* **201:**1096–1104.

Denou, E., Bruttin, A., Barretto, C., Ngom-Bru, C., Brüssow, H., and Zuber, S. (2009). T4 phages against *Escherichia coli* diarrhea: Potential and problems. *Virology* **388:**21–30.

Deplano, A., Denis, O., Poirel, L., Hocquet, D., Nonhoff, C., Byl, B., Nordmann, P., Vincent, J. L., and Struelens, M. J. (2005). Molecular characterization of an epidemic clone of panantibiotic-resistant *Pseudomonas aeruginosa. J. Clin. Microbiol.* **43:**1198–1204.

Disson, O., Grayo, S., Huillet, E., Nikitas, G., Langa-Vives, F., Dussurget, O., Ragon, M., Le Monnier, A., Babinet, C., Cossart, P., and Lecuit, M. (2008). Conjugated action of two species-specific invasion proteins for fetoplacental listeriosis. *Nature* **455:**1114–1118.

Falagas, M. E., Koletsi, P. K., and Bliziotis, I. A. (2006). The diversity of definitions of multidrug-resistant (MDR) and pandrug-resistant (PDR) *Acinetobacter baumannii* and *Pseudomonas aeruginosa. J. Med. Microbiol.* **55:**1619–1629.

Ferroni, A., Guillemot, D., Moumile, K., Bernede, C., Le Bourgeois, M., Waernessyckle, S., Descamps, P., Sermet-Gaudelus, I., Lenoir, G., Berche, P., and Taddei, F. (2009). Effect of mutator *P. aeruginosa* on antibiotic resistance acquisition and respiratory function in cystic fibrosis. *Pediatr. Pulmonol.* **44:**820–825.

Fichtenbaum, C. J., Woeltje, K. F., and Powderly, W. G. (1994). Serious *Pseudomonas aeruginosa* infections in patients infected with human immunodeficiency virus: A case-control study. *Clin. Infect. Dis.* **19:**417–422.

Fuhrman, J. A. (1999). Marine viruses and their biogeochemical and ecological effects. *Nature* **399:**541–548.

Graber Duvernay, B. (2002). *Pseudomonas aeruginosa* et infections bronchopulmonaires. *Presse Therm. Clim.* **139:**35–40.

Hagens, S., Habel, A., Von Ahsen, U., Von Gabain, A., and Blasi, U. (2004). Therapy of experimental *Pseudomonas* infections with a nonreplicating genetically modified phage. *Antimicrob. Agents Chemother.* **48:**3817.

Hawkins, C., Harper, D., Burch, D., Anggård, E., and Soothill, J. (2010). Topical treatment of *Pseudomonas aeruginosa* otitis of dogs with a bacteriophage mixture: A before/after clinical trial. *Vet. Microbiol.* **146:**309–313.

Heiniger, R. W., Winther-Larsen, H. C., Pickles, R. J., Koomey, M., and Wolfgang, M. C. (2010). Infection of human mucosal tissue by *Pseudomonas aeruginosa* requires sequential and mutually dependent virulence factors and a novel pilus-associated adhesin. *Cell. Microbiol.* **12:**1158–1173.

Heo, Y.-J., Lee, Y.-R., Jung, H.-H., Lee, J., Ko, G., and Cho, Y.-H. (2009). Antibacterial efficacy of phages against *Pseudomonas aeruginosa* infections in mice and *Drosophila melanogaster. Antimicrob. Agents Chemother.* **53:**2469–2474.

Kiska, D. L., Gilligan, P. H., and Murray, P. (2003). Pseudomonas. Manual of Clinical Microbiology, Vol. 1, pp. 719–728. Washington, DC, American Society for Microbiology Press.

Kumari, S., Harjai, K., and Chhibber, S. (2009a). Bacteriophage treatment of burn wound infection caused by *Pseudomonas aeruginosa* PAO in BALB/c mice. *Am. J. BioMed. Sci.* **1:**385–394.

Kumari, S., Harjai, K., and Chhibber, S. (2009b). Efficacy of bacteriophage treatment in murine burn wound infection induced by *Klebsiella pneumoniae. J. Microbiol. Biotechnol.* **19:**622–628.

Kumari, S., Harjai, K., and Chhibber, S. (2010). Evidence to support the therapeutic potential of bacteriophage Kpn5 in burn wound infection caused by *Klebsiella pneumoniae* in BALB/c mice. *J. Microbiol. Biotechnol.* **20:**935–941.

Kumari, S., Harjai, K., and Chhibber, S. (2011). Bacteriophage versus antimicrobial agents for the treatment of murine burn wound infection caused by *Klebsiella pneumoniae* B5055. *J. Med. Microbiol.* **60:**205–210.

Kutateladze, M., and Adamia, R. (2010). Bacteriophages as potential new therapeutics to replace or supplement antibiotics. *Trends Biotechnol.* **28:**591–595.

Lee, D. G., Urbach, J. M., Wu, G., Liberati, N. T., Feinbaum, R. L., Miyata, S., Diggins, L. T., He, J., Saucier, M., Déziel, E., Friedman, L., Li, L., *et al.* (2006). Genomic analysis reveals that *Pseudomonas aeruginosa* virulence is combinatorial. *Genome Biol.* **7:**R90–R.

Livermore, D. M. (2002). Multiple mechanisms of antimicrobial resistance in *Pseudomonas aeruginosa:* Our worst nightmare? *Clin. Infect. Dis.* **34:**634–640.

Lyczak, J. B., Cannon, C. L., and Pier, G. B. (2000). Establishment of *Pseudomonas aeruginosa* infection: Lessons from a versatile opportunist. *Microbes Infect.* **2:**1051–1060.

Lyczak, J. B., Cannon, C. L., and Pier, G. B. (2002). Lung infections associated with cystic fibrosis. *Clin. Microbiol. Rev.* **15:**194.

Markoishvili, K., Tsitlanadze, G., Katsarava, R., Morris, J. G., Jr., and Sulakvelidze, A. (2002). A novel sustained-release matrix based on biodegradable poly(ester amide)s and impregnated with bacteriophages and an antibiotic shows promise in management of infected venous stasis ulcers and other poorly healing wounds. *Int. J. Dermatol.* **41:**453–458.

Marza, J. A. S., Soothill, J. S., Boydell, P., and Collyns, T. A. (2006). Multiplication of therapeutically administered bacteriophages in *Pseudomonas aeruginosa* infected patients. *Burns* **32:**644–646.

Mathee, K., Narasimhan, G., Valdes, C., Qiu, X., Matewish, J. M., Koehrsen, M., Rokas, A., Yandava, C. N., Engels, R., Zeng, E., Olavarietta, R., Doud, M., *et al.* (2008). Dynamics of *Pseudomonas aeruginosa* genome evolution. *Proc. Natl. Acad. Sci. USA* **105:**3100–3105.

Mayhall, C. G. (2003). The epidemiology of burn wound infections: Then and now. *Clin. Infect. Dis.* **37:**543–550.

McVay, C. S., Velasquez, M., and Fralick, J. A. (2007). Phage therapy of *Pseudomonas aeruginosa* infection in a mouse burn wound model. *Antimicrob. Agents Chemother.* **51:**1934–1938.

Meitert, E., Petrovici, M., Sima, F., Costache, G., and Savulian, C. (1987). Investigation on the therapeutical efficiency of some adapted bacteriophages in experimental infection with *Pseudomonas aeruginosa. Arch. Roumaines Pathol. Exp. Microbiol.* **46:**17–26.

Merabishvili, M., Verhelst, R., Glonti, T., Chanishvili, N., Krylov, V., Cuvelier, C., Tediashvili, M., and Vaneechoutte, M. (2007). Digitized fluorescent RFLP analysis (fRFLP) as a universal method for comparing genomes of culturable dsDNA viruses: Application to bacteriophages. *Res. Microbiol.* **158:**572–581.

Morello, E., Saussereau, E., Maura, D., Huerre, M., Touqui, L., and Debarbieux, L. (2011). Pulmonary bacteriophage therapy on *Pseudomonas aeruginosa* cystic fibrosis strains: First steps towards treatment and prevention. *PloS One* **6:**e16963.

Morrison, A. J., Jr., and Wenzel, R. P. (1984). Epidemiology of infections due to *Pseudomonas aeruginosa. Rev. Infect. Dis.* **6**(Suppl. 3)**:**S627–S642.

Nordmann, P., Ronco, E., Naas, T., Duport, C., Michel-Briand, Y., and Labia, R. (1993). Characterization of a novel extended-spectrum beta-lactamase from *Pseudomonas aeruginosa. Antimicrob. Agents Chemother.* **37:**962.

Pier, G. B., and Ramphal, R. (2009). Pseudomonas aeruginosa. *In* "Mandell, Douglas, and Bennett's Principles and Practice of Infectious Diseases" (G. L. Mandell, ed.), 7th Ed, Vol. 2, pp. 2835–2860. Edinburgh, Churchill Livingstone.

Pirnay, J.-P., Bilocq, F., Pot, B., Cornelis, P., Zizi, M., Van Eldere, J., Deschaght, P., Vaneechoutte, M., Jennes, S., Pitt, T., and De Vos, D. (2009). *Pseudomonas aeruginosa* population structure revisited. *PloS One* **4:**e7740.

Pirnay, J. P., De Vos, D., Verbeken, G., Merabishvili, M., Chanishvili, N., Vaneechoutte, M., Zizi, M., Laire, G., Lavigne, R., Huys, I., Vanden Mooter, G., Buckling, A., *et al.* (2011). The phage therapy paradigm: Prêt-à-porter or sur-mesure? *Pharm. Res.*1–4.

Poirel, L., Carrër, A., Pitout, J. D., and Nordmann, P. (2009). Integron mobilization unit as a source of mobility of antibiotic resistance genes. *Antimicrob. Agents Chemother.* **53:**2492–2498.

Poole, K. (2005). Aminoglycoside resistance in *Pseudomonas aeruginosa. Antimicrob. Agents Chemother.* **49:**479–487.

Ramphal, R., Balloy, V., Huerre, M., Si-Tahar, M., and Chignard, M. (2005). TLRs 2 and 4 are not involved in hypersusceptibility to acute *Pseudomonas aeruginosa* lung infections. *J. Immunol.* **175:**3927–3934.

Ramphal, R., Balloy, V., Jyot, J., Verma, A., Si-Tahar, M., and Chignard, M. (2008). Control of *Pseudomonas aeruginosa* in the lung requires the recognition of either lipopolysaccharide or flagellin. *J. Immunol.* **181:**586–592.

Randall, T., and Irvin, R. T. (2008). Adherence of *Pseudomonas aeruginosa. In* "Pseudomonas: Model Organism, Pathogen, Cell Factory" (B. Rehm, ed.), pp. 45–84. Wiley-VCH.

Rhoads, D. D., Wolcott, R. D., Kuskowski, M. A., Wolcott, B. M., Ward, L. S., and Sulakvelidze, A. (2009). Bacteriophage therapy of venous leg ulcers in humans: Results of a phase I safety trial. *J. Wound Care* **18**(237–238)**:**240–243.

Rice, S. A., Tan, C. H., Mikkelsen, P. J., Kung, V., Woo, J., Tay, M., Hauser, A., McDougald, D., Webb, J. S., and Kjelleberg, S. (2008). The biofilm life cycle and virulence of *Pseudomonas aeruginosa* are dependent on a filamentous prophage. *ISME J.* **3:**271–282.

Roy, P. H., Tetu, S. G., Larouche, A., Elbourne, L., Tremblay, S., Ren, Q., Dodson, R., Harkins, D., Shay, R., Watkins, K., Mahamoud, Y., and Paulsen, I. T. (2010). Complete genome sequence of the multiresistant taxonomic outlier *Pseudomonas aeruginosa* PA7. *PLoS One* **5:**e8842.

Salzberg, S. L., Sommer, D. D., Puiu, D., and Lee, V. T. (2008). Gene-boosted assembly of a novel bacterial genome from very short reads. *PLoS Comp. Biol.* **4:**e1000186.

Soothill, J. S. (1992). Treatment of experimental infections of mice with bacteriophages. *J. Med. Microbiol.* **37:**258–261.

Stewart, R. M. K., Wiehlmann, L., Ashelford, K. E., Preston, S. J., Frimmersdorf, E., Campbell, B. J., Neal, T. J., Hall, N., Tuft, S., Kaye, S. B., and Winstanley, C. (2011). Genetic characterization indicates that a specific subpopulation of *Pseudomonas aeruginosa* is associated with keratitis infections. *J. Clin. Microbiol.* **49:**993–1003.

Stover, C. K., Pham, X. Q., Erwin, A. L., Mizoguchi, S. D., Warrener, P., Hickey, M. J., Brinkman, F. S., Hufnagle, W. O., Kowalik, D. J., Lagrou, M., Garber, R. L., Goltry, L., *et al.* (2000). Complete genome sequence of *Pseudomonas aeruginosa* PAO1, an opportunistic pathogen. *Nature* **406:**959–964.

Sulakvelidze, A., Alavidze, Z., and Morris, J. G., Jr. (2001). *Bacteriophage therapy. Antimicrob. Agents Chemother.* **45:**649.

Thiel, K. (2004). Old dogma, new tricks: 21st century phage therapy. *Nat. Biotechnol.* **22:**31–36.

Tinsley, C. R., Bille, E., and Nassif, X. (2006). Bacteriophages and pathogenicity: More than just providing a toxin? *Microbes Infect.* **8:**1365–1371.

Vaca-Pacheco, S., Paniagua-Contreras, G. L., García-González, O., and de la Garza, M. (1999). The clinically isolated FIZ15 bacteriophage causes lysogenic conversion in *Pseudomonas aeruginosa* PAO1. *Curr. Microbiol.* **38:**239–243.

Vento, S., Cainelli, F., and Temesgen, Z. (2008). Lung infections after cancer chemotherapy. *Lancet Oncol.* **9:**982–992.

Ventura, G. M., de, C., Balloy, V., Ramphal, R., Khun, H., Huerre, M., Ryffel, B., Plotkowski, M. C. M., Chignard, M., and Si-Tahar, M. (2009). Lack of MyD88 protects the immunodeficient host against fatal lung inflammation triggered by the opportunistic bacteria *Burkholderia cenocepacia. J. Immunol* **183:**670–676.

Verbeken, G., De Vos, D., Vaneechoutte, M., Merabishvili, M., Zizi, M., and Pirnay, J.-P. (2007). European regulatory conundrum of phage therapy. *Future Microbiol.* **2:**485–491.

Vinodkumar, C., Kalsurmath, S., and Neelagund, Y. (2008). Utility of lytic bacteriophage in the treatment of multidrug-resistant *Pseudomonas aeruginosa* septicemia in mice. *Indian J. Pathol. Microbiol.* **51:**360.

Wang, J., Hu, B., Xu, M., Yan, Q., Liu, S., Zhu, X., Sun, Z., Reed, E., Ding, L., Gong, J., Li, Q. Q., and Hu, J. (2006). Use of bacteriophage in the treatment of experimental animal bacteremia from imipenem-resistant *Pseudomonas aeruginosa. Int. J. Mol. Med.* **17:**309.

Watanabe, R., Matsumoto, T., Sano, G., Ishii, Y., Tateda, K., Sumiyama, Y., Uchiyama, J., Sakurai, S., Matsuzaki, S., Imai, S., and Yamaguchi, K. (2006). Efficacy of bacteriophage therapy against gut-derived sepsis caused by *Pseudomonas aeruginosa* in mice. *Antimicrob. Agents Chemother.* **51:**446–452.

Whiteley, M., Bangera, M. G., Bumgarner, R. E., Parsek, M. R., Teitzel, G. M., Lory, S., and Greenberg, E. P. (2001). Gene expression in *Pseudomonas aeruginosa* biofilms. *Nature* **413:**860–864.

Wibbenmeyer, L., Danks, R., Faucher, L., Amelon, M., Latenser, B., Kealey, G. P., and Herwaldt, L. A. (2006). Prospective analysis of nosocomial infection rates, antibiotic use, and patterns of resistance in a burn population. *J. Burn Care Res.* **27:**152–160.

Winstanley, C., Langille, M. G. I., Fothergill, J. L., Kukavica-Ibrulj, I., Paradis-Bleau, C., Sanschagrin, F., Thomson, N. R., Winsor, G. L., Quail, M. A., Lennard, N., Bignell, A., Clarke, L., *et al.* (2009). Newly introduced genomic prophage islands are critical determinants of *in vivo* competitiveness in the Liverpool epidemic strain of *Pseudomonas aeruginosa. Genome Res.* **19:**12–23.

Wisplinghoff, H., Bischoff, T., Tallent, S. M., Seifert, H., Wenzel, R. P., and Edmond, M. B. (2004). Nosocomial bloodstream infections in US hospitals: Analysis of 24,179 cases from a prospective nationwide surveillance study. *Clin. Infect. Dis.* **39:**309–317.

Wright, A., Hawkins, C. H., Anggård, E. E., and Harper, D. R. (2009). A controlled clinical trial of a therapeutic bacteriophage preparation in chronic otitis due to antibiotic-resistant *Pseudomonas aeruginosa*; a preliminary report of efficacy. *Clin. Otolaryngol.* **34:**349–357.

Zavascki, A. P., Carvalhaes, C. G., Picão, R. C., and Gales, A. C. (2010). Multidrug-resistant *Pseudomonas aeruginosa* and *Acinetobacter baumannii*: Resistance mechanisms and implications for therapy. *Expert Rev. Anti-Infective Ther.* **8:**71–93.

CHAPTER 5

Genomics of Staphylococcal Twort-like Phages - Potential Therapeutics of the Post-Antibiotic Era

Małgorzata Łobocka,[*,‡,1] **Monika S. Hejnowicz,**[*] **Kamil Dąbrowski,**[*,‡] **Agnieszka Gozdek,**[*] **Jarosław Kosakowski,**[*] **Magdalena Witkowska,**[*] **Magdalena I. Ulatowska,**[*,§] **Beata Weber-Dąbrowska,**[§] **Magdalena Kwiatek,**[¶] **Sylwia Parasion,**[¶] **Jan Gawor,**[†] **Helena Kosowska,**[†] and **Aleksandra Głowacka**[*,‡]

Contents

[*] Department of Microbial Biochemistry, Institute of Biochemistry and Biophysics, Polish Academy of Sciences, Warsaw, Poland
[†] Laboratory of DNA Sequencing and Oligonucleotide Synthesis, Institute of Biochemistry and Biophysics, Polish Academy of Sciences, Warsaw, Poland
[‡] Autonomous Department of Microbial Biology, Faculty of Agriculture and Biology, Warsaw University of Life Sciences, Warsaw, Poland
[§] Laboratory of Bacteriophages, Hirszfeld Institute of Immunology and Experimental Therapy, Polish Academy of Sciences, Wrocław, Poland
[¶] Military Institute of Hygiene and Epidemiology, Puławy, Poland
[1] Corresponding author, E-mail: lobocka@ibb.waw.pl; malgorzata_lobocka@sggw.pl

Advances in Virus Research, Volume 83
ISSN 0065-3527, DOI: 10.1016/B978-0-12-394438-2.00005-0

Abstract

Polyvalent bacteriophages of the genus Twort-like that infect clinically relevant *Staphylococcus* strains may be among the most promising phages with potential therapeutic applications. They are obligatorily lytic, infect the majority of *Staphylococcus* strains in clinical strain collections, propagate efficiently and do not transfer foreign DNA by transduction. Comparative genomic analysis of 11 *S. aureus/S. epidermidis* Twort-like phages, as presented in this chapter, emphasizes their strikingly high similarity and clear divergence from phage Twort of the same genus, which might have evolved in hosts of a different species group. Genetically, these phages form a relatively isolated group, which minimizes the risk of acquiring potentially harmful genes. The order of genes in core parts of their 127 to 140-kb genomes is conserved and resembles that found in related representatives of the *Spounavirinae* subfamily of myoviruses. Functions of certain conserved genes can be predicted based on their homology to prototypical genes of model spounavirus SPO1. Deletions in the genomes of certain phages mark genes that are dispensable for phage development. Nearly half of the genes of these phages have no known homologues. Unique genes are mostly located near termini of the virion DNA molecule and are expressed early in phage development as implied by analysis of their potential transcriptional signals. Thus, many of them are likely to play a role in host takeover. Single genes encode homologues of bacterial virulence-associated proteins. They were apparently acquired by a common ancestor of these phages by horizontal gene transfer but presumably evolved towards gaining functions that increase phage infectivity for bacteria or facilitate mature phage release. Major differences between the genomes of *S. aureus/S. epidermidis* Twort-like phages consist of single nucleotide polymorphisms and insertions/deletions of short stretches of nucleotides, single genes, or introns of group I. Although the number and location of introns may vary between particular phages, intron shuffling is unlikely to be a major factor responsible for specificity differences.

I. INTRODUCTION

The *Staphylococcus* genus groups several species of Gram-positive bacteria that inhabit human or animal organisms. Many of these species, especially *Staphylococcus aureus*, include highly pathogenic strains that can cause infections, which are manifested by various symptoms ranging from relatively mild to life threatening (reviewed by Tenover and Gorwitz, 2006). The treatment of such infections is becoming increasingly problematic due to the dissemination of methicillin-resistant *S. aureus*, as well as the emergence of vancomycin-resistant strains (Arias and Murray, 2009; Chambers and DeLeo, 2009; Howden *et al.*, 2010). Naturally occurring, virulent bacteriophages that infect and kill a wide range of staphylococcal strains may become an alternative in the treatment of otherwise incurable infections caused by antibiotic-resistant staphylococci (Borysowski *et al.*, 2011; Mann, 2008; Merabishvili *et al.*, 2009). A few of them were shown to be effective in the treatment of staphylococcal infections in animals or are used in phage therapy including clinical trials in humans (Capparelli *et al.*, 2007; Gill *et al.*, 2006; Gupta and Prasad, 2011; Jikia *et al.*, 2005; Kvachadze *et al.*, 2011; Markoishvili *et al.*, 2002; Merabishvili *et al.*, 2009; Paul *et al.*, 2011a; O'Flaherty *et al.*, 2005b; Rhoads *et al.*, 2009; Sulakvelidze *et al.*, 2001; Sulakvelidze and Kutter, 2005; Sunagar *et al.*, 2010; Wills *et al.*, 2005; Zimecki *et al.*, 2008, 2009, 2010). Additionally, proteins of these phages that are lethal to *Staphylococci* may serve as prototypes of antibacterials, mark potential drug targets in staphylococcal cells, or can be used as anti-staphylococcal agents by themselves (Fischetti, 2008; Gu *et al.*, 2011; Liu *et al.*, 2004; Pastagia *et al.*, 2011; Projan 2004; Rashel *et al.*, 2007). Phages of the most promising group from a therapeutic point of view belong to the Twort-like genus (Klumpp *et al.*, 2010). Although their history dates as far back as the history of phages, with their first representative, phage Twort, believed to have been isolated in 1915 (Lavigne *et al.*, 2009; Twort, 1915), little is yet known about the biology of these phages. This chapter provides the genome-wide comparison of 11 staphylococcal Twort-like phages, which are suitable for therapeutic applications. Some genomic features of these phages which appear to be responsible for their certain properties and wide strain specificity are also discussed.

II. STAPHYLOCOCCAL BACTERIOPHAGES: A SHORT OVERVIEW

Early interest in staphylococcal phages was stimulated by their potential therapeutic applications (reviewed by Alisky *et al.*, 1998; Sulakvelidze and Kutter, 2005; Sulakvelidze *et al.*, 2001). However, after the introduction of antibiotics, it was motivated mainly by the wide application of phages to differentiate clinical staphylococcal strains. The method of phage typing of staphylococci derived from this early work and originally developed by Wilson and Atkinson (1945) has been used, verified, and improved for decades (e.g., Aucken and Westwell, 2002; Blair and Williams, 1961; Davidson, 1972; Heczko *et al.*, 1977; Lundholm and Bergendahl, 1988; Marples and Rosdahl, 1997; Piechowicz *et al.*, 1999; Richardson *et al.*, 1999; Smith, 1948a; Wegener, 1993; Wildemauwe *et al.*, 2004; for a review of old literature, see Anderson and Williams, 1956; Wentworth, 1963). It has not been entirely abandoned even nowadays, when molecular typing methods have found a common use and appear superior for the differentiation of most strains (e.g., Andrasevic *et al.*, 1999; Emberger *et al.*, 2011; Grundmann *et al.*, 2002; Mehndiratta *et al.*, 2010; Piechowicz *et al.*, 2010; Shouval *et al.*, 2011; Tenover *et al.*, 1993; Wildemauwe *et al.*, 2010). Since the introduction of the routine bacteriophage set to type *S. aureus* strains by the Colindale laboratory (Williams and Rippon, 1952), other standard collections have also been established in different laboratories; these basic collections have often been supplemented with local isolates, resulting in hundreds of staphylococcal phages isolated around the world (e.g., Akatov *et al.*, 1991; Chistovich, 1960; Dmitrienko *et al.*, 2003; Dua *et al.*, 1982; Gershman *et al.*, 1988; Gibbs *et al.*, 1978; Holmberg, 1978; Kawano *et al.*, 1982; Martin-de-Nicolas *et al.*, 1990; Parisi *et al.*, 1978; Petrushina, 1975; Pillich *et al.*, 1978; Shimizu, 1977; Skahan and Parisi, 1977; Talbot and Parisi, 1976; Verhoef *et al.*, 1971; Vindel *et al.*, 1994; Wang, 1978; Zueva *et al.*, 1994, 1997; for a review of earlier literature see Wentworth, 1963).

Early studies with rabbit antisera led to the grouping of staphylococcal phages into 11 serological types (A, B, C, D, E, F, G, H, J, K, and L) (Blair and Williams, 1961; Doskar *et al.*, 2000; Rippon 1952, 1956; Rountree, 1949a). Phages that belong to a given serological type share many other features, including similar morphology, genome size and organization, number and size of major virion polypeptides, stability, and the ability to form lysogens or lyse strains of different origins (Ackermann, 1998; Ackermann and DuBow, 1987; Krzywy *et al.*, 1981; Kwan *et al.*, 2005; Lee and Stewart, 1985; Pariza and Iandolo, 1974; Rosenblum and Tyrone, 1964). The latter abilities became the basis for the division of staphylococcal phages into lytic groups. Temperate phages are the majority. A summary of their most important properties can be found in Kahankova *et al.* (2010). Those that belong to serological groups A, B, C, F, and L have been studied the most intensively. However, they are considered unsuitable for therapeutic applications due to safety concerns. An infection with these phages can either lead to phage development and cell lysis or initiate lysogeny (reviewed in Guttman *et al.*, 2005). Known staphylococcal temperate phages lysogenize cells by integration of their DNA (a prophage) into a certain attachment site in a host chromosome. Lysogeny appears to be common in staphylococci. Nearly all tested *S. aureus* isolates are lysogens, carrying from one to five prophages (Goerke *et al.*, 2009; Pantucek *et al.*, 2004; Rountree, 1949b). The prophage remains a part of the host chromosome as long as the phage repressor protein blocks transcription of its lytic genes and hence the initiation of lytic development. Additionally, the presence of phage repressor in lysogens makes them resistant to the infection with related phages. What is more, staphylococcal prophages can inactivate host genes for certain virulence factors, such as e.g. hemolysin or lipase, by integration at their attachment sites within those genes (Carroll *et al.*, 1993; Coleman *et al.*, 1991; Lee and Iandolo, 1986). However, as the evolutionary success of temperate phages depends on the selective advantage of lysogens over non-lysogens, temperate phages carry genes that encode bacterial adaptive functions, including toxins and other virulence determinants (Boyd and Brüssow, 2002; Brüssow *et al.*, 2004; Gill *et al.*, 2011). Genes for enterotoxin A and exfoliative toxin, Panton-Valentine leukocidin, and immune evasion factors are only some examples (Betley and Mekalanos, 1985; Coleman *et al.*, 1989; Kaneko *et al.*, 1998; van Wamel *et al.*, 2006; Yamaguchi *et al.*, 2000). Additionally, certain temperate phages can serve as carriers for staphylococcal pathogenicity islands and participate in the spread of the latter (reviewed by Novick *et al.*, 2010). Thus, despite successful attempts to use certain temperate phages for curing staphylococcal infections in animals (Matsuzaki *et al.*, 2003), the use of such phages for therapeutic purposes could potentially pose a risk of increasing the pathogenicity of the infecting bacterial strain.

Staphylococcal therapeutic phages of choice are obligatorily lytic and infect coagulase-positive staphylococci. They belong to serological groups D, G, and H. Several of their representatives are polyvalent and can lyse strains of human as well as bovine origin (Rippon, 1956). However, phages of lytic group H appear unstable and difficult to propagate, and hence only phages of groups D and G are potentially useful for therapeutic applications.

All of over 215 staphylococcal phages of different lytic groups that have been characterized morphologically belong to the order *Caudovirales*—phages of an icosahedral head, which contains linear, double-stranded DNA, and is attached by one vertex to a tube-like tail (Ackermann, 2007). Genomic sequences of only about 50 of them have been determined (Ackermann and Kropinsky, 2007 and references therein; Bae *et al.*, 2006; Christie *et al.*, 2010; García *et al.*, 2009; Hoshiba *et al.*, 2010; Lee *et al.*, 2011; Ma *et al.*, 2008; Narita *et al.*, 2001; Son *et al.*, 2010; see elsewhere in this chapter). They fall into three discrete size categories ($\approx$130, $\approx$40, and $\approx$18 kb) (Kwan *et al.*, 2005). Phages of medium-size genomes are a majority. They form the most diversified group, belong to the *Siphoviridae* family (phages with a long, noncontractile tail) and are temperate. Obligatorily lytic phages of serological groups D and G belong to those of the largest and the smallest genomes, respectively. They are representatives of the *Myoviridae* and *Podoviridae* families of tailed phages. The former have a long contractile tail, whereas the latter have a short noncontractile tail. Only one temperate phage of known genomic sequence appears to have a small genome (Kwan *et al.*, 2005). However, it is not related to obligatorily virulent phages of similar genome size.

Several features, in addition to the obligatory lytic lifestyle and polyvalency, make staphylococcal myo- and podoviruses of D and G serogroups suitable for therapeutic purposes. Homologies of these phages at the DNA level are mostly to phages of the same lineages, minimizing the risk of genetic exchange with prophages and bacterial chromosomes (Kwan *et al.*, 2005; see elsewhere in this chapter). The participation of these phages in horizontal gene transfer between bacteria is additionally limited by their inability to form transducing particles—virions that contain fragments of host DNA and could transfer such DNA to a new host (Anderson and Bodley, 1990; Bjornsti *et al.*, 1982; Klumpp *et al.*, 2008; Valpuesta *et al.*, 1993; Zimmer *et al.*, 2003; authors' unpublished results). This property is implied by the requirement of specific DNA sequences for phage DNA packaging and, in the case of serotype D phages, by the degradation of bacterial DNA at early stages of cell infection (Rees and Fry, 1981a). A fast and efficient adsorption of these phages to host cells, their short latent period as well as their ability to penetrate biofilms, contribute to their high therapeutic potential, whereas the optimized codon usage profiles in protein-coding genes may facilitate their fast propagation (Cerca *et al.*, 2007; Golec *et al.*, 2011; Pantucek *et al.*, 1998; Sau *et al.*, 2005; Son *et al.*, 2010; Vandersteegen *et al.*, 2011).

Staphylococcal podoviruses that are obligatorily lytic are rare in bacteriophage collections. Their host range is wide, but is limited to strains of coagulase-positive staphylococci (Rippon, 1956; Son *et al.*, 2010). The genomes of described isolates (44AHJD, P68, 66, and SAP-2) range in size between 16.7 and 18.2 kb and have highly homologous nucleotide sequences (Kwan *et al.*, 2005; Son *et al.*, 2010; Vybiral *et al.*, 2003). Phylogenetically, all these phages have been classified as a separate genus, AHJD-like, within the *Picovirinae* subfamily (Lavigne *et al.*, 2008). Their virion DNA molecules end with a few hundred base pair terminal repeats and encode 20–29 proteins. One of these proteins, endolysin of the SAP-2 phage, was shown to have the ability to digest the walls of various *Staphylococcus* species cells (Son *et al.*, 2010). Functions of only 11 other proteins can be predicted based on homologies of their amino acid sequences to sequences of functionally characterized proteins (Kwan *et al.*, 2005; Vybiral *et al.*, 2003). In the majority of cases, the homologies concern proteins of *Bacillus* phage ϕ29, a model representative of the *Picovirinae* subfamily, which belongs to a different genus, the ϕ29-like phages (Meijer *et al.*, 2001). By the extrapolation of relevant ϕ29 properties, one can predict that certain morphogenetic functions of AHJD-like phages and ϕ29 are similar, as well as that the replication of AHJD-like phages like that of ϕ29 is protein primed and relies on the phage type B DNA polymerase (an equivalent of *Escherichia coli* Pol II) (Meijer *et al.*, 2001; Vybiral *et al.*, 2003). Results of the authors' studies to be published elsewhere indicate a therapeutic efficacy of staphylococcal AHJD-like phages in an invertebrate model of *S. aureus* infection. Whether these phages can be effective in therapies of staphylococcal infections in other animals and in humans remains to be verified.

Polyvalent staphylococcal phages of serotype D, which belong to *Myoviridae*, have the widest host range with respect to staphylococcal strains and have been studied intensively. They infect the majority of *S. aureus* strains in different clinical strain collections and are also infective for several strains of coagulase-negative *Staphylococci* (Hsieh *et al.*, 2011; Kvachadze *et al.*, 2011; Kwiatek *et al.*, 2012; O'Flaherty *et al.*, 2005b; Pantucek *et al.*, 1998, Rippon, 1956; Rountree, 1949a; Synnot *et al.*, 2009). Over 30 staphylococcal myoviruses have been described (for a summary, see Hsieh *et al.*, 2011; Klumpp *et al.*, 2010 and references therein; Kwiatek *et al.*, 2012; Pantucek *et al.*, 1998; Synnott *et al.*, 2009). Six of them (K, G1, A5W, ISP, Sb-1, and Twort) have been characterized previously at the level of genomic sequence (Kvachadze *et al.*, 2011; Kwan *et al.*, 2005; O'Flaherty *et al.*, 2004; Ulatowska *et al.*, in prepatation; GenBank Acc. No. EU418428; Vandersteegen *et al.*, 2011). The genomic sequences of anadditional six (Staph1N, Fi200W, 676Ż, P4W, A3R, and MSA6) have been determined by the authors and are described in this chapter (Table I). All these phages have been studied with respect to potential therapeutic applications or have been used in phage therapy in humans (Dąbrowska *et al.*, 2010; Kvachadze *et al.*, 2011; Kwan *et al.*, 2005; Kwiatek *et al.*, 2012; Merabishvili *et al.*, 2009; Międzybrodzki *et al.*, 2012; O'Flaherty *et al.*, 2005b; Zimecki *et al.*, 2008, 2010). Some of them have also appeared effective therapeutically in the invertebrate staphylococcal infection model (authors' unpublished results). They have highly homologous DNA sequences and belong to the Twort-like genus of the *Spounavirinae* subfamily of myoviruses (Klumpp *et al.*, 2010; Lavigne *et al.*, 2009; see elsewhere in this chapter). Representatives of this subfamily have large genomes (127–140 kb), are strictly virulent, infect Gram-positive, low GC content bacteria, have a wide host range among strains of infected bacterial genus, and share considerable homologies at the level of protein sequences (Table I; Carlton *et al.*, 2005; Chibani-Chennoufi *et al.*, 2004; Kilcher *et al.*, 2010; Klumpp *et al.*, 2008; Uchiyama *et al.*, 2008; reviewed by Klumpp *et al.*, 2010). Their best-known member is the *Bacillus subtilis* phage SPO1, which has been classified to the SPO1-like genus of *Spounavirinae* (Lavigne *et al.*, 2009; for a review, see Stewart *et al.*, 2009a). A description, at the genomic level, of Twort-like *Spounavirinae* representatives that infect clinically

TABLE I Sequenced genomes of *Staphylococcus* Twort-like phages

Bacteriophage[a]	Genome size (bp)	LTR size (bp)[b]	%GC	Number of ORFs[c]	Number of tRNA genes	Number of introns[d]	GenBank Acc. No.	Source or reference
A5W	137,087	8455	30.47	214	4	3	EU418428	Ulatowska *et al.*, in preparation; this work
Staph1N	137,192	8455	30.46	214	4	3	JX080300	This work
Fi200W	140,079	8402	30.44	216	4	5+1*	JX080303	This work
P4W	139,173	8417	30.44	214	4	5+1*	JX080305	This work
676Ż	140,115	8447	30.45	216	4	5+1*	JX080302	This work
A3R	132,712	8306	30.53	196	4	5+1*	JX080301	This work
MSA6	140,194	8049	30.32	214	4	5	JX080304	This work
G1	138,715	ND (8432)	30.39	215	4	3	AY954969	Kwan *et al.*, 2005
ISP	138,339	ND (8422)	30.42	217	4	3	FR852584	Vandersteegen *et al.*, 2011
K	127,395	ND (5596)	30.60	184	4	3	AY176327	O'Flaherty *et al.*, 2004
Sb-1	127,188	ND (3959)	30.48	183	4	3	HQ163896	Kvachadze *et al.*, 2011
Twort	130,706	ND	30.26	196	1	7	NC_007021	Kwan *et al.*, 2005

[a] Bacteriophages Staph1N, Fi200W, P4W, 676Ż, and A3R are derived from the therapeutic phage collection of the Ludwik Hirszfeld Institute of Immunology and Experimental Therapy of the Polish Academy of Sciences, Wrocław, Poland (Międzybrodzki *et al.*, 2012). Phages A3R and Fi200W are descendants of phages A3 and Fi200, which were obtained from Gerhard Pulverer, Institute of Hygiene, University of Cologne, Germany, in 1986, and subsequently adapted to therapeutic applications by several rounds of passaging through certain clinical *S. aureus* strains, to broaden their range of specificity. Phage P4W corresponds to a clone of phage P4 that is a descendant of fi131-a phage originally isolated as infecting *S. epidermidis* and obtained from Hans Lodenkemper, Endoklinik, Hamburg, Germany in 1979. Phage 676Ż is a descendant of phage 676 obtained from Stefan Kryński, Department of Microbiology, Medical Academy, Gdańsk, Poland in 1978. P4W and 676Ż were adapted for therapeutic applications similarly as A3R and Fi200W. Phage Staph1N was obtained from a contaminated culture of PS80 strain. It is possibly a descendant of A5W, which acquired multiple point mutations. Phages that served as a source of DNA for sequencing were propagated in cells of the following host strains: PS80 (Staph1N), 6409 (Fi200W, P4W), R19930 (A3R), and Ż11788 (676Ż). The former two strains are standard strains for phage propagation, whereas the latter two strains are clinical *S. aureus* isolates. Preparation of DNA was as described elsewhere (Ulatowska *et al.*, in preparation). Pyrosequencing reads from Roche 454 Genome Sequencer GS FLX (with a 44-93-fold coverage), obtained at the Laboratory of Sequencing and Oligonucleotide Synthesis of the Institute of Biochemistry and Biophysics, PAS, Warsaw, Poland, served as a source of data for the assembly of crude versions of phage genomic sequences with the use of the Lasergene 9.1 program. Sequences of repeats and ambiguous sites were determined additionally by direct sequencing with specific primers using the ABI PRISM dye terminator cycle sequencing ready reaction kit with ApliTaq DNA polymerase, FS (Perkin Elmer), and ABI 377 automated sequencers (Perkin Elmer) for the separation of reaction products and reading the sequencing results. Physical maps of virion DNA to verify the correctness of sequence assembly were prepared using restriction enzymes (*Ava*I, *Ava*II, *Nde*I, *Nsb*I, *Kpn*I, *Pfl*MI, *Pst*I, *Sac*I, *Sma*I, and *Xho*I) by analysis of single or double digests.

[b] Sequences of terminal repeat regions of Staph1N, Fi200W, P4W, 676Ż, A3R, and MSA6 virion DNA molecules were determined as described elsewhere (Ulatowska *et al.*, in preparation). Terminal ≈9- to 12-kb restriction fragments of virion DNA molecules, identified by Bal31 nuclease digestion, separated by electrophoresis in 0.7% agarose gel from other fragments of virion DNA, and isolated from the gel, were used as templates for sequencing. Numbers in parentheses indicate the size of DNA regions (in bp) homologous to regions of known LTRs in other phages. ND, not described.

[c] Calculation of ORF number in the genomes of all phages, except phage Twort, is based on data in Table II. Calculation of ORF number in intron-containing regions was performed by the addition of intron-encoded ORFs and ORFs obtained after *in silico* splicing of relevant sequences.

[d] Asterisks indicate an intron-like sequence immediately downstream of protein-coding gene (*tmnA*).

relevant *Staphylococcus* strains, is the subject of the remaining part of this chapter.

III. TWORT-LIKE MYOVIRUSES OF *STAPHYLOCOCCUS*

A. Morphology

Morphologically, staphylococcal Twort-like phages resemble phages of the same genus that infect *Listeria monocytogenes* (A511, P100) and differ from *Bacillus* phage SPO1 by having longer tails (from 175–176 nm in ISP and 676Ż to 217–219 nm in K and P4W/Fi200W versus 140 nm in SPO1), which are also slightly more slender (Fig. 1; Duda *et al.*, 2006; Hotchin, 1954; Jarvis *et al.*, 1993; Klumpp *et al.*, 2008, 2010, and references therein; Kwiatek *et al.*, 2012; Ulatowska *et al.*, in preparation). However, other parts of their virions look alike. In both genera, phage heads are isometric and icosahedrons. Their diameters were estimated to be in the range of about 84.5 nm for SPO1 (with a maximum vertex-to-vertex diameter of 108 nm) and 90 nm for phage Twort. Whether slight differences between virion sizes of different Twort-like phages measured by different authors or on different preparations may reflect real differences is unclear. They may equally result from staining-related deformations of phage particles and differences in measurement techniques or microscope calibration, as was summarized by Ackermann (2009).

In SPO1 and Twort-like phages, the head is connected to a tail by a hollow neck. A characteristic feature of SPO1 and Twort-like phages is a baseplate at the end of the tail. On high-resolution electron micrographs of several of these phages, it was shown to undergo major structural rearrangements during tail contraction (Fig. 1; for a summary, see Klumpp *et al.*, 2008, 2010; Parker and Eiserling, 1983a). A slightly amorphous single layer plate with sixfold symmetry and protruding flexible fibers transforms into a double-layered plate that remains attached to the end of the contracted sheath and resembles the lacy ballet dancer's tutu. Each of its layers looks like a thin ring or centrally perforated disk that surrounds a significantly narrower sheath and is perpendicular to it. The diameter of the upper ring seems to be slightly smaller than that of the lower ring. Both rings are interconnected by a few thin vertical connectors, which extend further down from the bottom ring of the baseplate and form tail spikes. As a result of tail sheath contraction and the accompanying upward movement of the baseplate, the bottom part of the tail tube is exposed and can penetrate cell envelopes.

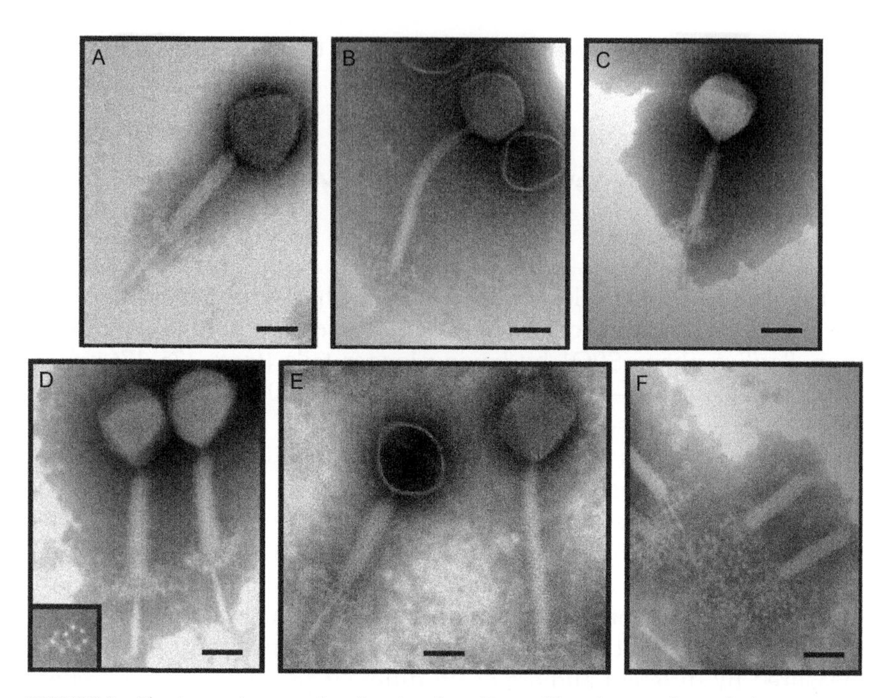

FIGURE 1 Electron micrographs of polyvalent Twort-like phages of *Staphylococcus*: (A) Staph1N, (B) P4W, (C) 676Ż, (D) A3R, and (E and F) Fi200W. (D, insert) A symmetric arrangement of six spikes in an upturned A3R baseplate. More baseplates are seen along the contracted tails of Fi200W (F). (B and E) Dark viral capsids are devoid of DNA that was apparently ejected during phage preparation. (E) The full phage head shows serrated edges indicating the presence of capsomers. Prior to electron microscopy, phages were washed twice with 0.1 *M* ammonium acetate (pH 7.2). A drop of phage suspension was applied to a formvar carbon coated copper grid and left for 5 min. Excess liquid was removed from the grid and phages were stained with 2% sodium phosphotungstate, as described elsewhere (Ulatowska *et al.*, in preparation). Grids were examined in a JEM 1400 (JEOL Co., Japan, 2008) transmission electron microscope equipped with a high-resolution digital camera (CCD MORADA, SiS-Olympus, Germany). Observations were performed in the Laboratory of Electron Microscopy at the Nencki Institute of Experimental Biology, Warsaw, Poland. Each scale bar represents 50 nm. Average head diameters and tail lengths were 95 and 215 for A3R, 89 and 217 for P4W, 94 and 213 for Staph1N, 87 and 217 for Fi200W, and 75 and 181 for 676Ż. Each size estimation is the average of measurements of at least 10 independent virions (error in neither case exceeds 10%). Necks and baseplates were not included in tail length estimation.

B. Overall genome organization and conserved regions

The similar head dimensions of different *Spounavirinae* representatives correlate with the similar lengths of their virion DNA molecules, which have to be packed to phage heads in the process of virion assembly ($\approx$127–148 kb; reviewed by Klumpp *et al.*, 2010). Typically, the ends of these molecules contain long, several thousand base pairs, exact (nonpermuted), direct terminal repeats (LTRs). Homologous recombination between the LTRs enables the circularization of infecting phage DNA (Okubo *et al.*, 1964; reviewed by Casjens and Gilcrease, 2009).

Virion DNA molecules of SPO1 and *Listeria monocytogenes* phage A511 have previously been shown to have LTRs of 13,185 and 3,125 bp, respectively (Klumpp *et al.*, 2008; Stewart *et al.*, 2009a). LTRs at virion DNA ends appear also a characteristic feature of staphylococcal Twort-like phages (Table I). The lengths of these repeats in virion DNA of the A5W phage (Ulatowska *et al.*, in preparation) and in DNA of the six *S. aureus* Twort-like phages of different origins described here are similar (from 8047 to 8455 bp).

The average GC content in the genomes of staphylococcal Twort-like phages is between 30.26 and 30.60%—slightly less than in the sequenced *S. aureus* and *S. epidermidis* genomes ($\approx$32.8 and 32.1%, respectively) (Table I; Avison *et al.*, 2002; Baba *et al.*, 2002; Holden *et al.*, 2004; Kuroda *et al.*, 2001; Rosenstein *et al.*, 2009). The distribution of GC pairs along the genomes is remarkably uniform, indicating an ancient origin of these phages and consistent with their affiliation to *Myoviridae*, which are mainly influenced by vertical evolution rather than by horizontal gene transfer (Lavigne *et al.*, 2009; Rohwer and Edwards, 2002).

Overall homologies between the genomes of staphylococcal Twort-like phages except phage Twort, are exceptionally high (88.3–99.9% identity at the DNA level; Fig. 2), implying a limited genetic exchange between the gene pool of these phages and phages of other families or of other hosts. Differences concern mostly single nucleotide substitutions or insertions/deletions. Genomes of only three phages, A3R, K and Sb-1, have been shortened substantially by deletions that include groups from 8 to 25 genes (Table II). The genome of phage K contains the largest number of genes that have no homologues in the genomes of other phages. Possibly, they were acquired to compensate for the functions of lost genes. The remaining insertions/deletions, in the genomes of these phages are short stretches of nucleotides, single protein-coding genes, or introns (see later in this chapter). Nucleotide substitutions and short insertions/deletions are scattered throughout the genomes. However, none of them influence the amino acid sequences of 80 predicted proteins, which are identical in all these phages (Fig. 3, Table II). The true number of identical proteins may be slightly larger, as suggested by ambiguous base designations or possible errors in the sequence files of certain phages deposited in GenBank.

Phage Twort shares only about 47.6–52.5% of identical genomic sequences with the genomes of other staphylococcal Twort-like phages. The homology at the protein sequence level is similarly low (53%; Table II). *S. hyicus*—a host of Twort, despite belonging to coagulase-positive staphylococci—exhibits a limited similarity to *S. aureus* subsp. *aureus* at the level of genomic sequence, as was found by SmaI restriction site mapping of both species' DNA (Pantucek *et al.*, 1996). Consistently, it has been classified by molecular typing methods in the *S. hyicus*/*S. intermedius* phylogenetic group, which is separate from the *S. aureus*/*S. epidermidis* group (Ghebremedhin *et al.*, 2008). Clearly, horizontal gene transfer between Twort-like phages that infect members of different species groups of the same genus is limited in staphylococci. Thus, in further parts of this chapter the authors have focused on those staphylococcal Twort-like phages which have evolved in clinically relevant *Staphylococcus* strains (phages: A5W, Staph1N, Fi200W, P4W, 676Ż, A3R, MSA6, G1, ISP, K and Sb-1). To distinguish them from phage Twort, they have been described here under the working name *S. aureus*/ *S. epidermidis* Twort-like phages, which does not imply their inability to infect cells of other *Staphylococcus* species strains.

Starting from the beginning of virion DNA molecule, genes of *S. aureus*/ *S. epidermidis* Twort-like phages can be grouped into three major clusters

A

PERCENT IDENTITY

	A5W	Staph1N	G1	Sb-1	MSA6	ISP	676Ż	A3R	P4W	FI200W	K	Twort
A5W		99.9	98.7	90.4	96.4	98.9	97.3	93.5	98.1	97.4	91.7	49.7
Staph1N	0.0		98.7	90.3	96.5	99.0	97.4	93.6	98.2	97.5	91.6	49.6
G1	0.3	0.3		91.5	97.0	99.6	97.6	92.5	97.2	97.8	90.6	49.6
Sb-1	0.4	0.4	0.1		89.1	91.1	89.4	89.1	88.9	89.5	96.0	52.1
MSA6	0.4	0.4	0.1	0.1		97.4	95.5	90.4	95.1	95.7	88.3	47.6
ISP	0.3	0.3	0.0	0.1	0.1		98.0	92.9	97.6	98.1	90.8	49.8
676Ż	0.5	0.5	0.6	0.6	0.6	0.6		94.6	98.9	99.5	89.6	50.2
A3R	0.6	0.6	0.7	0.8	0.8	0.7	0.4		94.9	94.4	90.7	51.8
P4W	0.4	0.3	0.4	0.4	0.4	0.4	0.4	0.6		99.1	90.0	50.0
FI200W	0.5	0.5	0.5	0.5	0.5	0.5	0.4	0.5	0.2		89.3	50.2
K	1.4	1.4	1.5	1.5	1.5	1.5	1.4	1.6	1.6	1.7		52.5
Twort	70.0	70.0	70.2	68.7	70.2	70.2	71.0	70.2	70.6	71.0	69.4	

DIVERGENCE

FIGURE 2 (Continued)

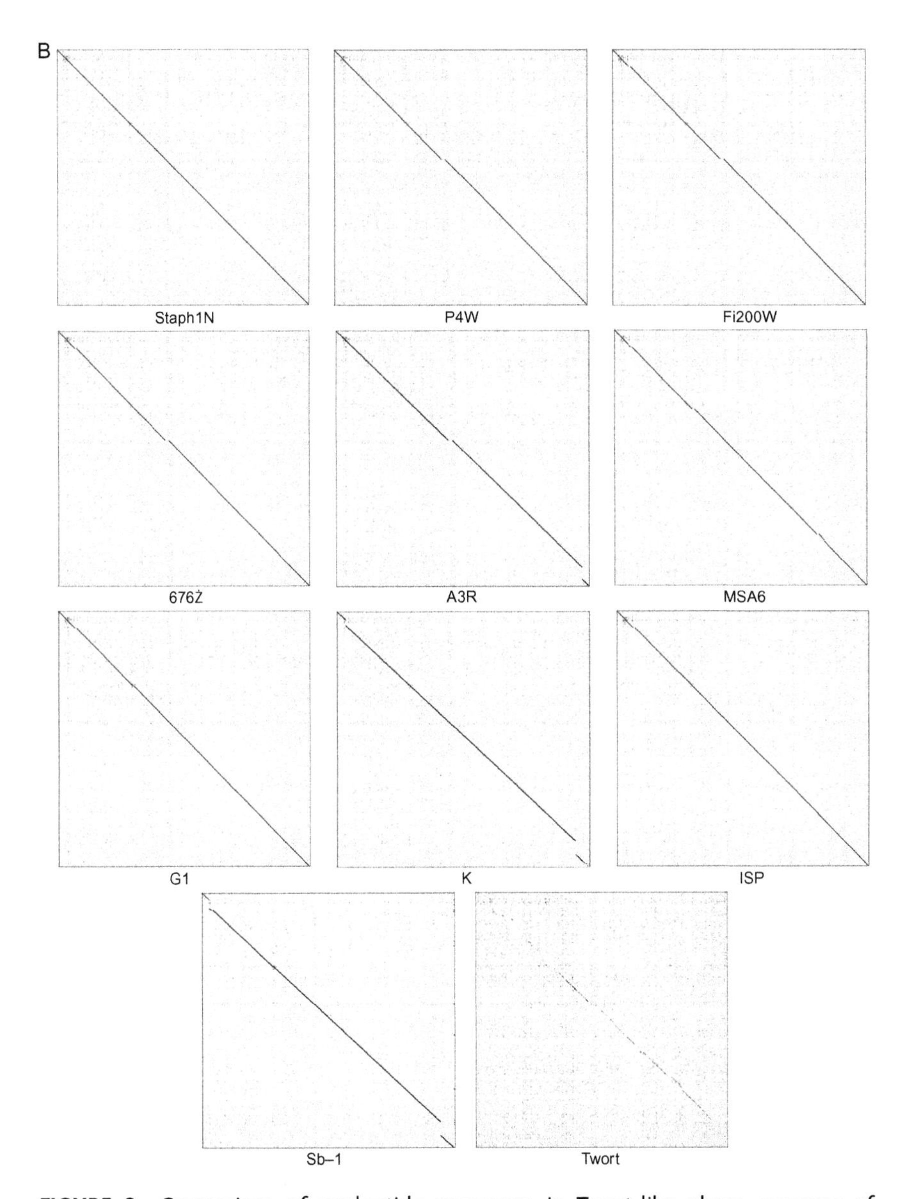

FIGURE 2 Comparison of nucleotide sequences in Twort-like phage genomes of *Staphylococcus*. (A) Pair-wise comparison. (B) Dot plot nucleotide comparisons of the A5W bacteriophage genomic sequence with genomic sequences of other staphylococcal Twort-like phages (regions of right LTRs were excluded from the comparisons). Beginnings and ends of K, G1, ISP, and Twort genomic sequences that were downloaded from GenBank were shifted to optimize the alignments.

TABLE II Known or predicted proteins encoded by *S. aureus*/*S. epidermidis* Twort–like phages[a]

A5W protein name [b]	Length (amino acid residues)[c]	Homologue encoded by phage[d]:										Degree of conservation[e] (%)	Presence of homo– logues in T, L, S, E[f]	Known or predicted function, or bases for name (references, comments)
		Staph 1N	Fi 200W	P4W	676Ż	A3R	MSA6	G1	ISP	K	Sb–1			
TreA	99	+	+	+	+	+	+	150	130	+	177	99–100	–	terminal repaet–encoded protein (7,11)
TreB	61	+	+	+	+	+	+	231	131	+	+	96–100	–	terminal repeat–encoded protein (7,11)
TreC	95–96	+	+	+	+	+	+	156	132	+	178	35–100	–	terminal repeat–encoded protein (7,11)
TreD	95–108	+	+	+	+	+	+	158	133	117	179	68–100	–	terminal repeat–encoded protein (7, 11)
TreE	97	+	+	+	+	+	+	154	134	+	180	97–100	–	terminal repeat–encoded protein (7,11)
TreF	85	+	+	+	+	+	+	175	135	+	181	99–100	–	terminal repeat–encoded protein (7,11)
TreG	79	+	+	+	+	+	+	183	136	+	182	99–100	–	terminal repeat–encoded protein (7,11)
TreH	115	+	+	+	+	+	+	125	137	118	183	100	–	terminal repeat–encoded protein (7,11)

TreI	112	+	+	+	+	+	+	128	138	+	–	100	T	terminal repeat–encoded protein (7,11)
TreJ	102	+	+	+	+	+	+	145	139	+	–	98–100	–	terminal repeat–encoded protein (7,11)
TreK	94–95	+	+	+	+	+	+	159	140	–	–	84–100	–	terminal repeat–encoded protein (7,11)
TreL	63	+	+	+	+	+	+	221	141	–	–	100	–	terminal repeat–encoded protein (7,11)
–	44	–	–	–	–	+	+	312	142	–	–	98–100	–	?
TreM	162	+	+	+	+	+	+	85	143	–	–	100	T, S	terminal repeat–encoded protein (7,11) homing endonuclease of HNH family (6)
TreN	52	+	+	+	+	+	+	+	144	–	–	98–100	T	terminal repeat–encoded protein (7,11)
TreO	43	+	+	+	+	–	+	297	145	–	–	98–100	–	terminal repeat–encoded protein (7,11)
TreP	107	+	+	+	+	+	+	135	146	–	–	94–100	–	terminal repeat–encoded protein (7,11)
TreQ	78	+	+	+	+	+	+	–	147	–	–	100	–	terminal repeat–encoded protein (7,11)
TreR	156	+	+	+	+	+	+	92	148	+	–	100	–	terminal repeat–encoded protein (7,11)

(*continued*)

TABLE II *(continued)*

A5W protein name[b]	Length (amino acid residues)[c]	Homologue encoded by phage[d]: Staph 1N	Fi 200W	P4W	676Ż	A3R	MSA6	G1	ISP	K	Sb-1	Degree of conservation[e] (%)	Presence of homo- logues in T, L, S, E[f]	Known or predicted function, or bases for name (references, comments)
TreS	57	+	+	+	+	+	–	234	149	+	–	100	–	terminal repeat-encoded protein (7,11)
+	56	+	+	+	+	+	–	247	+	+	–	100		?
TreT	41–89	+	+	+	+	+	+	166	150	+	1	85–100	–	terminal repeat-encoded protein (7,11)
TreU	59–73	+	+	+	+	+	+	226	151	+	2	28–100	–	terminal repeat-encoded protein (7,11)
+	26–33	+	+	+	+	+	+	450	+	+	+	97–100	–	?
BofL	78	+	+	+	+	+	+	184	152	+	3	97–100	T	border ORF L protein, function unknown (7)
–	161	–	UboC	–	UboC	–	UboC	88	153	–	4	100	T	upstream of *bof* (11)
	135	–	UboB	–	UboB	–	UboB	109	154	–	5	100	–	upstream of *bof* (11)
–	143	–	–	–	UboA	UboA	UboA	103	155	–	6	96–100	T	upstream of *bof* (11)
gpORF005	>56–57	+	+	+	–	–	–	–	–	+	–			
gpORF006	>84	+	+	–	–	–	–	–	–	+	–			
MbpT	63	+	+	–	+	+	+	224	156	+	7	98–100	–	putative membrane protein (7)
MbpA	161	+	+	+	+	+	+	87	157	1	8	99–100	–	membrane protein (7, 11)

gpORF008	143	+	+	+	+	+	+	104	158	2	9	100	T	
gpORF009	180	+	+	+	+	+	+	73	159	3	10	99–100	–	
gpORF010	162	+	+	+	+	+	+	86	160	4	11	100	L, E	
gpORF011	132	+	+	+	+	+	+	111	161	5	12	100	–	
gpORF012	235	+	+	+	+	+	+	51	162	6	13	100	T, L, E	serine/threonine protein phosphatase I, by homology (2); phosphoesterase (7)
gpORF013	184	+	+	+	+	+	+	70	163	7	14	100	–	
gpORF014	105	+	+	+	+	+	+	138	164	+	15	100	T	virion component (1), putative tail protein (7)
gpORF015	182	+	+	+	+	+	+	71	165	8	16	98–100	T	
gpORF016	72	+	+	+	+	+	+	201	166	+	17	99–100	–	
gpORF017	64	+	+	+	+	+	+	218	167	+	18	98–100	–	
gpORF018	245	+		+	+	+	+	50	168	9	19	99–100	–	
gpORF019	34	+	+	+	+	+	+	437	169	+	+	100	–	
gpORF020	82	+	+	+	+	+	+	180	170	+	20	99–100	–	
gpORF021	129	+	+	+	+	+	+	114	171	10	21	97–100	–	
gpORF022	57	+	+	+	+	+	+	245	172	+	22	100	T	
gpORF023	160	+	+	+	+	+	+	90	173	11	23	100	T	
gpORF024	180	+	+	+	+	+	+	72	174	12	24	100	–	
gpORF025	177	+	+	+	+	+	+	77	175	13	25	100	–	
MbpP	54	+	+	+	+	+	+	256	176	+	26	94–100	–	putative membrane protein (7)

(*continued*)

TABLE II *(continued)*

A5W protein name[b]	Length (amino acid residues)[c]	Homologue encoded by phage[d]: Staph 1N	Fi 200W	P4W	676Ż	A3R	MSA6	G1	ISP	K	Sb–1	Degree of conservation[e] (%)	Presence of homo– logues in T, L, S, E[f]	Known or predicted function, or bases for name (references, comments)
MbpR	91–92	+	+	+	+	+	+	163	177	+	27	96–100	–	putative membrane protein (7)
gpORF028	281	+	+	+	+	+	+	38	178	14	28	99–100	–	
gpORF029	372	+	+	+	+	+	+	24	179	+/–	29	98–100	L	AAA family ATPase (2)
gpORF030	71–108	+	+	+	+	+	+	134	180	+	30	97–100	T	
gpORF031	138–148	+	+	+	+	+	+	106	181	16	31	99–100	T	
HmzG	100	+	+	+	+	+	+	149	182	17	32	100	–	homologue of MazG (7)
gpORF034	62	+	+	+	+	+	+	228	183	+	33	98–100	T	
gpORF035	53	+	+	+	+	+	+	259	184	+	34	100	T	
gpORF036	682	+	+	+	+	+	+	7	185	18	35	76–100	L, E	
gpORF038	87	+	+	+	+	+	+	172	186	+	36	98–100	T	virion component (1,4)
gpORF039	57	+	+	+	+	+	+	+	+	+	37	98–100	T	
MbpB	192	+	+	+	+	+	+	68	187	19	38	99–100	T	membrane protein (11)
gpORF041	208	+	+	+	+	+	+	61	188	20	39	100	T	
Lig	298	+	+	+	+	+	+	32	189	21	40	98–100	L	DNA or RNA ligase
MbpU	74	+	+	+	+	+	+	+	+	+	41	100	T	putative membrane protein (7)

Phr	246	+	+	+	+	+	+	49	190	22	42	100	T	PhoH-related protein (2)
gpORF044	204	+	+	+	+	+	+	63	191	23	43	100	T	virion component (1)
Rbn	141	+	+	+	+	+	+	96	192	24	44	100	T	ribonuclease (2)
gpORF046	63	+	+	+	+	+	+	222	193	+	45	100	T	
gpORF047	213	+	+	+	+	+	+	57	194	25	46	100	T	
gpORF048	76	+	+	+	+	+	+	187	195	+	47	99–100	T, L, E	
gpORF049	75	+	+	+	+	+	+	190	196	+	48	100	T	
Tgl	210–230	+	+	+	+	+	+	54	197	26	49	100	T	putative transglycosylase (7)
gpORF051	211	+	+	+	+	+	+	58	198	27	50	100	T	
MbpS	263	+	+	+	+	+	+	44	199	28	51	100	–	putative membrane protein (7)
MbpV	102	+	+	+	+	+	+	146	200	29	52	100	–	putative membrane protein (7)
LysK(is)	495	+(is)	+(is)	+(is)	+(is)	+(is)	+(is)	60/42 (is)	201/ 203 (is)	30/32 (is)	53/55 (is)	99–100	T	endolysin, N–acetyl–muramoyl–L–alanine–amidase (2,10)
I-KsaI	166	+	+	+	+	+	+	84	202	31	54	100	T, S	intron–encoded HNH endonuclease I–KsaI (2)
HolA	167	+	+	+	+	+	+	83	204	33	56	100	T	holin (2)
UphA	61	+	+	+	+	+	+	233	205	+	57	100	T	upstream of *hol*

(continued)

TABLE II *(continued)*

A5W protein name [b]	Length (amino acid residues) [c]	Homologue encoded by phage [d]:										Degree of conservation [e] (%)	Presence of homo-logues in T, L, S, E [f]	Known or predicted function, or bases for name (references, comments)
		Staph 1N	Fi 200W	P4W	676Ż	A3R	MSA6	G1	ISP	K	Sb–1			
Iro	72	+	+	+	+	+	+	200	206	+	58	97–100	T, L	intergenic region ORF protein
DmcB	69	+	+	+	+	+	+	207	207	+	59	100	T	virion component (1), downstream of *mbpC* (7)
DmcA	110	+	+	+	+	+	+	209	208	+	60	99–100	T, L	downstream of *mbpC* (7)
MbpC	108	+	–	+	+	+	–	+	209	+	61	98–100	T, L	putative membrane protein (7)
MbpW	88	+	+	+	+	+	+	169	210	+	62A	100	–	putative membrane protein (7)
MbpD	128	+	+	+	+	+	+	168	211	+	62	100	T, L	putative membrane protein (7)
Dmd	92	+	+	+	+	+	+	161	212	+	63	100	T, L	downstream of *mbpD* (7)
gpORF057	136	+	+	+	+	+	+	133	213	34	64	100	T, L, E	
Ter	605	+	+	+	+	+	+ (is)	10	1	35	65	98–100	T, S, L, E	virion component, large terminase subunit (4)
–	323	–	–	–	–	–	I-MsaI	–	–	–	–	–	T, E	intron–encoded protein I-MsaI, function unknown (11)
gpORF059	273	+	+	+	+	+	+	40	2	36	66	98–100	T, L, E	virion component (4)

gpORF59A	60	+	+	+	+	+	+	235	3	+	67	100	–	
gpORF060	159	+	+	+	+	+	+	91	4	37	68	100	T, L, E	
MbpF	397–403	+	+	+	+	+	+	23	5	38	71	98–100	T	putative membrane protein (7); putative cell surface protein (2,6)
MbpE	116	+	+	+	+	+	+	120	6	39	73	99–100	T	membrane protein (11), see text
gpORF063	123	+	+	+	+	+	+	115	7	40	74	99–100	T, L, E	
Prt	563	+	+	+	+	+	+	14	8	41	75	99–100	T, L, S, E	virion component, portal protein (1,4)
Pro	257	+	+	+	+	+	+	48	9	42	76	100	T, S, L, E	virion component, prohead protease (1)
gpORF066	316–318	+	+	+	+	+	+	29	10	43	78	98–100	T, S, L	
Mcp	463	+	+	+	+	+	+	16	11	44	79	100	T, S, L, E	virion component, major capsid protein (1,2,4)
gpORF068	98	+	+	+	+	+	+	151	12	+	80	100	T, L, E	
gpORF069	302	+	+	+	+	+	+	30	13	45	81	99–100	T, S, L, E	virion component, putative tail fiber protein (4)
gpORF070	158–292	+	+	+	+	+	+	34	14	46	82	99–100	T, S, L, E	virion component (1)
gpORF071	206	+	+	+	+	+	+	62	15	47	83	100	T, S, L, E	
gpORF072	278	+	+	+	+	+	+	39	16	48	84	100	T, S, L, E	virion component, baseplate hub assembly protein (4)
gpORF073	71	+	+	+	+	+	+	202	17	+	85	99–100	T, L, E	

(*continued*)

TABLE II (*continued*)

A5W protein name [b]	Length (amino acid residues) [c]	Homologue encoded by phage [d]: Staph 1N	Fi 200W	P4W	676Ż	A3R	MSA6	G1	ISP	K	Sb–1	Degree of conservation [e] (%)	Presence of homo–logues in T, L, S, E [f]	Known or predicted function, or bases for name (references, comments)
Tsp	587	+	+	+	+	+	+	11	18	49	86	99–100	T, S, L, E	virion component, tail sheath protein (1,2,4)
TmpA	142	+	+(is)	+(is)	+(is)	+(is)	+	105	19	50	87	100	T, S, L, E	virion component, tail tube protein (1, 2, 4); tail morphogenetic protein (7)
–	467	–	I-812I	I-812I	I-812I	I-812I	–	–	–	–	–	99–100	T	homologue of putative intron-encoded nuclease of phage 812 host range mutant (11,13)
gpORF076	46	+	+	+	+	+	+	293	20	+	88	70–100	T	
gpORF077	152	+	+	+	+	+	+	93	21	51	89	99–100	T	
MbpX	64	+	+	+	+	+	+	215	22	+	89A	100	–	putative membrane protein (7)
gpORF078	103	+	+	+	+	+	+	141	23	52	90	100	T	virion component (1)
gpORF079	152	+	+	+	+	+	+	95	24	53	91	100	T, L, S, E	
TmpB	178	+	+	+	+	+	+	74	25	54	92	99–100	T, L, S, E	tail morphogenetic protein
TmpC	1351–1352	+	+	+	+	+	+	1	26	55	94	99–100	T, L, S, E	virion component (1,4); tape–measure protein (6); tail morphogenetic protein (7, 11)

TmpD	808	+	+	+	+	+	+	5	27	56	96	99–100	T, L, S, E	virion component, tail murein hydrolase TAME (5) tail morphogenetic protein (7)
TmpE	295	+	+	+	+	+	+	33	28	57	97	99–100	T, E	virion component, putative peptidoglycan hydrolase (4, 11), tail morphogenetic protein (7)
gpORF085	848	+	+	+	+	+	+	4	29	58	98	99–100	T, L, S, E	glycerophosphoryl diester phosphodies-terase, by homology (2,4,6)
gpORF086	263	+	+	+	+	+	+	43	30	59	99	100	T, L, S, E	virion component (1,4)
gpORF087	174	+	+	+	+	+	+	78	31	60	100	99–100	T, L, E	virion component (4)
BmpA	234	+	+	+	+	+	+	52	32	61	101	100	T, L, S, E	virion component, putative baseplate wedge subunit, by homology (4); baseplate morphogenetic protein (7)
BmpB	348	+	+	+	+	+	+	27	33	62	102	100	T, L, S, E	virion component (1,4); baseplate morphogenesis protein, by homology (7)

(*continued*)

TABLE II (*continued*)

A5W protein name [b]	Length (amino acid residues) [c]	Homologue encoded by phage [d]:										Degree of conservation [e] (%)	Presence of homologues in T, L, S, E [f]	Known or predicted function, or bases for name (references, comments)
		Staph 1N	Fi 200W	P4W	676Ż	A3R	MSA6	G1	ISP	K	Sb-1			
TmpF	1019	+	+	+	+	+	+	3	34	63	103	99–100	T, L, S, E	tail morphogenetic protein, by homology (7)
BmpC	173	+	+	+	+	+	+	79	35	64	104	100	T, L, S, E	virion component (1,4), baseplate morphogenetic protein, by homology (7)
TmpG	1152	+	+	+	+	+	+	2	36	65	105	99–100	T, L, S, E	virion component (1,4); tail morphogenetic protein, by homology (7, 11)
gpORF093	52	+	+	+	+	+	+	262	37	+	106	100	T, L, S, E	
gpORF094	640	+	+	+	+	+	+	8	38	66	107	99–100	T, L	virion component (1,4)
gpORF095	124	+	+	+	+	+	+	117	39	67	108	100	T	virion component (1)
gpORF096	458	+	+	+	+	+	+	17	40	68	109	99–100	T, L	virion component (1,4)
DhlA	582	+	+	+	+	+	+	12	41	69	110	99–100	T, L, S, E	DNA helicase A, by homology (2, 7)
gpORF098	537	+	+	+	+	+	+	13	42	70	111	100	T, L, E	Rep protein, by homology (2, 7)
DhlB	480	+	+	+	+	+	+	15	43	71	112	99–100	T, L, S, E	putative DNA helicase B, by homology (2, 7)

RncA	345	+	+	+	+	+	+	28	44	72	113	100	T, L, S, E	recombination nuclease A; similar to putative exonuclease SbcD (2, 7)
gpORF102	125	+	+	+	+	+	+	110	45	73	114	100	T	
RncB	639	+	+	+	+	+	+	9	46	74	115	100	T, L, S, E	recombination nuclease B; putative exonuclease, by homology (2, 7)
Asf	198	+	+	+	+	+	+	67	47	75	116	100	T, L, E	anti–sigma factor (8)
Pri	355	+	+	+	+	+	+	26	48	76	+/–	98–100	T, L, S, E	DNA primase, by homology (2)
gpORF106	112	+	+	+	+	+	+	127	49	+	119	100	–	
gpORF107	150	+	+	+	+	+	+	98	50	77	120	98–100	T, L, E	
gpORF108	202	+	+	+	+	+	+	64	51	78	121	99–100	T, L, S, E	homologue of T5 phage D14 protein (2), resolvase, by homology (4,7)
NrdI	143	+	+	+	+	+	+	102	52	79	122	100	T, L	ribonucleotide reductase stimulatory protein, NrdI, by homology (2,6)
NrdE	704	+	+	+	+	+	+	6	53	80	123	99–100	T, L, S, E	ribonucleotide reductase large (alpha) subunit, by homology (2,6)
NrdF	349	+	+	+	+	+	+	25	54	81	124	100	T, L, S, E	ribonucleotide reductase minor subunit, by homology (2,6)

(*continued*)

TABLE II *(continued)*

A5W protein name [b]	Length (amino acid residues)[c]	Homologue encoded by phage[d]:										Degree of conservation[e] (%)	Presence of homo– logues in T, L, S, E[f]	Known or predicted function, or bases for name (references, comments)
		Staph 1N	Fi 200W	P4W	676Ż	A3R	MSA6	G1	ISP	K	Sb–1			
gpORF112	109	+	+	+	+	+	+	130	55	82	125	98–100	–	
Tio	106	+	+	+	+	+	+	129	56	83	126	99–100	T, L	thioredoxin like protein, by homology (2)
gpORF114	198	+	+	+	+	+	+	66	+	84	127	100	T, L, E	
Trf	101	+	+	+	+	+	+	147	57	85	128	100	T, L, S, E	transcription factor (7,11)
PolA (2is)	1008–1072	+ (2is)	+ (2is)	+ (2is)	+ (2is)	+ (2is)	+ (2is))	35/19/ 37 (2is)	59/61/ 63 (2is)	86/88/ 90 (2is)	129/ 129b/ 129c (2is)	96–100	T, L, S, E	DNA–directed DNA polymerase A; related to bacterial PolA polymerases (2,6,7)
I–KsaII	170	+	+	+	+	+	+	81	60	87	130	100	–	intron–encoded protein I–KsaII (2), putative endonuclease (4)
I–KsaIII	269	+	+	+	+	+	+	41	62	89	132	99–100	T, S	intron–encoded protein I–KsaIII, putative HNH endonuclease (2)
gpORF121	80	+	+	+	+	+	+	181	64	+	135	99–100	–	
gpORF122	160	+	+	+	+	+	+	89	65	91	136	100	T, L, E	
gpORF123	296–423	+	+	+	+	+	+	20	66	92	137	100	T, L, E	

Rec	418	+	+	+	+	+	+ (is)	21	67	93	138	100	T, L, E	putative repair recombinase (2); homologue of bacterial RecA (6,7)
–	215	–	–	–	–	–	I-MsaII	–	–	–	–	–	–	putative intron-encoded endonuclease I-MsaII, by homology (11)
gpORF125	117	+	+	+	+	+	+	121	68	+	139	100	T, L, E	
Sig	220	+	+	+	+	+	+	56	69	94	140	100	T, L, S, E	RNA polymerase sigma factor, by homology (2,6)
gpORF127	210	+	+	+	+	+	+	59	70	95	141	100	T, L, E	virion component (1,2,4), putative tail morphogenetic protein (3,11)
TmpH	73–173	+	+	+	+	+	+	80	71	96	142	32–99	T, E	virion component (1,2,4); tail morphogenetic protein H, by homology (7,11), see text
TmpI	75	+	+	+	+	+	–	189	72	+	143	100	T	virion component (1,2); tail morphogenetic protein, by homology (7, 11); see text

(*continued*)

TABLE II *(continued)*

A5W protein name [b]	Length (amino acid residues) [c]	Homologue encoded by phage [d]: Staph 1N	Fi 200W	P4W	676Ż	A3R	MSA6	G1	ISP	K	Sb–1	Degree of conservation [e] (%)	Presence of homo–logues in T, L, S, E [f]	Known or predicted function, or bases for name (references, comments)
gpORF130	86	+	+	+	+	+	+	174	73	+	144	100	T	
gpORF131	251	+	+	+	+	+	+	46	74	97	–	97–100	T	
	>152	–	–	–	–	–	–	–	–	–	+			
	>106	–	–	–	–	–	–	–	–	–	+			
gpORF132	416	+	+	+	+	+	+	22	75	98	147	100	T, L, E	putative metallophos-phoesterase (6)
MbpG	122	+	+	+	+	+	+	118	76	99	148	100	T, L	putative membrane protein (7)
gpORF134	103	+	+	+	+	+	+	143	77	100	149	100	–	
gpORF135	178	+	+	+	+	+	+	75	78	+	150	100	T, L, E	
gpORF136	255	+	+	+	+	+	+	45	79	101	151	100	T, L, E	
gpORF137	148	+	+	+	+	+	+	99	80	102	152	100	T, L, E	
gpORF138	287	+	+	+	+	+	+	36	81	+	153	100	T, L, E	
gpORF139	243	+	+	+	+	+	+	47	82	103	154	100	T, L, S, E	
gpORF140	152	+	+	+	+	+	+	94	83	104	155	99–100	T, L, E	virion component (1)
gpORF141	147	+	+	+	+	+	+	100	84	105	156	100	T, L, E	
gpORF142	234	+	+	+	+	+	+	53	85	106	157	99–100	T	
MbpH	132	+	+	+	+	+	+	108	86	107	158	100	T	putative membrane protein (7)
PufA	20	+	+	+	+	+	+	+	87	+	+	100	–	peptide of unknown function (7,11)
gpORF144	80	+	+	+	+	+	+	182	88	+	+	100	–	
gpORF145	54–156	+	+	+	+	+	+	252	89	108	+	100	–	see text
gpORF146	59	+	+	+	+	+	+	+	+	–	–	98–100	–	

Sci	58	+	+	+	+	+	+	240	90	–	–	100	–	DNA sliding clump inhibitor, arrest of *S. aureus* DNA synthesis (7,9)
MbpJ	177	+	+	+	+	+	+	76	91	–	–	100	–	putative membrane protein (7)
MbpY	47–82	+	+	+	+	+	+	–	92	–	–	100	–	putative membrane protein (7)
gpORF149	58	+	+	+	+	+	+	241	93	–	–	91–100	T	
gpORF150	98	+	+	+	+	+	+	152	94	–	–	100	–	
MbpK	63	+	+	+	+	+	+	219	95	–	–	98–100	–	putative membrane protein (7)
gpORF152	122	+	+	+	+	+	+	119	96	–	–	97–100	–	
gpORF153	115	+	+	+	+	+	+	124	97	–	–	100	–	
MbpI	92	+	+	+	+	+	+	162	98	–	–	100	–	putative membrane protein (7)
gpORF155	101	+	+	+	+	+	+	140	99	–	–	100	T	
gpORF156	116	+	+	+	+	+	+	122	100	–	–	100	–	
gpORF157	200–381	+	+	+	+	–	+	65	101	–	–	49–100	–	see text
gpORF158	59	+	+	+	+	–	+	237	102	–	–	98–100	–	–
PufB	22	+	+	+	+	–	+	+	103	–	–	100	–	peptide of unknown function
PufC	24	+	+	+	+	–	+	+	104	–	–	100	–	peptide of unknown function
MbpM	133	+	+	+	+	–	+	107	105	–	–	100	–	putative membrane protein (7)
gpORF160	86–97	+	+	+	+	–	+	173	106	–	–	55–100	E	
MbpL	95	+	+	+	+	–	+	157	107	–	–	100	–	putative membrane protein (7)

(continued)

TABLE II *(continued)*

A5W protein name [b]	Length (amino acid residues)[c]	Homologue encoded by phage[d]: Staph 1N	Fi 200W	P4W	676Ż	A3R	MSA6	G1	ISP	K	Sb–1	Degree of conservation[e] (%)	Presence of homo– logues in T, L, S, E[f]	Known or predicted function, or bases for name (references, comments)
gpORF162	38	+	+	+	+	–	+	362	108	–	–	100	–	
gpORF163	87	+	+	+	+	–	+	170	109	–	–	100	–	
MbpZ	59	+	+	+	+	–	+	236	110	–	–	100	–	putative membrane protein (7)
gpORF164	87	+	+	+	+	–	+	171	111	–	–	100	–	
gpORF165	105	+	+	+	+	–	+	137	112	–	–	100	–	
gpORF166	226	+	+	+	+	–	+	55	113	–	–	100	L	see text
MbpN	52	+	+	+	+	–	+	263	114	+	160A	100	–	putative membrane protein (7)
gpORF168	66	+	+	+	+	–	+	211	115	+	161	100	–	
MbpO	96	+	+	+	+	–	+	155	116	+	162	100	–	putative membrane protein (7)
gpORF170	102	+	+	+	+	–	+	144	117	109	163	100	–	
Rpp	302	+	+	+	+	–	+	31	118	110	164	100	L	putative ribose–phosphate pyrophosphokinase, by homology (2, 7)
NadV	381–489	+	+	+	+	+	+	18	119	111	165	98–99	L	virion component (1); putative nicotinamide phosphoribosyl transferase NadV, see text (2,6,7)

gpORF173	81	+	+	+	+	+	+	178	120	+	+/–	86–100	–	
gpORF174	130–131	+	+	+	+	+	+	113	121	112	166	90–100	–	
gpORF175	73	+	+	+	+	+	+	194	122	–	167	100	–	
–	83	–	–	–	–	–	–	–	–	OsfA	–			ORF one hundred seventy five exchanger A (7)
–	98	–	–	–	–	–	–	–	–	OsfB	–			ORF one hundred seventy five exchanger B; homolog of TreA, TreC, TreD and TreE (7)
gpORF176	103	+	+	+	+	+	+	142	123	113	168	86–100	–	
gpORF177	169	+	+	+	+	+	+	82	124	–	169	100	T	
–	99	–	–	–	–	–	–	–	–	114	–			(homologue of *Bacillus* phage SP10, gpORF42) (11,12)
gpORF178	109	+	+	+	+	+	+	131	125	–	170	99–100	T	
–	75	–	–	–	–	–	–	–	–	Ose	–			ORF one hundred seventy eight exchanger (7)
gpORF179	64–69	+	+	+	+	+	+	–	126	+	171	86–100	–	
gpORF180	104	+	+	+	+	+	+	139	127	–	172	100	–	
–	117	–	–	–	–	–	–	–	–	115	–			

(*continued*)

TABLE II *(continued)*

A5W protein name [b]	Length (amino acid residues)[c]	Homologue encoded by phage[d]: Staph 1N	Fi 200W	P4W	676Ż	A3R	MSA6	G1	ISP	K	Sb–1	Degree of conservation[e] (%)	Presence of homo– logues in T, L, S, E[f]	Known or predicted function, or bases for name (references, comments)
–	128	–	–	–	–	–	–	–	–	116	–			
gpORF181	62	+	+	+	+	+	+	225	128	–	173	100	S	
BofR	33	+	+	+	+	+	+	445	129	–	+	100	–	Border ORF R protein, function unknown (7, 11)

[a] The function assignment of certain proteins is based on the following literature sources: 1, Eyer *et al.* (2007); 2, O'Flaherty *et al.* (2004); 3, Kilcher *et al.* (2010); 4, Vandersteegen *et al.* (2011); 5, Paul *et al.* (2011b); 6, Kwan *et al.* (2005); 7, Ulatowska *et al.* (in preparation); 8, Debhi *et al.* (2009); 9, Belley *et al.* (2006); 10, O'Flaherty *et al.* (2005a); 11, this work; 12, Yee *et al.* (2011); 13, Kasparek et al., (2007). Light-grey shaded rows indicate protein products of genes that are split in certain phages. Genes that are split by introns are indicated with (is). Proteins of some phages which are shorter homologues of corresponding proteins of other phages and are apparent products of split genes are listed below their longer counterparts and are indicated with >, prior to the number of their amino acid residues.

[b] Protein names and ORF numbers are as in Ulatowska *et al.* (in preparation). Names of proteins that are conserved in all *S. aureus*/*S. epidermidis* Twort–like phages are in bold.

[c] Ranges of protein lengths indicate the shortest and the longest version of each protein in *S. aureus*/*S. epidermidis* Twort–like phages.

[d] Homologues of A5W phage proteins that are designated by the names corresponding to those of A5W proteins, in the GenBank files of phages Staph1N, Fi200W, P4W, 676Ż, A3R, MSA6 (see Table 1) are designated with +. Proteins of phages G1, ISP, K and Sb–1 that are annotated in the GenBank files of genomic sequences of these phages are designated with the relevant ORF numbers or names. Homologues of A5W proteins of the latter phages that are not annotated in the relevant GenBank files are designated with +. Their identification was based on homology searches and on predictions of the MetaGeneAnnotator (Noguchi *et al.*, 2006, 2008). Predicted products of genes that do not match with MetaGeneAnnotator predictions are indicated with a question mark. Proteins that are missing in proteomes of certain phages are designated by – . Dark-grey shaded fields indicate genome regions that are deleted in certain phages. For the purpose of comparisons, amino acid sequences of homologous proteins that differed from each other due to annotation errors or differences in the assignments of start codons in DNA, despite identical gene sequences, were unified according to the annotations in the GenBank file of phage A5W sequence. Products of genes which may possibly contain errors in their sequences deposited in GenBank, are indicated with +/–. ORFs that are annotated in the sequences of phages G1 and ISP deposited in GenBank, but do not match with predictions of MetaGeneAnnotator were not included in the Table.

[e] Degree of conservation indicates the range of identities, in percent, between the least and the most homologous orthologues, as calculated based on the comparison of all orthologues with the shortest of them.

[f] T, L, S and E correspond to *Staphylococcus* phage Twort, *Listeria* phage A511 or P100, *Bacillus* phage SPO1 and *Enterococcus faecalis* phage phiEF24C, respectively. Homologies are indicated according to Stewart *et al.* (2009a) and Ulatowska *et al.* (in preparation).

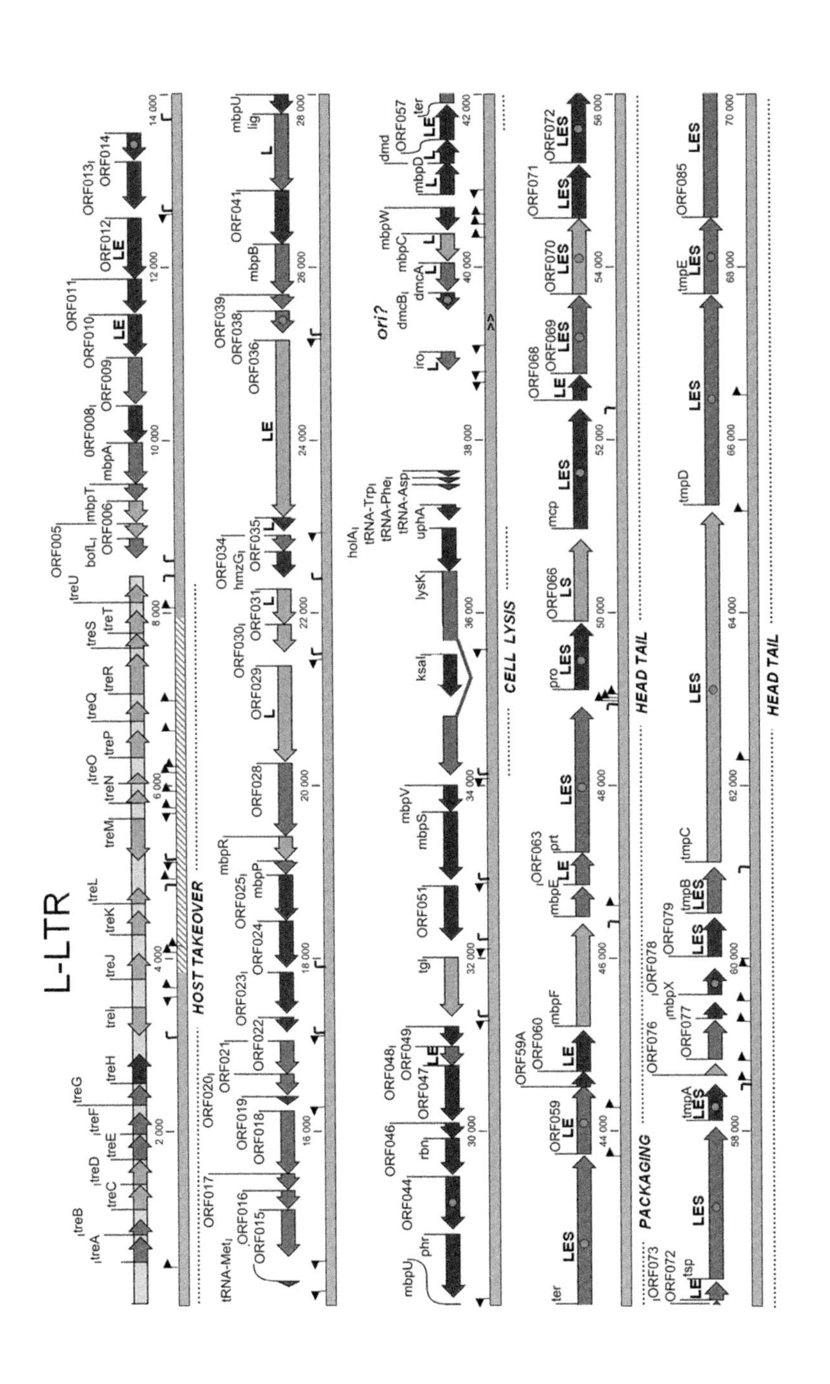
L-LTR
HOST TAKEOVER
CELL LYSIS
PACKAGING
HEAD TAIL
HEAD TAIL
ori?

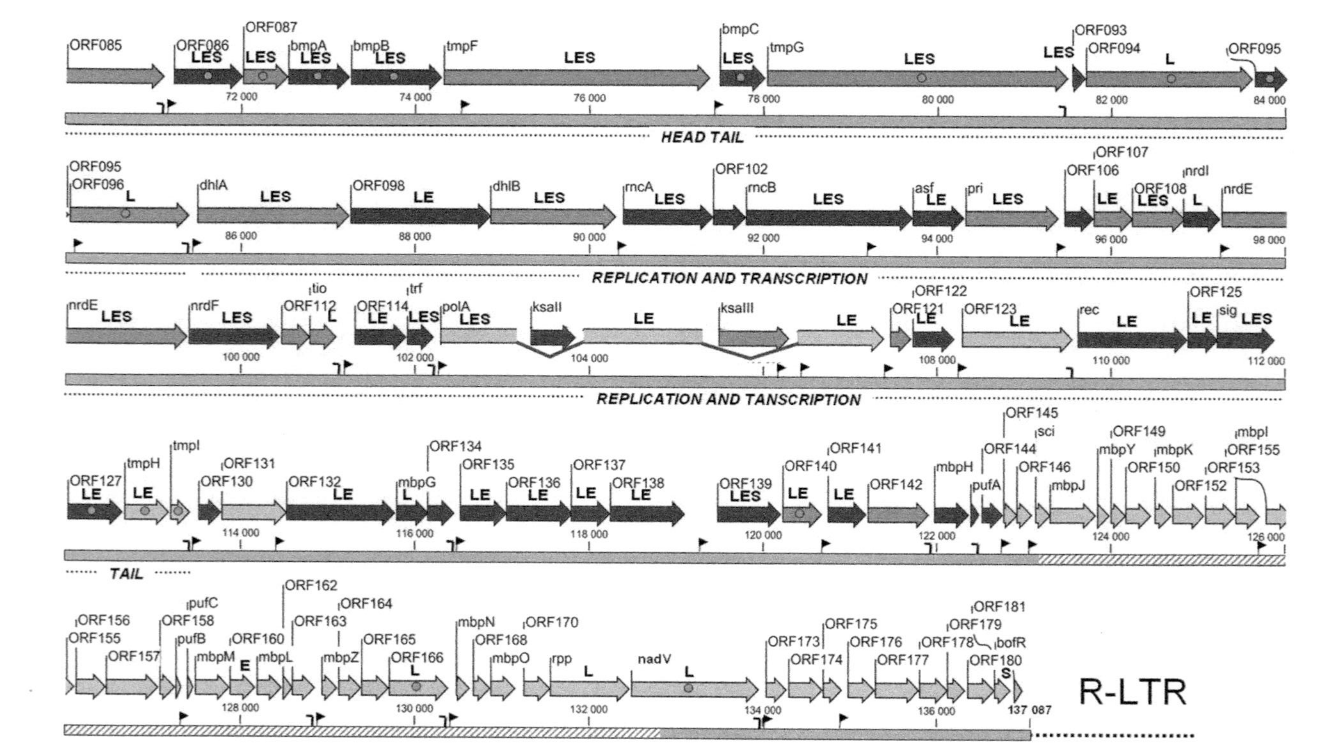

FIGURE 3 Genetic and physical organization of the A5W phage genome – an exemplary genome of *S. aureus/S. epidermidis* Twort-like phages. Arrows mark the position and orientation of genes. Red arrows indicate tRNA genes. Other arrows indicate protein-coding genes. Protein-coding genes are color coded according to the degree of conservation of their products in *S. aureus/S. epidermidis* Twort-like phages: dark green, 100% identity at the amino acid sequence level; light green, 90–99% and identical length; and olive green, below 90% identity or different length (see

according to their direction of transcription (Fig. 3; O'Flaherty *et al.*, 2004; Ulatowska *et al.*, in preparation). The first cluster consists of open reading frames (ORFs) of the left LTR region (ca. 8.5 kb). They are transcribed rightward, with the exception of two, which are transcribed leftward. However, at least one of the latter (*treM*), which encodes a putative homing endonuclease, represents a mobile genetic element and was presumably acquired by an ancestor of these phages via horizontal gene transfer. The second cluster of genes, which occupies about 32 kb of genomic DNA, consists of about 59 ORFs transcribed leftward. It contains two major genome regions that do not encode proteins. One of them, about 1 kb long, encodes one tRNA, while the other, about 1.6 kb long, encodes the three remaining tRNAs of these phages. The third cluster of genes, which is the major one and occupies 90-97 kb, consists of 107-132 ORFs transcribed rightward. In virion DNA, this cluster directly precedes the region of the right LTR. While the first cluster of genes (LTR) is separated from the second cluster by about a ≈0.3-kb noncoding region, spacing between the second and the third cluster is shorter, ≈0.14 kb. The genome of phage Twort is organized in a similar manner with respect to the direction of gene transcription (Kwan *et al.*, 2005). However, it encodes only one tRNA.

The aforementioned clusters of genes do not correspond to basic transcriptional units. Several putative promoter and terminator sequences have been identified by *in silico* analysis of K, ISP, and A5W genomic sequences (O'Flaherty *et al.*, 2004; Ulatowska *et al.*, in preparation; Vandersteegen *et al.*, 2011). Presumably, like in the case of *Bacillus* phage SPO1, proteins of these phages are translated from over 50 transcripts (Stewart *et al.*, 2009a).

The overall genome organization of staphylococcal Twort-like phages is typical of other *Spounavirinae* representatives. Each of them contains about 200 protein-coding genes and a few genes encoding tRNA (Tables I and II; Fig. 3; Kvachadze *et al.*, 2011; Kwan *et al.*, 2005; O'Flaherty *et al.*, 2004; Stewart *et al.*, 2009a; Ulatowska *et al.*, in preparation; Vandersteegen

---

Table II). Genes whose products are conserved in phage Twort (see Table I), in *Listeria* phage P100 or A511 (GenBank Acc. No. DQ004855.1; DQ003638 Carlton *et al.*, 2005; Klumpp *et al.*, 2008), in *Enterococcus faecalis* phage phiEF24C (GenBank Acc. No. NC_009904; Uchiyama *et al.*, 2008), and in *Bacillus* phage SPO1 (GenBank Acc. No. NC_011421.1; Stewart *et al.*, 2009a) are marked with capital T, L, E and S, respectively. Dark blue lines connect fragments of genes interrupted by introns. The grey line below the genes represents DNA. Direct repeats in the presumed origin of phage replication are marked as ≫. Striped fragments of the line that represents DNA mark regions that are absent from the genomes of phages A3R, K, or Sb-1. Predicted early phage promoters are represented by black flags on this line. They are pointing to the direction of transcription. Hooks indicate predicted Rho-independent transcription terminators. (See Pages 2 and 3 in Color Section at the back of the book.)

et al., 2011). Genes are tightly packed and occupy about 89% of the genomes. They are arranged in functional modules, but the organization of these modules is slightly different from that of staphylococcal temperate phages (O'Flaherty *et al.*, 2004). While modules in the genomes of the latter are clearly separated from each other and determine, in order, functions responsible for lysogeny, DNA replication and transcription, DNA packaging, phage morphogenesis, and cell lysis (Iandolo *et al.*, 2002), some modules in the genomes of staphylococcal Twort-like phages are not clearly separated, their order is different and certain genes are scrambled between modules. The lysis module is separated from the DNA packaging, and the head and tail morphogenesis module only by a short region that encodes tRNAs and a few proteins of unknown functions; it presumably contains an origin of phage replication (see further in the text). The head and tail morphogenesis module in turn seems to be split into a major and a minor part (containing only three genes), by insertion of the replication and transcription module. Additional genes that encode phage morphogenetic functions are scattered throughout different genome parts. The order of genes in central regions of the genomes is conserved substantially between staphylococcal Twort-like phages and other phages of the *Spounavirinae* subfamily (Stewart *et al.*, 2009a). Comparing the predicted gene products of the *Bacillus* SPO1 phage with those of staphylococcal phages G1, K, and Twort, *Listeria* phage P100, and *Lactobacillus plantarum* phage LP65, Stewart and colleagues (2009a) detected 73 known or putative proteins of G1, K, or Twort (34%) homologous to at least one protein of the remaining phages. Current comparisons, which include a new member of the *Spounavirinae* subfamily, *Enterococcus faecalis* phage phiEF24C (Uchiyama *et al.*, 2008), allow one to identify 83 conserved proteins (Table II). This list may be extended with the acquisition of crystallographic data concerning the remaining proteins. The closest relatives of *S. aureus*/*S. epidermidis* Twort-like phages among the Twort-like phages that infect nonstaphylococcal hosts appear to be *Listeria* phages P100 and A511. Seventy-eight proteins are highly conserved in these phages (Fig. 3, Table III). Sixty-six of them are also highly conserved in *Enterococcus* phage phiEF24C. One additional protein, represented by A5W gpORF160, is conserved in staphylococcal Twort-like phages and phiEF24C, but not in *Listeria* phage P100. Whether the homologue of ORF160 was deleted from the genome of P100 ancestor or acquired by the ancestor of staphylococcal Twort-like phages or phage phiEF24C by horizontal gene transfer is unclear.

The major cluster of conserved genes in the genomes of staphylococcal Twort-like phages is represented by the region from about 41 to 121 kb of the A5W genomic sequence (Fig. 3, Table II). Lavigne and colleagues (2009) identified in this region two clusters of genes shared between SPO1 and phages of the Twort-like genus: the cluster of morphogenesis

TABLE III Introns and their protein products (IEP) encoded by staphylococcal Twort-like phages[a]

Intron name (phage)[b]	Intron length (nucleotide)	Name of interrupted gene	Known or predicted function of interrupted gene	Name of IEP	Length of IEP (amino acids)	Known or predicted function of IEP[c]	Phages containing homologous intron at conserved DNA position (name of interrupted gene; length of conserved region if different; % identity if less than 100)[d]
Lys-I1 (K)	876	32&30 (*lysK*)	Lysin	I-KsaI	166	HNH endonuclease	G1 (60&42) ISP (201&203) Sb-1 (53&55) A5W (*lysK*) Staph1N (*lysK*) A3R (l*ysK*) Fi200W (*lysK*) 676Ż (*lysK*) P4W (*lysK*) 812 (*lyt*812) MSA6 (*lysK*)
Pol-I1 (K)	798	86&88 (*polA*)	DNA polymerase	I-KsaII	170	Two zinc fingers	G1 (33&19&37) ISP (59&61&63) Sb-1 (129&129b&129c) A5W (*polA*) Staph1N (*polA*) A3R (*polA*) Fi200W (*polA*) 676Ż (*polA*)

(continued)

TABLE III *(continued)*

Intron name (phage)[b]	Intron length (nucleotide)	Name of interrupted gene	Known or predicted function of interrupted gene	Name of IEP	Length of IEP (amino acids)	Known or predicted function of IEP[c]	Phages containing homologous intron at conserved DNA position (name of interrupted gene; length of conserved region if different; % identity if less than 100)[d]
Pol-I2 (K)	1082	88&90 (*polA*)	DNA polymerase	I-KsaIII	269	HNH endonuclease	P4W (*polA*) MSA6 (*polA*) G1 (33&19&37) ISP (59&61&63) Sb-1 (129&129b&129c) A5W (*polA*) Staph1N (*polA*; 1081; >99) A3R (*polA*) Fi200W (polA; 1081; >99) 676Ż (*polA*) P4W (*polA*; 1081; >99) MSA6 (*polA*)
nrd-I1 (Twort)	429	45&33 (*nrdI*)	Ribonucleotide reductase regulatory subunit	—	NA	NA	—
nrd-I2 (Twort)	1087	33&118 (*nrdE*)	Ribonucleotide reductase large subunit	I-TwoI	243	HNH endonuclease	—
tub-I1 (Twort)	325	106&NC (*tmpA*)	Tail tube	—	NA	NA	—
tub-I2 (Twort)	282	106&NC (*tmpA*)	Tail tube	—	NA	NA	—
	252		Tail tube	—	NA	NA	—

tub-I3 (Twort)		106&NC (*tmpA*)					
ter-I1 (Twort)	1514	151&59 (*ter*)	Large terminase	gp31	316	Unknown	—
ter-I2 (Twort)	1312	59&36 (*ter*)	Large terminase	gp119	320	Unknown	MSA6 (*ter*,221; 91)[e]
ter-I3 (MSA6)	1332	*ter*	Large terminase	I-MsaI	323	putative endonuclease	Twort (*ter*; 221; 91)[e]
rec-I1 (MSA6)	1376	*rec*	Recombinase	I-MsaII	322	putative endonuclease	812 (homologue of A5W *rec*; 1376; 92)
tmpA-I1 (676Ż)	305	*tmpA*	Tail tube	—	NA	NA	Fi200W (*tmpA*) P4W (*tmpA*) A3R (*tmpA*) 812 and U16 (homologue of A5W *tmpA*)
tmpA-I2 (676Ż)	256	*tmpA*	Tail tube	—	NA	NA	Fi200W (*tmpA*) P4W (*tmpA*) A3R (*tmpA*) 812 and U16 (homologue of A5W *tmpA*)

[a] Characteristic of introns and IEP-encoding genes in phages K, G1, Twort, and A5W is given according to Stewart *et al.* (2009a), Kasparek *et al.*, (2007) and Ulatowska *et al.* (in preparation). Characteristic of introns and IEP-encoding genes in other phages is based on homologies of relevant regions in DNA of these phages to known intron and IEP-encoding sequences. The lengths of the MSA6 *ter*-I3 and *rec*-I1 introns, as well as 676Ż *tmpA*-I1 and *tmpA*-I2 introns were predicted based on homologies of predicted translation products of intron-encoding and -flanking regions to *ter, rec* and *tmpA* gene products of other staphylococcal Twort-like phages, respectively. Coordinates of these introns are as in the relevant GenBank files (see Table I).

[b] The name of phage, in which a given intron was originally described, is in parentheses.

[c] Not applicable.

[d] Homologies of certain phage MSA6, 676Ż, P4W, Fi200, and A3R introns to introns of phage 812 are based on the comparison of DNA sequences of these phages with the fragments of 812 sequence that are deposited in GenBank (Acc. No. EF136586 and EF136582; Kasparek *et al.*, 2007).

[e] Homology concerns 5′ regions of Twort ter-I2 and MSA6-ter-I1 introns.

and DNA packaging genes (corresponding to A5W genes from *ter* to ORF093) and the cluster of DNA replication genes (including genes that encode helicase, exonuclease, primase and resolvase, and corresponding to A5W genes from *dhlA* to ORF108). Products of over half of the 33 further downstream genes until A5W ORF141 have homologues among P100, A511 and phiEF24C proteins, but only a few have homologues among proteins of the SPO1 phage. Genome regions close to the ends of virion DNA molecule are those of the least homology among the genomes of *Spounavirinae* that infect hosts of different species groups, indicating that mutational changes, as well as horizontal gene transfer events that led to the adaptation of these phages to their bacterial hosts, occurred mostly there. Remarkably, the pattern of gene conservation along the genomes of *Spounavirinae* representatives resembles that in T4-type bacteriophages. Perhaps, as was proposed in the case of the latter (Filée *et al.*, 2006), the conserved genome portions of *Spounavirinae* have been inherited as a block from a distant common ancestor, and selection for retaining their functions has limited the influence of horizontal gene transfer on them.

C. Gene designations

Despite the surprisingly high homology between genomes of Twort-like phages of *Staphylococcus*, there is no uniform system for designations of their genes. ORF numbering in the genomic sequences of phages K and G1, which were described first, is inconsistent and, in the case of G1, not related to the order of ORFs. Nearly 100 ORFs have not been annotated in the GenBank file of the phage K sequence and are not described in the relevant reference. Thus, to facilitate comparisons between genomes of different Twort-like phages, the authors use gene designations that were proposed elsewhere for the genes of phage A5W and were used in the annotations of genomic sequences of phages Staph1N, A3R, 676Ż, P4W, Fi200W, and MSA6 in the relevant GenBank files (Table I and II; Ulatowska *et al.*, in preparation). Genomic sequences of these phages that are deposited in GenBank, start with the beginning of the left LTR. LTR ORFs are designated as *tre* (terminal repeat-encoded). Certain other genes are designated by abbreviations that reflect the known or predicted function of their products, homology of their products to proteins of known function or the location of these genes in the genome. The remaining ORFs are designated by numbers, which increase rightward along the genome.

D. Long terminal repeat regions

The first genome region of staphylococcal Twort-like phages that reaches a host cell cytoplasm at infection is one of the LTRs. Whether all staphylococcal Twort-like phages have the LTRs of similar length is unclear. The

DNA sequences of G1 and ISP phages, which are deposited in GenBank contain ≈8.4-kb regions corresponding to the LTR regions of phages A5W, Staph1N, Fi200W, P4W, 676Ż, A3R and MSA6. In phage K, the LTR regions were estimated to occupy ≈20 kb of virion DNA molecule and represent the duplicated region corresponding to positions 124,000–127,395 and 1–5,000 of the current genome map (8.4 kb) (Klumpp *et al.*, 2008; O'Flaherty *et al.*, 2004). However, the exact borders between these LTRs and the rest of virion DNA have not been determined. Only 15 out of 21 LTR genes have their homologues in the K genomic sequence that has been deposited in GenBank (Table II). Together they occupy only about 5.6 kb. In the case of the phage Sb-1 sequence that is deposited in GenBank, the region homologous to LTRs of other phages spans only ≈4 kb. In both K and Sb-1 genomes the outermost sequences of the region that does not form LTRs in other phages are conserved. Whether these two phages have shorter LTRs or borders of LTRs shifted to compensate for the deleted regions is unclear. The colinearity of LTRs was shown previously for *Bacillus* phage SPO1 and related phages: SP82, 2 C, and ϕe (Hoet *et al.*, 1983). However, in the case of staphylococcal Twort-like phages deletions in the regions that correspond to LTRs restriction fragments of phage A5W DNA may be common. They have also been reported in certain host range mutants of A5W relative, phage 812 (Pantucek *et al.*, 1998). Additionally, LTRs of the MSA6 and A3R virion DNA appear to be shortened by about 0.4 and 0.2 kb deletions in the region of *treS* and *treO* genes, respectively (Table II).

The mechanism of redundant virion DNA end formation in *Spounavirinae* is not clearly understood. In SPO1, DNA concatamers formed in cells in the process of phage DNA replication contain only one copy of LTR between flanking copies of the remaining genome parts (reviewed by Stewart *et al.*, 2009a). Thus, the redundant copy of the LTR present in each virion DNA must be formed in concert with DNA packaging. Consistently, the nucleotide sequences of both LTRs in each phage are identical.

Comparison of border regions that separate the left and right LTR from the non redundant part of virion DNA in *S. aureus*/*S. epidermidis* Twort-like phages revealed 12-bp perfect inverted repeat (IR) sequences, which are present only there in the virion DNA molecules of these phages (Fig. 4). We presume that both these IRs are involved in the formation of redundant ends of virion DNA molecules during head packaging. What their exact role in this process is and whether it requires a product of any border region gene needs to be determined. Of ORFs that are adjacent to LTR borders, only *bofL* and *treA*, are conserved in *S. aureus/S. epidermidis* Twort-like phages. However, predicted products of both these ORFs are small, highly acidic proteins, which, by their nature, could not bind DNA directly (Ulatowska *et al.*, in preparation). Thus, it is more likely that one

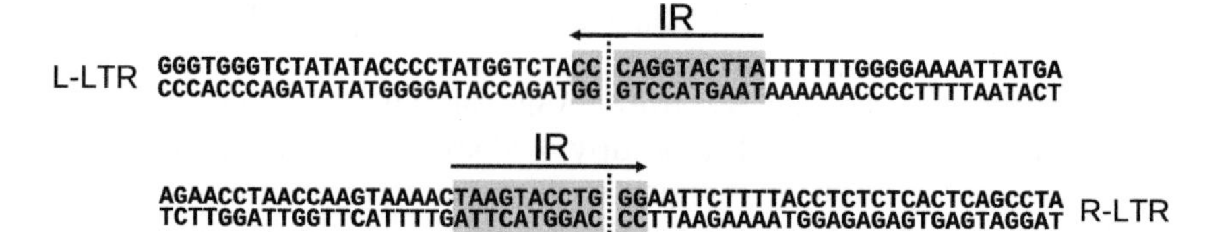

FIGURE 4 Nucleotide sequences of border regions between left and right LTRs and non-redundant parts of the genomes of *S. aureus/S. epidermidis* Twort-like phages. Regions of inverted repeats (IR) are indicated by arrows. Sequences that match perfectly in both repeats of all phages are highlighted in grey. Vertical dotted lines indicate borders between LTRs and non-redundant genome parts.

or both of them could mimic DNA, similarly to certain small, highly acidic phage proteins which assist in recombinational events (reviewed by Murphy, 2012).

The LTR region in *Bacillus* phage SPO1 encodes early phage proteins, whose primary role is subversion of the host's biosynthetic machinery to the purposes of the infecting bacteriophage (Perkus and Shub, 1985; Stewart *et al.*, 1998; Wei and Stewart, 1993). In general, their length does not exceed 100 amino acid residues. Their genes are preceded by strong promoters that are recognized by host DNA polymerase, and have ribosome binding sites characteristic of highly expressed genes. A cluster of early genes that occupy most of the terminal redundancy region in SPO1 phage DNA forms the so-called "host-takeover module". Mutations in certain genes in this region prevent or alter the shut off of host DNA and/or RNA synthesis (Sampath and Stewart, 2004; Stewart *et al.*, 2009a,b).

In *S. aureus/S. epidermidis* Twort-like phages, predicted products of LTR genes are short proteins of no significant homology to any protein of known function (Ulatowska *et al.*, in preparation). An exception is the *treM* gene, which is present in this region in most of the aforementioned phages and which encodes a putative homing endonuclease. Conceivably, some or all LTR genes play a role in host takeover, by anlogy to LTR genes of SPO1. Putative Shine–Dalgarno regions of 20 LTR genes are highly complementary to the 3′ end of staphylococcal 16 S rRNA, suggesting high translational efficiencies (Ulatowska *et al.*, in preparation). Not all LTR genes are essential for phage development, as indicated by the lack of up to eleven genes of this region in the genomes of certain phages (Table II).

E. Transcription of phage genes

Genes of most tailed phages, SPO1 among them, can be divided into early, middle, and late based on their order of expression in phage development (Stewart, 1993; Stewart *et al.*, 2009a). Some of them are expressed in more than one developmental stage at similar or various levels. Different timing of expression is associated with differences in transcriptional signals that precede genes of particular groups. Most of the early genes play a role in host takeover. Middle genes specify mostly DNA replication-associated functions, whereas late genes specify morphogenetic, DNA packaging, and lytic functions. Similar arrangement of genes in genome modules of SPO1 and Twort-like phages implies analogous regulatory patterns.

As many as 58 transcriptional units were proposed in the genome of phage SPO1 (Stewart *et al.*, 2009a). The transcriptional activity of 42 promoters scattered throughout the genome has been confirmed in *in vivo* and *in vitro* experiments (reviewed by Stewart *et al.*, 2009a). The presence of numerous putative promoters in connection with the identification of over 30 rho-independent terminators in the genomes of staphylococcal Twort-like phages also suggests complex transcriptional organization (O'Flaherty *et al.*, 2004; Ulatowska *et al.*, in preparation; Vandersteegen *et al.*, 2011).

Phages of the *Spounavirinae* subfamily do not encode their own RNA polymerase (RNAP) (O'Flaherty *et al.*, 2004; Stewart *et al.*, 2009a). Thus, they have to rely on host RNAP in transcription of all their genes and modify the host polymerase function to redirect it to the recognition of middle and late phage promoters at later phage developmental stages.

Four different σ factors which can associate with the RNAP holoenzyme, were indentified in staphylococcal cells. The primary σ, σ^{SA} - a member of the σ^{70} family, directs transcription of housekeeping genes during exponential cell growth (Deora and Misra, 1996). The results presented by Debhi and colleagues (2009) imply that this σ must also participate in the transcription initiation of early genes of staphylococcal Twort-like phages. In the case of *Bacillus* phage SPO1, transcription of early genes, both in *in vivo* and *in vitro* experiments, also depends entirely on the bacterial RNAP holoenzyme associated with the bacterial housekeeping σ factor (Heintz and Shub, 1982). Thus, promoters that are recognized by bacterial σ^{70} family RNAP σ factors are a hallmark of early transcripts in the SPO1 genome and in the genomes of staphylococcal Twort-like phages.

Numerous putative promoters of -35 and -10 regions characteristic of those recognized by bacterial σ^{70} family σ factors (Lonetto *et al.*, 1992) have been found in the sequences of phages K, ISP, and A5W using different *in silico* methods. Ulatowska and colleagues (in preparation) analyzed predicted transcriptional regulatory signals of LTR genes in phage A5W. All predicted LTR operons are preceded by putative σ^{70} family promoters of relatively high homology to the consensus sequence (Fig. 3). This implies a high level of expression of most of the LTR genes, as was found in the case of the LTR genes of phage SPO1 (Lee *et al.*, 1980; Romeo *et al.*, 1981; Stewart *et al.*, 1998, 2009a). In addition to the LTR region, putative σ^{SA}-recognized promoters were also identified in the upstream regions of several other K, A5W, and ISP genes (O'Flaherty *et al.*, 2004; Ulatowska *et al.*, in preparation; Vandersteegen *et al.*, 2011). At least eighteen of them are located in the regions preceding predicted operons in the 32-kb, leftward transcribed gene cluster, suggesting early expression of most of the genes that are grouped in this cluster, including tRNA genes. Three of the latter, grouped together (tRNA-Trp, tRNA-Phe, and tRNA-Asp), are preceded by two putative early promoters, while the remaining one (tRNA-Met), located separately, is preceded by one such promoter. All predicted early promoters of tRNA genes partially overlap with 14 nt sequences (TGTCAAGTTAATTT), which may be binding sites for a transcriptional regulatory factor. They are conserved in these regions in all *S. aureus*/*S. epidermidis* Twort-like phages. Comparison of the average codon usage frequencies between *S. aureus* and *S. aureus*/*S. epidermidis* Twort-like phages did not reveal major differences, suggesting that phage tRNAs' role is simply to support the host pool of tRNAs in phage mRNA translation.

One of the predicted early genes in the leftward transcribed gene cluster is *hmzG* - a homologue of *mazG*. MazG protein - the NTP pyrophasphatase, has been known as a "house-cleaning" enzyme which participates in the removal of non-canonical nucleoside triphosphates from a cell, and as a regulator of programmed cell death and ppGpp accumulation in *E. coli* (Galperin *et al.*, 2006; Gross *et al.*, 2006; Magnusson *et al.*, 2005; Traxler *et al.*, 2008). Its homologues are encoded by various phages and are signature genes of photosynthetic cyanomyophages. Phage-encoded MazG was proposed to modulate the cellular level of ppGpp pool to extend the period of cell survival under the stress of phage infection (Clockie and Mann, 2006; Weigele *et al.*, 2007). HmzG is likely to play a similar role.

At least fourty two putative σ^{SA}-recognized promoters are scattered throughout the large cluster (~96-kb) of rightward transcribed genes, most of which belong to DNA packaging and head-tail morphogenesis, and to replication and transcription modules. Eleven of them precede genes of the replication and transcription module. They might ensure an early start of phage DNA replication in infected cells. Eighteen predicted early promoters precede genes of the head-tail morphogenesis cluster, as it was observed similarly in the case of certain genes of this cluster in phage SPO1 (Stewart *et al.*, 2009a). Surprisingly, as many as three predicted early promoters are immediately upstream of the *pro* gene, which encodes the prohead protease (Ulatowska *et al.*, in preparation). One cannot exclude that the Pro protein, in addition to its function during head maturation, participates in the conversion of host metabolism at early stages of phage infection. Although the functionality of predicted σ^{SA}-recognized promoters in the genomes of staphylococcal Twort-like phages awaits verification, many of them are likely to be active in the initiation of transcription. In SPO1, 14 early promoters were identified in the non-redundant region of the SPO1 genome (summarized by Stewart *et al.*, 2009a). The functionality of three of them was confirmed.

One of the early genes of staphylococcal phage G1 that is not linked to the LTR region appears to be ORF240 (corresponding to A5W *sci*), which is preceded by a putative σ^{SA}-recognized promoter (Vandersteegen *et al.*, 2011). G1 ORF240 encodes a small, 58 amino acid protein that specifically inhibits *S. aureus* replicase via binding to a DNA sliding clump (Belley *et al.*, 2006). The binding prevents loading of the DNA sliding clump onto DNA and its interaction with DNA polymerase C. The relevant protein of phage Twort, gpORF168, has similar properties. Both G1 gpORF240 and phage Twort gpORF168 play a role in host takeover. Thus, it is surprising to find that homologues of these genes, as well as flanking genes, are absent from the genome of phage K and Sb-1. They are possibly replaced by additional genes specific for K and Sb-1 that are located at the right end of the genomes of these phages (Table II). Development of phage K in *S. aureus* cells was shown to cause inhibition of host DNA synthesis during the first 5 min following infection (Rees and Fry, 1981a). One cannot exclude that other proteins, in addition to gpORF240, may block host replication at phage infection as well. Alternatively, the deletion of *sci* gene region in phage K could occur recently, and differentiated the sequenced phage K clone from the one that was studied by Rees and Fry.

In the case of SPO1, after the early stage of phage development, the bacterial σ factor is replaced by SPO1-encoded factors. One of them, σ gp28, directs transcription of middle genes, and two, gp33 and σ gp34, direct transcription of late genes (Chelm *et al.*, 1982 and references therein; Costanzo and Pero, 1983, 1984; Fujita *et al.*, 1971; Gage and Geiduschek, 1971; Losick and Pero, 1981; Talkington and Pero, 1977). One can expect that staphylococcal Twort-like phages, like SPO1, encode their own sigmas or other factors redirecting the host polymerase holoenzyme from phage early, to middle, and then to late genes. However, no homologue of SPO1 gp28 could be found among predicted protein products of these phages. In SPO1, gp28 by itself appears insufficient enough in redirecting host RNA polymerase to explain the abrupt shut off of early gene transcription that accompanies the initiation of transcription of middle genes (Chelm *et al.*, 1982). At least three small proteins, products of genes 44, 50, and 51 of the "host-takeover module," are additionally involved in the regulation of early transcription shut off (Sampath and Stewart, 2004). Staphylococcal Twort-like phages do not encode an obvious homologue of any of these proteins. However, by searching for interactions of phage G1 proteins that inhibit growth of *S. aureus*, with proteins of *S. aureus* cells, Debhi and co-workers (2009) identified gpORF67 of phage G1 (equivalent to the product of A5W *asf*) as an anti-sigma factor that binds to the primary σ factor of *S. aureus* (σ^{SA}) and inhibits transcription at σ^{SA}-dependent promoters (Debhi *et al.*, 2009; Liu *et al.*, 2004). The binding of G1 gpORF67 to σ^{SA} and the resulting inhibition of host σ^{SA}-dependent RNAP activity was proposed to trigger a shift in the profile of transcription from early to middle phage genes.

Mechanisms of late transcription activation in staphylococcal Twort-like phages and SPO1 appear to have more in common. A protein that is a homologue of SPO1 σ gpORF34 is encoded by the genomes of all these phages (Sig of A5W, Table II). However, staphylococcal Twort-like phages do not encode a homologue of SPO1 gp33, which act in concert with gp34 to redirect host RNA polymerase to phage late genes. In the genome of SPO1, gene 33 is immediately upstream of gene 34. Whether the product of ORF125, which precedes the *sig* gene in *S. aureus*/*S. epidermdis* Twort-like phages, is involved in the action of Sig needs to be elucidated.

Of the additional phage SPO1 transcriptional regulators, staphylococcal Twort-like phages encode a homologue of Tf1 protein (A5W Trf; Fig. 3, Table II). SPO1 Tf1 is a histone-like transcription factor. It binds and bends double-stranded DNA, showing a preference for hydroxymethyluracil (HMU)-containing DNA (Geiduschek *et al.*, 1990; Schneider *et al.*, 1991). Additionally, it is required for shutoff of expression of certain middle genes and transcriptional activation of certain late genes (Sayre and Geiduschek, 1988). The Trf protein of staphylococcal Twort-like phages is likely to play a similar role in the regulation of transcription in these phages, despite the lack of HMU in their DNA.

F. Phage DNA replication

The kinetics of phage DNA synthesis was studied in detail in the case of phage K (Rees and Fry, 1981a,b; 1983). Infection of cells with phage K is followed by a cessation of bacterial DNA synthesis within 5 min and by the degradation of bacterial DNA. As soon as a replicative form of parental phage DNA appears in a cell, nucleotides from host DNA start to be incorporated into phage DNA. Mature phage DNA appears in cells in about 20 min post infection. The phage replicative complex is attached to the cell membrane. Some proteins in this complex must be host proteins, as formation of the complex is not completely inhibited by chloramphenicol.

Staphylococcal Twort-like phages, like other phages of the *Spounavirinae* subfamily, encode most of the proteins required for replication of their DNA (Klumpp *et al.*, 2008; O'Flaherty *et al.*, 2004; Stewart *et al.*, 2009a). In this respect they resemble T4 and other phages of the T4 superfamily whose replication relies mostly on their own proteins (Comeau *et al.*, 2007; Miller *et al.*, 2003).

At least 12 genes of the replication and transcription genome module of staphylococcal Twort-like phages can be unambiguously assigned functions related to DNA metabolism, replication, or repair (Table II). They encode DNA helicases, recombination nucleases, DNA-directed DNA polymerase, a homologue of plasmid replication protein, DNA primase, thioredoxin, homologues of DNA-binding proteins, and ribonucleotide reductase subunits. Neither the ligase gene, nor any of the numerous genes that encode potential membrane proteins are contained in this module.

Perhaps a protein that serves to attach the complex of replicating phage DNA to a cell membrane, analogous to membrane protein p16.7 of *Bacillus* phage φ29 (Alcorlo *et al.*, 2007), is encoded by a gene somewhere else in the genome. A candidate could be *mbpB* or *mbpU*, which are located close to the *lig* gene in the leftward transcribed gene cluster. Each of them encodes potential membrane protein.

The region of DNA replication origin in staphylococcal Twort-like phages has not been identified so far. However, according to the results of GC skew analysis in the genome of the A5W phage, it is likely to be located close to the region that encodes three tRNAs (Fig. 3; Ulatowska *et al.*, in preparation). Two direct repeats of the 28 nucleotide sequence in this region (AAAAAGTACGTATTTAGAAAATAAGGAG) could potentially serve as binding sites for the replication initiator protein. They are located at positions 39153 and 39343 of the A5W genomic sequence and are conserved in all *S. aureus*/*S. epidermidis* Twort-like phages (Fig. 3). Phages G1 and Sb-1 contain two additional repeats of this sequence in the relevant region due to the insertion of the 380 bp fragment, which was apparently generated by the duplication of neighboring DNA (unpublished results).

DNA polymerases of Twort-like phages and other spounaviruses are less closely related to DNA polymerases of most other phages than to bacterial type I DNA polymerases which are enzymes of low processivity (Lavigne *et al.*, 2009). In this respect they resemble the DNA polymerase of phage T7. The high processivity of T7 DNA polymerase is achieved through tight binding to its processivity factor, *E. coli* thioredoxin (Bedford *et al.*, 1997). Most likely, thioredoxin, which is encoded by staphylococcal Twort-like phages and other spounaviruses, also plays a role of a DNA polymerase processivity factor.

G. Morphogenetic and lytic functions

Thirty-three proteins of staphylococcal Twort-like phages have been identified as virion components (Fig. 3; Eyer *et al.*, 2007; O'Flaherty *et al.*, 2004; Paul *et al.*, 2011b; Vandersteegen *et al.*, 2011). Additional virion proteins still await identification, as implied by results of analysis of the *Bacillus* SPO1 phage, whose virion consists of 53 proteins (Parker and Eiserling, 1983b). However, homologies between proteins of SPO1 and proteins of staphylococcal Twort-like phages, which are encoded by genes of comparable location in their genomes, allow one to recognize presumable additional virion components of the latter (Table II).

Genes encoding 23 virion proteins of staphylococcal Twort-like phages are grouped together in the region of packaging and head–tail module that directly precedes the module of DNA replication and transcription genes (Fig. 3; Eyer *et al.*, 2007). They are occasionally interspaced with one to four genes, whose products either are not virion components or have not as yet been identified as such. Products of over half of the genes in this group are conserved in the *Spounavirinae* subfamily of myoviruses (Kwan *et al.* 2005; Stewart *et al.*, 2009a). The organization of genes in relevant genome regions is also conserved (Fig. 3). The first virion protein-specifying gene in this cluster, *ter*, encodes the large terminase subunit. The products of none of the two genes that are in the direct neighborhood of *ter* have homologies to known small subunits of phage terminases. However, the product of ORF59 (next to *ter*), has a homologue in the *Listeria* P100 phage and is a virion component.

Four genes of unknown function separate ORF59 from two other genes for virion proteins, *prt* and *pro*, which encode portal and prohead protease, respectively. Three of them—ORF60, *mbpF*, and ORF63—have similarly located homologues in P100, whereas *pro* and *prt* are also conserved in SPO1. A conserved gene of unknown function, ORF66, separates *pro* from a gene for a major capsid protein, *mcp*, which is highly conserved in all *Spounavirinae*.

Similarly, as in genomes of phages P100, A511, and SPO1, six ORFs separate the major capsid protein gene from the gene that encodes the major sheath protein, *tsp*. Three of them (ORF069, ORF070, and ORF072) encode virion components, while the function of the three others is unknown. The first of these ORFs, ORF068, although similar in length to ORFs of the same location in P100 and SPO1, encodes an unrelated protein, specific only for staphylococcal Twort-like phages. Products of the next four ORFs are conserved in the latter, as well as in P100 and SPO1. However, the ORF located immediately downstream has a homologue in P100, but is replaced by an unrelated ORF in SPO1. All proteins encoded by *Listeria* phage A511 genes in this region were predicted to be tail morphogenetic proteins (Loessner and Schrerer, 1995).

The *tsp* gene by itself is conserved in *Spounavirinae* representatives and is separated from the first gene of the replication and transcription module by 22 genes. Most, if not all, of these genes encode tail morphogenetic functions, as implied by homologies of their products to proteins of other *Spounavirinae* (Chibani-Chennoufi *et al.*, 2004; Klumpp *et al.*, 2008; Stewart *et al.*, 2009a). Products of 13 of them were identified in phage K, ISP, or 812 virions. The first seven of these genes separate *tsp* from the *tmpC* gene, which encodes the tape-measure protein—a tail length determinant. The product of the first of them, *tmpA*, a virion component, is conserved in *Spounavirinae*. It is a tail tube subunit, as implied by homology to the relevant protein of SPO1, gp10.1 (Stewart *et al.*, 2009a). In phage Twort, the corresponding gene (ORF106) is interrupted by three introns (Kwan *et al.*, 2005; Landthaler and Shub, 1999). The region of *tmpA* may be a common site of intron insertion. In phages Fi200W, P4W, A3R, and 676Ż, *tmpA* is also an intron-interrupted gene (Table III). The region of a few downstream genes is specific for staphylococcal Twort-like phages. It contains more genes than the relevant regions in the genomes of other

Spounavirinae representatives and varies between *S. aureus*/*S. epidermidis* Twort-like phages and phage Twort. In genomes of the former it contains four genes; one of them does not have a homologue in Twort. In the genome of the latter it contains five genes. Two are specific for Twort. One gene in this region, ORF78, encodes another virion component, perhaps also a tail protein. It is followed by two genes, ORF79 and *tmpB*, which precede *tmpC*. Their products were not detected in a virion of any staphylococcal Twort-like phage. However, they are conserved in phages of the *Spounavirinae* subfamily. In SPO1, a protein relevant to the product of *tmpB* was identified as a tail assembly chaperone. Thus, the homologous TmpB protein is likely to play a similar role in the assembly of phage tail. The location of the *tmpB* gene just upstream of a gene for the tape-measure protein *tmpC* is conserved in *Spounavirinae*.

Comparison of tape-measure proteins of *Spounavirinae* representatives fits with previous observations that in phages with a long flexible tail, the length of the tail is determined by a gene whose product corresponds in length to the tail length (Katsura and Hendrix, 1984; Pedulla *et al.*, 2003). The relevant tape measure proteins of staphylococcal Twort-like phages and *Listeria* phage P100, which have substantially longer tails than that of SPO1, are also substantially longer than the tape measure protein of the latter (1351–1352 and 1242 amino acids, respectively, versus 891 amino acids). Stewart and co-workers (2009a) noted that the ratios of tail lengths to the number of amino acids in the tape-measure proteins of these phages are consistent with the respective ratios in certain other phages ($\sim$1.5Å of tail length per amino acid) and are indicative of the α-helical structure of tape-measure proteins.

Of the 14 genes that are between *tmpC* and the first gene of the replication module, *dhlA*, only the 9 that are the closest to *tmpC* have homologues of a similar location in the genome of SPO1. The first of them, *tmpD*, encodes a virion component - tail-associated muralytic enzyme (Paul *et al.*, 2011b). Its product, TAME, which was purified from phage K virions (gpORF56), could hydrolyze the staphylococcal cell wall peptidoglycan. The domain of cysteine, histidine-dependent aminohydrolase/peptidase activity of this protein is located in its C-terminal moiety. When overproduced in *E. coli* cells, TAME undergoes autoproteolysis to shorter forms which retain muralytic activity.

Tail-associated peptidoglycan hydrolases are typical phage proteins (Briers *et al.*, 2006; Kao and McClain, 1980; Kenny *et al.*, 2004; Lehnherr *et al.*, 1998; Nakagawa *et al.*, 1985; Rashel *et al.*, 2008; Takac and Blassi, 2005; reviewed by Moak and Molineux, 2004). They are involved in the initial stages of infection by digesting bacterial murein at the point of phage adsorption and permitting penetration by the tail tube (Kanamaru *et al.*, 2002). The murein hydrolase of T4 myovirus is essential for infection (Nakagawa *et al.*, 1985), while the T7 podovirus or mycobacteriophage TM4 myovirus murein hydrolyses are not essential proteins (Moak and Molineaux, 2000; Piuri and

Hatfull, 2006). They facilitate the infection, especially at bacterial growth conditions, which lead to the increased cross-linking of peptidoglycan (e.g., in stationary phase cells). The decreased sensitivity of *S. epidermidis* stationary phase cells to infection with wild-type bacteriophage K suggests, that in Twort-like phages, TAME plays an essential role in infection rather than an accessory one (Cerca *et al.*, 2007). Conceivably, like tail lysins of certain other phages, it is also involved in the so-called "lysis from without" (Delbruck, 1940), which was observed in the case of these phages (Ralston, 1957, 1963; Ralston and McIvor, 1964). This phenomenon is manifested by rapid lysis of cells infected with a phage at a high multiplicity of infection (M.O.I.) or by sensitization of cells to low concentrations of lytic agents by adsorbed phages.

TAME is a virion component which contributes to a wide range of strain specificity of staphylococcal Twort-like phages. The C-terminal, muralytic domain of the enzyme lysed cells of all strains representing a global panel of clinical *S. aureus* isolates, when fused to the cell-wall binding domain of lysostaphin (Paul *et al.*, 2011b). However, TAME may not be the only baseplate petidoglycan-hydrolyzing enzyme of staphylococcal Twort-like phages and other spounaviruses. The product of the *tmpE* gene (next to *tmpD*) is likely to be a cell wall peptidase, as its C-terminal moiety contains a motif characteristic of these enzymes (Ulatowska *et al.*, in preparation). The TmpE protein is a virion component and is conserved in other spounaviruses.

Twelve genes between *tmpE* and *dhlA* are likely to encode other baseplate structural proteins or baseplate assembly functions. Products of 9 of them—ORF86, ORF87, *bmpA*, *bmpB*, *bmpC*, *tmpG*, ORF94, ORF95, and ORF96—are virion components, of which four—gpORF86, gpORF87, BmpA, and BmpB—have significant homologies to baseplate proteins of other phages (Stewart *et al.*, 2009a; Table II). A homologue of BmpC, the gp107 protein of *Bronchtrix* phage A9, a nonclassified spounavirus, was also identified as a virion component of this phage (Kilcher *et al.*, 2010). The product of *tmpG* (equivalent to K ORF65) is homologous in its C-terminal moiety to proteins of neuraminidase/sialidase activity (Table II; Eyer *et al.*, 2007). Enterobacterial phages of the K1 family, which infect *E. coli* K1, specifically recognize and degrade the polysialic capsule of their host bacteria utilizing tail spike endosialidases (Schulz *et al.*, 2010). Cells of numerous clinical *S. aureus* isolates are encapsulated or produce slime (Sompolinsky *et al.*, 1985; Sutter *et al.*, 2011). Sialic acid has not been described as a staphylococcal cell capsule component so far. However, Sakarya and colleagues (2004) showed that sialic acid may be a constituent molecule of slime that is produced by staphylococci and may be involved in bacterial adherence to inert surfaces. *Clostridium perfringens* neuraminidase significantly decreased slime production and adherence of certain strains of staphylococci to solid surfaces in a dose-dependent manner. Possibly, the TmpG protein of staphylococcal Twort-like phages facilitates the infection of certain strains with this phages by digesting sialic acid residues in their slime.

Nine of the 12 aforementioned genes for baseplate morphogenetic functions that are the most proximal to *tmpE* are conserved in phages P100 and SPO1. Of the remaining 3 genes, two, ORF094 and ORF96, are conserved only in P100, while the remaining one, ORF95, is specific for staphylococcal Twort-like phages. In the genome of SPO1, the region that corresponds to the region of these three genes was proposed to encode tail fiber proteins (Stewart *et al.*, 2009a). Ulatowska and co-workers (in preparation) noticed partial homologies of the predicted products of ORF94 and ORF95 to proteins of staphylococcal siphoviruses encoded by the genomic regions of these phages specifying tail structural components and located immediately upstream of lytic genes. Such a location in phages of the *Siphoviridae* family is characteristic of genes encoding tail fiber proteins (Brüssow and Desiere, 2001; Kwan *et al.*, 2005). Additionally, the product of ORF96 appears similar to a conserved tail structural protein of staphylococcal podoviruses responsible for the adsorption of these phages to host cells (Uchiyama *et al.*, Koshi University, Japan, manuscript in preparation). Conceivably, ORF94 and ORF95 encode tail fiber assembly functions or parts of tail fibers, whereas ORF96 encodes either the entire tail fiber or a part of tail fiber that interacts directly with teichoic acid on the surface of *Staphylococcus* cells. Results of studies by Xia and co-workers (2010, 2011) indicate that staphylococcal Twort-like phages adsorb to the anionic backbone of cell wall teichoic acid (WTA) in staphylococcal cells. Adsorption of phage 812 to cells of the *S. aureus* RN4220 mutant derivative, which lacks WTA, was severely impaired. However, mutants lacking α-GlcNAc residues in WTA or deficient in alanylation of teichoic acid were infected with this phage similarly as the wild-type strain, indicating that phage adsorption requires neither glycoepitope nor alanyl modification of WTA. This is in contrast with earlier results of Archibald and Coapes (1972) who showed that concanavalin A specifically blocks the adsorption of phage K to *S. aureus* cells containing α-N-acetyl-glycosaminyl but not β-N-acetyl-glucosaminyl substituents in WTA. The reasons for these discrepances are unclear. Possibly, staphylococcal Twort-like phages can use more than one receptor on the surface of bacterial cells.

Ten virion proteins of staphylococcal Twort-like phages are encoded by genes outside of the region of packaging and head–tail module. Genes for three of them, putative tail proteins gpORF127, TmpH, and TmpI, are clustered together and separated from the packaging and head–tail module by the replication and transcription module. Two of them, ORF127 and TmpH, have homologues encoded by the relevant genome regions of *Listeria* P100 and A511 phages and by certain other *Spounavirinae* representatives: *Enterococcus* phage phiEF24C, *Lactobacillus* phage Lb338-1, and *Bronchotrix* phage A9. Separation of these genes from other genes of head-tail module is likely to be a result of insertion of the replication and transcription module to the head-tail module in the genome of a common ancestor of these phages via horizontal gene transfer.

In *Bronchotrix* phage A9, the homologue of gpORF127 of phage A5W, gp154, was found in virions in a shorter and in a longer form (gp154L). The latter appeared to be a fusion protein that must be formed by translational frameshifting of ribosomes at the 3′ end of gene 154 mRNA to translate a message of the next gene, 155 (Kilcher *et al.*, 2010). Its N-terminal moiety is homologous to ORF127, while its C-terminal moiety is homologous to A5W TmpH protein. In the case of all *S. aureus / S. epidermidis* Twort-like phages, the end of ORF127 and the beginning of *treH* are separated from each other by a 23-bp DNA fragment, indicating that ORF127 and *treH* are translated separately.

TmpH and TmpI - a predicted product of the gene that is next to *tmpH* - have numerous homologues among tail proteins of various staphylococcal phages (data not shown). Homologous regions extend over the entire length of both proteins, suggesting that TmpH and TmpI may be tail components responsible for the specific interaction of Twort-like phages with cells of their hosts.

The amino acid sequence of the C-terminal moiety of TmpH, similar to its homologues in other spounaviruses, including the C-terminal end of *Bronchotrix* A9 phage gp154L, contains a motif characteristic of the bacterial and phage protein domains of the Ig-like fold of the Big2 sequence family (PFAM02368; E value: 1.55e-05; Eyer *et al.*, 2007). A similar motif was found in the C-terminal Ig-like domain of the bacteriophage λ tail tube protein, gpV, whose structure has been determined (Pell *et al.*, 2010), and in hundreds of proteins of tailed phages (Fraser *et al.*, 2006; Pell *et al.*, 2010 and references therein). Many phage Ig-like domains are added to larger proteins through programmed translational frameshifting and many are surface exposed in phage virions. It has been proposed that they play an accessory role in phage infection by interacting weakly with carbohydrates on the bacterial cell surface (Fraser *et al.*, 2007). In gpV, the Ig-like domain exists as a protrusion along a tail and is located at the surface of the virion (Roessner and Ihler, 1984). Plating efficiencies of lambda mutants that are devoid of this domain are several orders of magnitude lower than that of the wild-type phage (Katsura, 1981; Pell *et al.*, 2010). In phages of the *Myoviridae* family, Ig-like domains were identified within certain highly immunogenic outer capsid proteins (T4 Hoc and its homologues), fibritin and baseplate proteins (Fraser *et al.*, 2006, 2007). Although T4 Hoc is a nonessential protein, it was proposed to provide survival advantages to the virus by attaching the phage capsid loosely to bacterial surface molecules and allowing the virus to stay attached to the cell while its tail fibers find their receptors (reviewed by Black and Rao, 2012).

Whatever the role of TmpH Ig-like domain in the virions of staphylococcal Twort-like phages is, it is dispensable. Virions of phage MSA6 do not have this domain due to a deletion that encompasses the 3′ moiety of *tmpH* and nearly the entire *tmpI*. Instead, they may contain a fusion protein that is composed of the N-terminal fragment of TmpH and the 17 distal amino acid residues of TmpI. The N-terminal moiety of TmpH appears homologous to another virion protein, which was identified by Eyer and colleagues (2007) as the phage 812 homologue of G1 gpORF138 and is equivalent to A5W ORF014. This ORF is conserved in all staphylococcal Twort-like phages. It is located outside of the DNA packaging and head–tail genome module, like the *tmpH* gene, but separately from *tmpH*, in the cluster of leftward-oriented genes immediately downstream of left LTR (Fig. 3). Most likely it also encodes a tail protein. No other genes whose products were identified as virion components are linked to ORF014. However, the cluster of leftward-oriented genes in the genomes of Twort-like phages contains additional three genes that encode virion proteins and are scattered throughout this cluster (ORF038, ORF044, and *dmcB*; Fig. 3). Their predicted products have no homologues in databases.

The three remaining virion proteins, gpORF140, gpORF166, and NadV, are encoded by unlinked genes that are located in three distant regions of the about 20-kb terminal genome part of staphylococcal Twort-like phages. One of them, gpORF140, is conserved in phage Twort and certain other phages of the *Spounavirinae* subfamily. The second of these protein, gpORF166, is conserved only in A5W, Fi200W, P4W, MSA6, and G1, indicating that it is not an essential virion component. Phages K, Sb-1, and A3R, whose genomes are substantially smaller, do not have a part of the relevant genome region (Table II). In the case of K and Sb-1, the result of deletions are in-frame fusions between 25 or 26 3′ codons of ORF166, respectively, and all 54 codons of ORF145. Additionally, the fusion in the genome of phage K contains numerous codons derived from genes that are between ORF145 and ORF166. The third virion component encoded by the terminal genome region, NadV, contains amino acid sequence motifs characteristic of nicotinamide phosphoribosyl transferase - a principle enzyme of NAD^+ salvage pathway, which catalyzes the formation of nicotinamide mononucleotide, a NAD^+ precursor, from nicotinamide (O'Flaherty *et al.*, 2004; Ulatowska *et al.*, in preparation). The part of NadV protein preceding this motif is dispensable, as suggested by the deletion of the relevant DNA region from the genome of A3R phage. The deletion results in the formation of fusion protein, in which the N-terminal part of NadV is replaced by the N-terminal moiety of gpORF157. Homologues of NadV have been found among proteins of only a few phages so far (Lehman *et al.*, 2009; Miller *et al.*, 2003; Nolan *et al.*, 2006;

Yamada *et al.*, 2010). Possibly, as a virion component, NadV is injected to cells along with phage DNA at infection, ensuring the abundance of NAD^+ which is required as a cofactor for certain phage early enzymes, such as e.g. ribonucleotide reductase.

Certain virion components of staphylococcal Twort-like phages are synthesized as protein precursors that undergo proteolytic processing during virion maturation. One of them is a major capsid protein, Mcp. A form of this protein, which was isolated from the phage K virion, was truncated at its N-terminal end by 23 amino acid residues as compared to the predicted product of the relevant gene (O'Flaherty *et al.*, 2004). Proteolytic processing of major capsid proteins was also detected in the case of other *Spounavirinae* subfamily phages, such as SPO1 and A511 (Loessner and Scherer, 1995; Stewart *et al.*, 2009a). The region of proteolytic cleavage of the Mcp precursor is conserved in all these phages.

The major tail sheath protein of staphylococcal Twort-like phages, a product of the *tsp* gene, also appears to be processed during virion maturation. Its N-terminal and C-terminal moiety migrated as separate proteins in SDS–PAGE gels of denatured 812 virions (Eyer *et al.*, 2007). Tail sheath protein maturation by proteolyic cleavage may appear a conserved feature of *Spounavirinae* family representatives as well. In *Lactobacillus plantarum* Twort-like phage LP65, the tail sheath protein is synthesized in the form of a precursor, which undergoes proteolytic processing (Chibani-Chennoufi *et al.*, 2004).

Release of mature phages from a cell requires the action of two phage-encoded proteins: a peptidoglycan hydrolysing enzyme - endolysin, and a membrane pore-forming protein, holin (Young, 2005). Lesions in a cell membrane formed by holin pave a way for endolysin to access its substrate - the cell wall. A third protein, anti-holin, regulates the timing of lysis by inhibiting the holin action until mature phages are ready for release. Genes encoding holin and endolysin, *holA* and *lysK* (*plyTW* and *holTW* in phage Twort; Loessner *et al.*, 1998), were identified in the leftward-oriented gene cluster of staphylococcal Twort-like phage genomes (Fig. 3, Table II). The lytic potential of LysK protein was confirmed in *in vivo* and *in vitro* experiments (O'Flaherty *et al.*, 2005a; reviewed by Nelson *et al.*, 2012). Purified LysK has a wide spectrum of anti-staphylococcal activity. It can lyse untreated cells of different *S. aureus* strains as well as cells of coagulase-negative staphylococci (Becker *et al.*, 2009). How the time of lysis is controlled *in vivo* is incertain.

Staphylococcal Twort-like phages can encode more than one holin, analogously to enterobacteria phage P1 (Łobocka *et al.*, 2004); for example, the product of the *mpbE* gene of the head-tail module, a membrane domain-containing protein, triggers lysis of *E. coli* cells, producing phage λ endolysin (authors' unpublished results). An additional cell wall degrading enzyme may be the product of *tgl* gene, which encodes a putative transglycosylase. The *tgl* gene is located in the cluster of leftward transcribed genes, three genes upstream of *lysK* - a gene of the cell lysis module. The amino acid sequence of Tgl protein is homologous to amino acid sequences of staphylococcal surface exposed, putative lytic transglycosylases IsaA and SceD. Both these proteins are autolysins, and major *S. aureus* surface antigens, which participate in cell wall remodelling and are required for *S. aureus* virulence or nasal carriage (Burian *et al.*, 2010; Stapleton *et al.*, 2007). SceD is a candidate for a component of a vaccine against a carriage of *S. aureus*. The location of *tgl* gene in the direct neighborhood of the lytic module indicates that this apparently host-acquired gene might have evolved in the genomes of staphylococcal Twort-like phages to facilitate mature phage release from infected cells. Whether Tgl protein is indeed involved in cell lysis and why the *tgl* gene is preceded by a putative early phage promoter remains to be found (see Fig. 3).

H. Introns and inteins

A remarkable feature of staphylococcal Twort-like phages and other members of the *Spounavirinae* subfamily is the interruption of certain genes with sequences encoding group I introns or inteins - intervening RNA or proteins, respectively, which are able to splice in an autocatalytic manner from their precursor molecules (Landthaler and Shub, 1999; Landthaler *et al.*, 2002, 2004). An intein-coding gene was identified in the genome of phage Twort (Kwan *et al.*, 2005). Introns in the genomes of phages K, G1, and Twort have been analyzed in detail and are described elsewhere (Kwan *et al.*, 2005; O'Flaherty *et al.*, 2004; Stewart *et al.*, 2009a). None of the seven introns which have been identified in the genome of phage Twort have identical counterparts in the genome of any *S. aureus*/*S. epidermidis* Twort-like phages (Table III). However, an intron identified in the endolysis gene, as well as two introns identified in the DNA polymerase gene of phages K and G1, are conserved in all the latter phages. All these introns encode proteins (Tables II and III). Genome-wide comparison between all *S. aureus*/*S. epidermidis* Twort-like phages revealed the presence of additional insertions encoding intron-like sequences in the genomes of certain phages. The tail tube gene of phages 676Ż, Fi200W, P4W, and A3R, (*tmpA*) appears to contain introns which are the same as the identically located introns in the genomes of phage 812 and U16 host range mutants, whose complete DNA sequences have not been published yet (Kasparek *et al.*, 2007; author's unpublished results). None of them encodes any protein. An additional insertion that is identical to a part of the 812 and U16 mutants intron-like sequence is located immediately downstream of *tmpA*. It can encode a protein (I-812I, Table II) containing an amino acid sequence motif characteristic of intron-encoded nucleases. A verification of the intron borders in the *tmpA* gene requires further transcriptomic studies. Two additional introns are encoded by phage MSA6: rec-I1, in the *rec* gene for the recombinase, and ter-I1, in the *ter* gene for the large terminase. They both encode proteins, which are homologous to intron-encoded endonucleases (data not shown). A common deriviation of MSA6 ter-I1 intron and phage Twort intron in the terminase gene may be suggested by 221-bp regions of these introns that are in 91% identical and by the homology of their protein products. The detailed intron analysis of staphylococcal Twort-like phages is in progress and will be described elsewhere.

Shuffling of intron-encoding sequences appears to be responsible for certain differences between genomes of staphylococcal Twort-like phages and was proposed to be associated with spontaneous changes in the host range of particular phages (Kasparek *et al.*, 2007; Kwan *et al.*, 2005; Pantucek *et al.*, 1998; unpublished results). However, the variability of introns and their insertion sites in the genomes of the 11 *S. aureus*/ *S. epidermidis* Twort-like phages, which have been compared in this chapter, is too small to explain all differences in the host range between these phages. For example, phages 676Ż, Fi200W, P4W, and A3R have identical introns in their genomes but differ in specificity (data not shown). Conceivably, intron shuffling is only one of multiple mechanisms that can lead to host specificity changes in these phages.

I. Strategies of staphylococcal Twort-like phages to overcome host restriction systems

A major defense against invading DNA are restriction–modification (R-M) systems (Murray, 2000). They are also the main barrier limiting horizontal gene transfer between staphylococcal strains of different clonal lineages as well as the spread of staphylococcal phages (Lindsay, 2010; Novick, 1990; Waldron and Lindsay, 2006; Zueva *et al.*, 1990). Consistently, the distribution of prophages varies considerably between strains of dominant clonal lineages of human *S. aureus* strains, differing in their R-M specificity genes (Goerke *et al.*, 2009; Waldron and Lindsay, 2006). Inactivations of the R-M system were proposed as being responsible for the enhanced susceptibility of certain mutagenized *S. aureus* isolates to bacteriophage lysis (Iordanescu and Surdeanu, 1976; Stobberingh and Winkler, 1977). Additionally, *S. aureus* cells could acquire resistance to all 23 phages of the international *S. aureus* basic typing set upon lysogenization with phage ϕ42, which encodes its own R-M system, Sau42I (Dempsey *et al.*, 2005). Thus, an efficient protection from restriction must be an essential polyvalency determinant of staphylococcal Twort-like phages.

Bacteriophages have evolved various strategies to avoid host restriction (reviewed by Labrie *et al.*, 2010). In the prototypical *Spounavirinae* representative SPO1, thymine is replaced by hydroxymethyl uracil

(HMU) to avoid attacks from host restriction endonucleases that recognize sites containing the former. Staphylococcal Twort-like phages do not contain HMU in their DNA (O'Flaherty *et al.*, 2004). One of their strategies to overcome the activity of host R-Ms is avoiding host-recognized restriction sites in DNA. The genomes of these phages do not contain GATC sequences, which are the recognition sites for the *S. aureus* Sau3AI restriction endonuclease (O'Flaherty *et al.*, 2004; Sussenbach *et al.*, 1976; data not shown). Additionally, each of them contains only a single recognition sequence for the second *S. aureus* restrictase of known specificity, Sau96I (GGNCC; Sussenbach *et al.*, 1978; Sziläk *et al.*, 1990).

The protective efficacy of R-M systems is directly proportional to the number of recognition sites in invading DNA (Wilson and Murray, 1991). Thus, the absence or paucity of certain tetra- or pentanucleotides in DNA of large phages may serve as a guide in the discovery of new restriction enzyme specificities in bacteria that are their hosts. The following additional 4- or 5-bp sequences are absent or present in just a single copy in the DNA of staphylococcal Twort-like phages: CGCG, GGCC, CGWCG, GCWGC, CCGG, and CCSGG. One of them, CGCG - a recognition site for the *B. subtilis* endonuclease BsuE (Gaido *et al.*, 1988) - is also absent from the genome of *Bacillus* SPO1 phage. Possibly, staphylococcal cells encode an isoschizomer of BsuE. Specificities of at least a few *S. aureus* R-M systems still await identification (Dempsey *et al.*, 2005; Waldron and Lindsay, 2006).

The paucity of certain restriction sites in DNA cannot be the only strategy to protect staphylococcal Twort-like phages from their hosts' R-M systems. Too great a number of sites would have to be absent from the genomes of these phages to avoid various R-M type I systems present in *S. aureus* strains of different clonal lineages, as well as the systems of various types, that are encoded by staphylococcal mobilome (Dempsey *et al.*, 2005; Iordanescu and Surdeanu, 1976; Noto *et al.*, 2008; Sjöström *et al.*, 1978; Tsuru *et al.*, 2006; Waldron and Lindsay, 2006; Veiga and Pinho, 2009). Consistently, sequence CTYRAG, which is recognized by an isoschizomer of SmlI, detected in certain staphylococcal strains (Godany *et al.*, 2004), is abundant in the DNA of staphylococcal Twort-like phages.

Despite the remarkably wide range of strain specificity of *S. aureus*/ *S. epidermidis* Twort-like phages, certain *S. aureus*/*S. epidermidis* strains in large strain collections remain resistant to infection with the particular phages of this group (O'Flaherty *et al.*, 2005b; Pantucek *et al.*, 1998). In the majority of cases, R-M systems appear responsible for the observed resistance. Kelly and co-authors (2011) studied 29 strains resistant to phage K. They mixed a culture of each resistant strain with the supernatant of a culture of the same strain that was exposed to phage K for the time that usually suffices for cell lysis. In the case of 24 initially resistant strains, repetition of this procedure from two to five times resulted in the clearing of bacterial culture and release of progeny phages able to propagate efficiently in cells of such strains. Similar "adaptation" procedures of Twort-like phages to initially resistant bacterial hosts have been widely used in several laboratories (see e.g., O'Flaherty *et al.*, 2005b; Pantucek *et al.*, 1998; Ralston and Krueger, 1954).

Although R-M systems rarely provide complete protection to *S. aureus*/*S. epidermidis* strains against infection by relevant Twort-like phages, they may seriously limit the infection efficiency of certain strains that differ from the phage propagating strain. Ulatowska and co-workers (in preparation) studied the infection efficiency of 74 clinical *S. aureus* strains of 11 PFGE types and several subtypes with the A5W phage propagated in the *S. aureus* PS80 strain. Only in the case of 16 strains was the plating efficiency of A5W similar as in the case of the phage propagation strain. In other cases, plating efficiencies were two to seven orders of magnitude lower. The infection of cells from the first group of strains with A5W in liquid cultures (at M.O.I. >1) resulted in culture clearing within the first hour. Liquid cultures of certain strains of the second group also cleared completely when exposed to A5W at a similar M.O.I but with a significant delay. Apparently, phages released from a fraction of initially infected cells acquired the ability to infect the remaining cells in the culture efficiently by gaining their DNA methylation pattern. What makes these initial fractions of cells susceptible to phage infection, despite a foreign methylation pattern of infecting phage DNA, is unclear. One factor may be the leakiness of cellular R-M systems caused by variations in their activity or expression at different stages of the cell cycle or populational growth, as was found analogously in the case of *Bacillus staerothermophilus* R-M system BstVI (Gonzalez *et al.*, 1994). Additionally, staphylococcal Twort-like phages must encode efficient anti-restriction mechanisms. Many diverse anti-restriction systems of other phages have been described. Phage enzymes that inactivate host restriction endonucleases, modulate the function or expression of associated methyltransferases, or methylate phage DNA at infection are involved in their function (Atanasiu *et al.*, 2001; Davison and Brunel, 1979; Iida *et al.*, 1987; Łobocka *et al.*, 2004; McCorquodale & Warner, 1988). In the case of some phages, for example P1, anti-restriction proteins are injected into cells at infection together with phage DNA. Additionally, the activity of phage-encoded recombination enzymes or phage proteins that enhance host recombination functions may, at multiple infection, restore the disrupted phage DNA by recombinational repair. So far, no homologues of known anti-restriction proteins were identified among predicted proteins of staphylococcal Twort-like phages. Genes of these phages that encode anti-restriction functions have yet to be discovered by functional studies.

IV. CONCLUDING REMARKS

Genomic characterization of large obligatorily virulent bacteriophages on a mass scale became possible only recently with the advent of high-throughput cloning-independent sequencing technologies. The comparison of 11 phage genomic sequences presented in this chapter is only a small step toward understanding the biology of staphylococcal Twort-like phages and the biology of the *Spounavirinae* subfamily of myoviruses they belong to. Functional analysis of knockout mutants in genes of undetermined function is one of the obvious next steps to extend the knowledge on these phages and to accelerate their approval for medical use. Such analysis should allow for the identification of all genes crucial for phage specificity and the ability to propagate in *S. aureus* strains of different clonal lineages if performed in bacterial cells from collections of differentiated and molecularly characterized strains. Although this chapter touches only selected aspects of staphylococcal Twort-like phages biology, we believe that the results provided here will help direct further studies on these phages toward a deeper understanding of their physiology and molecular basics of polyvalency.

ACKNOWLEDGMENTS

The authors are grateful to Hans-Wolfgang Ackermann for help in the description of phage morphological features and to Wacław Szybalski for helpful comments on this manuscript. This work was supported by funds from the Operational Program 'Innovative Economy, 2007-2013' (Priority axis 1. Research and Development of Modern Technologies, Measure 1.3 Support for R&D projects for entrepreneurs carried out by scientific entities, Submeasure 1.3.1, Development Project No. POIG 01.03.01-02-003/08 entitled "Optimization of the production and characterization of bacteriophage preparations for therapeutic use"), targeted Grant No. PBZ-MNiSW-04/I/2007, funds from the Bacteriophage Biology and Biotechnology Network (1/E-35/BWSN/0082/2008), and statutory funds for the Faculty of Agriculture and Biology of the Warsaw University of Life Sciences, SGGW. Transmission electron microscopy observations were performed in the Laboratory of Electron Microscopy at the Nencki Institute of Experimetal Biology, Warsaw, Poland, using equipment that was installed within the project sponsored by the EU Structural Funds: Centre of Advanced Technology BIM—Equipment purchase for the Laboratory of Biological and Medical Imaging. Beata Weber-Dąbrowska is a coinventor on patent applications covering the therapeutic use of phages in bacterial infections filed by the Institute of Immunology and Experimental Therapy, Wrocław, Poland.

REFERENCES

Ackermann, H.-W. (2009). Basic phage electron microscopy. *Methods Mol. Biol.* **501:**113–126.

Ackermann, H.-W., and DuBow, M. (1987). Viruses of Prokaryotes II: Natural Groups of Bacteriophages. CRC Press, Boca Raton, FL.

Ackermann, H.-W. (1998). Tailed bacteriophages: The order caudovirales. *Adv. Virus Res.* **51:**135–201.

Ackermann, H.-W. (2007). 5500 Phages examined in the electron microscope. *Arch. Virol.* **152:**227–243.

Ackermann, H.-W., and Kropinski, A. M. (2007). Curated list of prokaryote viruses with fully sequenced genomes. *Res. Microbiol.* **158:**555–566.

Akatov, A. K., Zueva, V. S., and Dmitrenko, O. A. (1991). A new approach to establishing the set of phages for typing methicillin-resistant *Staphylococcus aureus. J. Chemother.* **3:**275–278.

Alisky, J., Iczkowski, K., Rapoport, A., and Troitsky, N. (1998). Bacteriophages show promise as antimicrobial agents. *J. Infect.* **36:**5–15.

Alcorlo, M., González-Huici, V., Hermoso, J. M., Meijer, W. J., and Salas, M. (2007). The phage ϕ29 membrane protein p16.7, involved in DNA replication, is required for efficient ejection of the viral genome. *J. Bacteriol.* **189:**5542–5549.

Anderson, D., and Bodley, J. W. (1990). Role of RNA in bacteriophage phi 29 DNA packaging. *J. Struct. Biol.* **104:**70–74.

Anderson, E. S., and Williams, R. E. (1956). Bacteriophage typing of enteric pathogens and staphylococci and its use in epidemiology. *J. Clin. Pathol.* **9:**94–127.

Andrasevic, A. T., Power, E. G., Anthony, R. M., Kalenic, S., and French, G. L. (1999). Failure of bacteriophage typing to detect an inter-hospital outbreak of methicillin-resistant *Staphylococcus aureus* (MRSA) in Zagreb subsequently identified by random amplification of polymorphic DNA (RAPD) and pulsed-field gel electrophoresis (PFGE). *Clin. Microbiol. Infect.* **10:**634–642.

Archibald, A. R., and Coapes, H. E. (1972). Blocking of bacteriophage receptor sites by Concanavalin A. *J. Gen. Microbiol.* **73:**581–585.

Arias, C. A., and Murray, B. E. (2009). Antibiotic-resistant bugs in the 21st century - a clinical super-challenge. *N. Engl. J. Med.* **360:**439–443.

Atanasiu, C., Byron, O., McMiken, H., Sturrock, S. S., and Dryden, D. T. F. (2001). Characterisation of the structure of ocr, the gene 0.3 protein of bacteriophage T7. *Nucleic Acids Res* **29:**3059–3068.

Aucken, H. M., and Westwell, K. (2002). Reaction difference rule for phage typing of *Staphylococcus aureus* at 100 times the routine test dilution. *J. Clin. Microbiol.* **40:**292–293.

Avison, M. B., Bennett, P. M., Howe, R. A., and Walsh, T. R. (2002). Preliminary analysis of the genetic basis for vancomycin resistance in *Staphylococcus aureus* strain Mu50. *J. Antimicrob. Chemother.* **49:**255–260.

Baba, T., Takeuchi, F., Kuroda, M., Yuzawa, H., Aoki, K., Oguchi, A., Nagai, Y., Iwama, N., Asano, K., Naimi, T., Kuroda, H., Cui, L., *et al.* (2002). Genome and virulence determinants of high virulence community-acquired MRSA. *Lancet* **359:**1819–1827.

Bae, T., Baba, T., Hiramatsu, K., and Schneewind, O. (2006). Prophages of *Staphylococcus aureus* Newman and their contribution to virulence. *Mol. Microbiol.* **62:**1035–1047.

Becker, S. C., Dong, S., Baker, J. R., Foster-Frey, J., Pritchard, D. G., and Donovan, D. M. (2009). LysK CHAP endopeptidase domain is required for lysis of live staphylococcal cells. *FEMS Microbiol. Lett.* **294:**52–60.

Bedford, E., Tabor, S., and Richardson, C. C. (1997). The thioredoxin binding domain of bacteriophage T7 DNA polymerase confers processivity on *Escherichia coli* DNA polymerase I. *Proc. Natl. Acad. Sci. USA.* **94:**479–484.

Belley, A., Callejo, M., Arhin, F., Dehbi, M., Fadhil, I., Liu, J., McKay, G., Srikumar, R., Bauda, P., Bergeron, D., Ha, N., Dubow, M., Gros, P., Pelletier, J., and Moeck, G. (2006). Competition of bacteriophage polypeptides with native replicase proteins for binding to the DNA sliding clamp reveals a novel mechanism for DNA replication arrest in *Staphylococcus aureus. Mol. Microbiol.* **62:**1132–1143.

Betley, M. J., and Mekalanos, J. J. (1985). Staphylococcal enterotoxin A is encoded by phage. *Science* **229:**185–187.

Bjornsti, M. A., Reilly, B. E., and Anderson, D. L. (1982). Morphogenesis of bacteriophage phi 29 of *Bacillus subtilis*: DNA-gp3 intermediate in *in vivo* and *in vitro* assembly. *J. Virol.* **41:**508–517.

Black, L. W., and Rao, V. B. (2012). Structure, assembly, and DNA packaging of the bacteriophage T4 head. *Adv. Virus Res.* **82:**119–153.
Blair, J. E., and Williams, R. E. (1961). Phage typing of staphylococci. *Bull. WHO* **24:**771–784.
Borysowski, J., Łobocka, M., Międzybrodzki, R., Weber-Dąbrowska, B., and Górski, A. (2011). Potential of bacteriophages and their lysins in the treatment of MRSA: current status and future perspectives. *BioDrugs* **25:**347–355.
Boyd, E. F., and Brüssow, H. (2002). Common themes among bacteriophage-encoded virulence factors and diversity among the bacteriophages involved. *Trends Microbiol.* **10:**521–529.
Briers, Y., Lavigne, R., Plessers, P., Hertveldt, K., Hanssens, I., Engelborghs, Y., and Volckaert, G. (2006). Stability analysis of the bacteriophage phiKMV lysin gp36C and its putative role during infection. *Cell Mol. Life Sci.* **63:**1899–1905.
Brüssow, H., Canchaya, C., and Hardt, W. D. (2004). Phages and the evolution of bacterial pathogens: From genomic rearrangements to lysogenic conversion. *Microbiol. Mol. Biol. Rev.* **68:**560–602.
Brüssow, H., and Desiere, F. (2001). Comparative phage genomics and the evolution of *Siphoviridae*: Insights from dairy phages. *Mol. Microbiol.* **39:**213–222.
Burian, M., Rautenberg, M., Kohler, T., Fritz, M., Krismer, B., Unger, C., Hoffmann, W. H., Peschel, A., Wolz, C., and Goerke, C. (2010). Temporal expression of adhesion factors and activity of global regulators during establishment of *Staphylococcus aureus* nasal colonization. *J. Infect. Dis.* **201:**1414–1421.
Capparelli, R., Parlato, M., Borriello, G., Salvatore, P., and Iannelli, D. (2007). Experimental phage therapy against Staphylococcus aureus in mice. *Antimicrob. Agents. Chemother.* **51:**2765–2773.
Carlton, R. M., Noordman, W. H., Biswas, B., de Meester, E. D., and Loessner, M. J. (2005). Bacteriophage P100 for control of *Listeria monocytogenes* in foods: Genome sequence, bioinformatic analyses, oral toxicity study, and application. *Regul. Toxicol. Pharmacol.* **43:**301–312.
Carroll, J. D., Cafferkey, M. T., and Coleman, D. C. (1993). Serotype F double- and triple-converting phage insertionally inactivate the *Staphylococcus aureus* beta-toxin determinant by a common molecular mechanism. *FEMS Microbiol. Lett.* **106:**147–155.
Casjens, S. R., and Gilcrease, E. B. (2009). Determining DNA packaging strategy by analysis of the termini of the chromosomes in tailed-bacteriophage virions. *Methods Mol. Biol.* **502:**91–111.
Cerca, N., Oliveira, R., and Azeredo, J. (2007). Susceptibility of *Staphylococcus epidermidis* planktonic cells and biofilms to the lytic action of staphylococcus bacteriophage K. *Lett. Appl. Microbiol.* **45:**313–317.
Chambers, H. F., and DeLeo, F. R. (2009). Waves of resistance: *Staphylococcus aureus* in the antibiotic era. *Nat. Rev. Microbiol.* **7:**629–641.
Chelm, B. K., Duffy, J. J., and Geiduschek, E. P. (1982). Interaction of *Bacillus subtilis* RNA polymerase core with two specificity-determining subunits: Competition between sigma and the SPO1 gene 28 protein. *J. Biol. Chem.* **257:**6501–6508.
Chibani-Chennoufi, S., Dillmann, M. L., Marvin-Guy, L., Rami-Shojaei, S., and Brüssow, H. (2004). *Lactobacillus plantarum* bacteriophage LP65: A new member of the SPO1-like genus of the family Myoviridae. *J. Bacteriol.* **186:**7069–7083.
Chistovich, G. N. (1960). Various problems in phagotyping of staphylococci. I. Method of obtaining a set of local strains of staphylophages and its use for typing the strains of different origin. *Zh Mikrobiol. Epidemiol. Immunobiol.* **31:**33–39[Russian].
Christie, G. E., Matthews, A. M., King, D. G., Lane, K. D., Olivarez, N. P., Tallent, S. M., Gill, S. R., and Novick, R. P. (2010). The complete genomes of *Staphylococcus aureus* bacteriophages 80 and 80α–implications for the specificity of SaPI mobilization. *Virology* **407:**381–390.

Clokie, M. R., and Mann, N. H. (2006). Marine cyanophages and light. *Environ. Microbiol.* **8:**2074–2082.

Coleman, D., Knights, J., Russell, R., Shanley, D., Birkbeck, T. H., Dougan, G., and Charles, I. (1991). Insertional inactivation of the *Staphylococcus aureus* beta-toxin by bacteriophage phi 13 occurs by site- and orientation-specific integration of the phi 13 genome. *Mol. Microbiol* **5:**933–939.

Coleman, D. C., Sullivan, D. J., Russell, R. J., Arbuthnott, J. P., Carey, B. F., and Pomeroy, H. M. (1989). *Staphylococcus aureus* bacteriophages mediating the simultaneous lysogenic conversion of beta-lysin, staphylokinase and enterotoxin A: Molecular mechanism of triple conversion. *J. Gen. Microbiol.* **135:**1679–1697.

Comeau, A. M., Bertrand, C., Letarov, A., Tétart, F., and Krisch, H. M. (2007). Modular architecture of the T4 phage superfamily: a conserved core genome and a plastic periphery. *Virology.* **362:**384–396.

Costanzo, M., and Pero, J. (1983). Structure of a *Bacillus subtilis* bacteriophage SPO1 gene encoding RNA polymerase sigma factor. *Proc. Natl. Acad. Sci. USA* **80:**1236–1240.

Costanzo, M., and Pero, J. (1984). Overproduction and purification of a bacteriophage SPO1-encoded RNA polymerase sigma factor. *J. Biol. Chem.* **259:**6681–6685.

Dabrowska, K., Skaradziński, G., Kurzepa, A., Owczarek, B., Zaczek, M., Weber-Dabrowska, B., Wietrzyk, J., Maciejewska, M., Budynek, P., and Górski, A. (2010). The effects of staphylococcal bacteriophage lysates on cancer cells *in vitro*. *Clin. Exp. Med.* **10:**81–85.

Davidson, I. (1972). A collaborative investigation of phages for typing bovine staphylococci. *Bull. WHO* **46:**81–98.

Davison, J., and Brunel, F. (1979). Restriction insensitivity in bacteriophage T5 I. Genetic characterization of mutants sensitive to EcoRI restriction. *J. Virol.* **29:**11–16.

Dehbi, M., Moeck, G., Arhin, F. F., Bauda, P., Bergeron, D., Kwan, T., Liu, J., McCarty, J., Dubow, M., and Pelletier, J. (2009). Inhibition of transcription in *Staphylococcus aureus* by a primary sigma factor-binding polypeptide from phage G1. *J. Bacteriol* **191:**3763–3771.

Delbruck, M. (1940). The growth of bacteriophage and lysis of the host. *J. Gen. Physiol* **23** (5)**:**643–660.

Dempsey, R. M., Carroll, D., Kong, H., Higgins, L., Keane, C. T., and Coleman, D. C. (2005). Sau42I, a BcgI-like restriction-modification system encoded by the *Staphylococcus aureus* quadruple-converting phage Phi42. *Microbiology* **151:**1301–1311.

Deora, R., and Misra, T. K. (1996). Characterization of the primary sigma factor of *Staphylococcus aureus*. *J. Biol. Chem.* **271:**21828–21834.

Dmitrienko, O. A., Sidorenko, S. V., Zhukhovitskiĭ, V. G., Terekhova, R. V., Karabak, V. I., Tarasevich, N. N., Vasil'eva, E. I., and Prokhorov, VIa (2003). Evaluation of the usefulness of three collections of bacteriophages for typing methicillin-resistant *Staphylococcus aureus*, isolated in Moscow hospitals. *Zh Mikrobiol. Epidemiol. Immunobiol.* **1:**3–9 [Russian].

Doskar, J., Pallova, P., Pantucek, R., Rosypal, S., Ruzickova, V., Pantuckova, P., Kailerova, J., Kleparnik, K., Mala, Z., and Bocek, P. (2000). Genomic relatedness of *Staphylococcus aureus* phages of the International Typing Set and detection of serogroup A, B, and F prophages in lysogenic strains. *Can. J. Microbiol.* **46:**1066–1076.

Dua, M., Agarwal, D. S., and Natarajan, R. (1982). Phage typing of *Staphylococcus aureus* using phages other than those of basic set and new methods. *Indian J. Med. Res.* **75:**348–354.

Duda, R. L., Hendrix, R. W., Huang, W. M., and Conway, J. F. (2006). Shared architecture of bacteriophage SPO1 and herpesvirus capsids. *Curr. Biol.* **16:**R11–R13Erratum in: Curr. Biol. 2006 16, 440.

Emberger, M., Koller, J., Laimer, M., Hell, M., Oender, K., Trost, A., Maass, M., Witte, W., Hintner, H., and Lechner, A. M. (2011). Nosocomial *Staphylococcal* scalded skin syndrome caused by intra-articular injection. *J. Eur. Acad. Dermatol. Venereol.* **25:**227–231.

Eyer, L., Pantůcek, R., Zdráhal, Z., Konecná, H., Kaspárek, P., Růzicková, V., Hernychová, L., Preisler, J., and Doskar, J. (2007). Structural protein analysis of the polyvalent staphylococcal bacteriophage 812. *Proteomics* **7**:64–72.
Filée, J., Bapteste, E., Susko, E., and Krisch, H. M. (2006). A selective barrier to horizontal gene transfer in the T4-type bacteriophages that has preserved a core genome with the viral replication and structural genes. *Mol. Biol. Evol.* **23**:1688–1696.
Fischetti, V. A. (2008). Bacteriophage lysins as effective antibacterials. *Curr. Opin. Microbiol.* **11**:393–400.
Fraser, J. S., Maxwell, K. L., and Davidson, A. R. (2007). Immunoglobulin-like domains on bacteriophage: Weapons of modest damage? *Curr. Opin. Microbiol.* **10**:382–387.
Fraser, J. S., Yu, Z., Maxwell, K. L., and Davidson, A. R. (2006). Ig-like domains on bacteriophages: A tale of promiscuity and deceit. *J. Mol. Biol.* **359**:496–507.
Fujita, D. J., Ohlsson-Wilhelm, B. M., and Geiduschek, E. P. (1971). Transcription during bacteriophage SPO1 development: Mutations affecting the program of viral transcription. *J. Mol. Biol.* **57**:301–317.
Gage, L. P., and Geiduschek, E. P. (1971). RNA synthesis during bacteriophage SPO1 development: Six classes of SPO1 DNA. *J. Mol. Biol.* **57**:279–300.
Gaido, M. L., Prostko, C. R., and Strobl, J. S. (1988). Isolation and characterization of BsuE methyltransferase, a CGCG specific DNA methyltransferase from *Bacillus subtilis*. *J. Biol. Chem.* **263**:4832–4836.
Galperin, M. Y., Moroz, O. V., Wilson, K. S., and Murzin, A. G. (2006). House cleaning, a part of good housekeeping. *Mol. Microbiol.* **59**:5–19.
García, P., Martínez, B., Obeso, J. M., Lavigne, R., Lurz, R., and Rodríguez, A. (2009). Functional genomic analysis of two *Staphylococcus aureus* phages isolated from the dairy environment. *Appl. Environ. Microbiol.* **75**:7663–7673.
Geiduschek, E. P., Schneider, G. J., and Sayre, M. H. (1990). TF1, a bacteriophage-specific DNA-binding and DNA bending protein. *J. Struct. Biol.* **104**:84–90.
Gershman, M., Hunter, J. A., Harmon, R. J., Wilson, R. A., and Markowsky, G. (1988). Phage typing set for differentiating *Staphylococcus epidermidis*. *Can. J. Microbiol.* **34**:1358–1361.
Ghebremedhin, B., Layer, F., König, W., and König, B. (2008). Genetic classification and distinguishing of *Staphylococcus* species based on different partial gap, 16S rRNA, *hsp60*, *rpoB*, *sodA*, and *tuf* gene sequences. *J. Clin. Microbiol.* **46**:1019–1025.
Gibbs, P. A., Patterson, J. T., and Thompson, J. K. (1978). Characterization of poultry isolates of *Staphylococcus aureus* by a new set of poultry phages. *J. Appl. Bacteriol.* **44**:387–400.
Gill, J. J., Pacan, J. C., Carson, M. E., Leslie, K. E., Griffiths, M. W., and Sabour, P. M. (2006). Efficacy and pharmacokinetics of bacteriophage therapy in treatment of subclinical Staphylococcus aureus mastitis in lactating dairy cattle. *Antimicrob. Agents. Chemother.* **50**:2912–2918.
Gill, S. R., McIntyre, L. M., Nelson, C. L., Remortel, B., Rude, T., Reller, L. B., and Fowler, V. G., Jr. (2011). Potential associations between severity of infection and the presence of virulence-associated genes in clinical strains of *Staphylococcus aureus*. *PloS One.* **6**:e18673.
Godány, A., Bukovská, G., Farkasovská, J., Brnáková, Z., Dmitriev, A., Tkáciková, E., Ayele, T., and Mikula, I. (2004). Characterization of a complex restriction-modification system detected in *Staphylococcus aureus* and *Streptococcus agalactiae* strains isolated from infections of domestic animals. *Folia Microbiol. (Praha)* **49**:307–314.
Goerke, C., Pantucek, R., Holtfreter, S., Schulte, B., Zink, M., Grumann, D., Bröker, B. M., Doskar, J., and Wolz, C. (2009). Diversity of prophages in dominant *Staphylococcus aureus* clonal lineages. *J. Bacteriol.* **191**:3462–3468.
Golec, P., Dąbrowski, K., Hejnowicz, M. S., Gozdek, A., Loś, J. M., Węgrzyn, G., Łobocka, M. B., and Łoś, M. A. (2011). A reliable method for storage of tailed phages. *J. Microbiol. Methods.* **84**:486–489.

González, E., Padilla, C., Saavedra, C., and Vásquez, C. (1994). The expression of the bstVIM gene from *Bacillus stearothermophilus* V is restricted to vegetative cell growth. *Microbiology* **140:**1337–1340.

Gross, M., Marianovsky, I., and Glaser, G. (2006). MazG – a regulator of programmed cell death in *Escherichia coli*. *Mol. Microbiol.* **59:**590–601.

Grundmann, H., Hori, S., Enright, M. C., Webster, C., Tami, A., Feil, E. J., and Pitt, T. (2002). Determining the genetic structure of the natural population of *Staphylococcus aureus*: A comparison of multilocus sequence typing with pulsed-field gel electrophoresis, randomly amplified polymorphic DNA analysis, and phage typing. *J. Clin. Microbiol.* **40:**4544–4546.

Gu, J., Xu, W., Lei, L., Huang, J., Feng, X., Sun, C., Du, C., Zuo, J., Li, Y., Du, T., Li, L., and Han, W. (2011). LysGH15, a novel bacteriophage lysin, protects a murine bacteremia model efficiently against lethal methicillin-resistant *Staphylococcus aureus* infection. *J. Clin. Microbiol.* **49:**111–117.

Gupta, R., and Prasad, Y. (2011). Efficacy of polyvalent bacteriophage P-27/HP to control multidrug resistant *Staphylococcus aureus* associated with human infections. *Curr. Microbiol.* **62:**255–260.

Guttman, B., Raya, P., and Kutter, E. (2005). Basic phage biology. *In* ''Bacteriophages Biology and Application'' (E. B. Kutter and A. Sulakvelidze, eds.), pp. 29–66. CRC Press, Boca Raton, FL.

Heczko, P. B., Pulverer, G., Kasprowicz, A., and Klein, A. (1977). Evaluation of a new bacteriophage set for typing of *Staphylococcus epidermidis* strains. *J. Clin. Microbiol.* **5:**573–577.

Heintz, N., and Shub, D. A. (1982). Transcriptional regulation of bacteriophage SPO1 protein synthesis *in vivo* and *in vitro*. *J. Virol.* **42:**951–962.

Hoet, P., Coene, M., and Cocito, C. (1983). Comparison of the physical maps and redundant ends of the chromosomes of phages 2C, SP01, SP82 and phi e. *Eur. J. Biochem.* **132:**63–67.

Holden, M. T., Feil, E. J., Lindsay, J. A., Peacock, S. J., Day, N. P., Enright, M. C., Foster, T. J., Moore, C. E., Hurst, L., Atkin, R., Barron, A., Bason, N., *et al.* (2004). Complete genomes of two clinical *Staphylococcus aureus* strains: Evidence for the rapid evolution of virulence and drug resistance. *Proc. Natl. Acad. Sci. USA* **101:**9786–9791.

Holmberg, O. (1978). Phage typing of coagulase-negative staphylococci. *Zentralbl. Bakteriol. Orig. A* **241:**68–71.

Hoshiba, H., Uchiyama, J., Kato, S., Ujihara, T., Muraoka, A., Daibata, M., Wakiguchi, H., and Matsuzaki, S. (2010). Isolation and characterization of a novel *Staphylococcus aureus* bacteriophage, phiMR25, and its therapeutic potential. *Arch. Virol.* **155:**545–552.

Hotchin, J. E. (1954). The purification and electron microscopical examination of the structure of staphylococcal bacteriophage K. *J. Gen. Microbiol.* **10:**250–260.

Howden, B. P., Seemann, T., Harrison, P. F., McEvoy, C. R., Stanton, J. A., Rand, C. J., Mason, C. W., Jensen, S. O., Firth, N., Davies, J. K., Johnson, P. D., and Stinear, T. P. (2010). Complete genome sequence of *Staphylococcus aureus* strain JKD6008, an ST239 clone of methicillin-resistant *Staphylococcus aureus* with intermediate-level vancomycin resistance. *J. Bacteriol.* **192:**5848–5849.

Hsieh, S. E., Lo, H. H., Chen, S. T., Lee, M. C., and Tseng, Y. H. (2011). Wide host range and strong lytic activity of *Staphylococcus aureus* lytic phage Stau2. *Appl. Environ. Microbiol.* **77:**756–761.

Iandolo, J. J., Worrell, V., Groicher, K. H., Qian, Y., Tian, R., Kenton, S., Dorman, A., Ji, H., Lin, S., Loh, P., Qi, S., Zhu, H., *et al.* (2002). Comparative analysis of the genomes of the temperate bacteriophages phi 11, phi 12 and phi 13 of *Staphylococcus aureus* 8325. *Gene* **289:**109–118.

Iida, S., Streiff, M. B., Bickle, T. A., and Arber, W. (1987). Two DNA antirestriction systems of bacteriophage P1, *darA*, and *darB*: Characterization of *darA*- phages. *Virology* **157:**156–166.

Iordanescu, S., and Surdeanu, M. (1976). Two restriction and modification systems in *Staphylococcus aureus* NCTC8325. *J. Gen. Microbiol.* **96:**277–281.

Jarvis, A. W., Collins, L. J., and Ackermann, H.-W. (1993). A study of five bacteriophages of the *Myoviridae* family which replicate on different gram-positive bacteria. *Arch. Virol.* **133:**75–84.

Jikia, D., Chkhaidze, N., Imedashvili, E., Mgaloblishvili, I., Tsitlanadze, G., Katsarava, R., Morris, J. G., Jr., and Sulakvelidze, A. (2005). The use of a novel biodegradable preparation capable of the sustained release of bacteriophages and ciprofloxacin, in the complex treatment of multidrug-resistant *Staphylococcus aureus*-infected local radiation injuries caused by exposure to Sr90. *Clin. Exp. Dermatol.* **30:**23–26.

Kahánková, J., Pantůček, R., Goerke, C., Růžičková, V., Holochová, P., and Doškař, J. (2010). Multilocus PCR typing strategy for differentiation of *Staphylococcus aureus* siphoviruses reflecting their modular genome structure. *Environ. Microbiol.* **12:**2527–2538.

Kanamaru, S., Leiman, P. G., Kostyuchenko, V. A., Chipman, P. R., Mesyanzhinov, V. V., Arisaka, F., and Rossmann, M. G. (2002). Structure of the cell-puncturing device of bacteriophage T4. *Nature* **415:**553–557.

Kaneko, J., Kimura, T., Narita, S., Tomita, T., and Kamio, Y. (1998). Complete nucleotide sequence and molecular characterization of the temperate staphylococcal bacteriophage phiPVL carrying Panton-Valentine leukocidin genes. *Gene* **215:**57–67.

Kao, S. H., and McClain, W. H. (1980). Roles of bacteriophage T4 gene 5 and gene s products in cell lysis. *J. Virol.* **34**(1)**:**104–107.

Kaspárek, P., Pantůcek, R., Kahánková, J., Růzicková, V., and Doskar, J. (2007). Genome rearrangements in host-range mutants of the polyvalent staphylococcal bacteriophage 812. *Folia Microbiol. (Praha)* **52:**331–338.

Katsura, I. (1981). Structure and function of the major tail protein of bacteriophage lambda: Mutants having small major tail protein molecules in their virion. *J. Mol. Biol.* **146:**493–512.

Katsura, I., and Hendrix, R. W. (1984). Length determination in bacteriophage lambda tails. *Cell* **39:**691–698.

Kawano, J., Shimizu, A., Kimura, S., and Blouse, L. (1982). Experimental bacteriophage set for typing *Staphylococcus intermedius*. *Zentralbl. Bakteriol. Mikrobiol Hyg. A* **253:**321–330.

Kelly, D., McAuliffe, O., Ross, R. P., O'Mahony, J., and Coffey, A. (2011). Development of a broad-host-range phage cocktail for biocontrol. *Bioeng. Bugs.* **2:**31–37.

Kenny, J. G., McGrath, S., Fitzgerald, G. F., and van Sinderen, D. V. (2004). Bacteriophage Tuc 2009 encodes a tail-associated cell wall degrading activity. *J. Bacteriol.* **186:**3480–3491.

Kilcher, S., Loessner, M. J., and Klumpp, J. (2010). *Brochothrix thermosphacta* bacteriophages feature heterogeneous and highly mosaic genomes and utilize unique prophage insertion sites. *J. Bacteriol.* **192:**5441–5453.

Klumpp, J., Dorscht, J., Lurz, R., Bielmann, R., Wieland, M., Zimmer, M., Calendar, R., and Loessner, M. J. (2008). The terminally redundant, nonpermuted genome of *Listeria* bacteriophage A511: A model for the SPO1-like myoviruses of gram-positive bacteria. *J. Bacteriol.* **190:**5753–5765.

Klumpp, J., Lavigne, R., Loessner, M. J., and Ackermann, H.-W. (2010). The SPO1-related bacteriophages. *Arch. Virol.* **155:**1547–1561.

Krzywy, T., Durlakowa, I., Kucharewicz-Krukowska, A., Krynski, S., and Slopek, S. (1981). Ultrastructure of bacteriophages used for typing of *Staphylococcus aureus*. *Zentralbl. Bakteriol. Mikrobiol. Hyg. A* **250:**287–295.

Kuroda, M., Ohta, T., Uchiyama, I., Baba, T., Yuzawa, H., Kobayashi, I., Cui, L., Oguchi, A., Aoki, K., Nagai, Y., Lian, J., Ito, T., *et al.* (2001). Whole genome sequencing of meticillin-resistant *Staphylococcus aureus*. *Lancet* **357:**1225–1240.

Kvachadze, L., Balarjishvili, N., Meskhi, T., Tevdoradze, E., Skhirtladze, N., Pataridze, T., Adamia, R., Topuria, T., Kutter, E., Rohde, C., and Kutateladze, M. (2011). Evaluation of lytic activity of staphylococcal bacteriophage Sb-1 against freshly isolated clinical pathogens. *Microb. Biotechnol.* **4:**643–650.

Kwan, T., Liu, J., DuBow, M., Gros, P., and Pelletier, J. (2005). The complete genomes and proteomes of 27 *Staphylococcus aureus* bacteriophages. *Proc. Natl. Acad. Sci. USA* **102:**5174–5179.

Kwiatek, M., Parasion, S., Mizak, L., Gryko, R., Bartoszcze, M., and Kocik, J. (2012). Characterization of a bacteriophage, isolated from a cow with mastitis, that is lytic against *Staphylococcus aureus* strains. *Arch. Virol.* **157:**225–234.

Labrie, S. J., Samson, J. E., and Moineau, S. (2010). Bacteriophage resistance mechanisms. *Nat. Rev. Microbiol.* **8:**317–327.

Landthaler, M., Begley, U., Lau, N. C., and Shub, D. A. (2002). Two self-splicing group I introns in the ribonucleotide reductase large subunit gene of *Staphylococcus aureus* phage Twort. *Nucleic Acids Res.* **30:**1935–1943.

Landthaler, M., Lau, N. C., and Shub, D. A. (2004). Group I intron homing in *Bacillus* phages SPO1 and SP82: A gene conversion event initiated by a nicking homing endonuclease. *J. Bacteriol.* **186:**4307–4314.

Landthaler, M., and Shub, D. A. (1999). Unexpected abundance of self-splicing introns in the genome of bacteriophage Twort: Introns in multiple genes, a single gene with three introns, and exon skipping by group I ribozymes. *Proc. Natl. Acad. Sci. USA* **96:**7005–7010.

Lavigne, R., Darius, P., Summer, E. J., Seto, D., Mahadevan, P., Nilsson, A. S., Ackermann, H.-W., and Kropinski, A. M. (2009). Classification of Myoviridae bacteriophages using protein sequence similarity. *BMC Microbiol.* **9:**224.

Lavigne, R., Seto, D., Mahadevan, P., Ackermann, H.-W., and Kropinski, A. M. (2008). Unifying classical and molecular taxonomic classification: Analysis of the Podoviridae using BLASTP-based tools. *Res. Microbiol.* **159:**406–414.

Lee, Y. D., Chang, H. I., and Park, J. H. (2011). Genomic sequence of temperate phage TEM126 isolated from wild type *S. aureus*. *Arch. Virol.* **156:**717–720.

Lee, G., Talkington, C., and Pero, J. (1980). Nucleotide sequence of a promoter recognized by *Bacillus subtilis* RNA polymerase. *Mol. Gen. Genet.* **180:**57–65.

Lee, J. S., and Stewart, P. R. (1985). The virion proteins and ultrastructure of *Staphylococcus aureus* bacteriophages. *J. Gen. Virol.* **66:**2017–2027.

Lee, C. Y., and Iandolo, J. J. (1986). Integration of staphylococcal phage L54a occurs by site-specific recombination: structural analysis of the attachment sites. *Proc. Natl. Acad. Sci. USA* **83:**5474–5478.

Lehman, S. M., Kropinski, A. M., Castle, A. J., and Svircev, A. M. (2009). Complete genome of the broad-host-range *Erwinia amylovora* phage phiEa21-4 and its relationship to *Salmonella* phage felix O1. *Appl. Environ. Microbiol.* **75:**2139–2147.

Lehnherr, H., Hansen, A.-M., and Ilyina, T. (1998). Penetration of the bacterial cell wall: A family of lytic transglycosylases in bacteriophages and conjugative plasmids. *Mol. Microbiol.* **30:**454–457.

Lindsay, J. A. (2010). Genomic variation and evolution of *Staphylococcus aureus*. *Int. J. Med. Microbiol.* **300:**98–103.

Liu, J., Dehbi, M., Moeck, G., Arhin, F., Bauda, P., Bergeron, D., Callejo, M., Ferretti, V., Ha, N., Kwan, T., McCarty, J., Srikumar, R., *et al.* (2004). Antimicrobial drug discovery through bacteriophage genomics. *Nat. Biotechnol.* **22:**185–191.

Loessner, M. J., Gaeng, S., Wendlinger, G., Maier, S. K., and Scherer, S. (1998). The two-component lysis system of *Staphylococcus aureus* bacteriophage Twort: a large TTG-start holin and an associated amidase endolysin. *FEMS Microbiol. Lett.* **162:**265–274.

Łobocka, M. B., Rose, D. J., Plunkett, G., 3 rd, Rusin, M., Samojedny, A., Lehnherr, H., Yarmolinsky, M. B., and Blattner, F. R. (2004). Genome of bacteriophage P1. *J. Bacteriol* **186:**7032–7068.

Loessner, M. J., and Scherer, S. (1995). Organization and transcriptional analysis of the *Listeria* phage A511 late gene region comprising the major capsid and tail sheath protein genes cps and tsh. *J. Bacteriol.* **177:**6601–6609.

Lonetto, M., Gribskov, M., and Gross, C. A. (1992). The σ^{70} family: Sequence conservation and evolutionary relationships. *J. Bacteriol.* **174:**3843–3849.
Losick, R., and Pero, J. (1981). Cascades of sigma factors. *Cell* **25:**582–584.
Lundholm, M., and Bergendahl, B. (1988). Heat treatment to increase phage typability of *Staphylococcus aureus*. *Eur. J. Clin. Microbiol. Infect. Dis.* **7:**300–302.
Ma, X. X., Ito, T., Kondo, Y., Cho, M., Yoshizawa, Y., Kaneko, J., Katai, A., Higashiide, M., Li, S., and Hiramatsu, K. (2008). Two different Panton-Valentine leukocidin phage lineages predominate in Japan. *J. Clin. Microbiol.* **46:**3246–3258.
Magnusson, L. U., Farewell, A., and Nystrom, T. (2005). ppGpp: a global regulator in *Escherichia coli*. *Trends Microbiol.* **13:**236–242.
Mann, N. H. (2008). The potential of phages to prevent MRSA infections. *Res. Microbiol.* **159:**400–405.
Markoishvili, K., Tsitlanadze, G., Katsarava, R., Morris, J. G., Jr., and Sulakvelidze, A. (2002). A novel sustained-release matrix based on biodegradable poly(ester amide)s and impregnated with bacteriophages and an antibiotic shows promise in management of infected venous stasis ulcers and other poorly healing wounds. *Int. J. Dermatol.* **41:**453–458.
Marples, R. R., and Rosdahl, V. T. (1997). International quality control of phage typing of *Staphylococcus aureus*. International Union of Microbial Societies Subcommittee. *J. Med. Microbiol.* **46:**511–516.
Martín-de-Nicolás, M. M., Vindel, A., and Sáez-Nieto, J. A. (1990). Development of a new set of phages as an epidemiological marker in *Staphylococcus epidermidis* causing nosocomial infections. *Epidemiol. Infect.* **104:**111–118.
Matsuzaki, S., Yasuda, M., Nishikawa, H., Kuroda, M., Ujihara, T., Shuin, T., Shen, Y., Jin, Z., Fujimoto, S., Nasimuzzaman, M. D., Wakiguchi, H., Sugihara, S., *et al.* (2003). Experimental protection of mice against lethal *Staphylococcus aureus* infection by novel bacteriophage phi MR11. *J. Infect. Dis.* **187:**613–624.
McCorquodale, J. D., and Warner, H. R. (1988). Bacteriophage T5 and related phages. *In* "The Bacteriophages" (R. Calendar, ed.), pp. 439–476. Plenum, New York.
Mehndiratta, P. L., Gur, R., Saini, S., and Bhalla, P. (2010). *Staphylococcus aureus* phage types and their correlation to antibiotic resistance. *Indian J. Pathol. Microbiol.* **53:**738–741.
Meijer, W. J., Horcajadas, J. A., and Salas, M. (2001). Phi29 family of phages. *Microbiol. Mol. Biol. Rev.* **65:**261–287.
Merabishvili, M., Pirnay, J. P., Verbeken, G., Chanishvili, N., Tediashvili, M., Lashkhi, N., Glonti, T., Krylov, V., Mast, J., Van Parys, L., Lavigne, R., Volckaert, G., Mattheus, W., Verween, G., De Corte, P., Rose, T., Jennes, S., Zizi, M., De Vos, D., and Vaneechoutte, M. (2009). Quality-controlled small-scale production of a well-defined bacteriophage cocktail for use in human clinical trials. *PLoS One* **4:**e4944.
Międzybrodzki, R., Borysowski, J., Weber-Dąbrowska, B., Fortuna, W., Letkiewicz, S., Szufnarowski, K., Pawełczyk, Z., Rogóż, P., Kłak, M., Wojtasik, E., and Górski, A. (2012). Clinical aspects of phage therapy. *Adv. Virus Res.* **83**.
Miller, E. S., Heidelberg, J. F., Eisen, J. A., Nelson, W. C., Durkin, A. S., Ciecko, A., Feldblyum, T. V., White, O., Paulsen, I. T., Nierman, W. C., Lee, J., Szczypinski, B., and Fraser, C. M. (2003). Complete genome sequence of the broad-host-range vibriophage KVP40: comparative genomics of a T4-related bacteriophage. *J. Bacteriol.* **185:**5220–5233.
Moak, M., and Molineux, I. J. (2000). Role of the Gp16 lytic transglycosylase motif in bacteriophage T7 virions at the initiation of infection. *Mol. Microbiol.* **37:**345–355.
Moak, M., and Molineux, I. J. (2004). Peptidoglycan hydrolytic activities associated with bacteriophage virions. *Mol. Microbiol.* **51:**1169–1183.
Murray, N. E. (2000). Type I restriction systems: Sophisticated molecular machines (a legacy of Bertani and Weigle). *Microbiol. Mol. Biol. Rev.* **64:**412–434.
Murphy, K. C. (2012). Phage recombinases and their applications. *Adv. Virus Res.* **83**.
Nakagawa, H., Arisaka, F., and Ishii, S. (1985). Isolation and characterization of the bacteriophage T4 tail-associated lysozyme. *J. Virol.* **54:**460–466.

Narita, S., Kaneko, J., Chiba, J., Piemont, Y., Jarraud, S., Etienne, J., and Kamio, Y. (2001). Phage conversion of Panton-Valentine leukocidin in *Staphylococcus aureus*: Molecular analysis of a PVL-converting phage, phiSLT. *Gene* **268:**195–206.

Nelson, D. C., Schmelcher, M., Rodriguez-Rubio, L., Klumpp, J., Pritchard, G. G., Dong, S., and Donovan, D. M. (2012). Endolysins as antibacterials. *Adv. Vir. Res.* **83:**297–364.

Noguchi, H., Park, J., and Takagi, T. (2006). MetaGene: prokaryotic gene finding from environmental genome shotgun sequences. *Nucleic Acids Res.* **34:**5623–5630.

Noguchi, H., Taniguchi, T., and Itoh, T. (2008). MetaGeneAnnotator: detecting species-specific patterns of ribosomal binding site for precise gene prediction in anonymous prokaryotic and phage genomes. *DNA Res.* **15:**387–396.

Nolan, J. M., Petrov, V., Bertrand, C., Krisch, H. M., and Karam, J. D. (2006). Genetic diversity among five T4-like bacteriophages. *Virol. J.* **3:**30.

Noto, M. J., Kreiswirth, B. N., Monk, A. B., and Archer, G. L. (2008). Gene acquisition at the insertion site for *SCCmec*, the genomic island conferring methicillin resistance in *Staphylococcus aureus. J. Bacteriol.* **190:**1276–1283.

Novick, R. P. (1990). The *Staphylococcus* às a molecular genetic system. *In* "Molecular Biology of the *Staphylococci*" (R. P. Novick, ed.). VCH, New York.

Novick, R. P., Christie, G. E., and Penadés, J. R. (2010). The phage-related chromosomal islands of Gram-positive bacteria. *Nat. Rev. Microbiol.* **8:**541–551.

O'Flaherty, S., Coffey, A., Edwards, R., Meaney, W., Fitzgerald, G. F., and Ross, R. P. (2004). Genome of staphylococcal phage K: A new lineage of Myoviridae infecting gram-positive bacteria with a low G+C content. *J. Bacteriol.* **186:**2862–2871.

O'Flaherty, S., Coffey, A., Meaney, W., Fitzgerald, G. F., and Ross, R. P. (2005a). The recombinant phage lysin LysK has a broad spectrum of lytic activity against clinically relevant staphylococci, including methicillin-resistant *Staphylococcus aureus. J. Bacteriol.* **187:**7161–7164.

O'Flaherty, S., Ross, R. P., Meaney, W., Fitzgerald, G. F., Elbreki, M. F., and Coffey, A. (2005b). Potential of the polyvalent anti-*Staphylococcus* bacteriophage K for control of antibiotic-resistant staphylococci from hospitals. *Appl. Environ. Microbiol.* **71:**1836–1842.

Okubo, S., Strauss, B., and Stodolsky, M. (1964). The possible role of recombination in the infection of competent *Bacillus subtilis* by bacteriophage deoxyribonucleic acid. *Virology* **24:**552–562.

Pantucek, R., Doskar, J., Růzicková, V., Kaspárek, P., Orácová, E., Kvardová, V., and Rosypal, S. (2004). Identification of bacteriophage types and their carriage in *Staphylococcus aureus. Arch. Virol.* **149:**1689–1703.

Pantucek, R., Götz, F., Doskar, J., and Rosypal, S. (1996). Genomic variability of *Staphylococcus aureus* and the other coagulase-positive *Staphylococcus* species estimated by macrorestriction analysis using pulsed-field gel electrophoresis. *Int. J. Syst. Bacteriol.* **46:**216–222.

Pantucek, R., Rosypalova, A., Doskar, J., Kailerova, J., Ruzickova, V., Borecka, P., Snopkova, S., Horvath, R., Gotz, F., and Rosypal, S. (1998). The polyvalent staphylococcal phage phi 812: Its host-range mutants and related phages. *Virology* **246:**241–252.

Parisi, J. T., Talbot, H. W., and Skahan, J. M. (1978). Development of a phage typing set for *Staphylococcus epidermidis* in the United States. *Zentralbl. Bakteriol. Orig. A* **241:**60–67.

Pariza, M. W., and Iandolo, J. J. (1974). Base ratio and deoxyribonucleic acid homology studies of six *Staphylococcus aureus* typing bacteriophages. *Appl. Microbiol.* **27** (2)**:**317–323.

Parker, M. L., and Eiserling, F. A. (1983a). Bacteriophage SPO1 structure and morphogenesis. I. Tail structure and length regulation. *J. Virol.* **46:**239–249.

Parker, M. L., and Eiserling, F. A. (1983b). Bacteriophage SPO1 structure and morphogenesis. III. SPO1 proteins and synthesis. *J. Virol.* **46:**260–269.

Pastagia, M., Euler, C., Chahales, P., Fuentes-Duculan, J., Krueger, J. G., and Fischetti, V. A. (2011). A novel chimeric lysin shows superiority to mupirocin for skin decolonization of methicillin-resistant and-sensitive *Staphylococcus aureus* strains. *Antimicrob. Agents. Chemother.* **55:**738–744.

Paul, V. D., Sundarrajan, S., Rajagopalan, S. S., Hariharan, S., Kempashanaiah, N., Padmanabhan, S., Sriram, B., and Ramachandran, J. (2011a). Lysis-deficient phages as novel therapeutic agents for controlling bacterial infection. *BMC Microbiol.* **11:**195.

Paul, V. D., Rajagopalan, S. S., Sundarrajan, S., George, S. E., Asrani, J. Y., Pillai, R., Chikkamadaiah, R., Durgaiah, M., Sriram, B., and Padmanabhan, S. (2011b). A novel bacteriophage Tail-Associated Muralytic Enzyme (TAME) from phage K and its development into a potent antistaphylococcal protein. *BMC Microbiol.* **11:**226.

Pedulla, M. L., Ford, M. E., Houtz, J. M., Karthikeyan, T., Wadsworth, C., Lewis, J. A., Jacobs-Sera, D., Falbo, J., Gross, J., Pannunzio, N. R., Brucker, W., Kumar, V., *et al.* (2003). Origins of highly mosaic mycobacteriophage genomes. *Cell* **113:**171–182.

Pell, L. G., Gasmi-Seabrook, G. M., Morais, M., Neudecker, P., Kanelis, V., Bona, D., Donaldson, L. W., Edwards, A. M., Howell, P. L., Davidson, A. R., and Maxwell, K. L. (2010). The solution structure of the C-terminal Ig-like domain of the bacteriophage λ tail tube protein. *J. Mol. Biol.* **403:**468–479.

Perkus, M. E., and Shub, D. A. (1985). Mapping the genes in the terminal redundancy of bacteriophage SPO1 with restriction endonucleases. *J. Virol.* **56:**40–48.

Petrushina, L. I. (1975). Results of typing staphylococci isolated from cows and their milk products using the basic set of phages and local phages. *Zh Mikrobiol. Epidemiol. Immunobiol.* **2:**77–89[Russian].

Piechowicz, L., Galiński, J., Garbacz, K., and Haras, K. (2010). Bacteriophage analysis of staphylokinase-negative *Staphylococcus aureus* strains isolated from people. *J. Basic Microbiol.* **50:**557–561.

Piechowicz, L., Wiśniewska, K., and Galiński, J. (1999). Evaluation of the usefulness of new international experimental phages for typing methicillin resistant *Staphylococcus aureus* (MRSA). *Med. Dosw. Mikrobiol.* **51:**31–36[Polish].

Pillich, J., Pulverer, G., Klein, A., and Vojtisková, M. (1978). A proposal for further modification of the phage-typing system for coagulase-negative staphylococci. *Zentralbl. Bakteriol. Orig. A* **241:**83–94.

Piuri, M., and Hatfull, G. F. (2006). A peptidoglycan hydrolase motif within the mycobacteriophage TM4 tape measure protein promotes efficient infection of stationary phase cells. *Mol. Microbiol.* **62:**1569–1585.

Projan, S. (2004). Phage-inspired antibiotics? *Nat. Biotechnol.* **22:**167–168.

Ralston, D. J. (1963). Staphylococcal sensitization: Specific biological effect of phage K on the bacterial cell wall in lysis from without. *J. Bacteriol.* **85:**1185–1193.

Ralston, D. J., Baer, B. S., Lieberman, M., and Krueger, A. P. (1957). Lysis-from-without of *S. aureus* K1 by the combined action of phage and virolysin. *J. Gen. Physiol.* **41:**343–358.

Ralston, D. J., and Krueger, A. P. (1954). The isolation of staphylococcal phage K variant susceptible to an unusual host control. *J. Gen. Physiol.* **37:**685–719.

Ralston, D. J., and McIvor, M. (1964). Lysis-from-without of *Staphylococcus aureus* strains by combination of specific phages and phage-inducedlytic enzymes. *J. Bacteriol.* **88:**676–681.

Rashel, M., Uchiyama, J., Takemura, I., Hoshiba, H., Ujihara, T., Takatsuji, H., Honke, K., and Matsuzaki, S. (2008). Tail-associated structural protein gp61 of *Staphylococcus aureus* phage MR11 has bifunctional lytic activity. *FEMS Microbiol. Lett.* **284:**9–16.

Rashel, M., Uchiyama, J., Ujihara, T., Uehara, Y., Kuramoto, S., Sugihara, S., Yagyu, K., Muraoka, A., Sugai, M., Hiramatsu, K., Honke, K., and Matsuzaki, S. (2007). Efficient elimination of multidrug-resistant *Staphylococcus aureus* by cloned lysin derived from bacteriophage phi MR11. *J. Infect. Dis.* **196:**1237–1247.

Rees, P. J., and Fry, B. A. (1981a). The morphology of staphylococcal bacteriophage K and DNA metabolism in infected *Staphylococcus aureus*. *J. Gen. Virol.* **53:**293–307.

Rees, P. J., and Fry, B. A. (1981b). Replication of bacteriophage K DNA in *Staphylococcus aureus*. *J. Gen. Virol.* **55:**41–51.

Rees, P. J., and Fry, B. A. (1983). Structure and properties of the rapidly sedimenting replicating complex of staphylococcal phage K DNA. *J. Gen. Virol.* **64:**191–198.

Rhoads, D. D., Wolcott, R. D., Kuskowski, M. A., Wolcott, B. M., Ward, L. S., and Sulakvelidze, A. (2009). Bacteriophage therapy of venous leg ulcers in humans: results of a phase I safety trial. *J. Wound Care* **18:**237–243.

Richardson, J. F., Rosdahl, V. T., van Leeuwen, W. J., Vickery, A. M., Vindel, A., and Witte, W. (1999). Phages for methicillin-resistant *Staphylococcus aureus*: An international trial. *Epidemiol. Infect.* **122:**227–233.

Rippon, J. E. (1952). A new serological division of *Staphylococcus aureus* bacteriophages: Group G. *Nature* **170:**287.

Rippon, J. E. (1956). The classification of bacteriophages lysing staphylococci. *J. Hyg. (Lond.)* **54:**213–226.

Roessner, C. A., and Ihler, G. M. (1984). Proteinase sensitivity of bacteriophage lambda tail proteins gpJ and pH in complexes with the lambda receptor. *J. Bacteriol.* **157:**165–170.

Rohwer, F., and Edwards, R. (2002). The phage proteomic tree: A genome-based taxonomy for phage. *J. Bacteriol.* **184:**4529–4535.

Romeo, J. M., Brennan, S. M., Chelm, B. K., and Geiduschek, E. P. (1981). A transcriptional map of the bacteriophage SPO1 genome. I. The major early promoters. *Virology.* **111:**588–603.

Rosenblum, E. D., and Tyrone, S. (1964). Serology, density and morphology of staphylococcal phages. *J. Bacteriol.* **88:**1737–1742.

Rosenstein, R., Nerz, C., Biswas, L., Resch, A., Raddatz, G., Schuster, S. C., and Götz, F. (2009). Genome analysis of the meat starter culture bacterium *Staphylococcus carnosus* TM300. *Appl. Environ. Microbiol.* **75:**811–822.

Rountree, P. M. (1949a). The serological differentiation of staphylococcal bacteriophages. *J. Gen. Microbiol.* **3:**164–173.

Rountree, P. M. (1949b). The phenomenon of lysogenicity in staphylococci. *J. Gen. Microbiol.* **3:**153–163.

Sakarya, S., Oncu, S., Oncu, S., Ozturk, B., Tuncer, G., and Sari, C. (2004). Neuraminidase produces dose-dependent decrease of slime production and adherence of slime-forming, coagulase-negative staphylococci. *Arch. Med. Res.* **35:**275–278.

Sampath, A., and Stewart, C. R. (2004). Roles of genes 44, 50, and 51 in regulating gene expression and host takeover during infection of *Bacillus subtilis* by bacteriophage SPO1. *J. Bacteriol.* **186:**1785–1792.

Sau, K., Gupta, S. K., Sau, S., and Ghosh, T. C. (2005). Synonymous codon usage bias in 16 *Staphylococcus aureus* phages: Implication in phage therapy. *Virus Res.* **113:**123–131.

Sayre, M. H., and Geiduschek, E. P. (1988). TF1, the bacteriophage SPO1-encoded type II DNA-binding protein, is essential for viral multiplication. *J. Virol.* **62:**3455–3462.

Schneider, G. J., Sayre, M. H., and Geiduschek, E. P. (1991). DNA-bending properties of TF1. *J. Mol. Biol.* **221:**777–794.

Schulz, E. C., Schwarzer, D., Frank, M., Stummeyer, K., Mühlenhoff, M., Dickmanns, A., Gerardy-Schahn, R., and Ficner, R. (2010). Structural basis for the recognition and cleavage of polysialic acid by the bacteriophage K1F tailspike protein EndoNF. *J. Mol. Biol.* **397:**341–351.

Shimizu, A. (1977). Establishment of a new bacteriophage set for typing avian staphylococci. *Am. J. Vet. Res.* **38:**1601–1605.

Shouval, D. S., Samra, Z., Shalit, I., Livni, G., Bilvasky, E., Ofir, O., Gadba, R., and Amir, J. (2011). Inducible clindamycin resistance among methicillin-sensitive *Staphylococcus aureus* infections in pediatric patients. *Isr. Med. Assoc. J.* **13:**605–608.

Sjöström, J. E., Löfdahl, S., and Philipson, L. (1978). Biological characteristics of a type I restriction-modification system in *Staphylococcus aureus*. *J. Bacteriol.* **133:**1144–1149.

Skahan, J. M., and Parisi, J. T. (1977). Development of a bacteriophage-typing set for *Staphylococcus epidermidis*. *J. Clin. Microbiol.* **6:**16–18.

Smith, H. W. (1948a). Investigations on the typing of staphylococci by means of bacteriophage; the origin and nature of lysogenic strains. *J. Hyg. (Lond.)* **46:**74–81.

Sompolinsky, D., Samra, Z., Karakawa, W. W., Vann, W. F., Schneerson, R., and Malik, Z. (1985). Encapsulation and capsular types in isolates of *Staphylococcus aureus* from different sources and relationship to phage types. *J. Clin. Microbiol* **22:**828–834.

Son, J. S., Lee, S. J., Jun, S. Y., Yoon, S. J., Kang, S. H., Paik, H. R., Kang, J. O., and Choi, Y. J. (2010). Antibacterial and biofilm removal activity of a *Podoviridae Staphylococcus aureus* bacteriophage SAP-2 and a derived recombinant cell-wall-degrading enzyme. *Appl. Microbiol. Biotechnol.* **86:**1439–1449.

Stapleton, M. R., Horsburgh, M. J., Hayhurst, E. J., Wright, L., Jonsson, I. M., Tarkowski, A., Kokai-Kun, J. F., Mond, J. J., and Foster, S. J. (2007). Characterization of IsaA and SceD, two putative lytic transglycosylases of *Staphylococcus aureus*. *J. Bacteriol.* **189:**7316–7325.

Stewart, C. R. (1993). SPO1 and related bacteriophages. *In* "*Bacillus subtilis* and Other Gram-Positive Bacteria: Biochemistry, Physiology, and Molecular Genetics" (A. L. Sonenshein, J. A. Hoch, and R. Losick, eds.), pp. 813–829. American Society for Microbiology, Washington, DC.

Stewart, C. R., Casjens, S. R., Cresawn, S. G., Houtz, J. M., Smith, A. L., Ford, M. E., Peebles, C. L., Hatfull, G. F., Hendrix, R.-W., Huang, W. M., and Pedulla, M. L. (2009a). The genome of *Bacillus subtilis* bacteriophage SPO1. *J. Mol. Biol.* **388:**48–70.

Stewart, C. R., Gaslightwala, I., Hinata, K., Krolikowski, K. A., Needleman, D. S., Peng, A. S.-Y., Peterman, M. A., Tobias, A., and Wei, P. (1998). Genes and regulatory sites of the "host-takeover module" in the terminal redundancy of *Bacillus subtilis* bacteriophage SPO1 *Virology* **246:**329–340.

Stewart, C. R., Yip, T. K., Myles, B., and Laughlin, L. (2009b). Roles of genes 38, 39, and 40 in shutoff of host biosyntheses during infection of Bacillus subtilis by bacteriophage SPO1. *Virology.* **392:**271–274.

Stobberingh, E., and Winkler, K. (1977). Restriction-deficient mutants of *Staphylococcus aureus*. *J. Gen. Microbiol.* **99:**359–367.

Sulakvelidze, A., Alavidze, Z., and Morris, J. G., Jr. (2001). Bacteriophage therapy. *Antimicrob. Agents Chemother.* **45:**649–659.

Sulakvelidze, A., and Kutter, E. (2005). Phage therapy in humans. *In* "Bacteriophages: Biology and Applications" (E. Kutter and A. Sulakvelidze, eds.), pp. 381–436. CRC Press, Boca Raton, FL.

Sunagar, R., Patil, S. A., and Chandrakanth, R. K. (2010). Bacteriophage therapy for *Staphylococcus aureus* bacteremia in streptozotocin-induced diabetic mice. *Res. Microbiol.* **161:**854–860.

Sussenbach, J. S., Monfoort, C. H., Schiphof, R., and Stobberingh, E. E. (1976). A restriction endonuclease from *Staphylococcus aureus*. *Nucleic Acids Res.* **3:**3193–3202.

Sussenbach, J. S., Steenbergh, P. H., Rost, J. A., van Leeuwen, W. J., and van Embden, J. D. (1978). A second site-specific restriction endonuclease from *Staphylococcus aureus*. *Nucleic Acids Res.* **5:**1153–1163.

Sutter, D. E., Summers, A. M., Keys, C. E., Taylor, K. L., Frasch, C. E., Braun, L. E., Fattom, A. I., and Bash, M. C. (2011). Capsular serotype of *Staphylococcus aureus* in the era of community-acquired MRSA. *FEMS Immunol. Med. Microbiol.* **63:**16–24.

Synnott, A. J., Kuang, Y., Kurimoto, M., Yamamichi, K., Iwano, H., and Tanji, Y. (2009). Isolation from sewage influent and characterization of novel *Staphylococcus aureus* bacteriophages with wide host ranges and potent lytic capabilities. *Appl. Environ. Microbiol.* **75:**4483–4490.

Sziläk, L., Venetianer, P., and Kiss, A. (1990). Cloning and nucleotide sequence of the genes coding for the Sau96I restriction and modification enzymes. *Nucleic Acids Res.* **18:**4659–4664.

Takac, M., and Blasi, U. (2005). Phage P68 virion-associated protein 17 displays activity against clinical Isolates of *Staphylococcus aureus. Antimicrob. Agents Chemother.* **49:**2934–2940.

Talbot, H. W., Jr., and Parisi, J. T. (1976). Phage typing of *Staphylococcus epidermidis. J. Clin. Microbiol.* **3:**519–523.

Talkington, C., and Pero, J. (1977). Restriction fragment analysis of temporal program of bacteriophage SPO1 transcription and its control by phage modified RNA polymerases. *Virology* **83:**365–379.

Tenover, F. C., Arbeit, R., Archer, G., Biddle, J., Byrne, S., Goering, R., Hancock, G., Hébert, G. A., Hill, B., Hollis, R., *et al.* (1993). Comparison of traditional and molecular methods of typing isolates of *Staphylococcus aureus. J. Clin. Microbiol.* **32:**407–415.

Tenover, F. C., and Gorwitz, R. J. (2006). The epidemiology of Staphylococcus infections. In ''Gram positive pathogens'' (V. A. Fischetti, R. P. Novick, J. J. Ferretti, D. A. Portnoy and J. I. Rood, eds.), pp. 256–235. ASM Press, Washington, DC. 526-535.

Traxler, M. F., Summers, S. M., Nguyen, H. T., Zacharia, V. M., Hightower, G. A., Smith, J. T., and Conway, T. (2008). The global, ppGpp-mediated stringent response to amino acid starvation in *Escherichia coli. Mol. Microbiol.* **68:**1128–1148.

Tsuru, T., Kawai, M., Mizutani-Ui, Y., Uchiyama, I., and Kobayashi, I. (2006). Evolution of paralogous genes: Reconstruction of genome rearrangements through comparison of multiple genomes within *Staphylococcus aureus. Mol. Biol. Evol.* **23:**1269–1285.

Twort, F. W. (1915). An investigation on the nature of the ultramicroscopic viruses. *Lancet* **189:**1241–1243.

Uchiyama, J., Rashel, M., Maeda, Y., Takemura, I., Sugihara, S., Akechi, K., Muraoka, A., Wakiguchi, H., and Matsuzaki, S. (2008). Isolation and characterization of a novel *Enterococcus faecalis* bacteriophage phiEF24C as a therapeutic candidate. *FEMS Microbiol. Lett.* **278:**200–206.

Valpuesta, J. M., Donate, L. E., Mier, C., Herranz, L., and Carrascosa, J. L. (1993). RNA-mediated specificity of DNA packaging into hybrid lambda/phi 29 proheads. *EMBO J.* **12:**4453–4459.

Vandersteegen, K., Mattheus, W., Ceyssens, P. J., Bilocq, F., De Vos, D., Pirnay, J. P., Noben, J. P., Merabishvili, M., Lipinska, U., Hermans, K., and Lavigne, R. (2011). Microbiological and molecular assessment of bacteriophage ISP for the control of *Staphylococcus aureus. PLoS One* **6:**e24418.

van Wamel, W. J., Rooijakkers, S. H., Ruyken, M., van Kessel, K. P., and van Strijp, J. A. (2006). The innate immune modulators staphylococcal complement inhibitor and chemotaxis inhibitory protein of *Staphylococcus aureus* are located on beta-hemolysin-converting bacteriophages. *J. Bacteriol.* **188:**1310–1305.

Veiga, H., and Pinho, M. G. (2009). Inactivation of the SauI type I restriction-modification system is not sufficient to generate *Staphylococcus aureus* strains capable of efficiently accepting foreign DNA. *Appl. Environ. Microbiol.* **75:**3034–3038.

Verhoef, J., Winkler, K. C., and van Boven, C. P. (1971). Characters of phages from coagulase-negative staphylococci. *J. Med. Microbiol.* **4:**413–424.

Vindel, A., Trincado, P., Gomez, E., Aparicio, P., Martin de Nicolas, M., Boquete, T., and Saez Nieto, J. A. (1994). An additional set of phages to characterize epidemic methicillin-resistant *Staphylococcus aureus* strains from Spain (1989-92). *Epidemiol. Infect.* **112:**299–306.

Vybiral, D., Takác, M., Loessner, M., Witte, A., von Ahsen, U., and Bläsi, U. (2003). Complete nucleotide sequence and molecular characterization of two lytic *Staphylococcus aureus* phages: 44AHJD and P68. *FEMS Microbiol. Lett.* **219:**275–283.
Waldron, D. E., and Lindsay, J. A. (2006). Sau1: A novel lineage-specific type I restriction-modification system that blocks horizontal gene transfer into *Staphylococcus aureus* and between *S. aureus* isolates of different lineages. *J. Bacteriol.* **188:**5578–5585.
Wang, C. T. (1978). Bacteriophage typing of canine staphylococci. II. Isolation of phage from lysogenic strains and establishment of a new phage set. *Nihon Juigaku Zasshi.* **40:**515–523.
Wegener, H. C. (1993). Development of a phage typing system for *Staphylococcus hyicus*. *Res. Microbiol.* **144:**237–244.
Wei, P., and Stewart, C. R. (1993). A cytotoxic early gene of *Bacillus subtilis* bacteriophage SPO1. *J. Bacteriol.* **175:**7887–7900.
Weigele, P. R., Pope, W. H., Pedulla, M. L., Houtz, J. M., Smith, A. L., Conway, J. F., King, J., Hatfull, G. F., Lawrence, J. G., and Hendrix, R. W. (2007). Genomic and structural analysis of Syn9, a cyanophage infecting marine *Prochlorococcus* and *Synechococcus*. *Environ. Microbiol.* **9:**1675–1695.
Wentworth, B. B. (1963). Bacteriophage typing of the staphylococci. *Bacteriol. Rev.* **27:**253–272.
Wildemauwe, C., De Brouwer, D., Godard, C., Buyssens, P., Dewit, J., Joseph, R., and Vanhoof, R. (2010). The use of spa and phage typing for characterization of a MRSA population in a Belgian hospital: Comparison between 2002 and 2007. *Pathol. Biol. (Paris)* **58:**70–72.
Wildemauwe, C., Godard, C., Verschraegen, G., Claeys, G., Duyck, M. C., De Beenhouwer, H., and Vanhoof, R. (2004). Ten years phage-typing of Belgian clinical methicillin-resistant *Staphylococcus aureus* isolates (1992-2001). *J. Hosp. Infect.* **56:**16–21.
Williams, R. E., and Rippon, J. E. (1952). Bacteriophage typing of *Staphylococcus aureus*. *J. Hyg. (Lond.)* **50:**320–353.
Wills, Q. F., Kerrigan, C., and Soothill, J. S. (2005). Experimental bacteriophage protection against *Staphylococcus aureus* abscesses in a rabbit model. *Antimicrob. Agents. Chemother.* **49:**1220–1221.
Wilson, G. G., and Murray, N. E. (1991). Restriction and modification systems. *Annu. Rev. Genet.* **25:**585–627.
Wilson, G. S., and Atkinson, J. D. (1945). Typing of staphylococci by the bacteriophage method. *Lancet* **1:**647.
Xia, G., Corrigan, R. M., Winstel, V., Goerke, C., Gründling, A., and Peschel, A. (2011). Wall teichoic acid-dependent adsorption of staphylococcal siphovirus and myovirus. *J. Bacteriol.* **193:**4006–4009.
Xia, G., Maier, L., Sanchez-Carballo, P., Li, M., Otto, M., Holst, O., and Peschel, A. (2010). Glycosylation of wall teichoic acid in *Staphylococcus aureus* by TarM. *J. Biol. Chem.* **285:**13405–13415.
Yamada, T., Satoh, S., Ishikawa, H., Fujiwara, A., Kawasaki, T., Fujie, M., and Ogata, H. (2010). A jumbo phage infecting the phytopathogen *Ralstonia solanacearum* defines a new lineage of the *Myoviridae* family. *Virology.* **398:**135–147.
Yamaguchi, T., Hayashi, T., Takami, H., Nakasone, K., Ohnishi, M., Nakayama, K., Yamada, S., Komatsuzawa, H., and Sugai, M. (2000). Phage conversion of exfoliative toxin A production in *Staphylococcus aureus*. *Mol. Microbiol.* **38:**694–705.
Yee, L. M., Matsumoto, T., Yano, K., Matsuoka, S., Sadaie, Y., Yoshikawa, H., and Asai, K. (2011). The genome of *Bacillus subtilis* phage SP10: a comparative analysis with phage SPO1. *Biosci. Biotechnol. Biochem.* **75:**944–952.
Young, R. (2005). Phage lysis. *In* "Phages. Their role in bacterial pathogenesis and Biotechnology" (M. K. Waldor, D. I. Friedman, and S. Adhya, eds.), pp. 92–113. ASM Press, Washington, D.C.

Zimecki, M., Artym, J., Kocieba, M., Weber-Dabrowska, B., Borysowski, J., and Górski, A. (2009). Effects of prophylactic administration of bacteriophages to immunosuppressed mice infected with *Staphylococcus aureus*. *BMC Microbiol.* **9**:169.

Zimecki, M., Artym, J., Kocieba, M., Weber-Dabrowska, B., Borysowski, J., and Górski, A. (2010). Prophylactic effect of bacteriophages on mice subjected to chemotherapy-induced immunosuppression and bone marrow transplant upon infection with *Staphylococcus aureus*. *Med. Microbiol. Immunol.* **199**:71–79.

Zimecki, M., Artym, J., Kocieba, M., Weber-Dabrowska, B., Lusiak-Szelachowska, M., and Górski, A. (2008). The concerted action of lactoferrin and bacteriophages in the clearance of bacteria in sublethally infected mice. *Postepy Hig. Med. Dosw* **62**:42–46.

Zimmer, M., Sattelberger, E., Inman, R. B., Calendar, R., and Loessner, M. J. (2003). Genome and proteome of *Listeria monocytogenes* phage PSA: An unusual case for programmed + 1 translational frameshifting in structural protein synthesis. *Mol. Microbiol.* **50**:303–317.

Zueva, V. S., Dmitrenko, O. A., Gladkova, K. K., and Zueva, E. A. (1994). A new collection of phages for typing methicillin resistant *Staphylococcus aureus*. *Zh. Mikrobiol. Epidemiol. Immunobiol.* **2**:20–23.

Zueva, V. S., Dmitrenko, O. A., Krupina, E. A., Belikov, N. G., and Nesterenko, L. N. (1990). The mechanism of the resistance of methicillin-resistant *Staphylococcus aureus* to phages from the International Collection. *Zh. Mikrobiol. Epidemiol. Immunobiol.* **10**:11–15.

Zueva, V. S., Dmitrenko, O. A., Shaginian, I. A., and Akatov, A. K. (1997). A system of step-by-step differentiation of methicillin-resistant *Staphylococcus aureus*. *Antibiot. Khimioter.* **42**:20–26.

Section 2
Interactions and Applications of Phage Proteins and Protein Complexes

CHAPTER 6

Bacteriophage Protein–Protein Interactions

Roman Häuser,[*,†] **Sonja Blasche,**[†] **Terje Dokland,**[‡] **Elisabeth Haggård-Ljungquist,**[§] **Albrecht von Brunn,**[||] **Margarita Salas,**[¶] **Sherwood Casjens,**[#] **Ian Molineux,**[**] and **Peter Uetz**[††,1]

Contents

* Institute of Toxicology and Genetics, Karlsruhe Institute of Technology, Karlsruhe, Germany
† Deutsches Krebsforschungszentrum, Heidelberg, Germany
‡ Department of Microbiology, University of Alabama at Birmingham, Birmingham, Alabama, USA
§ Department of Genetics, Microbiology and Toxicology, Stockholm University, SE-106 91 Stockholm, Sweden
|| Max-von-Pettenkofer-Institut, Lehrstuhl Virologie, Ludwig-Maximilians-Universität, München, Germany
¶ Centro de Biología Molecular "Severo Ochoa" (CSIC-UAM), Cantoblanco, Madrid, Spain
\# Division of Microbiology and Immunology, Pathology Department, University of Utah School of Medicine, Salt Lake City, Utah
** Molecular Genetics and Microbiology, Institute for Cell and Molecular Biology, University of Texas–Austin, Austin, Texas, USA
†† Center for the Study of Biological Complexity, Virginia Commonwealth University, Richmond, Virginia, USA
1 Corresponding author, E-mail: peter@uetz.us

Advances in Virus Research, Volume 83
ISSN 0065-3527, DOI: 10.1016/B978-0-12-394438-2.00006-2

Abstract Bacteriophages T7, λ, P22, and P2/P4 (from *Escherichia coli*), as well as ϕ29 (from *Bacillus subtilis*), are among the best-studied bacterial viruses. This chapter summarizes published protein interaction data of intraviral protein interactions, as well as known phage–host protein interactions of these phages retrieved from the literature. We also review the published results of comprehensive protein interaction analyses of *Pneumococcus* phages Dp-1 and Cp-1, as well as coliphages λ and T7. For example, the ≈55 proteins encoded by the T7 genome are connected by ≈43 interactions with another ≈15 between the phage and its host. The chapter compiles published interactions for the well-studied phages λ (33 intra-phage/22 phage-host), P22 (38/9), P2/P4 (14/3), and ϕ29 (20/2). We discuss whether different interaction patterns reflect different phage lifestyles or whether they may be artifacts of sampling. Phages that infect the same host can interact with different host target proteins, as exemplified by *E. coli* phage λ and T7. Despite decades of intensive investigation, only a fraction of these phage interactomes are known. Technical limitations and a lack of depth in many studies explain the gaps in our knowledge. Strategies to complete current interactome maps are described. Although limited space precludes detailed overviews of phage molecular biology, this compilation will allow future studies to put interaction data into the context of phage biology.

ABBREVIATIONS:

EM	electron microscopy
LC	liquid chromatography
MS	mass spectrometry
NMR	nuclear magnetic resonance
ORF	open reading frame
PPI	protein–protein interaction
Y2H	yeast two-hybrid

I. INTRODUCTION

Bacteriophages have been extremely important model systems in molecular biology. They were critical to many breakthrough discoveries, such as the nature of the genetic material (Hershey, 1952), the structure of genes and genomes and the first genome sequence (Sanger *et al.*, 1977), recombination (Kaiser and Jacob, 1957), transcriptional regulation (e.g., Roberts, 1969), and many others. In addition, phages have become important technical tools in molecular genetics and nanobiology, but these aspects are not reviewed here.

A great deal of work has been invested in learning the mechanistic details of the life cycles of a number of phages, including those described in this chapter. Although phages are tiny biological systems, surprisingly, not even the biology of main "model system phages" is completely understood. Despite a plethora of papers, few comprehensive summaries exist of the molecular interaction events that occur in both phage gene regulation and virion assembly. As in all organisms, protein–protein interactions (PPIs) play a fundamental role in phage biology. PPIs are important in all virus-related processes, including host infection, transcriptional regulation, DNA replication, virion assembly, and lysis. Current protein interaction databases (Lehne and Schlitt, 2009) have consistently neglected phage interactions. For instance, Intact, one of the most comprehenisve databases for protein–protein interactions, listed only 22 interactions for Caudovirales, the only phage group represented at the time of this writing (April 2011). We have thus attempted to compile such information for several phages that infect Gram-negative hosts, including λ, T7, P22, and P2/P4, as well as others that infect Gram-positive bacteria, specifically ϕ29, Dp-1, and Cp-1 (Table I). Coliphage

TABLE I Summary of phage interactomes described in this chapter[a]

Phage	Proteins	Phage–phage PPIs	Phage–host PPIs
T7	55	~43 (including 7 HTS[b])	15
λ	73	33 (+ 81 HTS)	26
P22	64	38	9
P2/P4	42/13	14	3
ϕ29	30	20	3
Dp-1	72	156 (HTS)	38 (HTS)
Cp-1	28	17 (HTS)	11 (HTS)

[a] Note that all interactions from this chapter are available as a downloadable supplement from http:/www.uetz.us.
[b] High-throughput data (see text for details).

T4 is one of the best understood of all the obligate lytic phages. T4 has been an outstanding model for phage biology, particularly for DNA replication, control of transcription, and virion assembly. Unfortunately, this chapter cannot discuss the extensive T4 PPIs here for reasons of space. Many of the small DNA phages, such as ϕX174 or M13, have been very well studied, but because of their tiny proteomes they do not have many protein–protein interactions. It would be highly desirable to establish their PPIs so that more comprehensive analyses of phage systems biology would be possible. Nevertheless, it is hoped that this chapter proves useful to the phage community and encourages similar compilations for other phages. The interactions collected for this chapter are available in Intact (http://www.ebi.ac.uk/intact/) for easy access.

II. METHODS USED TO DETECT PROTEIN–PROTEIN INTERACTIONS

A. Genetic detection of interactions

Methods used to study protein–protein interactions in phages have evolved over the decades. A few important methods are illustrated in Figure 1. In the beginnings of phage molecular biology, detection of genetic interactions dominated. For instance, certain mutants in the λ *J* gene abolished virion binding to host *Escherichia coli* cells. Similarly, mutants of the *E. coli* gene *lamB* made the cells resistant to λ infections. From such experiments it was concluded that J protein bound to LamB and that LamB was the λ receptor. This hypothesis has been confirmed by isolating compensatory mutations in both genes that regenerate the interaction (Werts *et al.*, 1994).

B. Electron microscopy

Another important method for the investigation of physical relationships between phage components is electron microscopy, especially in combination with genetic analysis. Mutants often have an aberrant morphology. If certain structures are missing in the mutant virion, it can be concluded that the mutated gene encodes this structure or is at least involved in its assembly. For instance, many laboratory strains of λ harbor a frameshift mutation in the *stf* gene, and these mutants lack side tail fibers (Hendrix and Duda, 1992). Many such studies not only showed which genes encode which proteins, but also indicated how the proteins interact (e.g., when particular structures could be "removed" from the virion by defined gene knockouts).

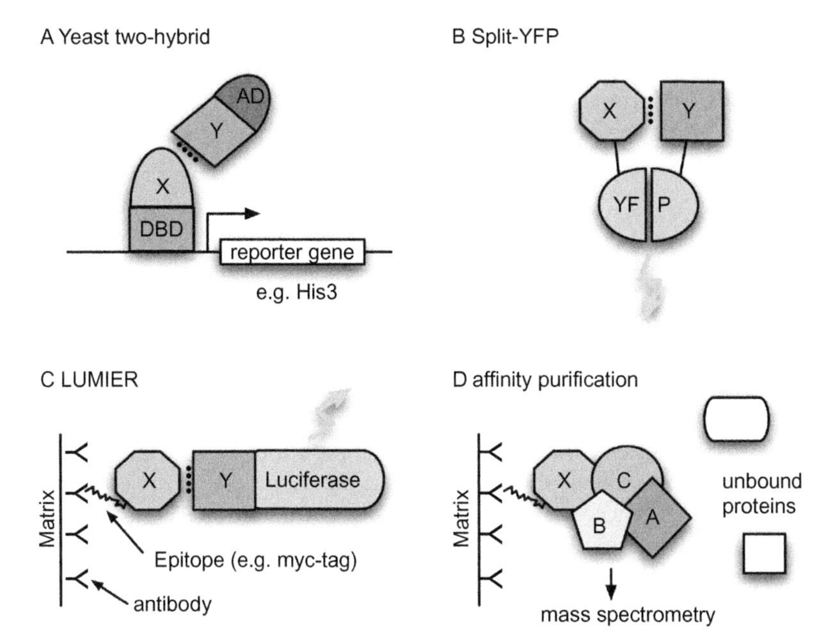

FIGURE 1 Selected methods for the study of protein–protein interactions. (A) The yeast two-hybrid (Y2H) system is based on the levels of two fusion proteins (typically in yeast but any cell type can be used). One of the proteins contains a DNA-binding domain (DBD), which can bind to the promoter of a reporter gene (here: HIS3), and a second protein X, the bait. The second fusion protein consists of a transcription-activation domain (AD) and a second protein, Y. If proteins X and Y interact, a transcription factor is formed and the reporter gene is activated. In this case, that means that the cell can grow on histidine-free medium. A yeast colony growing on such medium thus indicates an interaction of the two inserted proteins. (B) Protein complementation assay (PCA), e.g., split-YFP. As in the Y2H assay, two interacting proteins bring together two protein fragments that are inactive when separate but active when in close proximity. Here, fragments of yellow-fluorescent protein (YFP) reassociate and fluoresce when reassembled. Other fluorescent proteins, such as the green fluorescent protein (GFP), have been used in a similar way. (C) LUMIER (LUMInescence-based mammalian intER-actome). Two fusion proteins are purified by means of an epitope tag (here: FLAG tag), usually on an antibody-coated matrix. Interactions between X and Y can be detected using Luciferase, whose gene is fused to Y and which emits light when luciferin is added. (D) Affinity purification. Protein complexes can be purified from cellular lysates using an affinity epitope, as in (C) with a FLAG tag. Components of the complex can then be identified using mass spectrometry, using the unique mass of peptides when the protein is digested by trypsin. (See Page 4 in Color Section at the back of the book.)

More recently, many bacteriophage capsids and phage protein complexes have been analyzed by cryoelectron microscopy and three-dimensional reconstructions. This technique gives resolutions approaching that of X-ray crystallography, especially for proteins present in symmetric arrays in the virion, and has proven ideal for analyzing protein–protein interactions in large complexes, especially in combination with high-resolution structural data from other techniques, such as X-ray crystallography.

C. Yeast two-hybrid screens

More recently, the yeast two-hybrid system (Y2H) has become an important method to detect pairwise protein interactions. Interaction screens of phage proteins can be carried out easily by cloning all open reading frames in appropriate vectors and then testing these in a systematic pairwise fashion. This has been done for several phages, including coliphages T7 (Bartel *et al.*, 1996) and λ (Rajagopala *et al.*, 2011), as well as *Streptococcus pneumoniae* phages Cp-1 and Dp-1 (Häuser *et al.*, 2011; Sabri *et al.*, 2010). The whole process can also be automated.

One disadvantage of the Y2H system is that a typical pairwise screen, also called an "array screen" because of the systematic arrangement of clones in microtiter plates, may only detect 20–30% of all interactions. That is, the rate of false negatives is on the order of 70–80%. This can, in theory, be overcome by using multiple vectors that differ in the structure of the fusion proteins, for instance in the production of N- or C-terminal fusion proteins (Stellberger *et al.*, 2010). A combination of multiple vectors can achieve interaction detection rates of up to 80% (i.e., false negative rates as low as 20%) (Chen *et al.*, 2010).

Y2H also suffers from false positives that can only be understood with additional experimentation. Nonetheless, a multivector Y2H strategy can help in the recognition of false positives. For instance, if multiple vector combinations are used, interactions found in all or most cases are more likely to be "true positives," whereas interactions detected in a lower fraction of cases are less reliable and may be false positives.

D. Protein complex purification and mass spectrometry

Whole phage virions can be purified, typically by CsCl density gradient centrifugation. Proteins from purified phage can then be denatured and separated on polyacrylamide gels or subjected directly to trypsin digestion and subsequent liquid chromatography (LC) and mass spectrometry (MS). The mass of peptides determined by MS can usually identify the corresponding proteins unambiguously, and thus composition of the phage particles. Once all proteins in a virion are identified, at least some can be assigned easily to known structures, such as the icosahedral capsid or tail sheath. Typically, only a fraction of all phage-encoded proteins are represented in the mature virion (e.g., less than 20 of the $\approx$70 phage λ proteins). Especially for little-studied phages it can be extremely helpful to know which proteins are in the virion. For example, in an analysis of *Streptococcus* phage Dp-1, only 8 of the 72 predicted proteins were identified by LC/MS and were thus confirmed structural proteins (see later).

E. Biochemistry of protein complexes

Biochemical approaches often involve the purification of protein complexes and thus yield protein interaction information. This approach can be made technically easier if individual proteins are affinity tagged. For instance, phage-encoded proteins can often be fused to affinity column-binding proteins such as glutathione-*S*-transferase or oligohistidine without affecting their function. Such tagged proteins and proteins bound to them can then be purified using affinity reagents. The untagged proteins in the complex can then be identified by mass spectrometry or immunoblotting (if antibodies against specific proteins are available). A range of experimental techniques is available for the characterization of phage protein complexes and their activities; a comprehensive overview cannot be given here. For instance, chemical cross-linking of interacting proteins is used frequently to spatially locate physically associated proteins, and variable spacer lengths can determine the steric proximity of the binding partners. Other tests include activity assays (e.g., proteases or other modifying enzymes that may be required for phage assembly), replication, or other processes. Some examples of such activities are described here.

F. X-ray crystallography and nuclear magnetic resonance

Only atomic structures of all phage proteins and their interactions will give a mechanistic understanding of phage biology, and progress in this area has been accelerating. Individual proteins and protein complexes, if not the whole phage particle, can be crystallized and the structure of such proteins solved by X-ray analysis. Alternatively, protein structures can be solved by nuclear magnetic resonance (NMR) if high enough concentrations can be achieved. Unfortunately, size restrictions on NMR structural determinations limit the analysis of large protein complexes, but monomeric structures can be extremely useful in devising testable models for the interactions proteins may undergo. However, many phage proteins are refractory to structural analysis (especially those involved in virion assembly, as they often assemble only at high concentrations), and even in classical model systems such as phage λ less than a third of all phage-encoded proteins have been crystallized—most of these as pure proteins rather than as protein complexes (Rajagopala *et al.*, 2011;Yura *et al.*, 2006).

III. PROTEIN–PROTEIN INTERACTIONS OF BACTERIOPHAGES

A. Coliphage T7

Bacteriophage T7 was defined in 1945 as a member of the seven Type ("T") phages that grow lytically on *E. coli* B (Demerec and Fano, 1945), although it is probably identical to phage δ, used earlier by Delbrück; a close relative of T7 was likely studied by d'Herelle in the 1920s (d'Herelle, 1926). Relatives of T7 exhibit high synteny across their entire genomes, with the closest relatives differing primarily only by the number and location of homing endonucleases or other selfish elements not known to have any role in phage development. These phages therefore will exhibit life cycles and PPIs that parallel T7 itself, modified only by the resources available in their host organisms.

1. Growth physiology

T7 grows on rough strains of *E. coli* (i.e., those without full-length O-antigen polysaccharide on their surface) and some other enteric bacteria, but close relatives have been described that infect smooth and even capsulated strains (Molineux, 2006a). In such phages, tail fibers present on T7 are replaced with tailspikes with enzymatic activity that degrades the O- or K-antigens on the cell surface. T7 has a short latent period of 17 min at 37°C, which has resulted in most physiological studies being conducted at 30°C where infected cells lyse after 30 min. However, adaptation of T7 to high fitness with an excess of an *E. coli* K-12 strain growing under optimal conditions at 37°C in rich media results in a phage with a latent period of only ≈11 min. This adapted phage can undergo an effective expansion of its population by more than 10^{13} in 1 hr of growth (Heineman and Bull, 2007). There is no obvious subtlety in the T7 developmental pathway: the early region of its genome is transcribed efficiently by the host RNA polymerase, leading to synthesis of a phage-specific RNA polymerase and the concurrent inhibition of the host enzyme. Thus, T7 rapidly takes control of the transcriptional capacity of the infected cell. As *E. coli* mRNAs are also intrinsically unstable, T7 therefore also gains control of the translational capacity of the cell. Further, because T7 uses nucleotides derived from the host chromosome for its own replication, more than 80% of the DNA in progeny is derived from the bacterial chromosome and only a small burst is observed after infection of DNA-free minicells. T7 development is not known to be dependent on host chaperones, although it does code for two proteins that have been suggested to have chaperone activity in capsid and tail assembly. Aside from host components used for phage adsorption, plus general biosynthetic processes and the translational apparatus, T7 development is remarkably independent of host proteins: only *E. coli* RNA polymerase, thioredoxin—the cofactor for T7 DNA polymerase—and (deoxy)cytidine monophosphate kinase are known to be essential.

2. Early steps in infection

T7 is a member of *Podoviridae*, possessing only a short stubby tail that needs to be extended functionally in order to span the cell envelope, which thus allows its genome to be translocated into the cell. Tail extension is affected through internal head proteins that are ejected into the cell at the initiation of infection prior to or concomitant with transport of the leading end of the genome (Chang *et al.*, 2010; Kemp *et al.*, 2005). A schematic of the T7 virion is shown in Figure 2. In T7, the core consists of several copies each of three proteins (gp14, gp15, and gp16), which must interact, albeit perhaps loosely, with themselves and each other, in the phage capsid. However, during their ejection into the cell they must at least partially dissociate and also partially unfold in order to pass through the narrow channel through the portal and tail (Molineux, 2006b). The N-terminal domain of gp16 refolds spontaneously in the periplasm of the infected cell to form an enzyme with lytic transglycosylase activity (Moak and Molineux, 2000, 2004). This muralytic activity is only conditionally essential for T7 infection (Moak and Molineux, 2000).

The C-terminal residues of gp16, likely together with a more central portion of gp15, also refold in the periplasm and then insert spontaneously into the cytoplasmic membrane, completing the channel across the cell envelope for DNA translocation into the cell. Likely the last of the core proteins to exit the phage head, gp14 is found exclusively within the outer membrane of the infected cell. Using energy derived from the membrane potential, these proteins then ratchet the leading 1 kb of the T7 genome into the cell (Chang *et al.*, 2010; Kemp *et al.*, 2005). In both T7 and T3, genome internalization by this process stops, and the remainder of the DNA is normally pulled into the cytoplasm as a consequence of host and then phage RNA polymerase-catalyzed transcription (Garcia and Molineux, 1995, 1996; Kemp *et al.*, 2004). Several strong promoters lie on this leading 1-kb segment of the T7 genome.

3. Early genes

Transcription from T7 early promoters not only internalizes the remainder of the genome but also leads to expression of early T7 genes, some of which interact and inhibit or modulate the activity of host proteins (Molineux, 2006a). Of the early proteins, only T7 RNA polymerase is essential for phage growth. The first T7 protein to be synthesized is gp0.3, an anti-type I restriction protein (Studier, 1975). The structure of gp0.3 reveals that it resembles double-stranded DNA, and the protein binds to the specificity subunit of the heteropentameric EcoKI enzyme,

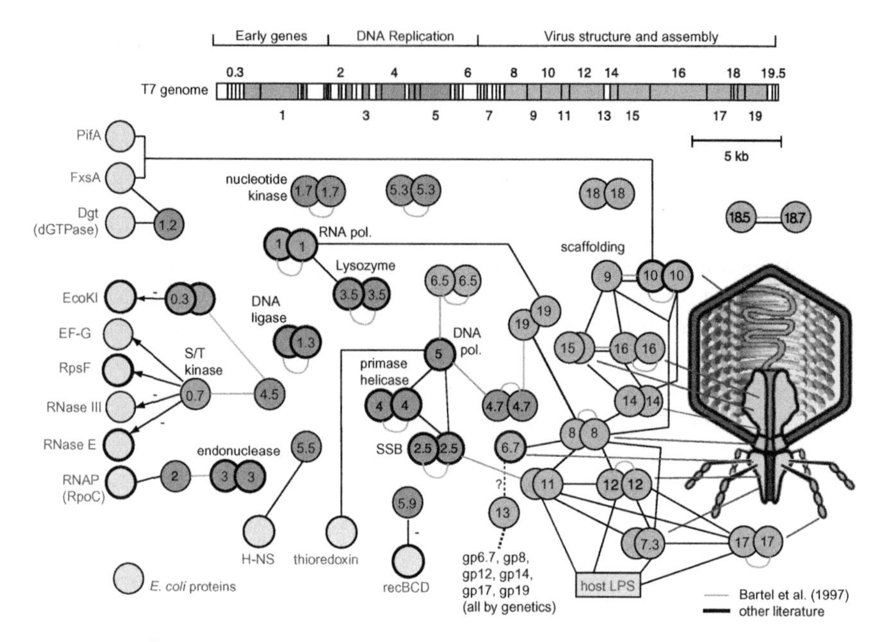

FIGURE 2 Phage T7: genome, virion, and interactome. ORFeome and protein interaction map of bacteriophage T7. Cloned open reading frames (ORFs) and random fragments were used to generate a two-dimensional interaction map. ORFeomes can also be used by structural genomics projects to derive three-dimensional structures. A combination of interaction maps and crystal structures often allows reconstruction of protein complexes as in viral particles or their subunits. (Top) T7 genome with each ORF represented as a box. ORFs indicated by integral numbers were identified originally as essential by genetic screens. Other genes have decimal numbers; of these, only *2.5* and *7.3* are essential. Proteins found to interact with other proteins are indicated by colored boxes; colors indicate a simplified assignment to functional classes as shown. Interactions between proteins are shown as lines, with self-interactions shown as dimers. Hatched lines indicate expected interactions that have not been found by two-hybrid analysis. Gray proteins correspond to host proteins (*E. coli*). Blue proteins are involved in virus assembly and structure, and, where known, their location in the virus particle is shown on the right. Proteins with thick borders have been crystallized and their structure determined. Genetic map modified after Dunn and Studier (1983). Figure modified after Uetz *et al.* (2004). Interactions based primarily on data in Table II (see references therein). (See Page 4 in Color Section at the back of the book.)

preventing both it and the cognate modification enzyme from recognizing its target sequence (Atanasiu *et al.*, 2001, 2002; Kennaway *et al.*, 2009; Murray, 2000; Stephanou *et al.*, 2009). T7 gp0.3 is active against all type I enzymes and thus T7 grows normally on B, C, and K-12 strains of *E. coli* regardless of its previous host. However, the protein affords no protection against type II and little, if any, against type III enzymes. Gp0.3 proteins of different T7-like phages fall into two distinct groups with no sequence similarity; both have anti-restriction activity but the group typified by T3 gp0.3 also has SAMase activity (Studier and Movva, 1976). The second major early protein whose function is known is gp0.7, a serine-threonine kinase (Rahmsdorf *et al.*, 1974) that is also responsible for the shut off of host transcription. The two activities are separable (Simon and Studier, 1973). Gp0.7 phosphorylates several host proteins, including RNA polymerase (Severinova and Severinov, 2006; Zillig *et al.*, 1975), the translational apparatus including EF-G and the small ribosomal subunit protein S6 (Robertson and Nicholson, 1990, 1992; Robertson *et al.*, 1994), and both RNase III (Marchand *et al.*, 2001) and RNase E (Mayer and Schweiger, 1983). T7 shuts down translation of host mRNAs but it remains unclear whether EF-G or S6 is involved directly. An unknown T7 late gene also modulates the specificity of RNase E (Savalia *et al.*, 2010). Although it is likely that others do so, only 1 other of the 10 early T7 proteins is currently known to interact with host proteins. Gp1.2 binds to *E. coli* dGTPase, converting it into a rGTP-binding protein (Huber *et al.*, 1988; Nakai and Richardson, 1990); gp1.2 also binds to the membrane protein FxsA and the F plasmid protein PifA (Cheng *et al.*, 2004; Schmitt and Molineux, 1991; Wang *et al.*, 1999).

4. T7 RNA polymerase

The hallmark of T7 is of course its phage-encoded RNA polymerase, which, together with its regulator T7 lysozyme, is responsible for most T7 gene expression. It is also essential for synthesizing RNA primers at the primary origin of replication (Fujiyama *et al.*, 1981) and for initiation of DNA packaging (Zhang and Studier, 1997, 2004). In addition to several complexes containing a promoter and product RNAs (Cheetham and Steitz, 1999; Cheetham *et al.*, 1999; Kennedy *et al.*, 2007; Tahirov *et al.*, 2002; Yin and Steitz, 2002, 2004), the crystal structure of the RNA polymerase–lysozyme complex has been determined (Jeruzalmi and Steitz, 1998). Many mutants in both RNA polymerase and lysozyme have been characterized that affect their functional interaction, but it is not obvious from the crystal structures how the mutations affect the regulation of transcription. Lysozyme is thought to target the initial transcribing complex in a promoter strength-dependent manner and have no effect on transcription elongation (Huang *et al.*, 1999; Stano and Patel, 2004; Zhang and Studier, 1997, 2004). Lysozyme is, however, important in pausing or terminating transcription at the class II terminator CJ, where it probably interacts directly with the T7 terminase protein gp19 and thus directs the packaging machinery to create the DNA terminus (Lyakhov *et al.*, 1997, 1998; Zhang and Studier, 2004).

After T7 RNA polymerase has been synthesized, there is no further requirement for the host polymerase and it actually becomes inhibitory to phage development. Inactivation is caused by the T7 gp2 protein, which binds to the β′ subunit and inhibits transcription by interfering with strand separation of the open complex at most promoters (Camara *et al.*, 2010; Nechaev and Severinov, 1999). *In vivo*, the only T7 promoter that must be inhibited by gp2 is A3 (Savalia *et al.*, 2010). Gp2 was found to interact with T7 endonuclease gp3 by a yeast two-hybrid screen (Bartel *et al.*, 1996). An interaction between gp2 and gp3 (albeit not necessarily direct) was suggested earlier (Mooney *et al.*, 1980), but the idea has not yet been critically evaluated experimentally.

5. DNA replication

T7 has been one of the primary model systems for understanding the biochemistry of DNA replication. Central to T7 replication is T7 DNA polymerase, which comprises a heterodimer of gp5 and the host thioredoxin. Binding of gp5 to thioredoxin is strong, with the latter serving as the processivity factor for DNA replication, increasing the length of DNA synthesized per binding event 100-fold. The interface between the two proteins also provides the docking site for binding of the other two proteins, SSB (gp2.5) and the primase–helicase gp4, which are central to T7 replication *in vivo*. The process of T7 replication *in vitro* has been reviewed recently (Hamdan and Richardson, 2009).

Two other proteins in the DNA replication cluster are known to interact with host proteins. The gene *5.5* product binds to the abundant nucleoid-associated protein H-NS (Ali *et al.*, 2011) and *in vitro* relieves its inhibition of both *E. coli* and T7 RNA polymerase-catalyzed transcription (Studier, 1981). *In vivo*, T7 *5.5* mutants direct reduced rates of DNA replication and exhibit a small plaque phenotype (Lin, 1992; Liu and Richardson, 1993; Studier, 1981). However, the phenotype persists in *hns*-defective strains, although an *hns* mutant that renders gene *5.5* essential has been isolated (Liu and Richardson, 1993). Gp5.5 also interacts with the λ proteins of RexAB system; it is not known whether this is a direct protein–protein interaction or whether gp5.5 interacts with a product of RexAB activity. A gene *5.5* missense, but not a null mutant, makes T7 sensitive to Rex-mediated exclusion. The product of gene *5.9* inhibits the *E. coli* RecBCD nuclease in a comparable manner to λ Gam protein; this interaction is also nonessential as mutants lacking gene *5.9* grow normally in *rec(BCD)*$^{+}$ hosts (Lin, 1992).

The assembly of T7 virions involves several symmetry mismatches and many protein–protein interactions, some of which may be helped by other T7 proteins that potentially have a chaperone-like activity (Kemp *et al.*, 2005). Note that, GenBank annotations notwithstanding, gp13 is not a structural component of the virion (Kemp *et al.*, 2005). Assembly is thought to initiate with oligomerization of the portal protein gp8 to form a dodecameric ring, although both 12- and 13-mers are observed following overexpression of gene *8* (Cerritelli and Studier, 1996b; Kemp *et al.*, 2005; Valpuesta *et al.*, 1992, 2000). The pathway of prohead formation may then involve assembly of the internal core (gp14, gp15, and gp16) on the inner surface of the portal, perhaps concomitant with the recruitment of capsid protein gp10 hexamers bound to scaffolding protein gp9. Subsequent recruitment of gp10 pentamers and hexamers results in shell formation around a scaffold of gp9 (Cerritelli *et al.*, 2003a). The 12-fold symmetrical portal protein faces a symmetry mismatch with the 5-fold vertex of the icosahedral capsid shell. The internal core of T7 contains 4 copies of gp16, 8 copies of gp15, and either 10 or 12 copies of gp14 (Agirrezabala *et al.*, 2007; Cerritelli *et al.*, 2003b; Kemp *et al.*, 2005). These proteins are ejected into the infected cell in order to transport the phage genome into the cytoplasm, and the protein–protein interactions in the mature virion are likely quite different than those found in the infected cell. Gp14 lies at the base of the internal core and is therefore juxtaposed to the 12-fold symmetrical portal protein. Complicating our understanding of the T7 virion is the identification of ≈18 copies of the small gp6.7 protein, which is ejected into the outer membrane of the infected cell at the initiation of infection (Kemp *et al.*, 2005). It is not known where gp6.7 is located; it likely forms part of the internal core or lies within the portal channel.

6. Tail

The outer face of the portal is the site of assembly of the sixfold symmetrical stubby tail. The tail is composed of three proteins—gp7.3, 11 and 12—to which six trimers of the tail fiber protein gp17 are attached. Each trimer consists of three domains: an N-terminal third that binds to the tail; this domain is conserved in phages related to T7 that have a different host range. This domain is followed by a triple-stranded coiled-coil of ≈120 residues and then a C-terminal half consisting of globular domains that interacts with the cell surface lipopolysaccharide (Steven *et al.*, 1988). The body of the tail consists of 12 copies of gp11, 6 copies of gp12, and ≈30 copies of the small protein gp7.3 (Kemp *et al.*, 2005). How these proteins are arranged is unknown. It has been suggested that gp7.3 functions as a tail assembly protein that remains with the mature virion; the protein is ejected into the outer membrane of the infected cell (Chang *et al.*, 2010; Kemp *et al*, 2005). Host range mutants of T7 suggest that all three proteins interact directly with the lipopolysaccharide of the host cell.

7. Interactome screening

T7 was the focus of the first genome-wide screen for phage protein–protein interactions using the yeast two-hybrid assay (Bartel *et al.*, 1996). As proof of principle, T7 was a good choice: the extensive genetic, molecular biology, and biochemical studies that had been performed on T7 provided a solid basis for evaluating the accuracy and completeness of the approach. Twenty-five interactions were identified in the screen, 15 of which were intramolecular. Many of these are to be expected as even monomeric polypeptides fold into a three-dimensional structure. Virion assembly clearly involves extensive protein–protein interactions, but as discussed by the authors, many of those that were known or plausibly expected from known structures were missed in the screen (Bartel *et al.*, 1996). Conversely, screen-predicted interactions involving gp1.7 that were confirmed subsequently by its purification as an oligomer (Tran *et al.*, 2010) and the predicted interactions between the spanin proteins gp18.5 and gp18.7 predated the discovery of a heterodimer even in phage λ (Casjens *et al.*, 1989). Six intermolecular interactions were predicted that have not yet found confirmatory support, and our current understanding of T7 suggests that some are implausible. Improved screening procedures should help resolve these outstanding issues and, if more intermolecular interactions can be found that are consistent with, in particular, the pathways of virion assembly, the utility of those screening procedures in predicting the protein interactome of lesser-studied phages will be enhanced greatly. All published T7 interactions are summarized in Table II.

B. Coliphage λ

Phage λ may be the best-studied temperate phage and has been under investigation since the early 1950s. There are probably thousands of papers that have been published on phage λ and its closer relatives N15, HK97, and P22. For instance, there are more than 400 citations in a review of lambdoid phages that do not even focus on the genetic switch of λ, the paradigm for gene regulation (Hendrix and Casjens, 2006). The switch has been described in detail by many other papers and in its own book (Court *et al.*, 2007; Ptashne, 2004). This section summarizes current knowledge of the interactions among phage λ proteins. Finally, a summary of interactions with its host, *E. coli*, is given. For a more detailed description of phage λ biology and regulation, the authors refer the reader to other reviews (Calendar, 2006; Court *et al.* 2007; Hendrix and Casjens, 2006; Hendrix *et al.*, 1983; Oppenheim *et al.*, 2005; Ptashne, 2004).

1. The λ virion

The phage λ genome and virion are shown in Figure 3. Note that the virion contains tail fibers that are not visible in most λ micrographs. The variant of λ used in most laboratories around the world, λPaPa, has a frameshift mutation in the structural *stf* gene and lacks the long "side tail fibers" of λ, Ur-λ, that comes directly from the original *E. coli* K-12 lysogen. It is still not entirely clear how many different proteins are in the λ particle but 12 are known with confidence and 2 more (gpM and gpL) are possibly in the particle in very low numbers. A current model of

TABLE II Protein–protein interactions of phage T7

Protein	Function	Interacts with	References
Class I			
gp0.3	B-DNA mimic	Host HsdRMS	Atanasiu *et al.*, 2001, 2002; Kennaway *et al.*, 2009; Murray, 2000; Stephanou *et al.*, 2009
gp0.3		gp0.3 (dimer)	Atanasiu *et al.*, 2001
gp0.3		gp4.5	Bartel *et al.*, 1996
gp0.7	Protein kinase	Host RNAP (activity change)	Rahmsdorf *et al.*, 1974; Severinova and Severinov, 2006; Zillig *et al.*, 1975
gp0.7		RNase III (activity change)	Mayer and Schweiger, 1983
gp0.7		RNase E (activity change)	Marchand *et al.*, 2001
gp0.7		Translation components	Robertson *et al.*, 1994; Robertson and Nicholson, 1990, 1992
gp0.7		gp4.5	Bartel *et al.*, 1996
gp1	T7 RNA polymerase	gp3.5	Jeruzalmi and Steitz, 1998; Zhang and Studier, 2004
gp1		gp19 (genetic)	Zhang and Studier, 1995, 2004
gp1.2	*E. coli* dGTPase inhibitor;	Host Dgt (dGTPase)	Huber *et al.*, 1988; Nakai and Richardson, 1990; Saito and Richardson, 1981
gp1.2	F-exclusion	F plasmid PifA	Cheng *et al.*, 2004 Schmitt and Molineux, 1991; Wang *et al.*, 1999
gp1.2		Host FxsA	Cheng *et al.*, 2004; Wang *et al.*, 1999
Class II			
gp1.7	Nucleotide kinase	gp1.7 (oligomer)	Tran *et al.*, 2010; Bartel *et al.*, 1996
gp2	*E. coli* RNAP inhibitor	Host RNAP	Camara *et al.*, 2010; Nechaev and Severinov, 1999
gp2		gp3	Bartel *et al.*, 1996

(*continued*)

TABLE II *(continued)*

Protein	Function	Interacts with	References
gp2.5	SSB	gp5 (with thioredoxin)	Hamdan *et al.*, 2005; He *et al.*, 2003 Ghosh *et al.*, 2008, 2010; Marintcheva *et al.*, 2009
gp2.5		gp4	Ghosh*et al.*, 2008; Hamdan*et al.*, 2005; He and Richardson, 2004
gp2.5		gp2.5 (dimer)	Rezende *et al.*, 2002; Bartel *et al.*, 1996
gp3	Endonuclease I, HJ resolvase	gp3 (dimer)	Hadden *et al.*, 2007; Parkinson and Lilley, 1997
gp3.5	Amidase (lysozyme)	gp1	Jeruzalmi and Steitz, 1998; Zhang and Studier, 2004
gp4	gp4A primase-helicase; gp4B helicase	gp5 (with thioredoxin)	Hamdan *et al.*, 2005; He and Richardson, 2004
gp4		gp2.5	Hamdan *et al.*, 2005; He and Richardson, 2004
gp4		gp4 (hexamer)	Singleton *et al.*, 2000; Bartel *et al.*, 1996
gp4.5		gp0.7	Bartel *et al.*, 1996
gp5	DNA polymerase	gp4 (with thioredoxin)	Ghosh *et al.*, 2008; Hamdan *et al.*, 2005; Bartel *et al.*, 1996
gp5		gp2.5 (with thioredoxin)	Ghosh *et al.*, 2008, 2010; He *et al.*, 2003; Marintcheva *et al.*, 2009; Hamdan*et al.*, 2005
gp5		gp4.7	Bartel *et al.*, 1996
gp5		gp6.5	Bartel *et al.*, 1996
gp5		Host thioredoxin	Ghosh *et al.*, 2008
gp5.3		gp5.3	Bartel *et al.*, 1996
gp5.5		Host H-NS	Lin, 1992; Liu and Richardson, 1993
gp5.9	Inhibits RecBCD nuclease	Host RecBCD	Lin, 1992

Class III			
gp6.7	Head completion; ejected virion protein	Other head proteins likely. gp13?	Kemp *et al.*, 2005
gp7.3	Tail chaperone; ejected structural virion protein	gp7.3, gp11, gp12	Kemp *et al.*, 2005
gp7.3		Host LPS	Kemp *et al.*, 2005
gp8	Head–tail connector (portal) protein	gp19	Fujisawa *et al.*, 1991; Morita *et al.*, 1995
gp8		gp9/10/gp6.7/7.3/11/12/14 (structure)	Agirrezabala *et al.*, 2005a, 2007; Cerritelli and Studier, 1996a; Liu *et al.*, 2010
gp8		gp8 (dodecamer) (structure)	Agirrezabala *et al.*, 2005b; Cerritelli and Studier, 1996b
gp9	Scaffolding protein	gp8, gp10 (structure)	Agirrezabala *et al.*, 2007; Cerritelli and Studier, 1996a,b
			Bartel *et al.*, 1996
gp9		gp14/15/16 (structure)	—
gp10	Major capsid protein (-1 frame-shift -> minor gp10B)	gp8, gp9, gp14 (structure)	Agirrezabala *et al.*, 2005a,b, 2007; Cerritelli and Studier, 1996a; Cerritelli and Studier, 1996b
		gp10 (pentamer/hexamer) (structure)	Cerritelli and Studier, 1996a; Bartel *et al.*, 1996
gp10		Host FxsA; F plasmid PifA	Cheng *et al.*, 2004; Schmitt and Molineux, 1991; Wang *et al.*, 1999
gp11	Tail protein	gp7.3, gp8, gp11, gp12 (structure)	Kemp *et al.*, 2005; Agirrezabala *et al.*, 2007; Liu *et al.*, 2010
gp11		Host LPS	Kemp *et al.*, 2005
gp11		gp2.5	Bartel *et al.*, 1996
gp12	Tail protein	gp7.3, gp8, gp11, gp12 (structure)	Kemp *et al.*, 2005; Liu *et al.*, 2010
gp12		Host LPS	Kemp *et al.*, 2005; Qimron *et al.*, 2006

(continued)

TABLE II *(continued)*

Protein	Function	Interacts with	References
gp13	Virion chaperone	gp6.7, gp8, gp12, gp14, gp17, gp19 (genetics)	Kemp *et al.*, 2005, Stone and Miller, 1985
gp14	Ejected core protein	gp8, gp14, gp15, gp16 (structure)	Agirrezabala *et al.*, 2007; Liu *et al.*, 2010
gp15	Ejected core protein	gp14, gp15, gp16 (structure)	Agirrezabala *et al.*, 2007; Chang *et al.*, 2010; Liu *et al.*, 2010; Bartel *et al.*, 1996
gp16	Ejected core protein	gp14, gp15, gp16 (structure)	Agirrezabala *et al.*, 2007; Chang *et al.*, 2010; Liu *et al.*, 2010
gp16		Peptidoglycan	Moak and Molineux, 2000, 2004
gp17	Tail fiber protein	gp7.3/11/12 (structure)	Agirrezabala *et al.*, 2007; Chang *et al.*, 2010; Liu *et al.*, 2010
gp17		gp17 (trimer)	Steven *et al.*, 1988; Bartel *et al.*, 1996
gp17		Host LPS	Steven *et al.*, 1988; Qimron *et al.*, 2006
gp18	Small terminase	gp18 (octamer)	White and Richardson, 1987; Hamada *et al.*, 1986
gp18.5-	gp18.7 Casjenset *al.*, 1989 Summer *et al.*, 2007; Bartel *et al.*, 1996	*Rz-Rz1* homologues; spanin	gp18.5–gp18.7
gp19	Large terminase	gp8	Fujisawa *et al.*, 1991; Hamada *et al.*, 1986; Morita *et al.*, 1994, 1995
gp19		gp19 (hexamer in the presence of prohead)	Fujisawa *et al.*, 1991
gp19		gp4.7	Bartel *et al.*, 1996

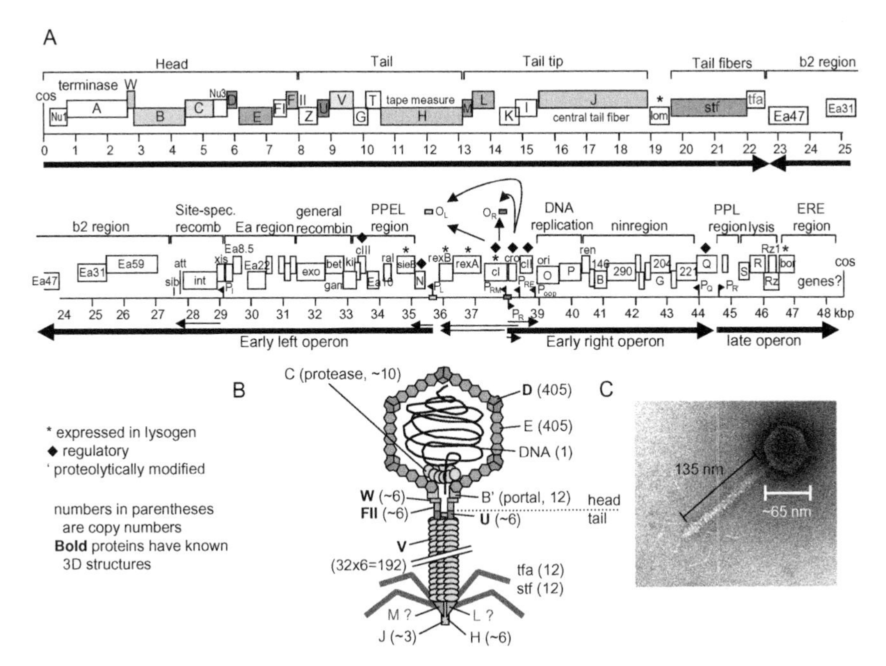

FIGURE 3 Genome and virion of phage λ. (A) Genome of phage λ. Colored ORFs correspond to colored proteins in (B). Main transcripts are shown as arrows. After Hendrix and Casjens in Calendar (2006). (B) Schematic model of λ virion. Numbers indicate the number of protein copies in the particle. It is unclear whether gpM and gpL proteins are in the final particle or only required for assembly. (C) Electron micrograph of phage λ. Modified after Hendrix and Casjens (2006). (See Page 5 in Color Section at the back of the book.)

the particle with its known proteins is shown in Figure 3. Based on this model, we can predict at least 23 PPIs in the virion (Table III), although several are inferred from genetic experiments and remain to be detected by biochemical or biophysical methods. The virion adsorbs to sensitive cells through an interaction between the tail tip protein gpJ and the host LamB outer membrane protein. DNA is then released through the tail into the cytoplasm.

2. Assembly of virion head (Fig. 4)

λ encodes 10 proteins required for head assembly, of which 6 end up in the mature virion. Head assembly starts with assembly of the portal, which requires the host GroEL/S chaperone for folding. It is thought that the portal serves as the nucleus for assembling the remaining capsid. However, the mechanistic details of assembly are not well understood. GpC is involved as a head maturation protease that becomes attached covalently to E, the main capsid protein. The resulting fusion proteins

TABLE III Published protein–protein interactions of phage λ

λ 1	λ 2	Notes	References
gpA	gpNu1	gpA (N-terminal) – gpNu1 (C-terminal)	Catalano, 2000; de Beer *et al.*, 2002; Hang *et al.*, 1999
gpA	gpB	gpA (C-terminal) – gpB (= portal)	Catalano, 2000; Duffy and Feiss, 2002
gpW	gpB	gpW (NMR, head–tail joining)	Casjens, 1974; Casjens *et al.*, 1972
gpW	gpFII	gpW required for gpFII binding	Casjens, 1974; Maxwell *et al.*, 2001
gpB	gpB	12-mer (22 amino acids removed from gpB N-terminal)	Kochan, Carrascosa and Murialdo, 1984; Tsui and Hendrix, 1980
gpC	gpE	Covalent PPI (in virion?)	Hendrix and Casjens, 1974;Hendrix and Duda, 1998
gpC′	gpB		Murialdo, 1979
gpC	gpNu3	gpC may degrade gpNu3 (before DNA packaging)	Hendrix and Casjens, 1975; Hohn *et al.*, 1975; Medina *et al.*, 2010; Murialdo and Becker, 1978
gpD	gpD	gpD forms trimers	Mikawa *et al.*, 1996; Sternberg and Hoess, 1995; Yang *et al.*, 2000
gpE	gpE	Major capsid protein	Casjens and Hendrix, 1974; Dokland and Murialdo, 1993; Lander *et al.*, 2008
gpU	gpU	"Probably a hexamer"	Katsura and Tsugita, 1977
gpD	gpE		Casjens and Hendrix, 1974; Dokland and Murialdo, 1993; Lander *et al.*, 2008
gpU	gpV	gpU is head-proximal tail protein	Pell *et al.*, 2009
gpV	gpV		Buchwald *et al.*, 1970a,b; Casjens and Hendrix, 1974; Katsura, 1983
gpO	gpP		Wickner and Zahn, 1986; Zahn and Blattner, 1985, 1987; Zahn and Landy, 1996)

gpO	gpO	gpO–gpO interactions when bound to ori DNA	Alfano and McMacken, 1989
gpB	gpA	Genetic evidence	Frackman *et al.*, 1985; Sippy and Feiss, 1992
gpFI	gpA	Genetic evidence	Catalano and Tomka, 1995
gpFI	gpE	Genetic evidence	Murialdo and Tzamtzis, 1997
gpH	gpG/ gpGT	gpG/gpGT holds gpH in an extended fashion	Hendrix and Casjens, 2006
gpV	gpGT	T domain binds soluble gpV	Hendrix and Casjens, 2006
gpH	gpV	gpV probably assembles around gpH, displacing gpG/gpGT	Xu *et al.*, 2004
CI	CI	Forms octamer that links O_R to O_L	Bell and Lewis, 2001; Dodd *et al.*, 2001
Exo	Bet		Radding *et al.*, 1971
SieB	Esc	Esc is encoded in frame in SieB and inhbits SieB	Ranade and Poteete, 1993a,b
CII	CII	Homotetramers	Ho and Rosenberg, 1982
Xis	Int		Sam *et al.*, 2002
Xis	Xis	Xis-Xis binding mediates cooperative DNA binding	Sam *et al.*, 2004
Int	Int	Dimer	Warren *et al.*, 2003
Stf	Stf	Likely homotrimer by homology with T4 fibers	
gpS	gpS	Large ring in inner membrane	Savva *et al.*, 2008
gpS	gpS′	gpS′ inhibits gpS ring formation	Grundling *et al.*, 2000
gpB	gpE	Copurify in procapsid	Hendrix and Casjens, 1975
gpNu3	pNu3	Monomer–dimer equilibrium	C. Catalano, personal communication
CIII	CIII	Dimer	Halder *et al.*, 2007
Cro	Cro	Dimer; X-ray str	Anderson *et al.*, 1979
gpRz	gpRz1	Heteromultimer supposed to span the periplasm	Summer *et al.*, 2007
gpNu3	gpB	gpNu3 required for gpB incorporation into procapsid	Ray and Murialdo, 1975

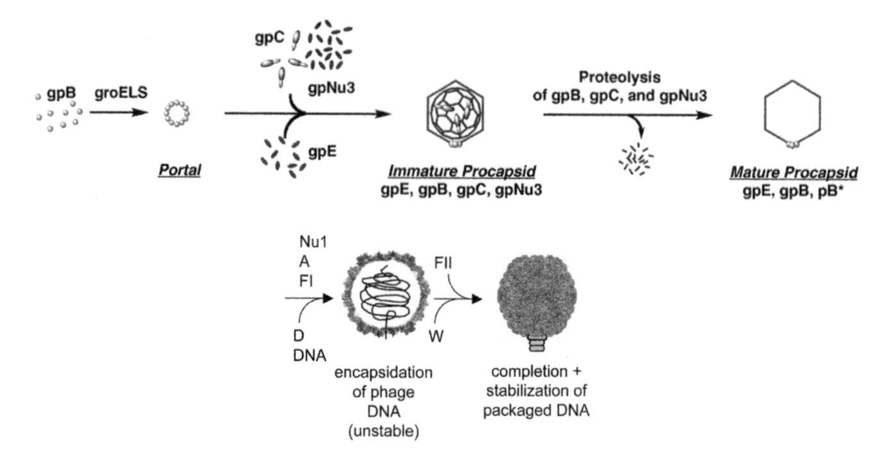

FIGURE 4 Assembly of the phage λ head. Head assembly has been subdivided into five steps, although most steps are not very well understood in mechanistic terms. Note that the tail is assembled independently. GpC protease, scaffolding protein gpNu3, and portal protein gpB form an ill-defined initiator structure. Protein gpE joins this complex in a step requiring the chaperonins GroES and GroEL. Proteins gpNu1, gpA, and gpFI are required for DNA packaging. GpD joins and stabilizes the capsid as a structural protein; gpFII and gpW are connecting the head to the tail that joins once the head is completed. Modified after Georgopoulos *et al.* (1983) and Lander *et al.* (2008). (See Page 5 in Color Section at the back of the book.)

subsequently get trimmed to make slightly shortened products named X1 and X2 (Hendrix and Casjens, 1974). The authors note, however, that attempts to identify X1 and X2 were unsuccessful and thus X1 and X2 may be artifacts (Medina *et al.*, 2010). GpNu3 acts as putative scaffolding protein that aids coat shell assembly and is removed from the capsid by proteolysis before DNA is packaged. Structures of the procapsid and mature λ capsid have been solved by cryo-EM, showing the T=7 arrangement of capsid protein gpE clustered into hexamers and pentamers (Dokland and Murialdo; Lander *et al.*, 2008). Trimers of the decoration protein gpD bind to the capsid at trivalent interaction points between the capsomers and stabilize the capsid (Sternberg and Weisberg, 1977). Comparison of the procapsid with the mature capsid reveals the considerable structural transitions that occur upon capsid maturation and DNA packaging, but this process is not well understood mechanistically.

3. Assembly of virion tail (Fig. 5)

Tail assembly begins with the antireceptor protein gpJ. Proteins gpI, gpK, gpL, and gpM assemble around gpJ in that order, although no mechanistic or structural details are known about this process. In parallel events, gpH, gpG, and gpG-T (the frameshifted form of gpG including the

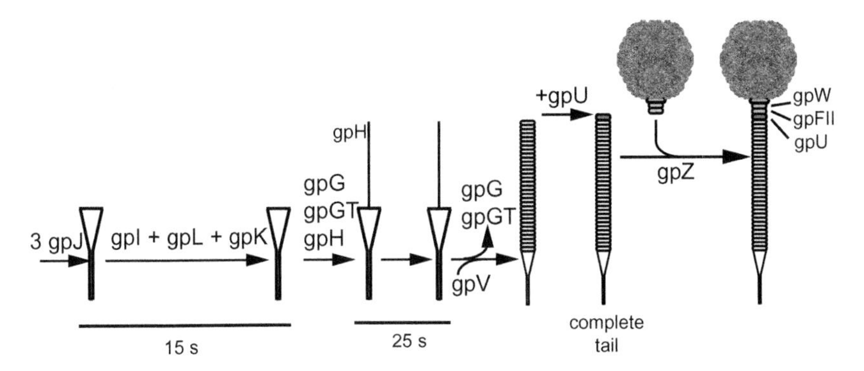

FIGURE 5 Assembly of the phage λ tail. The λ tail is made of at least six proteins (gpU, V, J, H, tfa, stf) with another seven required for assembly (gpI, M, L, K, G/T, Z). Assembly starts with protein gpJ, which then, in a poorly characterized fashion, requires proteins gpI, gpL, gpK, and gpG/T to add the tape measure protein H. GpM joins and then gpG and gpG/T leave the complex so that the main tail protein gpV can assemble on the gpJ/H scaffold. Finally, gpU is added to the head-proximal end of the tail. GpZ is required to connect the tail to the preassembled head. Protein gpH is cleaved by the action of gpU and gpZ (Tsui and Hendrix, 1983). It remains unclear if proteins gpM and gpL are part of the final particle (Hendrix and Casjens, 2006). From http://www.pitt.edu/~duda/lambdatail.html. (See Page 6 in Color Section at the back of the book.)

T reading frame) form a complex that binds the tail tip and the major tail protein (gpV). The tape meassure protein gpH determines the length of the phage tail (Katsura, 1990). Once the correct length is achieved, gpV assembly stops and the proximal tail protein gpU is added. Protein gpZ somehow facilitates head–tail joining without being assembled into the final virion.

4. Lysogeny and induction

After phage DNA ejection, circularization, and ligation by the host DNA ligase, the earliest genes to be expressed include *N*, *O*, and *P*, as well as *cII* and *cIII*, which play a key role in the lysis/lysogeny decision (Echols, 1986). Both CII and CIII can interact with and be cleaved by the bacterial FtsH (HflB) protease directly, which in turn is complexed with the host HflK and HflC proteins at the cytoplasmic membrane. HflD, a cytoplasmic membrane protein, also binds CII, making it both more susceptible to FtsH and less able to bind DNA (Kihara *et al.*, 2001; Parua *et al.*, 2010). As a substrate for FtsH, CIII acts as a competitive inhibitor of CII degradation and, as the half-life of CII protein increases, leads to an increase in CI repressor synthesis (Datta *et al.*, 2005a; Herman *et al.*, 1997; Kobiler *et al.*, 2007). Except when *cIII* is overexpressed, lysogeny is only established when the activity of these proteases (FtsH, Lon, and perhaps others) is low. Once established, lysogeny is a stable state until lysogens are exposed to ultraviolet light or other DNA-damaging agents that lead to the activation of RecA, which in turn accelerates autocleavage of the λ repressor CI. Loss of CI leads to prophage induction and the lytic cycle. The mechanism by which phage λ decides whether to grow lytically or become a lysogen is reviewed elsewhere (Court *et al.*, 2007; Friedman and Court, 2001; Gottesman and Weisberg, 2004; Ptashne, 2006). However, Table IV contains the host–phage protein–protein interactions critical for gene expression and thus the genetic switch of phage λ.

5. Anti-termination

Escherichia coli RNA polymerase transcribes λ DNA starting from the immediate early promoters P_L and P_R, which leads to expression of the *N* and *cro* genes, among others. As soon as the gpN-binding sites (*nut* sites) are transcribed, gpN modifies the transcription complex by recognizing a stem-loop structure in the mRNA and by binding to host Nus proteins [note that gpN interacts with *nut* RNA (not DNA)]. This modified complex is capable of overriding the termination signal that normally stops immediate early transcription. Because gpN is continually being degraded by the Lon protease, ongoing anti-termination requires continual transcription of the *N* gene (Fig. 6). In addition to the anti-terminator for early lytic genes (gpN), a second anti-terminator, gpQ, is required for late gene expression. GpQ anti-termination differs from that of gpN in that gpQ binds to the "Q binding element" (QBE, also known as the *qut* site) of the $P_{R'}$ promotor (i.e., unlike the gpN protein, gpQ binds DNA). When RNA polymerase (RNAP) initiates transcription at the $P_{R'}$ promotor, it almost immediately pauses at a pause-inducing sequence. Pausing allows gpQ to contact σ^{70} in the RNAP, which causes a conformational change in the enzyme that enables gpQ to bind the β subunit (RpoB). The gpQ-containing RNAP is then able to override the termination signal at $t_{R'}$ (Deighan *et al.*, 2008; Nickels *et al.*, 2002).

6. Replication

After λ gene *O* and *P* protein synthesis, replication of the phage genome is initiated (Fig. 6). Replication initiation starts with dimeric gpO bound to the four 18-bp direct repeats of the λ origin, located within the sequence of the *O* gene. Each repeat binds a gpO dimer, leading to a total of eight monomers bound directly to the DNA, in addition to non-DNA-bound gpO that associates with the complex solely by PPIs (Dodson *et al.*, 1986). This large nucleoprotein complex is called the O-some. GpP extracts the host DnaB helicase from its native complex with DnaC (Zylicz *et al.*, 1989). The DnaB helicase is inactive as long as it is tightly associated with gpP. The gpP–DnaB pair is then added to the O-some through interactions between gpP and gpO. Three host heat shock chaperone proteins—DnaK,

TABLE IV Published phage λ–host interactions[a]

λ	Host	Description	References
Transcription			
CI	RecA	RecA degrades CI[b]	Kobiler *et al.*, 2004
CI	RpoA		Kędzierska *et al.*, 2007
CI	RpoD		Li *et al.*, 1994; Nickels *et al.*, 2002
CII	ClpYQ	ClpYQ degrades CII *in vitro*	Kobiler *et al.*, 2004
CII	ClpAP	ClpAP degrades CII *in vitro*	Kobiler *et al.*, 2004
CII	HflD	HflD makes CII more vulnerable to FtsH (*hfl*, high-frequency lysogenization)	Kihara *et al.*, 2001
CII	HflB[c]	HflB (= FtsH) protease degrades cII	Kobiler *et al.*, 2002; Shotland *et al.*, 2000
CII	RpoA		Kędzierska *et al.*, 2004; Marr *et al.*, 2004
CII	RpoD		Kędzierska *et al.*, 2004
CIII	HflB	CIII inhibits HflB; HflB degrades CIII	Herman *et al.*, 1997; Kornitzer *et al.*, 1991
gpN	NusA	Transcriptional regulation	Friedman and Court, 2001
gpN	Lon	Lon degrades gpN	Kobiler *et al.*, 2004; Gottesman *et al.*, 1981
gpQ	σ^{70}	gpQ makes RNAP insensitive to *cis* terminal	Marr *et al.*, 2001;Nickels *et al.*, 2002; Roberts and Roberts, 1996; Roberts *et al.*, 1998
Head			
gpB	GroE	genetic interaction; *E. coli* GroE	Ang *et al.*, 2000; Georgopoulos *et al.*, 1973; Murialdo, 1979
gpE	GroE	genetic interaction	Georgopoulos *et al.*, 1973
Tail			
gpJ	LamB	LamB is the *E. coli* receptor	Buchwald and Siminovitch, 1969; Clement *et al.*, 1983; Mount *et al.*, 1968; Wang *et al.*, 2000a; Werts *et al.*, 1994
Stf	OmpC	OmpC is a secondary *E. coli* receptor	Hendrix and Duda, 1992

(continued)

TABLE IV *(continued)*

λ	Host	Description	References
Recombination			
Xis	Lon	Xis is degraded by *E. coli* Lon protease	Leffers and Gottesman, 1998
Xis	FtsH	Xis is degraded by *E. coli* FtsH protease	Leffers and Gottesman, 1998
Xis	Fis	both required for excision	Ball and Johnson, 1991; Cho *et al.*, 2002; Esposito and Gerard, 2003
Int	IHF	Both catalyze recombination at *attP*/*attB*	Campbell *et al.*, 2002; Crisona *et al.*, 1999
Gam	RecB	Gam inhibits RecBCD	Marsic *et al.*, 1993
NinB	SSB	NinB also binds ssDNA	Maxwell *et al.*, 2005
Replication			
gpO	ClpXP	ClpXP degrades gpO	Kobiler *et al.*, 2004
gpP	DnaA		Datta *et al.*, 2005b,c
gpP	DnaB		Mallory *et al.*, 1990

[a] CbpA (Wegrzyn *et al.*, 1996) perhaps interacts with DnaA and SeqA (Potrykus *et al.*, 2002; Wegrzyn and Wegrzyn, 2001, 2002).
[b] After DNA damage, RecA bound to single-stranded DNA stimulates autocleavage of the CI repressor (Kobiler *et al.*, 2004).
[c] HflB = FtsH.

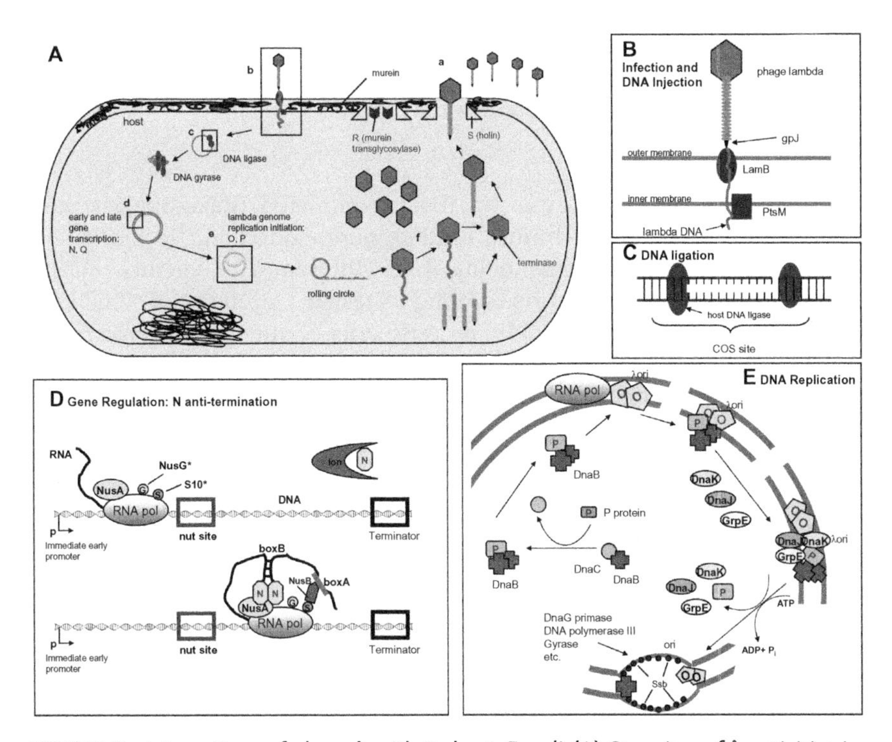

FIGURE 6 Interactions of phage λ with its host, *E. coli.* (A) Overview of λ activities in *E. coli*: (a) free phage, (b) binding and entry, (c) DNA circularization and supercoiling, (d) transcription, (e), replication, and (f) virion assembly. Details are shown in B–E. (B) Infection and DNA ejection, (C) DNA ligation, (D) gene regulation. An asterisk means weakly bound to the RNA polymerase. (E) DNA replication. Phage proteins are labeled in red, host proteins in black. Modified after Das (1992) and Mason and Greenblatt (1991). (See Page 6 in Color Section at the back of the book.)

DnaJ, and GrpE—are subsequently recruited (DnaK through direct interaction with gpP and gpO) and activate the DnaB helicase by liberating it from its inhibitory interaction with gpP. Finally gpP and the chaperones are released, and replication starts with the help of the host DnaG primase, DNA polymerase III, gyrase, and single-stranded DNA-binding (SSB) proteins (Dodson *et al.*, 1986; LeBowitz and McMacken, 1986). The RNA polymerase β subunit was shown to interact directly with gpO (Szambowska *et al.*, 2010). Even though not yet fully understood, how *E. coli* RNA polymerase activates λ *ori* seems to be far more complex than the original idea that it simply changed DNA topology.

Studies on λ *ori*-dependent plasmids have revealed interactions with even more host proteins, including the DnaJ homolog CbpA (Wegrzyn *et al.*, 1996), and through transcriptional activation of origin activity, perhaps also with DnaA and SeqA (Potrykus *et al.*, 2002; Wegrzyn and

Wegrzyn, 2001, 2002). As λ *ori*-dependent plasmid maintenance and copy number are very sensitive to changes in host physiology, interactions of the O-some with additional proteins may yet be revealed.

7. The λ interactome

A λ Y2H interactome analysis has been completed (Rajagopala *et al.*, 2011). All λ open reading frames (ORFs) were cloned and tested in all possible pairs for interactions, using three different Y2H vectors. These screens identified 97 interactions among 68 tested λ proteins, of which 16 were known previously. Notably, because only 33 interactions had been reported previously, this screen found more than 50% of all previously published interactions in a single experiment! At least 18 of the newly observed Y2H interactions seem to be plausible, based on the functions of the interacting proteins, indicating that even in λ many new discoveries can be made. Interestingly, relatively few interactions involving morphogenetic proteins were revealed, and possible reasons for the still significant fraction of apparent false negatives are discussed later.

C. *Salmonella* phage P22

P22 infects *Salmonella enterica* serovar Typhimurium and was originally isolated by Zinder and Lederberg (1952). It is a temperate member of *Podoviridae*. Its virion proteins are not particularly closely related to other tailed phage groups, but its early genes and overall lifestyle are very closely related to phage λ. For this reason, it is often placed in the "lambdoid" phage group. Phage P22 is also a very important tool for those who use *Salmonella* genetics, as it is particularly easy to handle in the laboratory and is a very good generalized transducing phage (i.e., it packages a significant amount of host DNA by accident and can be used to move host alleles from one *Salmonella* strain to another).

1. P22 life cycle

The early operons of P22 are organizationally similar to those of λ and carry parallel prophage repressor, Cro, recombination, DNA replication, and transcription anti-termination functions (Susskind and Botstein, 1978). The P22 repressor has a different DNA operator target specificity from that of λ repressor and encodes nonhomologous types of recombination and replication proteins, but its late operon anti-termination protein is 97% identical to that of λ and has the same target specificity. Unlike λ, in addition to its prophage repressor (C2) and lysogeny control proteins (C1, C3, Cro), P22 carries an "*immunity I*" region that encodes an anti-repressor protein, Ant, that binds to the prophage repressor and blocks its ability to bind its operator. Two additional repressors, Mnt and Arc, regulate Ant synthesis (Susskind and Botstein, 1975; Vershon *et al.*, 1987a,b).

Also, instead of a protein that recruits the host DnaB protein to the replication origin, P22 encodes its own DnaB homologue (Backhaus and Petri, 1984; Wickner, 1984). The repressed P22 prophage encodes several "lysogenic conversion" genes, several of which encode a set of three Gtr (glucose transfer) proteins that modify the host O-antigen polysaccharide by glucosylation (Broadbent *et al.*, 2010; Makela, 1973; Vander Byl and Kropinski, 2000). Finally, P22 has been studied as a prototypical headful DNA packaging phage, which is different from the packaging strategies used by the other phages discussed here; unlike these other phages that package DNA between two specific nucleotide sequences in the DNA substrate, P22 packages DNA until the head is full and the nucleotide sequence at the termination point does not matter (Casjens and Hayden, 1988; Jackson *et al.*, 1978; Tye *et al.*, 1974).

2. P22 virion assembly

The P22 virion is assembled by proteins encoded by 14 genes in its late operon (Fig. 7) (Botstein *et al.*, 1973; Eppler *et al.*, 1991; King *et al.*, 1973; Poteete and King, 1977; Youderian and Susskind, 1980). Some P22-like phages, for example, phages ε34 and L, have a 15th virion assembly gene, *dec*, at the promoter proximal end of the cluster, whose encoded protein "decorates" the outside of the head and stabilizes it (Gilcrease *et al.*, 2005; Tang *et al.*, 2006; Villafane *et al.*, 2008). The P22 late operon encodes lysis proteins very similar to those of λ; however, the virion assembly genes are not closely related to λ or to any other tailed phage type. P22 virion assembly is best known for the fact that its head-assembling scaffolding protein was the first to be discovered and studied (Cortines *et al.*, 2011; King and Casjens, 1974; Prevelige *et al.*, 1988; Weigele *et al.*, 2005). The P22 procapsid assembles as a complex of 415 coat protein molecules, 12 portal ring subunits, and about 250 scaffolding protein molecules. At about the time when DNA is packaged, all 250 scaffolding proteins are released intact from the procapsid and are reused with newly synthesized coat protein in subsequent rounds of procapsid assembly. Thus, the P22 scaffolding protein acts catalytically in head assembly and individual molecules can participate in at least five rounds of procapsid assembly. A particularly high-resolution asymmetric cryo-EM reconstruction of the virion has been determined (Lander *et al.*, 2006; Tang *et al.*, 2011), and because assembly of the four proteins of its short tail has also been quite well studied [X-ray structures are known for three tail proteins and the portal protein (Olia *et al.*, 2006, 2007; Steinbacher *et al.*, 1997)], the P22 virion is perhaps the best understood of all tailed phage virions. The details of P22 virion assembly have been reviewed in light of its evolution and genome mosaicism (Casjens and Thuman-Commike, 2011), and the authors refer the reader to this review for further details. The published PPIs of phage P22 are summarized in Table V. Interactions between P22

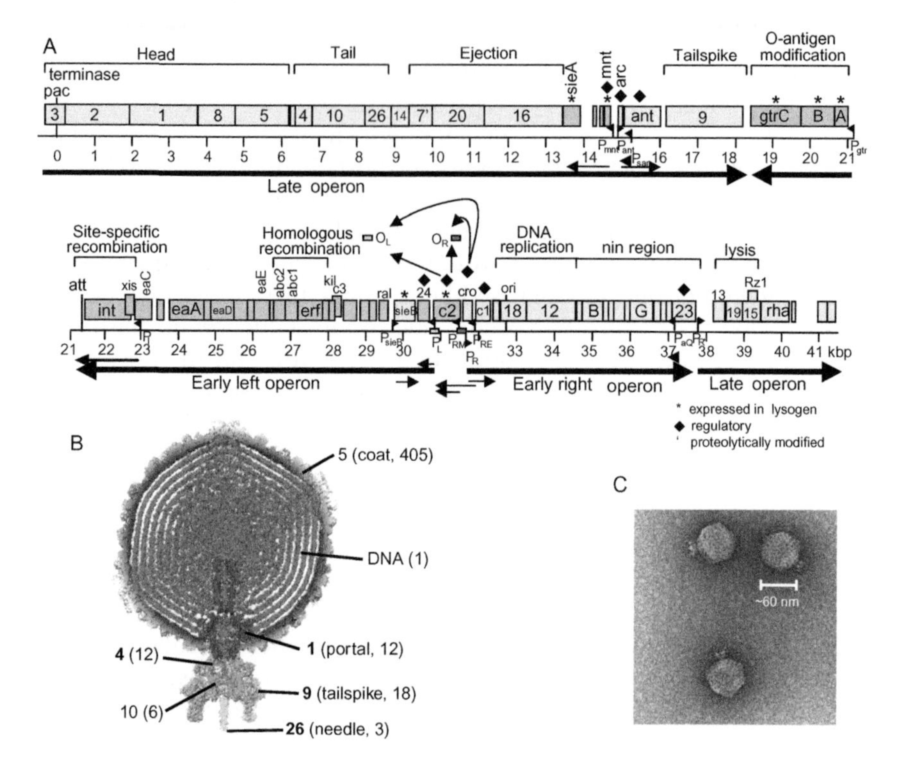

FIGURE 7 Genome and virion of phage P22. (A) Genome of phage P22 with a scale in kbp below. Green open reading frames (ORFs; rectangles) are transcribed left to right; red ORFs are transcribed right to left. Known RNA polymerase promoters and transcripts are shown as black flags and arrows, respectively. (B) The asymmetric (not icosahedrally averaged) P22 virion three-dimensional cryo-EM reconstruction from Tang *et al.* (2011). Numbers in parentheses indicate molecules/virion, and bold numbers indicate proteins for which X-ray structures are known. (C). Negatively stained electron micrograph of P22 virions. (See Page 7 in Color Section at the back of the book.)

and its host are much less well studied than those of λ and are mostly understood by homology arguments in the case of proteins that are similar in the two phages. Known P22–host interactions are listed in Table VI.

3. Comparison of P22 and λ interactomes

As noted earlier, comparison of P22 and λ is a particularly informative way to understand the kinds of relationships among the proteins of different tailed phages. The "lambdoid" phages all encode similar functions and have similar gene organization, but these "similar functions" can be of several different types. Because these phages undergo a

TABLE V P22 intraphage protein interactions

P22 1	P22 2	Description	References
gp3	gp3	Small terminase subunit; hommonomer in solution	Nemecek *et al.*, 2007, 2008
gp3	gp2	Purify as complex; bind each other *in vitro*	Nemecek *et al.*, 2007; Poteete and Botstein, 1979
gp2	gp1	gp2 is large terminase subunit; assembles in packaging motor to gp1 by analogy with other phages (gp2 by itself is monomer in solution); evolutionary argument for P22 gp2–gp1 interaction	Casjens, 2011
gp1	gp1	Portal protein; homododecamer; EM of 12-mer; EM of virion; X-ray structure	Bazinet *et al.*, 1988; Lander *et al.*, 2006; Olia *et al.*, 2011
gp1	gp5	Purify as complex (procapsids); three-dimensional (3D) reconstructions of virion and procapsid	Botstein *et al.*, 1973; Chen *et al.*, 2011; Lander *et al.*, 2006
gp1	gp8	gp8 required for assembly of gp1 into procapsid; interaction seen in procapsid 3D reconstruction	Chen *et al.*, 1989; Greene and King, 1996; Weigele *et al.*, 2005
gp8	gp8	Scaffolding protein; forms homodimers and tetramers in solution	Parker *et al.*, 1997
gp8	gp5	Purify as complex (procapsid); bind *in vitro*	Fuller and King, 1982; Greene and King, 1994; Parker *et al.*, 1998
gp5	gp5	Coat (major capsid) protein; forms T=7 *l* icosahedral shell of virion	Botstein *et al.*, 1973; Casjens, 1979; Lander *et al.*, 2006
gp5	gp1	Purify as complex (procapsids); interaction seen in 3D reconstruction	Chen *et al.*, 2011; Lander *et al.*, 2006
gp4	gp4	Tail protein; monomer in solution (naturally unfolded?); gp4s barely touching each other in dodecamer ring in virion (i.e., contacts not large); EM, X-ray structure	Olia *et al.*, 2011; Tang *et al.*, 2011
gp4	gp5	Contact in high-resolution 3D reconstruction of virion	Olia *et al.*, 2006, 2011; Tang *et al.*, 2011
gp4	gp1	Order of assembly, EM, 3D reconstruction; X-ray cocrystal structure	Olia *et al.*, 2006, 2011; Tang *et al.*, 2011

(continued)

TABLE V *(continued)*

P22 1	P22 2	Description	References
gp4	gp10	12 gp4s contact 6 gp10s; order of assembly, EM, 3D reconstruction	Lander *et al.*, 2006; Olia *et al.*, 2007
gp4	gp9	gp4 ring contacts 6 gp9 trimers; 3D reconstruction	Lander *et al.*, 2006
gp10	gp10	Tail protein; probably hexamer in solution; 3D reconstruction	Lander *et al.*, 2006; Olia *et al.*, 2007
gp10	gp9	6 gp4s contact 6 gp9 trimers; 3D reconstruction	Lander *et al.*, 2006
gp10	gp26	6 gp10s contact one gp26 trimer	Lander*et al.*, 2006
gp26	gp26	Tail needle protein; trimer in solution - crystal structure; one trimer in virion	Olia*et al.*, 2007
gp16	gp8	Mutants of gp8 do not incorporate gp16 into virion	Chen *et al.*, 1989; Greene and King, 1996; Weigele *et al.*, 2005
C1	C1	C1 is homologue of λ CII; homotetramer	Ho *et al.*, 1992
Mnt	Mnt	Repressor of antirepressor gene; tetramer	Nooren *et al.*, 1999
Arc	Arc	Repressor of antirepressor gene; dimer in solution, tetramer on DNA	Bowie and Sauer, 1989; Breg *et al.*, 1990; Brown, Bowie and Sauer, 1990
Ant	C2	Antirepressor; binds to and inhibits C2 repressor	Susskind and Botstein, 1975
gp9	gp9	Tailspike protein; trimer in solution, 6 trimers in virion; crystal structure; binds to polysaccharide on cell surface	Lander *et al.*, 2006; Steinbacher *et al.*, 1994, 1997
Int	Xis	Form heteromultimer	Cho *et al.*, 2002
Xis	Xis	Excisionase; homomultimers	Mattis *et al.*, 2008
Erf	Erf	Recombination protein; homomultimeric ring of 10–14 subunits	Poteete *et al.*, 1983
C3	C3	C3 is homologue of λ CIII; λ CIII is dimer by disulfide bond; P22 interaction by homology with λ CIII	Halder *et al.*, 2007
SieB	Esc	Similar to λ but not studied in detail	Ranade and Poteete, 1993a
C2	C2	Prophage repressor homologue of λ CI; homodimer	De Anda *et al.*, 1983
Cro	Cro	Dimer; solution studies and crystal structure	Darling *et al.*, 2000; Newlove *et al.*, 2004

gp18	gp12	Origin binding protein gp18 recruits DnaB-like gp12 to replication origin by analogy with λ (note that gp12 is not a homologue of λ P protein; gp18 is homologous to λ O protein only in N-terminal domain)	Backhaus and Petri, 1984
Rz	Rz1	Rz and Rz1 are supposed to bind to one another and form a bridge that spans the periplasm in phage λ; P22 gp15 and Rz1 should do the same by homology.	Summer *et al.*, 2007
gp13	gp13	Holin forms rings of large diameter in λ; by homology same should be true of P22 holin	Savva *et al.*, 2008
gp13	gp13′	Antiholin–holin interactions delay the formation of holin rings in λ; by homology same should be true in P22	Rennell and Poteete, 1985
Dec	Dec	Trimer; Note that P22 has no Dec, but its very close relative phage L does and L Dec binds to P22 capsids.	Tang *et al.*, 2006
Dec	gp5	Dec occupies a subset of the quasi-equivalent local threefold positions on the shell	Tang *et al.*, 2006

TABLE VI P22–host protein interactions

P22	Host	Description	References
gp5	GroEL	Solution studies	de Beus *et al.*, 2000
gp24	NusA? etc?	Homologue/analogue of λ gpN; likely binds same host factors as λ gpN	Franklin, 1985; Hendrix and Casjens, 2006
C3	FtsH(HflB)	C3 is homologue of λ CIII; inhibition of FstH by homology with λ CIII	Semerjian *et al.*, 1989
C1	FtsH(HflB)	C1 is homologue of λ CII; CII is degraded by FtsH	Backhaus and Petri, 1984
C2	RecA	P22 repressor autocleavage stimulated by RecA binding like λ	Prell and Harvey, 1983; Tokuno and Gough, 1976
Xis, gp18, gp24, C1	Host protein degradation machinery	These may be degraded analogously to their λ homologues, but direct information is not available	Backhaus and Petri, 1984
gp23	RNA polymerase?	By homology to λ; gp23 is nearly identical to λ gpQ	Pedulla *et al.*, 2003
gp12	Host replication machinery	Possible interaction with host DnaG primase	Wickner, 1984
Abc2	RecC	Solution studies	Murphy, 2000

significant amount of horizontal transfer of genetic information (especially if they are "closely" related), their functions can be homologous or even nearly identical (Casjens, 2005; Hendrix, 2002). For example, the two "gpQ" late operon control anti-termination proteins of the two phages are nearly identical and can replace one another in function. A second kind of relationship is represented by the prophage repressors. These proteins from the two phages have weak but recognizable homology and similar folds, and probably interact with host proteins the same way, but have different operator-binding specificity. Thus, one repressor cannot functionally replace the other without additional changes in the

DNA-binding sites of the protein. A third kind of relationship is that of the virion coat, terminase, and portal proteins, λ gpE, gpA, and gpB and P22 gp5, gp2, and gp1, respectively. These proteins have the same functions in the two phages—coat protein and DNA packaging nuclease/motor ATPase and procapsid DNA entry channel, respectively (summarized in Hendrix and Casjens, 2006). These cognate proteins are not recognizably similar in overall amino acid sequence between the two phages, although the two ATPase proteins have similar ATP-binding motifs. However, the two proteins almost certainly have the same basic fold. Thus, gpE and gp5 and gpA and gp2, as well as gpB and gp1, are all very likely ancient homologous pairs that have diverged beyond the point of having recognizable sequence similarity. Finally, similar functions are performed by nonhomologues in the two phages. Examples of this relationship are the (i) nonhomologous Exo (λ) and Erf (P22) proteins that catalyze homologous recombination in completely different ways (Little, 1967; Poteete and Fenton, 1983); (ii) the endolysin proteins gpR and gp19, where the λ protein is a transglycosylase and the P22 protein is a true lysozyme that both cleave the same peptidoglycan bond (Bienkowska-Szewczyk *et al.*, 1981; Rennell and Poteete, 1989); (iii) the gpW and gp4 proteins, which bind between the portal ring of the head and the tail and yet have no recognizable homology or structural similarity; and (iv) the λ receptor binding "fiber" (gpJ and STF) and P22 tailspike (gp9), which have no homology, recognize different structures, and act in different ways [the P22 gp9 enzyme cleaves its polysaccharide receptor (Iwashita and Kanegasaki, 1976)]. Thus, there are potentially a number of different ways in which parallel λ and P22 functions might interact with the hosts of these two phages, and it will be interesting to understand these in detail in both cases. But are there very similar phage proteins that have different host interactions? Do very different phage proteins that have the same function have equally different host interactions? Although no such situations have been identified to date, it remains possible. A comparison of PPIs in λ and P22 is given in Tables VII and VIII.

D. Enterobacteriophage P2 and its satellite phage, P4

Bacteriophage P2 was originally isolated from the Lisbonne & Carrère strain of *E. coli* by Bertani in 1951 and is a member of the *Myoviridae* family of viruses. It has an icosahedral head and a contractile tail and is a 33.6-kb double-stranded DNA genome (Bertani, 1951; Bertani and Six, 1988; Nilsson and Haggård-Ljungquist, 2006). P2-like prophages are common in the environment (Breitbart *et al.*, 2002) and are present in about 30% of strains in the *E. coli* reference collection (Nilsson *et al.*, 2004).

Bacteriophage P2 has attracted much interest as a "helper" for bacteriophage P4, an 11.6-kb replicon that can exist either as a plasmid or integrated

TABLE VII Comparison of λ and P22 interactions

λ 1	λ 2	P22 1	P22 2	Description
gpA	gpB	gp2	gp1	Likely similar contacts—by evolutionary argument[a]
gpA	gpNu1	gp2	gp3	Likely similar contacts—by evolutionary argument [a]
gpNu1	gpNu1	gp3	gp3	gpNu1 N-terminal fragment is dimer; gp3 is nonomer; not known if they have the same fold (i.e., are ancient homologues)
gpW	gpB	gp4	gp1	gpW and gp4 have no known homology, but are analogous in that they are the first proteins to bind the underside of the the portal ring; gp4 has no structural homology to **λ** gpFII or gpW (Cardarelli *et al.*, 2010; Olia *et al.*, 2011)
gpW	gpFII			No clear gpFII homologue in P22 (gp4 and gp10 of P22 have no sequence similarity)
gpB	gpB	gp1	gp1	Both form dodeamer portal rings[a]
gpC	gpE			No gpC analogue in P22
gpC′	gpB			No gpC analogue in P22
gpC	gpNu3			No gpC analogue in P22
gpNu3	gpNu3	gp8	gp8	Monomer/dimer/tetramer for gp8; monomer/dimer for gpNu3
gpB	gpE	gp1	gp5	Seen in reconstructions[a]
gpNu3	gpB	gp8	gp1	Scaffolds required for portal assembly into procapsid
		gp8	gp5	Scaffold–coat interaction not studied for λ
gpD	gpD	Dec	Dec	Both trimers; bind slightly different sites.[a] Note that P22 has no Dec, but its very close relative phage L does and L Dec binds to P22 capsids
gpE	gpE	gp5	gp5	Both make icosahedral T=7 head shells[a]
gpU	gpU			No analogue in P22
gpD	gpE	Dec	gp5	Dec is more discriminatory than D, i.e., unlike D, Dec does not occupy all of the quasi-equivalent local threefold positions on the shell
		gp4	gp5	Unkown in λ
		gp4	gp10	No gp10 analogue in λ (but see gpFII above)
		gp4	gp9	No gp9 analogue in λ (but see Stf below?)
		gp10	gp9	No gp9 analogue in λ (but see Stf below?)
		gp10	gp10	No gp10 analogue in λ (but see gpFII above)
		gp10	gp26	No gp10 or gp26 (?) analogue in λ

		gp26	gp26	No gp26 analogue in λ (maybe gpJ?)
Stf	Stf	gp9	gp9	Both trimers (Stf by homology to T4 fibers); analogues, not homologues
gpU	gpV			No analogues in P22
gpV	gpV			No analogue in P22
gpO	gpP	gp18	gp12	gpO and gp18 are homologous but gpP and gp12 are not; nonetheless, a good argument can be made that gp18 and gp12 must interact similarly to gpO and gpP to get replication started
gpW	gpW	gp4	gp4	Analogues—first proteins to bind below the portal ring (see gpW–gpB interaction above)
gpFI	gpA			No gpFI analogue in P22
gpFI	gpE			No gpFI analogue in P22
gpH	gpG/gpGT			No analogues in P22
gpV	gpGT			No analogues in P22
gpH	gpV			No analogues in P22
		Arc	Arc	No analogue in λ
		Ant	C2	No Ant in λ
		Mnt	Mnt	No analogue in λ
CI	CI	C2	C2	Homologues (Sevilla-Sierra *et al.*, 1994)
Exo	Bet	Erf	Erf	λ Exo and P22 Erf proteins are not homologues and function in basically different ways, but they are both essential for homologous recombination
SieB	Esc	SieB	Esc	Good homologues in P22 but they are unstudied
CII	CII	C1	C1	Homotetramers in both cases
Xis	Int	Xis	Int	Xis proteins are quite different
Xis?	Xis?	Xis	Xis	Monomer in λ, but binds cooperatively so could interact on DNA (Abbani *et al.*, 2007); multimer in P22
CIII	CIII	C3	C3	C3 is homologue of λ CIII; λ CIII is dimer (Halder *et al.*, 2007); P22 interaction by homology with λ CIII
Cro	Cro	Cro	Cro	Both homodimers
gpRz	gpRz1	gp15	gp15	Predicted for P22 by homology

[a] Not recognizably similar in amino acid sequence but *very* likely ancient homologues.

TABLE VIII Comparison of phage λ–host and P22–host interactions

λ	Host	P22	Host	Notes
CII	HflB	C1	HflB	Should be true for P22 by homology
CIII	HflB	C3	HflB	Should be true for P22 by homology
gpJ	LamB			P22 tails recognize polysaccharide so no analogue in P22
Stf	OmpC			P22 tails recognize polysaccharide so no analogue in P22
Xis	Lon	Xis	Lon?	Might be true for P22 by homology
Xis	FtsH	Xis	FtsH?	Might be true for P22 by homology
gpN	RNAP	gp24	RNAP?	Likely true for P22 by similarity of function
Int	IHF			P22 does not interact with IHF (Cho *et al.*, 1999)
Xis	Fis			Unknown for P22
gpQ	σ^{70}	gp23	σ^{70}	Likely true for P22 by similarity of function
P	DnaB			No analogue of P in P22
		gp12	DnaG?	No analogue of 12 in λ
gpE	GroE	gp5	GroE	
gpB	GroE			Not known for P22
gpN	Lon	gp24	Lon	Not known for P22
CI	RecA	C2	RecA	Might be true for P22 by homology and induction by ultraviolet light
CII	ClpYQ	C1	ClpYQ	Might be true for P22 by homology
CII	ClpAP	C1	ClpAP	Might be true for P22 by homology
gpO	ClpXP	gp18	ClpXP	Might be true for P22 by homology
gpN	NusA	gp24	NusA	Likely true by similarity of function
CII	HflD	C1	HflD	Might be true for P22 by homology
Gam	RecB			No Gam in P22
		Abc2	RecC	No Abc2 in λ

into the host genome as a prophage (Briani *et al.*, 2001; Deho and Ghisotti, 2006; Lindqvist *et al.*, 1993). P4 lacks genes encoding major structural proteins, but can utilize the structural gene products encoded by P2 for assembly of its own capsid (Christie and Calendar, 1990; Lindqvist *et al.*, 1993; Six, 1975). An overview of the P2 structural biology and a summary of its interactions are shown in Figure 8 and Table IX.

1. Gene control circuits

The decision whether to grow lytically or form a lysogen after P2 infection is dependent on the levels of two repressors: the immunity repressor gpC, which blocks lytic growth, and Cox, which blocks the expression of the

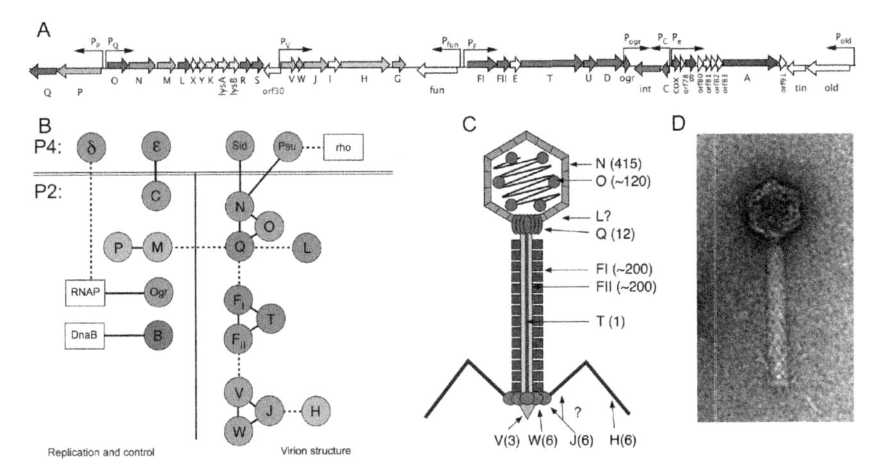

FIGURE 8 (A) Schematic diagram of the 33.6-kb P2 genome. Open reading frames (ORFs) are indicated by arrows that reflect their direction and size and are color coded as follow: red/orange, capsid-associated proteins (gpQ, O, N, L); blue, terminase proteins (gpP, M); yellow, lysis proteins (gpY, K, lysA, lysB); green, tail-related proteins (gpR, S, FI, FII, T, U, D); blue, base plate and tail fiber-related proteins (gpV, W, J, H, G); pink, transcriptional control proteins (Ogr, C, Cox); brown, integrase (Int); and purple, replication-related proteins (gpA,B). Nonessential genes and ORFs of unknown function are white. Promoters are indicated by arrows above ORFs. (B) Diagram of protein–protein interactions listed in Table IX. Each P2 and P4 protein is shown as a circle, colored as given earlier. Host proteins are shown as white boxes. (C) Schematic diagram of the P2 virion, color coded as given earlier. Copy numbers of proteins, when known, are indicated in parentheses. (D) Electron micrograph of a P2 virion, stained negatively with uranyl acetate. (See Page 7 in Color Section at the back of the book.)

genes required for lysogenization. GpC acts as a symmetric dimer, and its three-dimensional structure has been determined (Eriksson *et al.*, 2000; Massad *et al.*, 2010). Cox is multifunctional and acts as a repressor, as well as an excisionase, thereby coupling the transcriptional switch that controls lytic growth versus formation of lysogeny with the integration/excision of the phage genome in or out of the host chromosome. Cox is a tetramer in solution that self-associates to octamers in the absence of DNA (Eriksson and Haggard-Ljungquist, 2000). The transcriptional activator Ogr is required to turn on late gene transcription. Ogr is a Zn finger-containing protein that interacts with the α subunit of the *E. coli* RNA polymerase, recruiting it to the late promoters (Ayers *et al.*, 1994; Fujiki *et al.*, 1976; King *et al.*, 1992; Sunshine and Sauer, 1975; Wood *et al.*, 1997).

Some P2 prophage genes are not blocked by the action of gpC and instead are expressed from the prophage providing the host cell with new properties. Among these "lysogenic conversion" genes, P2 has three genes, *old*, *fun*, and *tin*, that block superinfection of unrelated phages,

TABLE IX Protein–protein interaction of phages P2 and P4

Protein	Function	Interactions	Notes	References
P2–P2 proteins				
gpC	Immunity repressor	Dimer	Y2H	Eriksson *et al.*, 2000; Massad *et al.*, 2010
			3D structure	
Cox	Transcriptional repressor, excisionase	Tetramer that can self-associate to octamers	*In vitro*; cross-linking and gel filtration	Eriksson and Haggard-Ljungquist, 2000
Int	Integrase	Dimer	*In vitro*; cross-linking	Frumerie *et al.*, 2005
			Y2H	
gpF_{I}	Tail sheath	Polymer, six molecules per sheath striation	Electron microscopy	Lengyel *et al.*, 1974
gpF_{II}	Tail tube	Polymer, six molecules per tube striation	Electron microscopy	Lengyel *et al.*, 1974
gpM	Small terminase subunit	Polymer, 8- to 10-fold ring	Biochemistry	Dokland, unpublished data
gpM	Small terminase subunit	gpP	Biochemistry	Bowden and Modrich, 1985
gpN	Major capsid protein	T=7 icosahedron (415 copies)	Cryo-EM and 3D reconstruction	Dokland *et al.*, 1992
		gpO	Biochemistry	Chang *et al.*, 2008, 2009
gpO	Internal scaffolding protein	Dimer	Biochemistry, chromatography	Chang *et al.*, 2008, 2009
		gpN		
gpP	Large terminase subunit	gpP	Biochemistry	Bowden and Modrich, 1985
gpQ	Connector, portal protein	Dodecameric ring	EM, crystallography	Doan and Dokland, 2007; Rishovd *et al.*, 1994

gpT	Tape measure	gpF_{II} (gpF_{I} ?)	Inside tail tube	
gpV	Tailspike	Tetramer	Sedimentation velocity and molecular mass determination	Haggard-Ljungquist *et al.*, 1995; Kageyama *et al.*, 2009
gpV	Tailspike	gpW, base plate component. Belongs to the T4 gene 25-like superfamily	Copurification	Haggard-Ljungquist *et al.*, 1995
P2–other phage proteins				
Tin	Block superinfection of T-even phages	T4 SSB protein (gp32)	Genetic evidence	Mosig *et al.*, 1997
gpC	Immunity repressor	P4 anti-repressor ε	Genetic evidence. Y2H	Geisselsoder *et al.*, 1981; Liu *et al.*, 1998; Eriksson *et al.*, 2000
gpN	Major capsid protein	P4 Sid; size determinator, external scaffold	Cryo-EM and 3D reconstruction	Marvik *et al.*, 1995; Wang *et al.*, 2000; Dokland *et al.* , 2002
gpN	Major capsid protein	P4 Psu; polarity suppressor; decoration protein	Cryo-EM and 3D reconstruction	Dokland *et al.*, 1993
P2–host proteins				
gpB	Helicase loader. Required for lagging strand synthesis	*E. coli* helicase DnaB	Genetic evidence. Electron micoscopy. *In vitro* binding.	Sunshine *et al.*, 1975; Funnell and Inman, 1983; Odegrip *et al.*, 2000
Ogr	Transcriptional activator of late operons	*E. coli* RNA polymerase α C-terminal domain	Genetic evidence	Sunshine and Sauer, 1975; Fujiki *et al.*, 1976; King *et al.*, 1992; Ayers *et al.*, 1994; Wood *et al.*, 1997
P4–host proteins				
Psu	Polarity suppressor; anti-terminator	Rho	Genetics; chemical cross-linking	Sauer *et al.*, 1981; Pani *et al.*, 2009
Delta	Transcriptional activator, contains two Ogr domains	*E. coli* RNA polymerase	Genetics; homology to Ogr	Julien *et al.*, 1997

but only gpTin is known to have a PPI. P2 gpTin excludes phage T4 by binding to and poisoning T4 gp32 (SSB), thereby inhibiting DNA replication (Mosig *et al.*, 1997).

Satellite phage P4 has to get access to the morphogenetic genes of its helper P2 to grow lytically. This is accomplished by P4-encoded functions during coinfection with P2. The P4 antirepressor Epsilon derepresses the P2 prophage by interacting with gpC, leading to the expression of P2 lytic genes (Eriksson *et al.*, 2000; Geisselsoder *et al.*, 1981; Liu *et al.*, 1998). P4 also has the capacity to *trans*-activate P2 late genes in the absence of Ogr using the transcriptional activator Delta. Delta contains two Ogr-like domains and recognizes the same DNA sequence as Ogr. Like Ogr, Delta interacts with the α subunit of *E. coli* RNA polymerase (Christie *et al.*, 2003; Julien *et al.*, 1997; Souza *et al.*, 1977).

2. P2 DNA replication

After infection, the linear DNA molecule is circularized by ligation of the cohesive (*cos*) ends by the host DNA ligase, and if P2 enters the lytic pathway, DNA replication is initiated by the *cis*-acting gpA that creates a single-stranded cut at the origin. GpA remains linked covalently to the 5′ end of the cleaved strand (Liu and Haggard-Ljungquist, 1994). Replication occurs via a modified rolling-circle mechanism that generates a double-stranded monomeric circle that is the packaging substrate (Pruss *et al.*, 1975). The cleavage/joining reactions are promoted by two tyrosine residues interspaced by three amino acids in gpA (Odegrip and Haggard-Ljungquist, 2001). The host Rep helicase is required for all DNA replication (Calendar *et al.*, 1970), while the phage-encoded gpB is a helicase loader that interacts with the *E. coli* DnaB helicase required for lagging strand synthesis (Funnell and Inman, 1983; Odegrip *et al.*, 2000).

3. Head structure and assembly

The main structural proteins involved in P2 capsid assembly are gpN, capsid protein; gpO, internal scaffolding protein; and gpQ, connector or portal protein (Chang *et al.*, 2008; Lengyel *et al.*, 1973; Linderoth *et al.*, 1991). P2 capsids are assembled as empty precursor procapsids consisting of 415 copies of the gpN capsid protein arranged with T=7 icosahedral symmetry and clustered in hexamers and pentamers that make trivalent interactions (Dokland *et al.*, 1992) (Fig. 9). GpQ forms a dodecameric ring that is incorporated into the capsid at one fivefold vertex and through which the DNA is subsequently packaged (Doan and Dokland, 2007; Rishovd *et al.*, 1994).

GpO plays a dual role in the assembly process as both a protease and a scaffolding protein (Chang *et al.*, 2009; Dokland, 2011). The full-length, 284 residue gpO protein is associated with procapsids at early times (Marvik *et al.*, 1994), but is cleaved autoproteolytically between residues

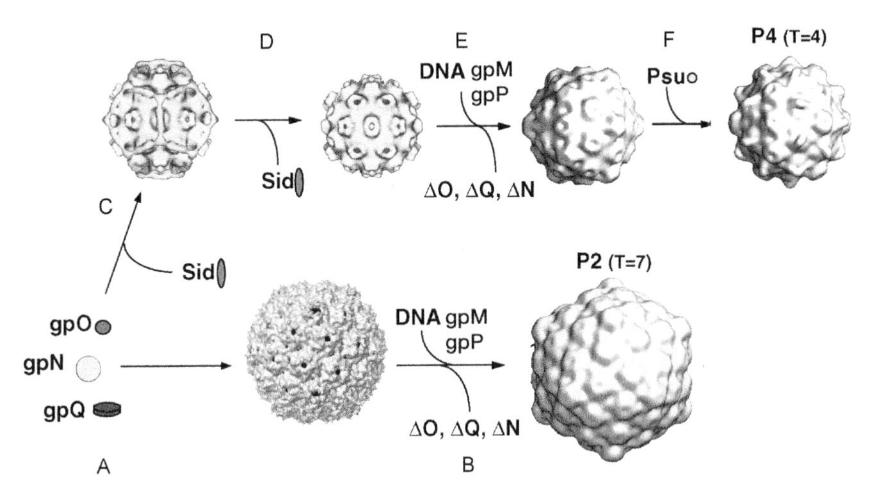

FIGURE 9 P2 assembly. Schematic diagram of the P2/P4 capsid assembly pathway. (A) Capsid protein gpN, scaffolding protein gpO, and connector (portal) protein gpQ are assembled into the T=7 P2 procapsid. (B) GpN, gpO, and gpQ are then processed to their mature forms, N*, O*, and Q*. DNA is packaged into the procapsid by the P2 terminase complex (gpM and gpP), accompanied by expansion into the mature, angular capsid (B). (C) In the presence of the P4-encoded external scaffolding protein Sid, P2 structural proteins are assembled into a small, T=4 P4 procapsid. Loss of Sid (D), protein processing, and DNA packaging lead to the mature, expanded P4 capsid (E). The decoration protein Psu is added to the mature capsid (F). (See Page 8 in Color Section at the back of the book.)

141 and 142, leaving an N-terminal fragment, gpO* (formerly known as h7), that remains inside the mature capsid (Chang *et al.*, 2008; Lengyel *et al.*, 1973). Concomitantly, gpN and gpQ are cleaved—presumably by gpO—to their mature forms, gpN* and gpQ*, by removal of 31 and 26 residues, respectively, from the N terminus (Rishovd and Lindqvist, 1992; Rishovd *et al.*, 1994). The protease activity of gpO resides in its N-terminal domain, which binds tightly to gpN and constitutes a serine protease similar to that of the herpesviruses (Chang *et al.*, 2009; Cheng *et al.*, 2004; Dokland, 2011). The C-terminal 90 amino acids comprise a scaffolding domain, which is necessary and sufficient for promoting capsid assembly (Chang *et al.*, 2008, 2009; Wang *et al.*, 2006). It too presumably binds to gpN during assembly, but any interaction is transient (Chang *et al.*, 2009). Both N- and C-terminal domains form dimers in solution (Chang *et al.*, 2009).

Circular P2 genomic DNA substrates are packaged into empty procapsids by the terminase complex, which consists of the small and large terminase proteins (gpM and gpP, respectively). Terminase recognizes and cleaves P2 DNA at the *cos* sites (Bowden and Modrich, 1985; Ziermann and Calendar, 1990). DNA packaging is associated with major structural transitions that lead to a larger, more angular and thin-shelled, mature capsid (Chang *et al.*, 2008; Dokland *et al.*, 1992). A head completion protein, gpL, is added (Chang *et al.*, 2008), and tails, which are assembled via a separate pathway, are attached to the unique gpQ-containing portal vertex (Lengyel *et al.*, 1974).

4. Tail structure

The 135-nm-long tail is composed of an inner tube and a contractile sheath (Lengyel *et al.*, 1974). The head-proximal end of the free tail has two disc-like structures of different thickness and diameter. In the complete phage, only the thinner disc is seen outside the head. The distal end of the tail contains a baseplate with a spike and six fibers. Sixteen essential tail genes have been identified located in three transcription units. GpFI and gpFII constitute the tail sheath and tail tube, respectively, and there are about 200 copies of each protein per tail (Temple *et al.*, 1991). RpR and gpS are involved in tail completion, as *amber* mutations in the respective gene produce giant tails and giant tail sheaths (Linderoth *et al.*, 1994). GpV is the tail spike, which forms a trimer in solution (Haggard-Ljungquist *et al.*, 1995; Kageyama *et al.*, 2009). GpW copurifies with gpV and is thus thought to be part of the baseplate, while gpJ localizes to the edge of the baseplate or proximal part of the tail fiber by immunoelectron microscopy (Haggard-Ljungquist *et al.*, 1995). GpH is believed to be the distal part of the tail fiber because it contains regions similar to tail fiber proteins of phages Mu, P1, λ, K3, and T2 (Haggard-Ljungquist *et al.*, 1992). A region of gene *H* is highly similar to the variable part of the *S* gene (*Sv*) of phage Mu and the variable part of gene *19* (*19v*) of phage P1, in both cases in the (+) orientation of the invertible segments. P2 does not have an invertible segment, but it contains a sequence identical to the left repeat of the Mu invertible G segment located within the *H* gene and upstream of the region with a high similarity to Mu *Sv* and P1 *19v*, indicating that the P2 *H* gene has been frozen in the (+) orientation, probably by a deletion covering the right inverted repeat and the invertase. This region is therefore inferred to constitute the receptor-binding tip of the tail fiber. By its similarities to gpU and gpU′ of phages Mu and P1, and gp38 of phages T4, Tula, and Tulb, gpG is probably required for fiber assembly. GpT is likely the tape measure, as it is the only protein long enough to span the length of the tail shaft (Christie *et al.*, 2002). Function of the products of genes *X*, *I*, *E*, *E′* (programmed -1 frameshift extension of E), *U*, and *D* remains unknown.

5. P4-induced redirection of P2 capsid assembly

Satellite phage P4 cannot form phage particles by itself, but in the presence of phage P2, P4 genomes are packaged into phage particles using structural proteins supplied by P2 (Six, 1975). However, the P4 capsid is smaller than P2 and contains only 235 copies of gpN, arranged on a T=4 icosahedral lattice (Dokland *et al.*, 1992) (Fig. 9). This size control is affected by the P4 gene product Sid (size determination)(Barrett *et al.*, 1976), which forms an dodecahedral cage-like scaffold around the outside of P4 procapsids (Marvik *et al.*, 1995). Sid self-associates into trimers and dimers that interact with gpN only in two hexamer subunits near the twofold symmetry axes (Wang *et al.*, 2000; Dokland *et al.*, 2002). Gene *N* mutants, called *sir* (*sid* responsiveness), render the capsid protein unable to form small capsids even in the presence of Sid (Six *et al.*, 1991). Conversely, the so-called super-*sid* or *nms* (*N* mutation sensitive) mutations in *sid* act as second-site suppressors of the *sir* mutations and enable the formation of small capsids (Kim *et al.*, 2001). Although well-formed small procapsids can be assembled from Sid and gpN alone (Dokland *et al.*, 2002), gpO is required for the formation of viable P4 phage (Six, 1975). Presumably this is because its protease activity is essential, and gpO may also facilitate incorporation of the portal. Indeed, *sid* complements the P2 *Oam*279 mutant, which produces a 237 residue N-terminal fragment of gpO that includes the protease domain but lacks scaffolding activity (Agarwal *et al.*, 1990). The P4 genome contains P2 *cos* sites and is packaged into small capsids using the P2 terminase (Ziermann and Calendar, 1990). The P4-encoded protein Psu is a suppressor of Rho-dependent transcription termination (Pani *et al.*, 2009; Sauer *et al.*, 1981); it also functions as a decoration protein that binds to gpN hexamers and stabilizes the P4 capsid (Dokland *et al.*, 1993; Isaksen *et al.*, 1993).

E. *Pneumococcus* phage Dp-1 and Cp-1

The lytic bacteriophages Dp-1 and Cp-1 infect the Gram-positive bacterium *Streptococcus pneumoniae*, which colonizes the mucosal surface of the upper respiratory tract and causes serious invasive infections such as pneumonia, meningitis, and sepsis (van der Poll and Opal, 2009). Dp-1 was the first isolated *Streptococcus* phage and is an unassigned member of the *Siphoviridae* (Fig. 10) (McDonnell, Lain and Tomasz, 1975). It was reported that the Dp-1 virion surprisingly contains lipids that might be arranged as a bilayer surrounding the capsid (Lopez *et al.*, 1977), a feature not reported for any other phage belonging to the *Caudovirales*. The lytic enzyme Pal of Dp-1 has been studied intensively because it kills virulent streptococci efficiently and has been used as an enzybiotic to cure pneumococcal infections in animals (Loeffler *et al.*, 2001; Rodriguez-Cerrato *et al.*, 2007). The podovirus Cp-1 lytic enzyme Cpl1 was also demonstrated to be an efficient enzybiotic (Loeffler *et al.*, 2003). The Cp-1 genome is a linear, 20-kbp dsDNA predicted to encode 28 proteins and a packaging RNA (pRNA). Like ϕ29-like viruses, Cp-1 uses protein priming for the initiation of DNA replication (Martin *et al.*, 1996). The biology of Dp-1 and Cp-1 is otherwise only poorly understood.

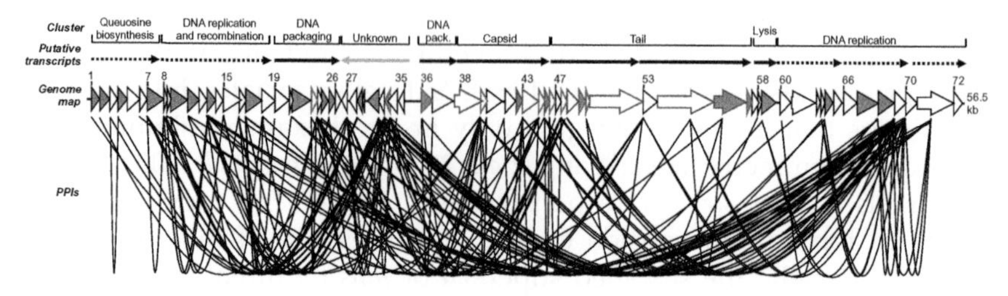

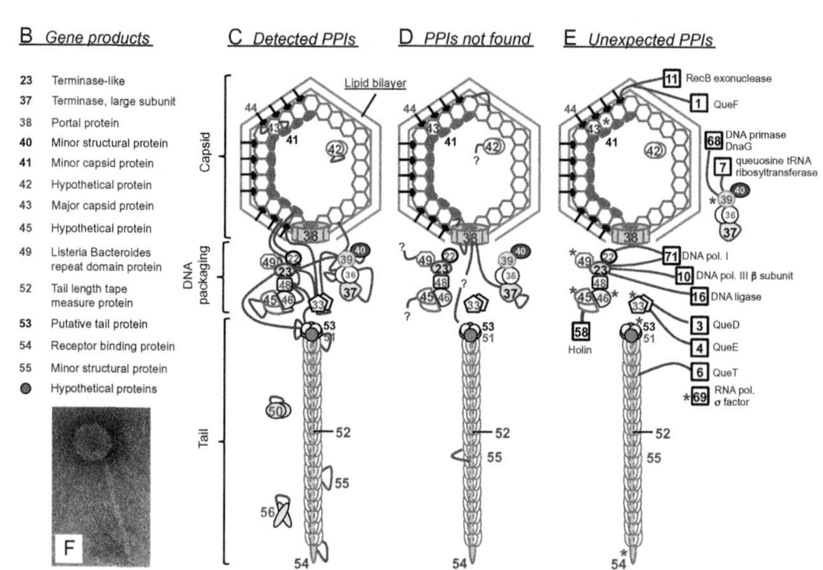

FIGURE 10 The Dp-1 genome and interactome. (A) Dp-1 genome map. Open reading frames (ORFs) corresponding to their transcription direction are shown as arrows (genome map). Predicted transcripts and functional gene clusters are indicated. All protein–protein interactions (PPIs) identified by systematic Y2H screens among Dp-1 proteins are shown as lines that connect the corresponding ORF symbols. Proteins that bind themselves (homomers) are highlighted by gray ORF symbols. Other ORF colors: refer to Figure C-E. (B–E) A Dp-1 virion model based on identified binary PPIs, mass spectrometry analysis (MS) of mature Dp-1 virions, and homology predictions. (B) Dp-1 virion proteins, including structural proteins identified by MS analysis (in red), or proteins with a homology-based annotation that indicates a virion-related function (in black). (C) Core model of the Dp-1 virion: proteins highlighted in red were identified by MS analysis as structural components. Proteins highlighted in blue represent hypothetical gene products that interact with structural components, thus presumably playing a transient role in virion assembly as they were not identified by MS analysis of Dp-1 virions. Red edges that connect the proteins were shown to interact in Y2H tests. (D) Y2H screens do not identify all expected PPIs. Red lines represent PPIs that were expected to occur among the corresponding proteins known from homologues but were not identified in the systematic Y2H screens. (E) Systematic Y2H screens reveal unexpected PPIs.

The 56.5-kbp genome of Dp-1 is predicted to encode 72 proteins (Fig. 10A) (Sabri *et al.*, 2011). Approximately half of the gene products could be functionally annotated by homology predictions. However, in this and hundreds of other completely sequenced phage genomes, more than half of phage gene products could not be annotated. Proteome-wide Y2H screens, coupled with mass spectrometry analysis (MS) of mature Dp-1 virions, identified 156 unique PPIs among 57 Dp-1 proteins [a detailed list can be found in Sabri *et al.* (2011)], and MS revealed that the mature virion is composed of eight structural proteins. A Dp-1 virion model (Figs. 10B and 10C) physically links many proteins with unknown function to the virion. Because they bind structural proteins but are absent from the mature virion, these Y2H PPIs either indicate novel structural components (when found by MS of the virion particles) or virion assembly facilitators. For instance, homology comparisons of gp40, gp41, and gp53 predict that these proteins are structural (Fig. 10B). However, MS analysis did not detect them as components of the virion. Even though functional predictions by homology comparisons are extremely helpful in proteome annotation, a certain fraction of proteins may be wrongly annotated in the absence of other experimental data (i.e., whether a protein is annotated as a structural component or an assembly facilitator can have an important impact on future experiments).

Such a comprehensive analysis also reveals unexpected interactions that normally would not be tested in hypothesis-driven experiments (Fig. 10E). For example, Dp-1 proteins believed to be involved in DNA replication and recombination, transcriptional regulation, and other pathways exhibit interactions with structural components. This indicates that DNA packaging and DNA replication/repair might be coupled (as in phage T4), for example, by resolving intermediate DNA structures in parallel with packaging. Although the identified PPIs were helpful in obtaining the first hints about the function of the proteins, not all expected PPIs between the structural proteins were detected (Fig. 10D). Nevertheless, this study demonstrated that homology-based genome annotation complemented by proteomic approaches is a useful strategy to gain novel insights into the biology of poorly understood bacteriophages.

Proteins are included that were shown to bind with structural proteins and putative morphogenetic factors, e.g., DNA metabolism (DNA polymerase subunits, RecB, DNA ligase, DnaG), queuosine biosynthesis (Que proteins), lysis (Holin), and transcription (sigma factor). Interactions are symbolized by edges that connect the corresponding proteins (in the case of sigma factor gp69, interaction partners are highlighted by an asterisk). (F) Electron micrograph of a mature Dp-1 virion. C–F from Sabri *et al.* (2010), with permission. (See Page 8 in Color Section at the back of the book.)

The same combination of genome annotation and proteomics was repeated for the 28 putative Cp-1 proteins (Häuser *et al.*, 2011). They detected a total of 12 structural proteins by MS analysis. More than half the proteins showed at least one protein–protein interaction with 17 binary interactions in total. Notably, nearly all were detected between proteins encoded in the structural gene cluster, reflecting a highly modular organization of its interactome. In contrast, Dp-1, with its larger proteome/interactome, shows more cross-communication between proteins belonging to different pathways, although a functional enrichment of interactions between proteins belonging to related cellular processes was still detected (Sabri *et al.*, 2011). It is likely that larger phage proteomes/interaction networks routinely exhibit a higher frequency of protein cross-communication at the protein interaction level due to regulation between various pathways (i.e., the larger the proteome, the more promiscuous the interaction network).

F. *Bacillus subtilis* phage ϕ29-like viruses

ϕ29-like phages belong to the *Picovirinae* subfamily of the *Podoviridae*. Members include ϕ29, ϕ15, PZA, BS32, B103, M2Y (M2), Nf, and GA-1. All are lytic; most infect *B. subtilis*, but some grow in *B. pumilus*, *B. amyloliquefaciens*, and *B. licheniformis* (Meijer *et al.*, 2001a). The characteristics of their virions are a prolate capsid (e.g., ϕ29 is 41.5 by 31.5 nm) with attached head fibers, a connector, a short noncontractile tail (e.g., ϕ29 with 32.5 nm in length), and the presence of tail appendages (Anderson *et al.*, 1966) (Fig. 11). GA-1 and M2Y lack the gene for head fibers (Yoshikawa *et al.*, 1986), which may stabilize the capsid (Tao *et al.*, 1998). The approximate 20-kbp genome of ϕ29-like phages is linear and encodes about 25 proteins. The 5′ ends of the genome are attached covalently to a terminal protein (TP). ϕ29-like viruses are among the smallest dsDNA phages isolated so far and thus represent minimal model systems (Meijer *et al.*, 2001a).

1. Protein–protein interactions

A literature and database search on ϕ29-like viruses retrieved 30 PPIs (Tables X and XI) with evidence mainly from directed experiments and structure determination studies. Twenty-three are attributed to phage ϕ29 itself, five to GA-1, and two to Nf, emphasizing the role of ϕ29 as a model for the whole group. The biology of other ϕ29-like phages is poorly understood, and only a few proteins have been characterized experimentally. In some cases, ϕ29 PPIs are conserved among other ϕ29-like homologues, supporting a close relationship of these phages. However, there are also differences that reveal interesting insights into their evolution. Known ϕ29 protein–protein interactions include proteins involved in

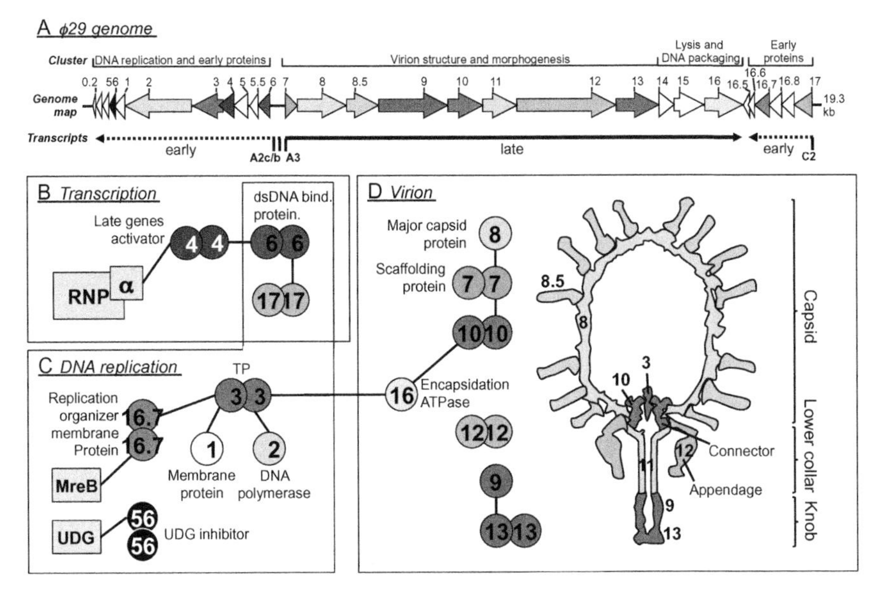

FIGURE 11 Genome map and literature curated φ29 interactome. (A) Genome and simplified transcriptional map of φ29 (after Meijer *et al.*, 2001a). Major promoters are indicated by their names, and major transcripts are shown as dashed (early) and solid (late) arrows. Other promoters are not indicated for clarity. Open reading frame (ORF) arrows indicate transcription direction. The color of ORF symbols corresponds to the color used for protein symbols in B, C, and D. PPIs known for φ29, organized by their cellular function (transcription, DNA replication, or virion). *B. subtilis* proteins are shown as rectangles, phage proteins as circles. Equal protein symbols contacting each other indicate homomers. Heteromeric PPIs are highlighted by protein symbols that are connected by lines. (D) Schematic cross view of the mature φ29 virion (after Xiang *et al.*, 2008). UDG, uracil–DNA glycosylase; RNAP, RNA polymerase; TP, terminal protein. (See Page 9 in Color Section at the back of the book.)

transcriptional regulation, DNA replication, and virion structure. Only three PPIs between φ29 and *B. subtilis* proteins have been identified so far.

2. Protein-primed DNA replication

Phage φ29 DNA polymerase (DNAP, gp2) interacts with terminal protein (TP, gp3) to initiate genome replication. Interactions have also been shown between the corresponding GA-1 and Nf orthologs (Table XI). The structure of the φ29 TP–DNAP complex has been determined and suggests that the priming domain of TP occupies the dsDNA-binding tunnel, allowing the DNAP to bind negatively charged amino acid residues of TP (Kamtekar *et al.*, 2006).

Phage φ29 replication depends on the presence of a single dAMP residue linked covalently via its 5′-phosphoryl group to the hydroxyl group of Ser^{232} of TP. Formation of the phosphoester bond is catalyzed

TABLE X Protein–protein interactions of phage ϕ29 including phage–host interactions[a]

ϕ29 1	Function 1	ϕ29 2	Function 2	References
gp2	DNA polymerase	gp3	DNA terminal protein TP	Kamtekar *et al.*, 2006; Blanco *et al.*, 1987; Dufour *et al.*, 2000; González-Huici *et al.*, 2000
gp3	DNA terminal protein TP	gp1	Membrane-associated early protein	Bravo *et al.*, 2000
gp3	DNA terminal protein TP	gp3	DNA terminal protein TP	Serna-Rico *et al.*, 2003
gp3	DNA terminal protein	gp16	Packaging ATPase	Guo *et al.*, 1987 Grimes and Anderson, 1997
gp4	Late gene activator	gp4	Late gene activator	Badia *et al.*, 2006 Mencía *et al.*, 1996b
gp4	Late gene activator	gp6	Histone-like, nonspecific DNA-binding protein	Calles *et al.*, 2002; Camacho and Salas, 2001
gp6	Histone-like, nonspecific dsDNA-binding protein	gp6	Histone-like, nonspecific dsDNA-binding protein	Abril *et al.*, 1999; Pastrana *et al.*, 1985
gp7	Scaffolding protein	gp7	Scaffolding protein	Morais *et al.*, 2003
gp7	Scaffolding protein	gp8	Major head protein	Lee and Guo, 1995
gp7	Scaffolding protein	gp10	Upper collar/connector protein	Guo *et al.*, 1991; Lee and Guo, 1995
gp10	Upper collar/connector protein	gp10	Upper collar/connector protein	Guasch *et al.*, 2002; Simpson *et al.*, 2001
gp12	Preneck appendage protein (anti-receptor)	gp12	Preneck appendage protein (anti-receptor)	Xiang *et al.*, 2009
gp13	Tail-associated, peptidoglycan-degrading enzyme	gp9	Tail knob protein	García *et al.*, 1983

gp13	Tail-associated, peptidoglycan-degrading enzyme	gp13	Tail-associated, peptidoglycan-degrading enzyme	Xiang *et al.*, 2008
gp16.7	Replication organizer; DNA-binding membrane protein p16.7	gp16.7	Replication organizer; DNA-binding membrane protein p16.7	Asensio *et al.*, 2005; Albert *et al.*, 2005; Crucitti *et al.*, 1998; Serna-Rico *et al.*, 2003; Meijer *et al.*, 2001b
gp16.7	Replication organizer; DNA-binding membrane protein p16.7	gp3	DNA terminal protein TP	Serna-Rico *et al.*, 2003
gp17	Early protein involved in early DNA replication	gp6	Histone-like, nonspecific dsDNA-binding protein	Crucitti *et al.* 2003
gp17	Early protein involved in early DNA replication	gp17	Early protein involved in early DNA replication	Crucitti *et al.*, 2003
gp56 = gp0.8	Host uracil-DNA glycosylase inhibitor	gp56 = gp0.8	Host uracil-DNA glycosylase inhibitor	Serrano-Heras *et al.*, 2007)
ϕ29–*Bacillus subtilis* interactions				
gp4	Late gene activator (3)	RpoA	RNA polymerase subunit α	Mencía *et al.*, 1996a; Monsalve *et al.*, 1996b
gp16.7	Replication organizer, DNA-binding membrane protein gp16.7 (2)	MreB	Actin-like cytoskeleton protein	Muñoz-Espín *et al.*, 2009
gp56 = gp0.8	UDG inhibitor (1)	UDG	Uracil-DNA glycosylase	Serrano-Heras *et al.*, 2006

[a] Data from BOND, IntACT, and PDB (ϕ29, Taxon id 186846). All proteins are termed "gp" (gene product), although they are also termed "p" (protein), e.g., gp4 = p4. In the phage–host interaction list, the first protein is from ϕ29, the second from *B. subtilis*. (1) Host RNAP binds via α subunit to p4 (i) bound at the late promoter PA3, thus activating late transcription and (ii) bound to the promoter PA2c, thus blocking transcription through promoter clearance inhibition. (2) PPI plays a role in the organization of ϕ29 DNA replication as peripheral helix-like structures. (3) Inhibition of host UDG; avoids removal of uracil from ϕ29 DNA via UDG-base excision repair.

TABLE XI Protein–protein interactions of ϕ29-related phages[a]

Phage	Protein 1		Protein 2		References
		Function		Function	
GA-1	gp6	Histone-like, nonspecific dsDNA binding protein	gp6	Histone-like, nonspecific dsDNA-binding protein	Freire *et al.*, 1996
Nf	gp6	dsDNA binding protein	gp6	dsDNA-binding protein	Freire *et al.*, 1996
GA-1	gp2	DNA polymerase	gp3	DNA terminal protein	Illana *et al.*, 1996; Longás *et al.*, 2006
Nf	gp2	DNA polymerase	gp3	DNA terminal protein	Longás *et al.*, 2006; González-Huici *et al.*, 2000
GA-1	gp5	Single-stranded DNA-binding protein SSB (1)	gp5	Single-stranded DNA-binding protein SSB	Gascón *et al.*, 2000
GA-1	gp12	Neck appendage protein (2)	gp12	Neck appendage protein	Schulz *et al.*, 2010
GA-1	$gp4_G$	Transcriptional regulator (3)	$gp4_G$	Transcriptional regulator	Horcajadas *et al.*, 1999

[a] Data from BOND, IntACT, and PDB (searched for ϕ29, Taxon id 186846). (1) ϕ29 and Nf SSB were shown to be stable monomers in solution, whereas GA-1 SSB forms homohexamers. (2) Homotrimer; autocatalytic cleavage similar to ϕ29 gp12. PDB: 3GUD. (3) Homodimer; negative regulator of early transcription; was shown not to interact with RNAP α subunit as the homologue gp4 of ϕ29.

by the DNAP itself (Blanco and Salas, 1984; Hermoso *et al.*, 1985); thus formation of a stable DNAP–TP heterodimeric protein complex is required (Blanco *et al.*, 1987). The TP–dAMP is elongated by the same DNAP (Méndez *et al.*, 1992). After the synthesis of six to nine nucleotides, DNAP dissociates from TP and starts continuous, linear synthesis (Méndez *et al.*, 1997). This mechanism of protein-dependent DNA replication initiation was shown for Nf and GA-1 (Illana *et al.*, 1996; Longás *et al.*, 2008), *S. pneumoniae* phage Cp-1, *E. coli* phage PRD1, adenoviruses, and some plasmids and bacterial genomes (e.g., *Streptomyces*), although there may be differences in the various priming mechanisms (de Jong and van der Vliet, 1999; Martin *et al.*, 1996; Salas, 1991).

The TP–DNAP interaction plays a central role in ϕ29 DNA replication initiation. However, other PPIs with TP are known, and thus protein-primed DNA replication may be more complicated than is currently known (Figs. 11C and Fig. 12). Chemical cross-linking shows that TP forms homodimers (Serna-Rico *et al.*, 2003). Analysis of TP mutants suggests that the parental TP (TP bound to the parental genome) recruits the free TP–DNAP complex via a TP–TP interaction (Illana *et al.*, 1999), but an additional interaction between parental TP and DNAP might be important for the specificity of replication initiation (González-Huici *et al.*, 2000).

TP interacts with the replication organizer gp16.7 (Serna-Rico *et al.*, 2003) (Fig. 12). Gp16.7, which has the potential to bind to both ssDNA and dsDNA, is an integral membrane protein consisting of an N-terminal membrane anchor and a C-terminal cytoplasmic domain. It localizes ϕ29 DNA replication to the membrane (Meijer *et al.*, 2000, 2001b; Serna-Rico *et al.*, 2002, 2003). The C-terminal domain of gp16.7 dimerizes in solution but forms larger filaments in the presence of ssDNA (Asensio *et al.*, 2005; Serna-Rico *et al.*, 2003). Replication, which starts at both DNA ends, first produces a so-called type I replication intermediate consisting of dsDNA with ssDNA regions caused by displacement of the non-template DNA parental strand. Later, when the two replication forks have merged, the parental DNA strands become separated into two type II replication intermediates, still containing ssDNA regions in the parental template. A crystal structure of the gp16.7 DNA-binding C-terminal domain complexed with dsDNA reveals three gp16.7 dimers forming a deep, positively charged longitudinal cavity that interacts nonspecifically with the DNA phosphate backbone (Albert *et al.*, 2005). Rp16.7 localizes in peripheral helix-like structures in a bacterial MreB cytoskeleton-dependent manner (Muñoz-Espín *et al.*, 2010). In fact, protein gp16.7 interacts directly with MreB; this interaction is required for efficient ϕ29 DNA replication, allowing simultaneous replication of multiple templates at various peripheral sites.

Phage ϕ29 gp1 is another early protein that binds TP (Bravo *et al.*, 2000). Rp1, like gp16.7, is membrane associated and self-associates into long filamentous structures (Bravo and Salas, 1997, 1998). Its membrane

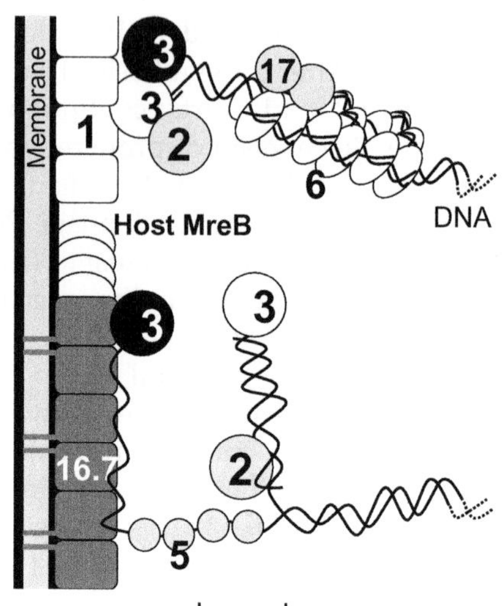

Legend

p**1** Membrane associated early protein
p**2** DNA polymerase
p**3** Terminal protein
p**5** Single-stranded DNA-binding protein
p**6** Histone-like, nonspecific dsDNA binding protein
p**16.7** Replication organizer; ssDNA-binding membrane protein
p**17** Early protein; enhances early DNA replication

FIGURE 12 Interactions in ϕ29 genome replication and membrane localization. Proteins that contact each other have been shown to interact. For clarity, only one genome end is shown. Parental TP is highlighted by a black symbol, whereas primer TP is in white. The upper part of the figure includes PPIs that preferentially play a role in replication initiation. The lower part of the figure indicates elongated DNA with the associated ssDNA regions. Because p16.7 and p5 (SSB) both bind ssDNA, they could either redirect the parental ssDNA strand to the membrane via TP-p16.7 and p16.7–ssDNA interactions or stabilize it in the cytoplasm via SSB alone (Meijer *et al.*, 2001b). After Bravo *et al.* (2000) and Serna-Rico *et al.* (2003).

association and polymerization are mediated through a C-terminal hydrophobic sequence, whereas binding to TP involves an N-terminal region. In contrast to gp16.7, gp1 does not bind DNA (Fig. 12) and is thought to be another component that directs ϕ29 DNA replication to the membrane (Bravo *et al.*, 2000). Any interplay between gp16.7 and gp1 in this process is unknown.

The ϕ29 histone-like, nonspecific dsDNA-binding protein gp6 forms a nucleoprotein complex with ϕ29 DNA that covers most of the genome, implicating it in genome organization (Serrano *et al.*, 1994). Rp6 either unwinds the genome termini or interacts with TP directly, thus making a TP complex competent to bind the genome ends (Blanco *et al.*, 1986; Serrano *et al.*, 1994). Rp6 forms long oligomeric protein filaments thought to function as a protein backbone for DNA binding (Abril *et al.*, 1999). GA-1 and Nf gp6 orthologues also self-associate (Freire *et al.*, 1996). Rp6 also binds gp17, an early protein that stimulates the first viral replication round *in vivo* when other replication proteins are present at low levels (Crucitti *et al.*, 1998, 2003). In contrast to gp6, gp17 does not bind DNA but its interaction with gp6 enhances gp6 binding to the replication origins, thereby stimulating early rounds of replication.

A homomeric interaction was demonstrated for the single-stranded DNA-binding protein of GA-1. GA-1 SSB was surprisingly found as a homohexamer in solution, whereas SSBs of Nf and ϕ29 (p5) appear to be stable monomers (Gascón *et al.*, 2000). The homohexamer binds to ssDNA with higher affinity than the monomer and exhibits a higher stimulatory effect on DNA replication.

Bacteriophages have evolved different mechanisms to protect their genomes from degradation by host nucleases in the cell. One protective strategy includes the incorporation of noncanonical nucleotides into their DNA. For instance, *B. subtilis* phage PBS2 uses uracil instead of thymine (Takahashi and Marmur, 1963). PBS2 achieves this unusual goal by increasing the dUTP pool, relative to dTTP, after infection. In addition, PBS2 encodes Ugi, a protein that specifically inhibits the host uracil–DNA glycosylase (UDG)(Cone *et al.*, 1980), which is involved in the first step of removing incorporated uracil residues via the base excision repair pathway. Normally, this takes place to avoid mutational effects in the host genome but it would impair the incorporation of uracil into the PBS2 genome. Cocrystallization of Ugi and UDG revealed that Ugi mimics uracil-containing duplex DNA, thereby blocking efficient DNA binding of UDG (Putnam *et al.*, 1999). ϕ29 blocks the host UDG by an interaction with the early protein gp56 (gp0.8) that competes with the DNA-binding ability of UDG (Serrano-Heras *et al.*, 2006, 2007, 2008). Although the ϕ29 genome does not contain uracil (Serrano-Heras *et al.*, 2006), ϕ29 DNAP is able to incorporate dUMP with a catalytic activity only twofold lower than for that of dTMP. Because ϕ29 DNAP does not discriminate stringently between dTMP and dUMP, UDG inhibition via gp56 minimizes uracil incorporation.

3. Interactions during transcriptional regulation

Transcriptional regulation of the ϕ29 genome has been studied extensively (Meijer *et al.*, 2001a; Rojo *et al.*, 1998). The transition from early to late gene expression is controlled by gp4 (Fig. 11). Gp4 is a homodimer in solution (Mencía *et al.*, 1996b); it forms a tetramer when bound to the intergenic region between gene *6* and gene *7*, which contains the A2b,

A2c, and A3 promoters (Fig. 11A). One binding site is upstream of A3, the late promoter that lacks a -35 box (Barthelemy and Salas, 1989). Once gp4 is bound, it recruits, via a PPI, the host RNAP, leading both to its stabilization on the promoter and transcriptional activation (Nuez *et al.*, 1992). Although gp4 activates late gene expression, it represses early gene expression from the A2b and A2c promoters (Monsalve *et al.*, 1997; Rojo and Salas, 1991). Interestingly, repression of A2c and activation of A3 involve the same residues of gp4 with the C-terminal domain of the RNAP α subunit (Mencía *et al.*, 1996a; Monsalve *et al.*, 1996a,b). Rp6 also interacts with gp4 in regulating the transcription of promoters A2b, A2c, and A3. Gp6 binds gp4 cooperatively to form a repression complex at promoters A2b and A2c while stimulating A3 (Calles *et al.*, 2002; Elías-Arnanz and Salas, 1999).

4. Virion structure and morphogenesis

Cryo-EM analyses have revealed many details about the ϕ29 virion structure (Morais *et al.*, 2003, 2005; Tao *et al.*, 1998; Xiang *et al.*, 2008, 2009), and many structures of virion proteins and factors are known (Table X). Assembly starts from a connector core particle consisting of 12 subunits of gp10 on which the scaffolding protein gp7 binds, followed by recruitment of the major capsid protein gp8 (Guo *et al.*, 1991; Lee and Guo, 1995). The scaffolding protein is needed to create a lattice of concentric shells (Morais *et al.*, 2003) and helps to organize the single capsomers built up by gp8 except for the basal connector capsomer. ϕ29 proheads were shown to consist of 235 copies of gp8 arranged as hexameric and pentameric capsomeres, indicating that direct interactions between gp8 monomers occur (Morais *et al.*, 2005). The capsid fiber protein gp8.5, which is present in ϕ29 proheads, is presumably connected via gp8 to the prohead and is suggested to be organized as homotrimers via interactions of a C-terminal coiled-coil region (Morais *et al.*, 2005). In addition to gp8, gp8.5, and gp10, ϕ29 proheads contain a homopentameric packaging RNA (pRNA) that binds to the connector, forming a basket-like complex (Simpson *et al.*, 2000). pRNA is needed to translocate DNA into the prohead. Furthermore, the DNA translocation machinery needs the ATPase gp16 to drive the packaging reaction (see also chapter 9 of this volume). Rp16 was shown to bind to the connector protein (Ibarra *et al.*, 2001). While other phages use specific DNA sequences to recognize and recruit the genome to the procapsids, ϕ29 DNA packaging initiation is dependent on the TP gp3 present at the genome ends, indicating that an interaction between gp16 and DNA-bound gp3 recruits the DNA to proheads and initiates packaging (Guo *et al.*, 1987). Release of the scaffolding protein is associated with DNA packaging and pRNA and gp16 are released when there is no more DNA to package.

An interesting aspect of ϕ29 is that the virion contains several enzymes that are associated with structural proteins. For instance, gp12 is a homotrimer that participates in host cell recognition and entry. It also shows autocatalytic cleavage, although it remains unclear what the function of that reaction is. Similarly, gp13 is bound as a dimer at the distal end of the ϕ29 tail knob to gp9; gp13 cleaves both the polysaccharide backbone and peptide cross-links of the cell wall during infection.

IV. CONCLUSIONS

A. High-throughput data vs small-scale interactions

It took several decades to collect the interactions of phage λ, even though it has only a few dozen genes. These interactions have been studied in painstaking individual experiments. Since the mid-1990s, new strategies have been developed to identify and analyze PPIs on a larger scale, particularly two-hybrid-related methods and techniques involving protein complex purification and mass spectrometry. Despite the pioneering efforts of (Bartel *et al.*, 1997) to map the interactome of phage T7 in a "single" experiment, their results remain somewhat contentious. Why is that? Studies show that each two-hybrid system produces system-specific data, and each system may only be able to identify as few as 20–30% of all interactions (Chen *et al.*, 2010). It has been shown that the application of up to 10 different Y2H systems can identify up to 80% of all interactions, but there are still some that go undetected (Chen *et al.*, 2010). This limited sensitivity can be explained by differences between the individual Y2H systems, expression levels, the way fusion proteins are constructed (e.g., N-terminal fusions vs C-terminal fusions), or linkers between fusion proteins. Some interactions may be completely undetectable by Y2H assays as these experiments are typically done in yeast, which may not have the necessary chaperones for bacterial or phage proteins to fold properly or the proteins may localize differently than in their normal organism.

Similar limitations apply to strategies that involve protein complex purification and mass spectrometry. Because many purifications are sensitive to small changes in protocols, even when a complex is purified within the same laboratory different preparations often result in different protein compositions in subsequent MS analyses (Gavin *et al.*, 2006). If two different laboratories purify the same protein complexes using different methods, the results may differ even more dramatically (Hu *et al.*, 2009).

Even if large-scale screens identify the majority of interactions, they will always also identify false positives. For instance, a screen of λPPIs yielded 97 interactions among λ proteins (Rajagopala *et al.*, 2011). While the authors detected about half of the 33 published interactions listed in

Table III, they also found many new interactions. Many do not seem plausible and are likely false positives. However, there is no way of knowing which ones are in fact false. We can only make educated guesses and rank all interactions by their plausibility, selecting only the best candidates for more detailed studies.

B. Phage of the same host has different interactomes

Where it is known, all bacteria are infected by numerous unrelated phages. While unrelated phages cannot have identical intraphage interactions, their proteins may interact with the same targets in their host cells. While interaction databases do not have enough information about phage–host interactions to draw general conclusions, several lines of evidence indicate that this is not generally true. For instance, the well-studied phages T7 and λ interact with different host proteins. They seem to share only one common interaction target, namely the RecBCD recombinase complex (see Tables II and IV). λ certainly has about twice as many host–phage interactions than T7, but this is certainly due to λ being temperate. T4 probably does interact somewhat more extensively with the host than T7, and one could argue that this is due to genome size. Similarly, in *Streptococcus* phage Dp-1 the authors found 38 interactions among the 72 phage proteins and its host, while the 28 (unrelated) proteins of phage Cp1 showed only 11 such interations (Häuser *et al.*, 2011). Together, Dp-1 and Cp-1 target a total of 36 host proteins, of which only 2 are shared. More host–phage interactomes are needed to draw more general conclusions though.

C. Evolution of interactions [λ, P22]

We expect that related phage have similar interaction patterns, both among phage proteins and between phage and host proteins. This is certainly true for very closely related phages. However, phages evolve much faster than any other living system and thus may lose or gain interactions faster too. Unfortunately, there is not enough interaction data to make detailed comparisons, even among the well-studied lambdoid phages. However, many individual cases of proteins are not obviously homologous by sequence but which are believed to share a common ancestor, for example, the capsid proteins of λ and HK97. These proteins also have similar structures and make similar contacts in the capsid (Hendrix, 2005). Evolutionary relationships are also supported by the extensive similarities in parts of their genomes, that is, genome mosaicism (Casjens, 2011; Casjens and Thuman-Commike, 2011; Hendrix and Casjens, 2006). It remains to be seen how many interactions are conserved and how successful phages are in inventing new interactions (or discarding them).

More systematic studies of multiple phages in parallel will be required to obtain comparable data sets in which differences in interaction patterns are not based on the use of noncomparable methods.

ACKNOWLEDGMENTS

Work on this chapter was partially funded by NIH Grant R01GM79710 to P.U., the European Union (Grant HEALTH-F3-2009-223101)(P.U., R.H., and S.B.), NIH Grant RO1 AI074825 (S.R.C), and the Spanish Ministry of Science and Innovation (Grant BFU2008-00215) to M.S.

REFERENCES

Abbani, M. A., Papagiannis, C. V., Sam, M. D., Cascio, D., Johnson, R. C., and Clubb, R. T. (2007). Structure of the cooperative Xis-DNA complex reveals a micronucleoprotein filament that regulates phage lambda intasome assembly. *Proc. Natl. Acad. Sci. USA* **104:**2109–2114.

Abril, A. M., Marco, S., Carrascosa, J. L., Salas, M., and Hermoso, J. M. (1999). Oligomeric structures of the phage phi29 histone-like protein p6. *J. Mol. Biol.* **292:**581–588.

Agarwal, M., Arthur, M., Arbeit, R. D., and Goldstein, R. (1990). Regulation of icosahedral virion capsid size by the *in vivo* activity of a cloned gene product. *Proc. Natl. Acad. Sci. USA* **87:**2428–2432.

Agirrezabala, X., Martin-Benito, J., Caston, J. R., Miranda, R., Valpuesta, J. M., and Carrascosa, J. L. (2005a). Maturation of phage T7 involves structural modification of both shell and inner core components. *EMBO J.* **24:**3820–3829.

Agirrezabala, X., Martin-Benito, J., Valle, M., Gonzalez, J. M., Valencia, A., Valpuesta, J. M., and Carrascosa, J. L. (2005b). Structure of the connector of bacteriophage T7 at 8A resolution: Structural homologies of a basic component of a DNA translocating machinery. *J. Mol. Biol.* **347:**895–902.

Agirrezabala, X., Velazquez-Muriel, J. A., Gomez-Puertas, P., Scheres, S. H., Carazo, J. M., and Carrascosa, J. L. (2007). Quasi-atomic model of bacteriophage t7 procapsid shell: insights into the structure and evolution of a basic fold. *Structure* **15:**461–472.

Albert, A., Munoz-Espin, D., Jimenez, M., Asensio, J. L., Hermoso, J. A., Salas, M., and Meijer, W. J. (2005). Structural basis for membrane anchorage of viral phi29 DNA during replication. *J. Biol. Chem.* **280:**42486–42488.

Alfano, C., and McMacken, R. (1989). Ordered assembly of nucleoprotein structures at the bacteriophage lambda replication origin during the initiation of DNA replication. *J. Biol. Chem.* **264:**10699–10708.

Ali, S. S., Beckett, E., Bae, S. J., and Navarre, W. W. (2011). The 5.5 protein of phage T7 inhibits H-NS through interactions with the central oligomerization domain. *J. Bacteriol* **193:**4881–4892.

Anderson, D. L., Hickman, D. D., and Reilly, B. E. (1966). Structure of *Bacillus subtilis* bacteriophage phi 29 and the length of phi 29 deoxyribonucleic acid. *J. Bacteriol.* **91:**2081–2089.

Anderson, W. F., Takeda, Y., Echols, H., and Matthews, B. W. (1979). The structure of a repressor: Crystallographic data for the Cro regulatory protein of bacteriophage lambda. *J. Mol. Biol.* **130:**507–510.

Ang, D., Keppel, F., Klein, G., Richardson, A., and Georgopoulos, C. (2000). Genetic analysis of bacteriophage-encoded cochaperonins. *Annu. Rev. Genet.* **34:**439–456.

Asensio, J. L., Albert, A., Muñoz-Espín, D., González, C., Hermoso, J., Villar, L., Jiménez-Barbero, J., Salas, M., and Meijer, W. J. (2005). Structure of the functional domain of phi29 replication organizer: Insights into oligomerization and DNA binding. *J. Biol. Chem.* **280:**20730–20739.

Atanasiu, C., Byron, O., McMiken, H., Sturrock, S. S., and Dryden, D. T. (2001). Characterisation of the structure of ocr, the gene 0.3 protein of bacteriophage T7. *Nucleic Acids Res.* **29:**3059–3068.

Atanasiu, C., Su, T. J., Sturrock, S. S., and Dryden, D. T. (2002). Interaction of the ocr gene 0.3 protein of bacteriophage T7 with EcoKI restriction/modification enzyme. *Nucleic Acids Res* **30:**3936–3944.

Ayers, D. J., Sunshine, M. G., Six, E. W., and Christie, G. E. (1994). Mutations affecting two adjacent amino acid residues in the alpha subunit of RNA polymerase block transcriptional activation by the bacteriophage P2 Ogr protein. *J. Bacteriol.* **176:**7430–7438.

Backhaus, H., and Petri, J. B. (1984). Sequence analysis of a region from the early right operon in phage P22 including the replication genes 18 and 12. *Gene* **32:**289–303.

Badia, D., Camacho, A., Pérez-Lago, L., Escandón, C., Salas, M., and Coll, M. (2006). The structure of phage phi29 transcription regulator p4-DNA complex reveals an N-hook motif for DNA. *Mol. Cell.* **22:**73–81.

Ball, C. A., and Johnson, R. C. (1991). Efficient excision of phage lambda from the *Escherichia coli* chromosome requires the Fis protein. *J. Bacteriol.* **173:**4027–4031.

Barrett, K. J., Marsh, M. L., and Calendar, R. (1976). Interactions between a satellite bacteriophage and its helper. *J. Mol. Biol.* **106:**683–707.

Bartel, P. L., Roecklein, J. A., SenGupta, D., and Fields, S. (1996). A protein linkage map of Escherichia coli bacteriophage T7. *Nat. Genet.* **12:**72–77.

Barthelemy, I., and Salas, M. (1989). Characterization of a new prokaryotic transcriptional activator and its DNA recognition site. *J. Mol. Biol.* **208:**225–232.

Bazinet, C., Benbasat, J., King, J., Carazo, J. M., and Carrascosa, J. L. (1988). Purification and organization of the gene 1 portal protein required for phage P22 DNA packaging. *Biochemistry* **27:**1849–1856.

Bell, C. E., and Lewis, M. (2001). Crystal structure of the lambda repressor C-terminal domain octamer. *J. Mol. Biol.* **314:**1127–1136.

Bertani, G. (1951). Studies on lysogenesis. I. The mode of phage liberation by lysogenic *Escherichia coli. J. Bacteriol.* **62:**293–300.

Bertani, L. E., and Six, E. (1988). The P2-like phages and their parasite, P4. (R. Calendar, ed.), pp. 73-143. Plenum Press, New York.

Bienkowska-Szewczyk, K., Lipinska, B., and Taylor, A. (1981). The R gene product of bacteriophage lambda is the murein transglycosylase. *Mol. Gen. Genet.* **184:**111–114.

Blanco, L., Gutierrez, J., Lazaro, J. M., Bernad, A., and Salas, M. (1986). Replication of phage phi 29 DNA *in vitro*: Role of the viral protein p6 in initiation and elongation. *Nucleic Acids Res.* **14:**4923–4937.

Blanco, L., Prieto, I., Gutierrez, J., Bernad, A., Lazaro, J. M., Hermoso, J. M., and Salas, M. (1987). Effect of NH4+ ions on phi 29 DNA-protein p3 replication: Formation of a complex between the terminal protein and the DNA polymerase. *J. Virol.* **61:**3983–3991.

Blanco, L., and Salas, M. (1984). Characterization and purification of a phage phi 29-encoded DNA polymerase required for the initiation of replication. *Proc. Natl. Acad. Sci. USA* **81:**5325–5329.

Botstein, D., Waddell, C. H., and King, J. (1973). Mechanism of head assembly and DNA encapsulation in *Salmonella* phage P22. I. Genes, proteins, structures and DNA maturation. *J. Mol. Biol* **80:**669–695.

Bowden, D. W., and Modrich, P. (1985). *In vitro* maturation of circular bacteriophage P2 DNA: Purification of ter components and characterization of the reaction. *J. Biol. Chem.* **260:**6999–7007.

Bowie, J. U., and Sauer, R. T. (1989). Equilibrium dissociation and unfolding of the Arc repressor dimer. *Biochemistry* **28:**7139–7143.
Bravo, A., Illana, B., and Salas, M. (2000). Compartmentalization of phage phi29 DNA replication: Interaction between the primer terminal protein and the membrane-associated protein p1. *EMBO J.* **19:**5575–5584.
Bravo, A., and Salas, M. (1997). Initiation of bacteriophage phi29 DNA replication *in vivo*: assembly of a membrane-associated multiprotein complex. *J. Mol. Biol.* **269:**102–112.
Bravo, A., and Salas, M. (1998). Polymerization of bacteriophage phi 29 replication protein p1 into protofilament sheets. *EMBO J.* **17:**6096–6105.
Breg, J. N., van Opheusden, J. H., Burgering, M. J., Boelens, R., and Kaptein, R. (1990). Structure of Arc repressor in solution: Evidence for a family of beta- sheet DNA-binding proteins. *Nature* **346:**586–589.
Breitbart, M., Salamon, P., Andresen, B., Mahaffy, J. M., Segall, A. M., Mead, D., Azam, F., and Rohwer, F. (2002). Genomic analysis of uncultured marine viral communities. *Proc. Natl Acad. Sci. USA* **99:**14250–14255.
Briani, F., Deho, G., Forti, F., and Ghisotti, D. (2001). The plasmid status of satellite bacteriophage P4. *Plasmid* **45:**1–17.
Broadbent, S. E., Davies, M. R., and van der Woude, M. W. (2010). Phase variation controls expression of Salmonella lipopolysaccharide modification genes by a DNA methylation-dependent mechanism. *Mol. Microbiol.* **77:**337–353.
Brown, B. M., Bowie, J. U., and Sauer, R. T. (1990). Arc repressor is tetrameric when bound to operator DNA. *Biochemistry* **29:**11189–11195.
Buchwald, M., Murialdo, H., and Siminovitch, L. (1970a). The morphogenesis of bacteriophage lambda. II. Identification of the principal structural proteins. *Virology* **42:**390–400.
Buchwald, M., and Siminovitch, L. (1969). Production of serum-blocking material by mutants of the left arm of the lambda chromosome. *Virology* **38:**1–7.
Buchwald, M., Steed-Glaister, P., and Siminovitch, L. (1970b). The morphogenesis of bacteriophage lambda. I. Purification and characterization of lambda heads and lambda tails. *Virology* **42:**375–389.
Calendar, R. (2006). "The Bacteriophages" 2nd Ed. Oxford University Press, Oxford.
Calendar, R., Lindqvist, B., Sironi, G., and Clark, A. (1970). Characterization of REP- mutants and their interaction with P2 phage. *Virology* **40:**72–83.
Calles, B., Salas, M., and Rojo, F. (2002). The phi29 transcriptional regulator contacts the nucleoid protein p6 to organize a repression complex. *EMBO J.* **21:**6185–6194.
Camacho, A., and Salas, M. (2001). Mechanism for the switch of phi29 DNA early to late transcription by regulatory protein p4 and histone-like protein p6. *EMBO J.* **20:**6060–6070.
Camara, B., Liu, M., Reynolds, J., Shadrin, A., Liu, B., Kwok, K., Simpson, P., Weinzierl, R., Severinov, K., Cota, E., Matthews, S., and Wigneshweraraj, S. R. (2010). T7 phage protein Gp2 inhibits the *Escherichia coli* RNA polymerase by antagonizing stable DNA strand separation near the transcription start site. *Proc. Natl. Acad. Sci. USA* **107:**2247–2252.
Campbell, A., del-Campillo-Campbell, A., and Ginsberg, M. L. (2002). Specificity in DNA recognition by phage integrases. *Gene* **300:**13–18.
Cardarelli, L., Lam, R., Tuite, A., Baker, L. A., Sadowski, P. D., Radford, D. R., Rubinstein, J. L., Battaile, K. P., Chirgadze, N., Maxwell, K. L., and Davidson, A. R. (2010). The crystal structure of bacteriophage HK97 gp6: Defining a large family of head-tail connector proteins. *J. Mol. Biol.* **395:**754–768.
Casjens, S. (1974). Bacteriophage lambda FII gene protein: Role in head assembly. *J. Mol. Biol.* **90:**1–20.
Casjens, S. (1979). Molecular organization of the bacteriophage P22 coat protein shell. *J. Mol. Biol.* **131:**1–14.
Casjens, S. (2011). Evolution of mosaically related tailed bacteriophage genomes seen through the lens of phage P22 virion assembly. Virology, in press.

Casjens, S., Eppler, K., Parr, R., and Poteete, A. R. (1989). Nucleotide sequence of the bacteriophage P22 gene 19 to 3 region: Identification of a new gene required for lysis. *Virology* **171:**588–598.

Casjens, S., and Hayden, M. (1988). Analysis *in vivo* of the bacteriophage P22 headful nuclease. *J. Mol. Biol.* **199:**467–474.

Casjens, S., Horn, T., and Kaiser, A. D. (1972). Head assembly steps controlled by genes F and W in bacteriophage lambda. *J. Mol. Biol.* **64:**551–563.

Casjens, S. R. (2005). Comparative genomics and evolution of the tailed-bacteriophages. *Curr. Opin. Microbiol.* **8:**451–458.

Casjens, S. R., and Hendrix, R. W. (1974). Locations and amounts of major structural proteins in bacteriophage lambda. *J. Mol. Biol.* **88:**535–545.

Casjens, S. R., and Thuman-Commike, P. A. (2011). Evolution of mosaically related tailed bacteriophage genomes seen through the lens of phage P22 virion assembly. *Virology* **411:**393–415.

Catalano, C. E. (2000). The terminase enzyme from bacteriophage lambda: A DNA-packaging machine. *Cell. Mol. Life Sci.* **57:**128–148.

Catalano, C. E., and Tomka, M. A. (1995). Role of gpFI protein in DNA packaging by bacteriophage lambda. *Biochemistry* **34:**10036–10042.

Cerritelli, M. E., Conway, J. F., Cheng, N., Trus, B. L., and Steven, A. C. (2003a). Molecular mechanisms in bacteriophage T7 procapsid assembly, maturation, and DNA containment. *Adv. Protein Chem.* **64:**301–323.

Cerritelli, M. E., and Studier, F. W. (1996a). Assembly of T7 capsids from independently expressed and purified head protein and scaffolding protein. *J. Mol. Biol.* **258:**286–298.

Cerritelli, M. E., and Studier, F. W. (1996b). Purification and characterization of T7 head-tail connectors expressed from the cloned gene. *J. Mol. Biol.* **258:**299–307.

Cerritelli, M. E., Trus, B. L., Smith, C. S., Cheng, N., Conway, J. F., and Steven, A. C. (2003b). A second symmetry mismatch at the portal vertex of bacteriophage T7: 8-fold symmetry in the procapsid core. *J. Mol. Biol.* **327:**1–6.

Chang, C. Y., Kemp, P., and Molineux, I. J. (2010). Gp15 and gp16 cooperate in translocating bacteriophage T7 DNA into the infected cell. *Virology* **398:**176–186.

Chang, J. R., Poliakov, A., Prevelige, P. E., Mobley, J. A., and Dokland, T. (2008). Incorporation of scaffolding protein gpO in bacteriophages P2 and P4. *Virology* **370:**352–361.

Chang, J. R., Spilman, M. S., Rodenburg, C. M., and Dokland, T. (2009). Functional domains of the bacteriophage P2 scaffolding protein: Identification of residues involved in assembly and protease activity. *Virology* **384:**144–150.

Cheetham, G. M., Jeruzalmi, D., and Steitz, T. A. (1999). Structural basis for initiation of transcription from an RNA polymerase-promoter complex. *Nature* **399:**80–83.

Cheetham, G. M., and Steitz, T. A. (1999). Structure of a transcribing T7 RNA polymerase initiation complex. *Science* **286:**2305–2309.

Chen, D. H., Baker, M. L., Hryc, C. F., DiMaio, F., Jakana, J., Wu, W., Dougherty, M., Haase-Pettingell, C., Schmid, M. F., Jiang, W., Baker, D., King, J. A., *et al.* (2011). Structural basis for scaffolding-mediated assembly and maturation of a dsDNA virus. *Proc. Natl. Acad. Sci. USA* **108:**1355–1360.

Chen, K. C., Vannais, D. B., Jones, C., Patterson, D., and Davidson, J. N. (1989). Mapping of the gene encoding the multifunctional protein carrying out the first three steps of pyrimidine biosynthesis to human chromosome 2. *Hum. Genet.* **82:**40–44.

Chen, Y. C., Rajagopala, S. V., Stellberger, T., and Uetz, P. (2010). Exhaustive benchmarking of the yeast two-hybrid system. *Nature Methods* **7:**667–668.

Cheng, H., Shen, N., Pei, J., and Grishin, N. V. (2004). Double-stranded DNA bacteriophage prohead protease is homologous to herpesvirus protease. *Protein Sci.* **13:**2260–2269.

Cheng, X., Wang, W., and Molineux, I. J. (2004). F exclusion of bacteriophage T7 occurs at the cell membrane. *Virology* **326:**340–352.

Cho, E. H., Gumport, R. I., and Gardner, J. F. (2002). Interactions between integrase and excisionase in the phage lambda excisive nucleoprotein complex. *J. Bacteriol.* **184:**5200–5203.

Cho, E. H., Nam, C. E., Alcaraz, R., Jr., and Gardner, J. F. (1999). Site-specific recombination of bacteriophage P22 does not require integration host factor. *J. Bacteriol.* **181:**4245–4249.

Christie, G., Anders, D., McAlister, V., Goodwin, T., Julien, B., and Calendar, R. (2003). Identification of upstream sequences essential for activation of a bacteriophage P2 late promoter. *J Bacteriol* **185:**4609–4614.

Christie, G., Temple, L., Bartlett, B., and Goodwin, T. (2002). Programmed translational frameshift in the bacteriophage P2 FETUD tail gene operon. *J Bacteriol* **184:**6522–6531.

Christie, G. E., and Calendar, R. (1990). Interactions between satellite bacteriophage P4 and its helpers. *Annu. Rev. Genet.* **24:**465–490.

Clement, J. M., Lepouce, E., Marchal, C., and Hofnung, M. (1983). Genetic study of a membrane protein: DNA sequence alterations due to 17 lamB point mutations affecting adsorption of phage lambda. *EMBO J.* **2:**77–80.

Cone, R., Bonura, T., and Friedberg, E. C. (1980). Inhibitor of uracil-DNA glycosylase induced by bacteriophage PBS2: Purification and preliminary characterization. *J. Biol. Chem.* **255:**10354–10358.

Cortines, J. R., Weigele, P. R., Gilcrease, E. B., Casjens, S. R., and Teschke, C. M. (2011). Decoding bacteriophage P22 assembly: Identification of two charged residues in scaffolding protein responsible for coat protein interaction. *Virology* **421:**1–11.

Court, D. L., Oppenheim, A. B., and Adhya, S. L. (2007). A new look at bacteriophage lambda genetic networks. *J. Bacteriol.* **189:**298–304.

Crisona, N. J., Weinberg, R. L., Peter, B. J., Sumners, D. W., and Cozzarelli, N. R. (1999). The topological mechanism of phage lambda integrase. *J. Mol. Biol.* **289:**747–775.

Crucitti, P., Abril, A. M., and Salas, M. (2003). Bacteriophage phi 29 early protein p17. Self-association and hetero-association with the viral histone-like protein p6. *J. Biol. Chem* **278:**4906–4911.

Crucitti, P., Lazaro, J. M., Benes, V., and Salas, M. (1998). Bacteriophage phi29 early protein p17 is conditionally required for the first rounds of viral DNA replication. *Gene* **223:**135–142.

d'Herelle, F. (1926). The Bacteriophage and Its Behavior. Williams & Wilkins, Baltimore, MD: 490–497.

Darling, P. J., Holt, J. M., and Ackers, G. K. (2000). Coupled energetics of lambda cro repressor self-assembly and site-specific DNA operator binding I: Analysis of cro dimerization from nanomolar to micromolar concentrations. *Biochemistry* **39:**11500–11507.

Das, A. (1992). How the phage lambda N gene product suppresses transcription termination: communication of RNA polymerase with regulatory proteins mediated by signals in nascent RNA. *J. Bacteriol.* **174:**6711–6716.

Datta, A. B., Roy, S., and Parrack, P. (2005a). Role of C-terminal residues in oligomerization and stability of lambda CII: Implications for lysis-lysogeny decision of the phage. *J. Mol. Biol.* **345:**315–324.

Datta, I., Banik-Maiti, S., Adhikari, L., Sau, S., Das, N., and Mandal, N. C. (2005b). The mutation that makes *Escherichia coli* resistant to lambda P gene-mediated host lethality is located within the DNA initiator Gene dnaA of the bacterium. *J. Biochem. Mol. Biol.* **38:**89–96.

Datta, I., Sau, S., Sil, A. K., and Mandal, N. C. (2005c). The bacteriophage lambda DNA replication protein P inhibits the oriC DNA- and ATP-binding functions of the DNA replication initiator protein DnaA of *Escherichia coli*. *J. Biochem. Mol. Biol.* **38:**97–103.

De Anda, J., Poteete, A. R., and Sauer, R. T. (1983). P22 c2 repressor. Domain structure and function. *J. Biol. Chem* **258:**10536–10542.

de Beer, T., Fang, J., Ortega, M., Yang, Q., Maes, L., Duffy, C., Berton, N., Sippy, J., Overduin, M., Feiss, M., and Catalano, C. E. (2002). Insights into specific DNA recognition during the assembly of a viral genome packaging machine. *Mol. Cell.* **9:**981–991.

de Beus, M. D., Doyle, S. M., and Teschke, C. M. (2000). GroEL binds a late folding intermediate of phage P22 coat protein. *Cell Stress Chaperones* **5:**163–172.

de Jong, R. N., and van der Vliet, P. C. (1999). Mechanism of DNA replication in eukaryotic cells: Cellular host factors stimulating adenovirus DNA replication. *Gene* **236:**1–12.

Deho, G., and Ghisotti, D. (2006). The satellite phage P4. (R. Calendar, ed.), pp. 391-408. Oxford University Press, New York.

Deighan, P., Diez, C. M., Leibman, M., Hochschild, A., and Nickels, B. E. (2008). The bacteriophage lambda Q antiterminator protein contacts the beta-flap domain of RNA polymerase. *Proc. Natl. Acad. Sci. USA* **105:**15305–15310.

Demerec, M., and Fano, U. (1945). Bacteriophage-resistant mutants in *Escherichia coli. Genetics* **30:**119–136.

Doan, D. N. P., and Dokland, T. (2007). The gpQ portal protein of bacteriophage P2 forms dodecameric connctors in crystals. *J. Struct. Biol.* **157:**432–436.

Dodd, I. B., Perkins, A. J., Tsemitsidis, D., and Egan, J. B. (2001). Octamerization of lambda CI repressor is needed for effective repression of P(RM) and efficient switching from lysogeny. *Genes Dev.* **15:**3013–3022.

Dodson, M., Echols, H., Wickner, S., Alfano, C., Mensa-Wilmot, K., Gomes, B., LeBowitz, J., Roberts, J. D., and McMacken, R. (1986). Specialized nucleoprotein structures at the origin of replication of bacteriophage lambda: Localized unwinding of duplex DNA by a six-protein reaction. *Proc. Natl. Acad. Sci. USA* **83:**7638–7642.

Dokland, T. (2011). gpO peptidase (Enterobacteria phage P2). *In* "Handbook of proteolytic enzymes" (N. D. Rawlings and G. Salvesen, eds.). Elsevier, Oxford.

Dokland, T., Isaksen, M. L., Fuller, S. D., and Lindqvist, B. H. (1993). Capsid localization of the bacteriophage P4 Psu protein. *Virology* **194:**682–687.

Dokland, T., Lindqvist, B. H., and Fuller, S. D. (1992). Image reconstruction from cryo-electron micrographs reveals the morphopoietic mechanism in the P2-P4 bacteriophage system. *EMBO J. Eur. Mol. Biol. Organ. J* **11:**839–846.

Dokland, T., and Murialdo, H. (1993). Structural transitions during maturation of bacteriophage lambda capsids. *J. Mol. Biol.* **233:**682–694.

Dokland, T., Wang, S., and Lindqvist, B. H. (2002). The structure of P4 procapsids produced by coexpression of capsid and external scaffolding proteins. *Virology* **298:**224–231.

Duffy, C., and Feiss, M. (2002). The large subunit of bacteriophage lambda's terminase plays a role in DNA translocation and packaging termination. *J. Mol. Biol.* **316:**547–561.

Dufour, E., Méndez, J., Lázaro, J. M., de Vega, M., Blanco, L., and Salas, M. (2000). An aspartic acid residue in TPR-1, a specific region of protein-priming DNA polymerases, is required for the functional interaction with primer terminal protein. *J. Mol. Biol.* **304:**289–300.

Dunn, J. J., and Studier, F. W. (1983). Complete nucleotide sequence of bacteriophage T7 DNA and the locations of T7 genetic elements. *J. Mol. Biol.* **166:**477–535.

Echols, H. (1986). Multiple DNA-protein interactions governing high-precision DNA transactions. *Science* **233:**1050–1056.

Elías-Arnanz, M., and Salas, M. (1999). Functional interactions between a phage histone-like protein and a transcriptional factor in regulation of phi29 early-late transcriptional switch. *Genes Dev.* **13:**2502–2513.

Eppler, K., Wyckoff, E., Goates, J., Parr, R., and Casjens, S. (1991). Nucleotide sequence of the bacteriophage P22 genes required for DNA packaging. *Virology* **183:**519–538.

Eriksson, J. M., and Haggard-Ljungquist, E. (2000). The multifunctional bacteriophage P2 cox protein requires oligomerization for biological activity. *J. Bacteriol.* **182:**6714–6723.

Eriksson, S. K., Liu, T., and Haggard-Ljungquist, E. (2000). Interacting interfaces of the P4 antirepressor E and the P2 immunity repressor C. *Mol. Microbiol.* **36:**1148–1155.

Esposito, D., and Gerard, G. F. (2003). The *Escherichia coli* Fis protein stimulates bacteriophage lambda integrative recombination *in vitro*. *J. Bacteriol.* **185:**3076–3080.
Frackman, S., Siegele, D. A., and Feiss, M. (1985). The terminase of bacteriophage lambda: Functional domains for cosB binding and multimer assembly. *J. Mol. Biol.* **183:**225–238.
Franklin, N. C. (1985). Conservation of genome form but not sequence in the transcription antitermination determinants of bacteriophages lambda, phi 21 and P22. *J. Mol. Biol.* **181:**75–84.
Freire, R., Serrano, M., Salas, M., and Hermoso, J. M. (1996). Activation of replication origins in phi29-related phages requires the recognition of initiation proteins to specific nucleoprotein complexes. *J. Biol. Chem.* **271:**31000–31007.
Friedman, D. I., and Court, D. L. (2001). Bacteriophage lambda: Alive and well and still doing its thing. *Curr. Opin. Microbiol.* **4:**201–207.
Frumerie, C. M., Eriksson, J., Dugast, M., and Haggard-Ljungquist, E. (2005). Dimerization of bacteriophage P2 integrase is not required for binding to its DNA target but for its biological activity. *Gene* **344:**221–231.
Fujiki, H., Palm, P., Zillig, W., Calendar, R., and Sunshine, M. (1976). Identification of a mutation within the structural gene for the a subunit of DNA-dependent RNA polymerase of *E. coli. Mol. Gen. Genet* **145:**19–22.
Fujisawa, H., Shibata, H., and Kato, H. (1991). Analysis of interactions among factors involved in the bacteriophage T3 DNA packaging reaction in a defined *in vitro* system. *Virology* **185:**788–794.
Fujiyama, A., Kohara, Y., and Okazaki, T. (1981). Initiation sites for discontinuous DNA synthesis of bacteriophage T7. *Proc. Natl. Acad. Sci. USA* **78:**903–907.
Fuller, M. T., and King, J. (1982). Assembly *in vitro* of bacteriophage P22 procapsids from purified coat and scaffolding subunits. *J. Mol. Biol.* **156:**633–665.
Funnell, B. E., and Inman, R. B. (1983). Bacteriophage P2 DNA replication: Characterization of the requirement of the gene B protein *in vivo*. *J. Mol. Biol.* **167:**311–334.
García, J. A., Carrascosa, J. L., and Salas, M. (1983). Assembly of the tail protein of the *Bacillus subtilis* phage phi 29. *Virology* **125:**18–30.
Garcia, L. R., and Molineux, I. J. (1995). Rate of translocation of bacteriophage T7 DNA across the membranes of *Escherichia coli*. *J. Bacteriol.* **177:**4066–4076.
Garcia, L. R., and Molineux, I. J. (1996). Transcription-independent DNA translocation of bacteriophage T7 DNA into *Escherichia coli*. *J. Bacteriol.* **178:**6921–6929.
Gascón, I., Gutiérrez, C., and Salas, M. (2000). Structural and functional comparative study of the complexes formed by viral o29, Nf and GA-1 SSB proteins with DNA. *J. Mol. Biol.* **296:**989–999.
Gavin, A. C., Aloy, P., Grandi, P., Krause, R., Boesche, M., Marzioch, M., Rau, C., Jensen, L. J., Bastuck, S., Dumpelfeld, B., Edelmann, A., Heurtier, M. A., *et al.* (2006). Proteome survey reveals modularity of the yeast cell machinery. *Nature* **440:**631–636.
Geisselsoder, J., Youdarian, P., Deho, G., Chidambaram, M., Goldstein, R., and Ljungquist, E. (1981). Mutants of satellite virus P4 that cannot derepress their bacteriophage P2 helper. *J. Mol. Biol.* **148:**1–19.
Georgopoulos, C., Tilly, K., and Casjens, S. (1983). Lambdoid phage head assembly. *In* "Lambda II" (R. Hendrix, J. Roberts, F. W. Stahl, and R. Weisberg, eds.), pp. 279–304. Cold Spring Harbor Laboratory, Cold Spring Harbor, NY.
Georgopoulos, C. P., Hendrix, R. W., Casjens, S. R., and Kaiser, A. D. (1973). Host participation in bacteriophage lambda head assembly. *J. Mol. Biol.* **76:**45–60.
Ghosh, S., Hamdan, S. M., Cook, T. E., and Richardson, C. C. (2008). Interactions of *Escherichia coli* thioredoxin, the processivity factor, with bacteriophage T7 DNA polymerase and helicase. *J. Biol. Chem.* **283:**32077–32084.
Ghosh, S., Hamdan, S. M., and Richardson, C. C. (2010). Two modes of interaction of the single-stranded DNA-binding protein of bacteriophage T7 with the DNA polymerase-thioredoxin complex. *J. Biol. Chem.* **285:**18103–18112.

Gilcrease, E. B., Winn-Stapley, D. A., Hewitt, F. C., Joss, L., and Casjens, S. R. (2005). Nucleotide sequence of the head assembly gene cluster of bacteriophage L and decoration protein characterization. *J. Bacteriol.* **187:**2050–2057.

González-Huici, V., Lázaro, J. M., Salas, M., and Hermoso, J. M. (2000). Specific recognition of parental terminal protein by DNA polymerase for initiation of protein-primed DNA replication. *J. Biol. Chem.* **275:**14678–14683.

Gottesman, M. E., and Weisberg, R. A. (2004). Little lambda, who made thee? *Microbiol. Mol. Biol. Rev.* **68:**796–813.

Gottesman, S., Gottesman, M., Shaw, J. E., and Pearson, M. L. (1981). Protein degradation in *E. coli*: The lon mutation and bacteriophage lambda N and cII protein stability. *Cell* **24:**225–233.

Greene, B., and King, J. (1994). Binding of scaffolding subunits within the P22 procapsid lattice. *Virology* **205:**188–197.

Greene, B., and King, J. (1996). Scaffolding mutants identifying domains required for P22 procapsid assembly and maturation. *Virology* **225:**82–96.

Grimes, S., and Anderson, D. (1997). The bacteriophage phi29 packaging proteins supercoil the DNA ends. *J. Mol. Biol.* **266:**901–914.

Grundling, A., Smith, D. L., Blasi, U., and Young, R. (2000). Dimerization between the holin and holin inhibitor of phage lambda. *J. Bacteriol.* **182:**6075–6081.

Guasch, A., Pous, J., Ibarra, B., Gomis-Ruth, F. X., Valpuesta, J. M., Sousa, N., Carrascosa, J. L., and Coll, M. (2002). Detailed architecture of a DNA translocating machine: The high-resolution structure of the bacteriophage phi29 connector particle. *J. Mol. Biol.* **315:**663–676.

Guo, P., Peterson, C., and Anderson, D. (1987). Initiation events in *in vitro* packaging of bacteriophage phi 29 DNA-gp3. *J. Mol. Biol.* **197:**219–228.

Guo, P. X., Erickson, S., Xu, W., Olson, N., Baker, T. S., and Anderson, D. (1991). Regulation of the phage phi 29 prohead shape and size by the portal vertex. *Virology* **183:**366–373.

Hadden, J. M., Declais, A. C., Carr, S. B., Lilley, D. M., and Phillips, S. E. (2007). The structural basis of Holliday junction resolution by T7 endonuclease I. *Nature* **449:**621–624.

Haggard-Ljungquist, E., Halling, C., and Calendar, R. (1992). DNA sequences of the tail fiber genes of bacteriophage P2: evidence for horizontal transfer of tail fiber genes among unrelated bacteriophages. *J Bacteriol* **174:**1462–1477.

Haggard-Ljungquist, E., Jacobsen, E., Rishovd, S., Six, E. W., Nilssen, O., Sunshine, M. G., Lindqvist, B. H., Kim, K. J., Barreiro, V., Koonin, E. V., *et al.* (1995). Bacteriophage P2: Genes involved in baseplate assembly. *Virology* **213:**109–121.

Halder, S., Datta, A. B., and Parrack, P. (2007). Probing the antiprotease activity of lambda-CIII, an inhibitor of the *Escherichia coli* metalloprotease HflB (FtsH). *J. Bacteriol.* **189:**8130–8138.

Hamada, K., Fujisawa, H., and Minagawa, T. (1986). Overproduction and purification of the products of bacteriophage T3 genes 18 and 19, two genes involved in DNA packaging. *Virology* **151:**110–118 .

Hamdan, S. M., Marintcheva, B., Cook, T., Lee, S. J., Tabor, S., and Richardson, C. C. (2005). A unique loop in T7 DNA polymerase mediates the binding of helicase-primase, DNA binding protein, and processivity factor. *Proc. Natl. Acad. Sci. USA* **102:**5096–5101.

Hamdan, S. M., and Richardson, C. C. (2009). Motors, switches, and contacts in the replisome. *Annu. Rev. Biochem.* **78:**205–243.

Hang, Q., Woods, L., Feiss, M., and Catalano, C. E. (1999). Cloning, expression, and biochemical characterization of hexahistidine-tagged terminase proteins. *J. Biol. Chem.* **274:**15305–15314.

Häuser, R., Sabri, M., Moineau, S., and Uetz, P. (2011). The proteome and interactome of Streptococcus pneumoniae phage Cp-1. *J. Bacteriol.* **193:**3135–3138.

He, Z. G., Rezende, L. F., Willcox, S., Griffith, J. D., and Richardson, C. C. (2003). The carboxyl-terminal domain of bacteriophage T7 single-stranded DNA-binding protein modulates DNA binding and interaction with T7 DNA polymerase. *J. Biol. Chem.* **278:**29538–29545.

He, Z. G., and Richardson, C. C. (2004). Effect of single-stranded DNA-binding proteins on the helicase and primase activities of the bacteriophage T7 gene 4 protein. *J. Biol. Chem.* **279:**22190–22197.

Heineman, R. H., and Bull, J. J. (2007). Testing optimality with experimental evolution: Lysis time in a bacteriophage. *Evolution* **61:**1695–1709.

Hendrix, R., and Casjens, S. (2006). Bacteriophage lambda and its genetic neighborhood. Chapter 27. *In* "The Bacteriophages" (R. Calendar, ed.), pp. 409–445. Oxford University Press, Oxford.

Hendrix, R., Roberts, J., Stahl, F. W., and Weisberg, R. (1983). Lambda II. p. 694. CSHL Press, Cold Spring Harbor, NY.

Hendrix, R. W. (2002). Bacteriophages: Evolution of the majority. *Theor. Popul. Biol.* **61:**471–480.

Hendrix, R. W. (2005). Bacteriophage HK97: Assembly of the capsid and evolutionary connections. *Adv. Virus Res.* **64:**1–14.

Hendrix, R. W., and Casjens, S. R. (1974). Protein fusion: A novel reaction in bacteriophage lambda head assembly. *Proc. Natl. Acad. Sci. USA* **71:**1451–1455.

Hendrix, R. W., and Casjens, S. R. (1975). Assembly of bacteriophage lambda heads: Protein processing and its genetic control in petit lambda assembly. *J. Mol. Biol.* **91:**187–199.

Hendrix, R. W., and Duda, R. L. (1992). Bacteriophage lambda PaPa: Not the mother of all lambda phages. *Science* **258:**1145–1148.

Hendrix, R. W., and Duda, R. L. (1998). Bacteriophage HK97 head assembly: A protein ballet. *Adv. Virus Res.* **50:**235–288.

Herman, C., Thevenet, D., D'Ari, R., and Bouloc, P. (1997). The HflB protease of *Escherichia coli* degrades its inhibitor lambda cIII. *J. Bacteriol* **179:**358–363.

Hermoso, J. M., Méndez, E., Soriano, F., and Salas, M. (1985). Location of the serine residue involved in the linkage between the terminal protein and the DNA of phage phi 29. *Nucleic Acids Res.* **13:**7715–7728.

Hershey, A. D. (1952). Inheritance in bacteriophage. *Ann. N.Y. Acad. Sci* **54:**960–962.

Ho, Y., and Rosenberg, M. (1982). Characterization of the phage lambda regulatory protein cII. *Ann. Microbiol.* **133:**215–218.

Ho, Y. S., Pfarr, D., Strickler, J., and Rosenberg, M. (1992). Characterization of the transcription activator protein C1 of bacteriophage P22. *J. Biol. Chem.* **267:**14388–14397.

Hohn, T., Flick, H., and Hohn, B. (1975). Petit lambda, a family of particles from coliphage lambda infected cells. *J. Mol. Biol.* **98:**107–120.

Horcajadas, J. A., Monsalve, M., Rojo, F., and Salas, M. (1999). The switch from early to late transcription in phage GA-1: Characterization of the regulatory protein p4G. *J. Mol. Biol.* **290:**917–928.

Hu, P., Janga, S. C., Babu, M., Diaz-Mejia, J. J., Butland, G., Yang, W., Pogoutse, O., Guo, X., Phanse, S., Wong, P., Chandran, S., Christopoulos, C., *et al.* (2009). Global functional atlas of *Escherichia coli* encompassing previously uncharacterized proteins. *PLoS Biol.* **7:**e96.

Huang, J., Villemain, J., Padilla, R., and Sousa, R. (1999). Mechanisms by which T7 lysozyme specifically regulates T7 RNA polymerase during different phases of transcription. *J. Mol. Biol.* **293:**457–475.

Huber, H. E., Beauchamp, B. B., and Richardson, C. C. (1988). *Escherichia coli* dGTP triphosphohydrolase is inhibited by gene 1.2 protein of bacteriophage T7. *J. Biol. Chem* **263:**13549–13556.

Ibarra, B., Valpuesta, J. M., and Carrascosa, J. L. (2001). Purification and functional characterization of p16, the ATPase of the bacteriophage phi29 packaging machinery. *Nucleic Acids Res.* **29:**4264–4273.

Illana, B., Blanco, L., and Salas, M. (1996). Functional characterization of the genes coding for the terminal protein and DNA polymerase from bacteriophage GA-1: Evidence for a sliding-back mechanism during protein-primed GA-1 DNA replication. *J. Mol. Biol.* **264**:453–464.

Illana, B., Lázaro, J. M., Gutierrez, C., Meijer, W. J., Blanco, L., and Salas, M. (1999). Phage phi29 terminal protein residues Asn80 and Tyr82 are recognition elements of the replication origins. *J. Biol. Chem.* **274**:15073–15079.

Isaksen, M. L., Dokland, T., and Lindqvist, B. H. (1993). Characterization of the capsid associating activity of bacteriophage P4's Psu protein. *Virology* **194**:647–681.

Iwashita, S., and Kanegasaki, S. (1976). Enzymic and molecular properties of base-plate parts of bacteriophage P22. *Eur. J. Biochem.* **65**:87–94.

Jackson, E. N., Jackson, D. A., and Deans, R. J. (1978). EcoRI analysis of bacteriophage P22 DNA packaging. *J. Mol. Biol.* **118**:365–388.

Jeruzalmi, D., and Steitz, T. A. (1998). Structure of T7 RNA polymerase complexed to the transcriptional inhibitor T7 lysozyme. *EMBO J.* **17**:4101–4113.

Julien, B., Lefevre, P., and Calendar, R. (1997). The two P2 Ogr-like domains of the delta protein from bacteriophage P4 are required for activity. *Virology* **230**:292–299.

Kageyama, Y., Murayama, M., Onodera, T., Yamada, S., Fukada, H., Kudou, M., Tsumoto, K., Toyama, Y., Kado, S., Kubota, K., and Takeda, S. (2009). Observation of the membrane binding activity and domain structure of gpV, which comprises the tail spike of bacteriophage P2. *Biochemistry* **48**:10129–10135.

Kaiser, A. D., and Jacob, F. (1957). Recombination between related bacteriophages and the genetic control of immunity and prophage localization. *Virology* **4**:509–517.

Kamtekar, S., Berman, A. J., Wang, J., Lázaro, J. M., de Vega, M., Blanco, L., Salas, M., and Steitz, T. A. (2006). The phi29 DNA polymerase:protein-primer structure suggests a model for the initiation to elongation transition. *EMBO J.* **25**:1335–1343.

Katsura, I. (1983). Tail assembly and injection. *In* ""Lambda II" (R. Hendrix, J. Roberts, F. W. Stahl, and R. Weisberg, eds.), pp. 331–346. Cold Spring Harbor Laboratory, Cold Spring Harbor, NY.

Katsura, I. (1990). Mechanism of length determination in bacteriophage lambda tails. *Adv. Biophys.* **26**:1–18.

Katsura, I., and Tsugita, A. (1977). Purification and characterization of the major protein and the terminator protein of the bacteriophage lambda tail. *Virology* **76**:129–145.

Kędzierska, B., Lee, D. J., Węgrzyn, G., Busby, S. J., and Thomas, M. S. (2004). Role of the RNA polymerase alpha subunits in CII-dependent activation of the bacteriophage lambda pE promoter: identification of important residues and positioning of the alpha C-terminal domains. *Nucleic Acids Res.* **32**:834–841.

Kędzierska, B., Szambowska, A., Herman-Antosiewicz, A., Lee, D. J., Busby, S. J., Węgrzyn, G., and Thomas, M. S. (2007). The C-terminal domain of the *Escherichia coli* RNA polymerase alpha subunit plays a role in the CI-dependent activation of the bacteriophage lambda pM promoter. *Nucleic Acids Res.* **35**:2311–2320.

Kemp, P., Garcia, L. R., and Molineux, I. J. (2005). Changes in bacteriophage T7 virion structure at the initiation of infection. *Virology* **340**:307–317.

Kemp, P., Gupta, M., and Molineux, I. J. (2004). Bacteriophage T7 DNA ejection into cells is initiated by an enzyme-like mechanism. *Mol. Microbiol.* **53**:1251–1265.

Kennaway, C. K., Obarska-Kosinska, A., White, J. H., Tuszynska, I., Cooper, L. P., Bujnicki, J. M., Trinick, J., and Dryden, D. T. (2009). The structure of M.EcoKI Type I DNA methyltransferase with a DNA mimic antirestriction protein. *Nucleic Acids Res* **37**:762–770.

Kennedy, W. P., Momand, J. R., and Yin, Y. W. (2007). Mechanism for de novo RNA synthesis and initiating nucleotide specificity by t7 RNA polymerase. *J. Mol. Biol.* **370**:256–268.

Kihara, A., Akiyama, Y., and Ito, K. (2001). Revisiting the lysogenization control of bacteriophage lambda: Identification and characterization of a new host component. *HflD. J. Biol. Chem.* **276:**13695–13700.
Kim, K.-J., Sunshine, M. G., Lindqvist, B. H., and Six, E. W. (2001). Capsid size determination in the P2–P4 bacteriophage system: suppression of sir mutations in P2's capsid gene N by supersid mutations in P4's external scaffold gene sid. *Virology* **283:**49–58.
King, J., and Casjens, S. (1974). Catalytic head assembling protein in virus morphogenesis. *Nature* **251:**112–119.
King, J., Lenk, E. V., and Botstein, D. (1973). Mechanism of head assembly and DNA encapsulation in Salmonella phage P22. *II. Morphogenetic pathway. J. Mol. Biol.* **80:**697–731.
King, R. A., Anders, D. L., and Christie, G. E. (1992). Site-directed mutagenesis of an amino acid residue in the bacteriophage P2 ogr protein implicated in interaction with *Escherichia coli* RNA polymerase. *Mol. Microbiol.* **6:**3313–3320.
Kobiler, O., Koby, S., Teff, D., Court, D., and Oppenheim, A. B. (2002). The phage lambda CII transcriptional activator carries a C-terminal domain signaling for rapid proteolysis. *Proc. Natl. Acad. Sci. USA* **99:**14964–14969.
Kobiler, O., Oppenheim, A. B., and Herman, C. (2004). Recruitment of host ATP-dependent proteases by bacteriophage lambda. *J. Struct. Biol.* **146:**72–78.
Kobiler, O., Rokney, A., and Oppenheim, A. B. (2007). Phage lambda CIII: A protease inhibitor regulating the lysis-lysogeny decision. *PLoS One* **2:**e363.
Kochan, J., Carrascosa, J. L., and Murialdo, H. (1984). Bacteriophage lambda preconnectors. *Purification and structure. J. Mol. Biol.* **174:**433–447.
Kornitzer, D., Altuvia, S., and Oppenheim, A. B. (1991). The activity of the CIII regulator of lambdoid bacteriophages resides within a 24-amino acid protein domain. *Proc. Natl. Acad. Sci. USA* **88:**5217–5221.
Lander, G. C., Evilevitch, A., Jeembaeva, M., Potter, C. S., Carragher, B., and Johnson, J. E. (2008). Bacteriophage lambda stabilization by auxiliary protein gpD: Timing, location, and mechanism of attachment determined by cryo-EM. *Structure* **16:**1399–1406.
Lander, G. C., Tang, L., Casjens, S. R., Gilcrease, E. B., Prevelige, P., Poliakov, A., Potter, C. S., Carragher, B., and Johnson, J. E. (2006). The structure of an infectious P22 virion shows the signal for headful DNA packaging. *Science* **312:**1791–1795.
LeBowitz, J. H., and McMacken, R. (1986). The *Escherichia coli* dnaB replication protein is a DNA helicase. *J. Biol. Chem.* **261:**4738–4748.
Lee, C. S., and Guo, P. (1995). Sequential interactions of structural proteins in phage phi 29 procapsid assembly. *J. Virol.* **69:**5024–5032.
Leffers, G. G., Jr., and Gottesman, S. (1998). Lambda Xis degradation *in vivo* by Lon and FtsH. *J. Bacteriol.* **180:**1573–1577.
Lehne, B., and Schlitt, T. (2009). Protein-protein interaction databases: Keeping up with growing interactomes. *Hum. Genom.* **3:**291–297.
Lengyel, J. A., Goldstein, R. N., Marsh, M., and Calendar, R. (1974). Structure of the bacteriophage P2 tail. *Virology* **62:**161–174.
Lengyel, J. A., Goldstein, R. N., Marsh, M., Sunshine, M. G., and Calendar, R. (1973). Bacteriophage P2 head morphogenesis: cleavage of the major capsid protein. *Virology* **53:**1–23.
Li, M., Moyle, H., and Susskind, M. M. (1994). Target of the transcriptional activation function of phage lambda cI protein. *Science* **263:**75–77.
Lin, L. (1992). "Study of Bacteriophage T7 Gene 5.9 and Gene 5.5" SUNY, Stonybrook, NY.
Linderoth, N. A., Julien, B., Flick, K. E., Calendar, R., and Christie, G. (1994). Molecular cloning and characterization of bacteriophage P2 genes R and S involved in tail completion. *Virology* **200:**347–359.

Linderoth, N. A., Ziermann, R., Haggård-Ljungquist, E., Christie, G. E., and Calendar, R. (1991). Nucleotide sequence of the DNA packaging and capsid synthesis genes of bacteriophage P2. *Nucl. Acids Res.* **19:**7207–7214.

Lindqvist, B. H., Deho, G., and Calendar, R. (1993). Mechanisms of genome propagation and helper exploitation by satellite phage P4. *Microbiol. Rev.* **57:**683–702.

Little, J. W. (1967). An exonuclease induced by bacteriophage lambda. II. Nature of the enzymatic reaction. *J. Biol. Chem* **242:**679–686.

Liu, Q., and Richardson, C. C. (1993). Gene 5.5 protein of bacteriophage T7 inhibits the nucleoid protein H-NS of *Escherichia coli*. *Proc. Natl. Acad. Sci.*USA **90:**1761–1765.

Liu, T., Renberg, S. K., and Haggard-Ljungquist, E. (1998). The E protein of satellite phage P4 acts as an anti-repressor by binding to the C protein of helper phage P2. *Mol. Microbiol.* **30:**1041–1050.

Liu, X., Zhang, Q., Murata, K., Baker, M. L., Sullivan, M. B., Fu, C., Dougherty, M. T., Schmid, M. F., Osburne, M. S., Chisholm, S. W., and Chiu, W. (2010). Structural changes in a marine podovirus associated with release of its genome into Prochlorococcus. *Nat. Struct. Mol. Biol.* **17:**830–836.

Liu, Y., and Haggard-Ljungquist, E. (1994). Studies of bacteriophage P2 DNA replication: localization of the cleavage site of the A protein. *Nucleic Acids Res.* **22:**5204–5210.

Loeffler, J. M., Djurkovic, S., and Fischetti, V. A. (2003). Phage lytic enzyme Cpl-1 as a novel antimicrobial for pneumococcal bacteremia. *Infect. Immun.* **71:**6199–6204.

Loeffler, J. M., Nelson, D., and Fischetti, V. A. (2001). Rapid killing of *Streptococcus pneumoniae* with a bacteriophage cell wall hydrolase. *Science* **294:**2170–2172.

Longás, E., de Vega, M., Lázaro, J. M., and Salas, M. (2006). Functional characterization of highly processive protein-primed DNA polymerases from phages Nf and GA-1, endowed with a potent strand displacement capacity. *Nucleic Acids Res.* **34:**6051–6063.

Longás, E., Villar, L., Lázaro, J. M., de Vega, M., and Salas, M. (2008). Phage phi29 and Nf terminal protein-priming domain specifies the internal template nucleotide to initiate DNA replication. *Proc. Natl. Acad. Sci. USA* **105:**18290–18295.

Lopez, R., Ronda, C., Tomasz, A., and Portoles, A. (1977). Properties of "diplophage": A lipid-containing bacteriophage *J. Virol.* **24:**201–210.

Lyakhov, D. L., He, B., Zhang, X., Studier, F. W., Dunn, J. J., and McAllister, W. T. (1997). Mutant bacteriophage T7 RNA polymerases with altered termination properties. *J. Mol. Biol.* **269:**28–40.

Lyakhov, D. L., He, B., Zhang, X., Studier, F. W., Dunn, J. J., and McAllister, W. T. (1998). Pausing and termination by bacteriophage T7 RNA polymerase. *J. Mol. Biol.* **280:**201–213.

Makela, P. H. (1973). Glucosylation of lipopolysaccharide in Salmonella: Mutants negative for O antigen factor 1221. *J. Bacteriol.* **116:**847–856.

Mallory, J. B., Alfano, C., and McMacken, R. (1990). Host virus interactions in the initiation of bacteriophage lambda DNA replication: Recruitment of *Escherichia coli* DnaB helicase by lambda P replication protein. *J. Biol. Chem.* **265:**13297–13307.

Marchand, I., Nicholson, A. W., and Dreyfus, M. (2001). Bacteriophage T7 protein kinase phosphorylates RNase E and stabilizes mRNAs synthesized by T7 RNA polymerase. *Mol. Microbiol.* **42:**767–776.

Marintcheva, B., Qimron, U., Yu, Y., Tabor, S., and Richardson, C. C. (2009). Mutations in the gene 5 DNA polymerase of bacteriophage T7 suppress the dominant lethal phenotype of gene 2.5 ssDNA binding protein lacking the C-terminal phenylalanine. *Mol. Microbiol* **72:**869–880.

Marr, M. T., Datwyler, S. A., Meares, C. F., and Roberts, J. W. (2001). Restructuring of an RNA polymerase holoenzyme elongation complex by lambdoid phage Q proteins. *Proc. Natl. Acad. Sci. USA* **98:**8972–8978.

Marr, M. T., Roberts, J. W., Brown, S. E., Klee, M., and Gussin, G. N. (2004). Interactions among CII protein, RNA polymerase and the lambda PRE promoter: Contacts between

RNA polymerase and the -35 region of PRE are identical in the presence and absence of CII protein. *Nucleic Acids Res.* **32:**1083–1090.

Marsic, N., Roje, S., Stojiljkovic, I., Salaj-Smic, E., and Trgovcevic, Z. (1993). *In vivo* studies on the interaction of RecBCD enzyme and lambda Gam protein. *J. Bacteriol.* **175:**4738–4743.

Martin, A. C., Blanco, L., Garcia, P., Salas, M., and Mendez, J. (1996). *In vitro* protein-primed initiation of pneumococcal phage Cp-1 DNA replication occurs at the third 3′ nucleotide of the linear template: A stepwise sliding-back mechanism. *J. Mol. Biol.* **260:**369–377.

Marvik, O. J., Dokland, T., Nøkling, R. H., Jacobsen, E., Larsen, T., and Lindqvist, B. H. (1995). The capsid size-determining protein Sid forms an external scaffold on phage P4 procapsids. *J. Mol. Biol.* **251:**59–75.

Marvik, O. J., Jacobsen, E., Dokland, T., and Lindqvist, B. H. (1994). Bacteriophage P2 and P4 morphogenesis: assembly precedes proteolytic processing of the capsid proteins. *Virology* **205:**51–65.

Mason, S. W., and Greenblatt, J. (1991). Assembly of transcription elongation complexes containing the N protein of phage lambda and the *Escherichia coli* elongation factors NusA, NusB, NusG, and S10. *Genes Dev.* **5:**1504–1512.

Massad, T., Skaar, K., Nilsson, H., Damberg, P., Henriksson-Peltola, P., Haggard-Ljungquist, E., Hogbom, M., and Stenmark, P. (2010). Crystal structure of the P2 C-repressor: A binder of non-palindromic direct DNA repeats. *Nucleic Acids Res.* **38:**7778–7790.

Mattis, A. N., Gumport, R. I., and Gardner, J. F. (2008). Purification and characterization of bacteriophage P22 Xis protein. *J. Bacteriol.* **190:**5781–5796.

Maxwell, K. L., Reed, P., Zhang, R. G., Beasley, S., Walmsley, A. R., Curtis, F. A., Joachimiak, A., Edwards, A. M., and Sharples, G. J. (2005). Functional similarities between phage lambda Orf and *Escherichia coli* RecFOR in initiation of genetic exchange. *Proc. Natl. Acad. Sci. USA* **102:**11260–11265.

Maxwell, K. L., Yee, A. A., Booth, V., Arrowsmith, C. H., Gold, M., and Davidson, A. R. (2001). The solution structure of bacteriophage lambda protein W, a small morphogenetic protein possessing a novel fold. *J. Mol. Biol.* **308:**9–14.

Mayer, J. E., and Schweiger, M. (1983). RNase III is positively regulated by T7 protein kinase. *J. Biol. Chem.* **258:**5340–5343.

McDonnell, M., Lain, R., and Tomasz, A. (1975). "Diplophage": A bacteriophage of Diplococcus pneumoniae *Virology* **63:**577–582.

Medina, E., Wieczorek, D., Medina, E. M., Yang, Q., Feiss, M., and Catalano, C. E. (2010). Assembly and maturation of the bacteriophage lambda procapsid: gpC is the viral protease. *J. Mol. Biol.* **401:**813–830.

Meijer, W. J., Horcajadas, J. A., and Salas, M. (2001a). Phi29 family of phages. *Microbiol. Mol. Biol. Rev.* **65:**261–287.

Meijer, W. J., Lewis, P. J., Errington, J., and Salas, M. (2000). Dynamic relocalization of phage phi 29 DNA during replication and the role of the viral protein p16.7. *EMBO J* **19:**4182–4190.

Meijer, W. J., Serna-Rico, A., and Salas, M. (2001b). Characterization of the bacteriophage phi29-encoded protein p16.7: A membrane protein involved in phage DNA replication. *Mol. Microbiol* **39:**731–746.

Mencía, M., Monsalve, M., Rojo, F., and Salas, M. (1996a). Transcription activation by phage phi29 protein p4 is mediated by interaction with the alpha subunit of *Bacillus subtilis* RNA polymerase. *Proc. Natl. Acad. Sci. USA* **93:**6616–6620.

Mencía, M., Monsalve, M., Salas, M., and Rojo, F. (1996b). Transcriptional activator of phage phi 29 late promoter: mapping of residues involved in interaction with RNA polymerase and in DNA bending. *Mol. Microbiol.* **20:**273–282.

Méndez, J., Blanco, L., Esteban, J. A., Bernad, A., and Salas, M. (1992). Initiation of phi 29 DNA replication occurs at the second 3′ nucleotide of the linear template: A sliding-back mechanism for protein-primed DNA replication. *Proc. Natl. Acad. Sci. USA* **89:**9579–9583.

Méndez, J., Blanco, L., and Salas, M. (1997). Protein-primed DNA replication: A transition between two modes of priming by a unique DNA polymerase. *EMBO J.* **16:**2519–2527.
Mikawa, Y. G., Maruyama, I. N., and Brenner, S. (1996). Surface display of proteins on bacteriophage lambda heads. *J. Mol. Biol.* **262:**21–30.
Moak, M., and Molineux, I. J. (2000). Role of the Gp16 lytic transglycosylase motif in bacteriophage T7 virions at the initiation of infection. *Mol. Microbiol.* **37:**345–355.
Moak, M., and Molineux, I. J. (2004). Peptidoglycan hydrolytic activities associated with bacteriophage virions. *Mol. Microbiol.* **51:**1169–1183.
Molineux, I. J. (2006a). The T7 group. Chapter 20*In* "The Bacteriophages" (R. Calendar, ed.), pp. 277–301. Oxford University Press, Oxford.
Molineux, I. J. (2006b). Fifty-three years since Hershey and Chase; much ado about pressure but which pressure is it? *Virology* **344:**221–229.
Monsalve, M., Calles, B., Mencía, M., Salas, M., and Rojo, F. (1997). Transcription activation or repression by phage psi 29 protein p4 depends on the strength of the RNA polymerase-promoter interactions. *Mol. Cell.* **1:**99–107.
Monsalve, M., Mencía, M., Rojo, F., and Salas, M. (1996a). Activation and repression of transcription at two different phage phi29 promoters are mediated by interaction of the same residues of regulatory protein p4 with RNA polymerase. *EMBO J.* **15:**383–391.
Monsalve, M., Mencía, M., Salas, M., and Rojo, F. (1996b). Protein p4 represses phage phi 29 A2c promoter by interacting with the alpha subunit of *Bacillus subtilis* RNA polymerase. *Proc. Natl. Acad. Sci. USA* **93:**8913–8918.
Mooney, P. Q., North, R., and Molineux, I. J. (1980). The role of bacteriophage T7 gene 2 protein in DNA replication. *Nucleic Acids Res.* **8:**3043–3053.
Morais, M. C., Choi, K. H., Koti, J. S., Chipman, P. R., Anderson, D. L., and Rossmann, M. G. (2005). Conservation of the capsid structure in tailed dsDNA bacteriophages: The pseudoatomic structure of phi29. *Mol. Cell.* **18:**149–159.
Morais, M. C., Kanamaru, S., Badasso, M. O., Koti, J. S., Owen, B. A., McMurray, C. T., Anderson, D. L., and Rossmann, M. G. (2003). Bacteriophage phi29 scaffolding protein gp7 before and after prohead assembly. *Nat. Struct. Biol.* **10:**572–576.
Morita, M., Tasaka, M., and Fujisawa, H. (1994). Analysis of functional domains of the packaging proteins of bacteriophage T3 by site-directed mutagenesis. *J. Mol. Biol.* **235:**248–259.
Morita, M., Tasaka, M., and Fujisawa, H. (1995). Structural and functional domains of the large subunit of the bacteriophage T3 DNA packaging enzyme: Importance of the C-terminal region in prohead binding. *J. Mol. Biol.* **245:**635–644.
Mosig, G., Yu, S., Myung, H., Haggard-Ljungquist, E., Davenport, L., Carlson, K., and Calendar, R. (1997). A novel mechanism of virus-virus interactions: Bacteriophage P2 Tin protein inhibits phage T4 DNA synthesis by poisoning the T4 single-stranded DNA binding protein, gp32. *Virology* **230:**72–81.
Mount, D. W., Harris, A. W., Fuerst, C. R., and Siminovitch, L. (1968). Mutations in bacteriophage lambda affecting particle morphogenesis. *Virology* **35:**134–149.
Muñoz-Espín, D., Daniel, R., Kawai, Y., Carballido-Lopez, R., Castilla-Llorente, V., Errington, J., Meijer, W. J., and Salas, M. (2009). The actin-like MreB cytoskeleton organizes viral DNA replication in bacteria. Proc. Natl. Acad. Sci, USA.
Muñoz-Espín, D., Holguera, I., Ballesteros-Plaza, D., Carballido-Lopez, R., and Salas, M. (2010). Viral terminal protein directs early organization of phage DNA replication at the bacterial nucleoid. *Proc. Natl. Acad. Sci. USA* **107:**16548–16553.
Murialdo, H. (1979). Early intermediates in bacteriophage lambda prohead assembly. *Virology* **96:**341–367.
Murialdo, H., and Becker, A. (1978). Head morphogenesis of complex double-stranded deoxyribonucleic acid bacteriophages. *Microbiol. Rev.* **42:**529–576.

Murialdo, H., and Tzamtzis, D. (1997). Mutations of the coat protein gene of bacteriophage lambda that overcome the necessity for the Fl gene; the EFi domain. *Mol. Microbiol.* **24:**341–353.

Murphy, K. C. (2000). Bacteriophage P22 Abc2 protein binds to RecC increases the 5′ strand nicking activity of RecBCD and together with lambda bet, promotes Chi- independent recombination. *J. Mol. Biol.* **296:**385–401.

Murray, N. E. (2000). Type I restriction systems: Sophisticated molecular machines (a legacy of Bertani and Weigle). *MMBR* **64:**412–434.

Nakai, H., and Richardson, C. C. (1990). The gene 1.2 protein of bacteriophage T7 interacts with the *Escherichia coli* dGTP triphosphohydrolase to form a GTP-binding protein. *J. Biol. Chem* **265:**4411–4419.

Nechaev, S., and Severinov, K. (1999). Inhibition of *Escherichia coli* RNA polymerase by bacteriophage T7 gene 2 protein. *J. Mol. Biol.* **289:**815–826.

Nemecek, D., Gilcrease, E. B., Kang, S., Prevelige, P. E., Jr., Casjens, S., and Thomas, G. J., Jr. (2007). Subunit conformations and assembly states of a DNA-translocating motor: The terminase of bacteriophage P22. *J. Mol. Biol.* **374:**817–836.

Nemecek, D., Lander, G. C., Johnson, J. E., Casjens, S. R., and Thomas, G. J., Jr. (2008). Assembly architecture and DNA binding of the bacteriophage P22 terminase small subunit. *J. Mol. Biol.* **383:**494–501.

Newlove, T., Konieczka, J. H., and Cordes, M. H. (2004). Secondary structure switching in Cro protein evolution. *Structure (Camb.)* **12:**569–581.

Nickels, B. E., Roberts, C. W., Sun, H., Roberts, J. W., and Hochschild, A. (2002). The sigma (70) subunit of RNA polymerase is contacted by the (lambda)Q antiterminator during early elongation. *Mol. Cell.* **10:**611–622.

Nilsson, A. S., and Haggård-Ljungquist, E. (2006). The P2-like bacteriophages. (R. Calendar, ed.), pp. 365-390. Oxford University Press, New York.

Nilsson, A. S., Karlsson, J. L., and Haggård-Ljungquist, E. (2004). Site-specific recombination links the evolution of P2-like coliphages and pathogenic enterobacteria. *Mol. Biol. Evol.* **21:**1–13.

Nooren, I. M., Kaptein, R., Sauer, R. T., and Boelens, R. (1999). The tetramerization domain of the Mnt repressor consists of two right-handed coiled coils. *Nat. Struct. Biol.* **6:**755–759.

Nuez, B., Rojo, F., and Salas, M. (1992). Phage phi 29 regulatory protein p4 stabilizes the binding of the RNA polymerase to the late promoter in a process involving direct protein-protein contacts. *Proc. Natl. Acad. Sci. USA* **89:**11401–11405.

Odegrip, R., and Haggard-Ljungquist, E. (2001). The two active-site tyrosine residues of the a protein play non-equivalent roles during initiation of rolling circle replication of bacteriophage p2. *J Mol Biol* **308:**147–163.

Odegrip, R., Schoen, S., Haggard-Ljungquist, E., Park, K., and Chattoraj, D. K. (2000). The interaction of bacteriophage P2 B protein with *Escherichia coli* DnaB helicase. *J. Virol.* **74:**4057–4063.

Olia, A., Prevelige, P., Jr., Johnson, J., and Cingolani, G. (2011). Three-dimensional structure of a viral genome-delivery portal vertex. *Nat. Struct. Mol. Biol* in press.

Olia, A. S., Al-Bassam, J., Winn-Stapley, D. A., Joss, L., Casjens, S. R., and Cingolani, G. (2006). Binding-induced stabilization and assembly of the phage P22 tail accessory factor gp4. *J. Mol. Biol.* **363:**558–576.

Olia, A. S., Bhardwaj, A., Joss, L., Casjens, S., and Cingolani, G. (2007). Role of gene 10 protein in the hierarchical assembly of the bacteriophage P22 portal vertex structure. *Biochemistry* **46:**8776–8784.

Oppenheim, A. B., Kobiler, O., Stavans, J., Court, D. L., and Adhya, S. (2005). Switches in bacteriophage lambda development. *Annu. Rev. Genet.* **39:**409–429.

Pani, B., Ranjan, A., and Sen, R. (2009). Interaction surface of bacteriophage P4 protein Psu required for complex formation with the transcription terminator Rho. *J. Mol. Biol.* **389:**647–660.

Parker, M. H., Casjens, S., and Prevelige, P. E., Jr. (1998). Functional domains of bacteriophage P22 scaffolding protein. *J. Mol. Biol.* **281:**69–79.

Parker, M. H., Stafford, W. F., 3rd, and Prevelige, P. E., Jr. (1997). Bacteriophage P22 scaffolding protein forms oligomers in solution. *J. Mol. Biol.* **268:**655–665.

Parkinson, M. J., and Lilley, D. M. (1997). The junction-resolving enzyme T7 endonuclease I: Quaternary structure and interaction with DNA. *J. Mol. Biol.* **270:**169–178.

Parua, P. K., Mondal, A., and Parrack, P. (2010). HflD, an *Escherichia coli* protein involved in the lambda lysis-lysogeny switch, impairs transcription activation by lambdaCII. *Arch. Biochem. Biophys.* **493:**175–183.

Pastrana, R., Lazaro, J. M., Blanco, L., Garcia, J. A., Mendez, E., and Salas, M. (1985). Overproduction and purification of protein P6 of *Bacillus subtilis* phage phi 29: Role in the initiation of DNA replication. *Nucleic Acids Res.* **13:**3083–3100.

Pedulla, M. L., Ford, M. E., Karthikeyan, T., Houtz, J. M., Hendrix, R. W., Hatfull, G. F., Poteete, A. R., Gilcrease, E. B., Winn-Stapley, D. A., and Casjens, S. R. (2003). Corrected sequence of the bacteriophage p22 genome. *J. Bacteriol.* **185:**1475–1477.

Pell, L. G., Liu, A., Edmonds, L., Donaldson, L. W., Howell, P. L., and Davidson, A. R. (2009). The X-ray crystal structure of the phage lambda tail terminator protein reveals the biologically relevant hexameric ring structure and demonstrates a conserved mechanism of tail termination among diverse long-tailed phages. *J. Mol. Biol.* **389:**938–951.

Poteete, A. R., and Botstein, D. (1979). Purification and properties of proteins essential to DNA encapsulation by phage P22. *Virology* **95:**565–573.

Poteete, A. R., and Fenton, A. C. (1983). DNA-binding properties of the Erf protein of bacteriophage P22. *J. Mol. Biol.* **163:**257–275.

Poteete, A. R., and King, J. (1977). Functions of two new genes in Salmonella phage P22 assembly. *Virology* **76:**725–739.

Poteete, A. R., Sauer, R. T., and Hendrix, R. W. (1983). Domain structure and quaternary organization of the bacteriophage P22 Erf protein. *J. Mol. Biol.* **171:**401–418.

Potrykus, K., Baranska, S., Wegrzyn, A., and Wegrzyn, G. (2002). Composition of the lambda plasmid heritable replication complex. *Biochem. J.* **364:**857–862.

Prell, H. H., and Harvey, A. M. (1983). P22 antirepressor protein prevents *in vivo* recA-dependent proteolysis of P22 repressor. *Mol. Gen. Genet.* **190:**427–431.

Prevelige, P. E., Jr., Thomas, D., and King, J. (1988). Scaffolding protein regulates the polymerization of P22 coat subunits into icosahedral shells *in vitro*. *J. Mol. Biol.* **202:**743–757.

Pruss, G. J., Wang, J. C., and Calendar, R. (1975). In vitro packaging of covalently-closed circular monomers of bacteriophage DNA. *J. Mol. Biol.* **98:**465–478.

Ptashne, M. (2004). A Genetic Switch: Phage Lambda Revisited. Third Ed. Cold Spring Harbor Laboratory Press, Cold Spring Harbor, NY.

Ptashne, M. (2006). A Genetic Switch. Phage Lambda Revisited. Cold Spring Harbor Laboratory Press, Cold Spring Harbor, NY.

Putnam, C. D., Shroyer, M. J., Lundquist, A. J., Mol, C. D., Arvai, A. S., Mosbaugh, D. W., and Tainer, J. A. (1999). Protein mimicry of DNA from crystal structures of the uracil-DNA glycosylase inhibitor protein and its complex with *Escherichia coli* uracil-DNA glycosylase. *J. Mol. Biol.* **287:**331–346.

Qimron, U., Marintcheva, B., Tabor, S., and Richardson, C. C. (2006). Genomewide screens for *Escherichia coli* genes affecting growth of T7 bacteriophage. *Proc. Natl. Acad. Sci. USA* **103:**19039–19044.

Radding, C. M., Rosenzweig, J., Richards, J., and Cassuto, E. (1971). Appendix: Separation and characterization of exonuclease, β protein and a complex of both. *J. Biol. Chem.* **146:**2510–2512.

Rahmsdorf, H. J., Pai, S. H., Ponta, H., Herrlich, P., Roskoski, R., Jr., Schweiger, M., and Studier, F. W. (1974). Protein kinase induction in *Escherichia coli* by bacteriophage T7. *Proc. Natl. Acad. Sci. USA* **71:**586–589.

Rajagopala, S. V., Casjens, S., and Uetz, P. (2011). The protein interaction map of bacteriophage lambda. *BMC Microbiol.* **11:**213.

Ranade, K., and Poteete, A. R. (1993a). Superinfection exclusion (sieB) genes of bacteriophages P22 and lambda. *J. Bacteriol.* **175:**4712–4718.

Ranade, K., and Poteete, A. R. (1993b). A switch in translation mediated by an antisense RNA. *Genes Dev.* **7:**1498–1507.

Ray, P., and Murialdo, H. (1975). The role of gene Nu3 in bacteriophage lambda head morphogenesis. *Virology* **64:**247–263.

Rennell, D., and Poteete, A. R. (1985). Phage P22 lysis genes: Nucleotide sequences and functional relationships with T4 and lambda genes. *Virology* **143:**280–289.

Rennell, D., and Poteete, A. R. (1989). Genetic analysis of bacteriophage P22 lysozyme structure. *Genetics* **123:**431–440.

Rezende, L. F., Hollis, T., Ellenberger, T., and Richardson, C. C. (2002). Essential amino acid residues in the single-stranded DNA-binding protein of bacteriophage T7: Identification of the dimer interface. *J. Biol. Chem.* **277:**50643–50653.

Rishovd, S., and Lindqvist, B. H. (1992). Bacteriophage P2 and P4 morphogenesis: protein processing and capsid size determination. *Virology* **187:**548–554.

Rishovd, S., Marvik, O. J., Jacobsen, E., and Lindqvist, B. H. (1994). Bacteriophage P2 and P4 morphogenesis: Identification and characterization of the portal protein. *Virology* **200:**744–751.

Roberts, C. W., and Roberts, J. W. (1996). Base-specific recognition of the nontemplate strand of promoter DNA by *E. coli* RNA polymerase. *Cell* **86:**495–501.

Roberts, J. W. (1969). *Termination factor for RNA synthesis. Nature* **224:**1168–1174.

Roberts, J. W., Yarnell, W., Bartlett, E., Guo, J., Marr, M., Ko, D. C., Sun, H., and Roberts, C. W. (1998). Antitermination by bacteriophage lambda Q protein. *Cold Spring Harb. Symp. Quant. Biol.* **63:**319–325.

Robertson, E. S., Aggison, L. A., and Nicholson, A. W. (1994). Phosphorylation of elongation factor G and ribosomal protein S6 in bacteriophage T7-infected *Escherichia coli. Mol. Microbiol.* **11:**1045–1057.

Robertson, E. S., and Nicholson, A. W. (1990). Protein kinase of bacteriophage T7 induces the phosphorylation of only a small number of proteins in the infected cell. *Virology* **175:**525–534.

Robertson, E. S., and Nicholson, A. W. (1992). Phosphorylation of *Escherichia coli* translation initiation factors by the bacteriophage T7 protein kinase. *Biochemistry* **31:**4822–4827.

Rodriguez-Cerrato, V., Garcia, P., Del Prado, G., Garcia, E., Gracia, M., Huelves, L., Ponte, C., Lopez, R., and Soriano, F. (2007). *In vitro* interactions of LytA, the major pneumococcal autolysin, with two bacteriophage lytic enzymes (Cpl-1 and Pal), cefotaxime and moxifloxacin against antibiotic-susceptible and -resistant *Streptococcus pneumoniae* strains. *J. Antimicrob. Chemother.* **60:**1159–1162.

Rojo, F., Mencia, M., Monsalve, M., and Salas, M. (1998). Transcription activation and repression by interaction of a regulator with the alpha subunit of RNA polymerase: The model of phage phi 29 protein p4. *Prog. Nucleic Acid Res. Mol. Biol.* **60:**29–46.

Rojo, F., and Salas, M. (1991). A DNA curvature can substitute phage phi 29 regulatory protein p4 when acting as a transcriptional repressor. *EMBO J.* **10:**3429–3438.

Sabri, M., Häuser, R., Ouellette, M., Liu, J., Dehbi, M., Moeck, G., Garcia, E., Titz, B., Uetz, P., and Moineau, S. (2011). Genome annotation and intraviral interactome for the Streptococcus pneumoniae virulent phage Dp-1. *J. Bacteriol.* **193:**551–562.

Saito, H., and Richardson, C. C. (1981). Processing of mRNA by ribonuclease III regulates expression of gene 1.2 of bacteriophage T7. *Cell* **27:**533–542.

Salas, M. (1991). Protein-priming of DNA replication. *Annu. Rev. Biochem.* **60:**39–71.

Sam, M. D., Papagiannis, C. V., Connolly, K. M., Corselli, L., Iwahara, J., Lee, J., Phillips, M., Wojciak, J. M., Johnson, R. C., and Clubb, R. T. (2002). Regulation of directionality in bacteriophage lambda site-specific recombination: structure of the Xis protein. *J. Mol. Biol.* **324:**791–805.

Sam, M. D., Cascio, D., Johnson, R. C., and Clubb, R. T. (2004). Crystal structure of the excisionase-DNA complex from bacteriophage lambda. *J. Mol. Biol.* **338:**229–240.

Sanger, F., Air, G. M., Barrell, B. G., Brown, N. L., Coulson, A. R., Fiddes, C. A., Hutchison, C. A., Slocombe, P. M., and Smith, M. (1977). Nucleotide sequence of bacteriophage phi X174 DNA. *Nature* **265:**687–695.

Sauer, B., Ow, D., Ling, L., and Calendar, R. (1981). Mutants of satellite bacteriophage P4 that are defective in the suppression of transcriptional polarity. *J. Mol. Biol.* **145:**29–46.

Savalia, D., Robins, W., Nechaev, S., Molineux, I., and Severinov, K. (2010). The role of the T7 Gp2 inhibitor of host RNA polymerase in phage development. *J. Mol. Biol.* **402:**118–126.

Savva, C. G., Dewey, J. S., Deaton, J., White, R. L., Struck, D. K., Holzenburg, A., and Young, R. (2008). The holin of bacteriophage lambda forms rings with large diameter. *Mol. Microbiol.* **69:**784–793.

Schmitt, C. K., and Molineux, I. J. (1991). Expression of gene 1.2 and gene 10 of bacteriophage T7 is lethal to F plasmid-containing *Escherichia coli*. *J. Bacteriol.* **173:**1536–1543.

Schulz, E. C., Dickmanns, A., Urlaub, H., Schmitt, A., Muhlenhoff, M., Stummeyer, K., Schwarzer, D., Gerardy-Schahn, R., and Ficner, R. (2010). Crystal structure of an intramolecular chaperone mediating triple-beta-helix folding. *Nat. Struct. Mol. Biol.* **17:**210–215.

Semerjian, A. V., Malloy, D. C., and Poteete, A. R. (1989). Genetic structure of the bacteriophage P22 PL operon. *J. Mol. Biol.* **207:**1–13.

Serna-Rico, A., Munoz-Espin, D., Villar, L., Salas, M., and Meijer, W. J. (2003). The integral membrane protein p16.7 organizes *in vivo* phi29 DNA replication through interaction with both the terminal protein and ssDNA. *EMBO J* **22:**2297–2306.

Serna-Rico, A., Salas, M., and Meijer, W. J. (2002). The *Bacillus subtilis* phage phi 29 protein p16.7, involved in phi 29 DNA replication, is a membrane-localized single-stranded DNA-binding protein. *J. Biol. Chem* **277:**6733–6742.

Serrano, M., Gutierrez, C., Freire, R., Bravo, A., Salas, M., and Hermoso, J. M. (1994). Phage phi 29 protein p6: A viral histone-like protein. *Biochimie* **76:**981–991.

Serrano-Heras, G., Bravo, A., and Salas, M. (2008). Phage phi29 protein p56 prevents viral DNA replication impairment caused by uracil excision activity of uracil-DNA glycosylase. *Proc. Natl. Acad. Sci. USA* **105:**19044–19049.

Serrano-Heras, G., Ruiz-Maso, J. A., del Solar, G., Espinosa, M., Bravo, A., and Salas, M. (2007). Protein p56 from the *Bacillus subtilis* phage phi29 inhibits DNA-binding ability of uracil-DNA glycosylase. *Nucleic Acids Res.* **35:**5393–5401.

Serrano-Heras, G., Salas, M., and Bravo, A. (2006). A uracil-DNA glycosylase inhibitor encoded by a non-uracil containing viral DNA. *J. Biol. Chem.* **281:**7068–7074.

Severinova, E., and Severinov, K. (2006). Localization of the *Escherichia coli* RNA polymerase beta′ subunit residue phosphorylated by bacteriophage T7 kinase Gp0.7. *J. Bacteriol* **188:**3470–3476.

Sevilla-Sierra, P., Otting, G., and Wuthrich, K. (1994). Determination of the nuclear magnetic resonance structure of the DNA-binding domain of the P22 c2 repressor (1 to 76) in solution and comparison with the DNA-binding domain of the 434 repressor. *J. Mol. Biol.* **235:**1003–1020.

Shotland, Y., Shifrin, A., Ziv, T., Teff, D., Koby, S., Kobiler, O., and Oppenheim, A. B. (2000). Proteolysis of bacteriophage lambda CII by Escherichia coli FtsH (HflB). *J. Bacteriol.* **182:**3111–3116.
Simon, M. N., and Studier, F. W. (1973). Physical mapping of the early region of bacteriophage T7 DNA. *J. Mol. Biol.* **79:**249–265.
Simpson, A. A., Leiman, P. G., Tao, Y., He, Y., Badasso, M. O., Jardine, P. J., Anderson, D. L., and Rossmann, M. G. (2001). Structure determination of the head-tail connector of bacteriophage phi29. *Acta Crystallogr. D Biol. Crystallogr.* **57:**1260–1269.
Simpson, A. A., Tao, Y., Leiman, P. G., Badasso, M. O., He, Y., Jardine, P. J., Olson, N. H., Morais, M. C., Grimes, S., Anderson, D. L., Baker, T. S., and Rossmann, M. G. (2000). Structure of the bacteriophage phi29 DNA packaging motor. *Nature* **408:**745–750.
Singleton, M. R., Sawaya, M. R., Ellenberger, T., and Wigley, D. B. (2000). Crystal structure of T7 gene 4 ring helicase indicates a mechanism for sequential hydrolysis of nucleotides. *Cell* **101:**589–600.
Sippy, J., and Feiss, M. (1992). Analysis of a mutation affecting the specificity domain for prohead binding of the bacteriophage lambda terminase. *J. Bacteriol.* **174:**850–856.
Six, E. W. (1975). The helper dependence of satellite bacteriophage P4: which gene functions of bacteriophage P2 are needed by P4?. *Virology* **67:**249–263.
Six, E. W., Sunshine, M. G., Williams, J., Haggård-Ljungquist, E., and Lindqvist, B. H. (1991). Morphopoietic switch mutations of bacteriophage P2. *Virology* **182:**34–46.
Souza, L., Calendar, R., Six, E., and Lindqvist, B. (1977). A transactivation mutant of satellite phage P4. *Virology* **81:**81–90.
Stano, N. M., and Patel, S. S. (2004). T7 lysozyme represses T7 RNA polymerase transcription by destabilizing the open complex during initiation. *J. Biol. Chem.* **279:**16136–16143.
Steinbacher, S., Miller, S., Baxa, U., Budisa, N., Weintraub, A., Seckler, R., and Huber, R. (1997). Phage P22 tailspike protein: Crystal structure of the head-binding domain at 2.3 A, fully refined structure of the endorhamnosidase at 1.56 A resolution, and the molecular basis of O-antigen recognition and cleavage. *J. Mol. Biol* **267:**865–880.
Steinbacher, S., Seckler, R., Miller, S., Steipe, B., Huber, R., and Reinemer, P. (1994). Crystal structure of P22 tailspike protein: Interdigitated subunits in a thermostable trimer. *Science* **265:**383–386.
Stellberger, T., Häuser, R., Baiker, A., Pothineni, V. R., Haas, J., and Uetz, P. (2010). Improving the yeast two-hybrid system with permutated fusions proteins: The Varicella Zoster Virus interactome. *Proteome Sci.* **8:**8.
Stephanou, A. S., Roberts, G. A., Cooper, L. P., Clarke, D. J., Thomson, A. R., MacKay, C. L., Nutley, M., Cooper, A., and Dryden, D. T. (2009). Dissection of the DNA mimicry of the bacteriophage T7 Ocr protein using chemical modification. *J. Mol. Biol.* **391:**565–576.
Sternberg, N., and Hoess, R. H. (1995). Display of peptides and proteins on the surface of bacteriophage lambda. *Proc. Natl. Acad. Sci. USA* **92:**1609–1613.
Sternberg, N., and Weisberg, R. (1977). Packaging of coliphage lambda DNA. II. The role of the gene D protein. *J. Mol. Biol* **117:**733–759.
Steven, A. C., Trus, B. L., Maizel, J. V., Unser, M., Parry, D. A., Wall, J. S., Hainfeld, J. F., and Studier, F. W. (1988). Molecular substructure of a viral receptor-recognition protein. The gp17 tail-fiber of bacteriophage T7. *J. Mol. Biol* **200:**351–365.
Stone, J. C., and Miller, R. C., Jr. (1985). Spontaneous temperature-sensitive mutations in bacteriophage T7. *J. Virol.* **54:**886–888.
Studier, F. W. (1975). Gene 0.3 of bacteriophage T7 acts to overcome the DNA restriction system of the host. *J. Mol. Biol* **94:**283–295.
Studier, F. W., and Movva, N. R. (1976). SAMase gene of bacteriophage T3 is responsible for overcoming host restriction. *J. Virol.* **19:**136–145.
Studier, F. W. (1981). Identification and mapping of five new genes in bacteriophage T7. *J. Mol. Biol.* **153:**493–502.

Summer, E. J., Berry, J., Tran, T. A., Niu, L., Struck, D. K., and Young, R. (2007). Rz/Rz1 lysis gene equivalents in phages of Gram-negative hosts. *J. Mol. Biol.* **373:**1098–1112.

Sunshine, M., Usher, D., and Calendar, R. (1975). Interaction of P2 bacteriophage with the dnaB gene of *Escherichia coli*. *J. Virol.* **16:**284–289.

Sunshine, M. G., and Sauer, B. (1975). A bacterial mutation blocking P2 phage late gene expression. *Proc. Natl. Acad. Sci. USA* **72:**2770–2774.

Susskind, M. M., and Botstein, D. (1975). Mechanism of action of Salmonella phage P22 antirepressor. *J. Mol. Biol.* **98:**413–424.

Susskind, M. M., and Botstein, D. (1978). Molecular genetics of bacteriophage P22. *Microbiol. Rev.* **42:**385–413.

Szambowska, A., Pierechod, M., Węgrzyn, G., and Glinkowska, M. (2010). Coupling of transcription and replication machineries in λ DNA replication initiation: Evidence for direct interaction of *Escherichia coli* RNA polymerase and the λ O protein. *Nucleic Acids Res.* **39:**168–177.

Tahirov, T. H., Temiakov, D., Anikin, M., Patlan, V., McAllister, W. T., Vassylyev, D. G., and Yokoyama, S. (2002). Structure of a T7 RNA polymerase elongation complex at 2.9 A resolution. *Nature* **420:**43–50.

Takahashi, I., and Marmur, J. (1963). Replacement of thymidylic acid by deoxyuridylic acid in the deoxyribonucleic acid of a transducing phage for *Bacillus subtilis*. *Nature* **197:**794–795.

Tang, J., Lander, G., Olia, A., Li, R., Casjens, S., Prevelige, P., Cingolani, G., Baker, T., and Johnson, J. (2011). Peering down the barrel of a bacteriophage portal: The genome packaging and release valve in P22. *Structure* **19:**496–502.

Tang, L., Gilcrease, E. B., Casjens, S. R., and Johnson, J. E. (2006). Highly discriminatory binding of capsid-cementing proteins in bacteriophage L. *Structure* **14:**837–845.

Tao, Y., Olson, N. H., Xu, W., Anderson, D. L., Rossmann, M. G., and Baker, T. S. (1998). Assembly of a tailed bacterial virus and its genome release studied in three dimensions. *Cell* **95:**431–437.

Temple, L., Forsburg, S., Calendar, R., and Christie, G. (1991). Nucleotide sequence of the genes encoding the major tail sheath and tail tube proteins of bacteriophage P2. *Virology* **181:**353–358.

Tokuno, S. I., and Gough, M. (1976). UV sensitivity of a nonrepressor regulatory protein of bacteriophage P22. *J. Virol.* **18:**65–70.

Tran, N. Q., Lee, S. J., Richardson, C. C., and Tabor, S. (2010). A novel nucleotide kinase encoded by gene 1.7 of bacteriophage T7. *Mol. Microbiol* **77:**492–504.

Tsui, L., and Hendrix, R. W. (1980). Head-tail connector of bacteriophage lambda. *J. Mol. Biol.* **142:**419–438.

Tsui, L. C., and Hendrix, R. W. (1983). Proteolytic processing of phage lambda tail protein gpH: Timing of the cleavage. *Virology* **125:**257–264.

Tye, B. K., Huberman, J. A., and Botstein, D. (1974). Non-random circular permutation of phage P22 DNA. *J. Mol. Biol.* **85:**501–528.

Uetz, P., Rajagopala, S. V., Dong, Y. A., and Haas, J. (2004). From ORFeomes to protein interaction maps in viruses. *Genome Res.* **14:**2029–2033.

Valpuesta, J. M., Fujisawa, H., Marco, S., Carazo, J. M., and Carrascosa, J. L. (1992). Three-dimensional structure of T3 connector purified from overexpressing bacteria. *J. Mol. Biol.* **224:**103–112.

Valpuesta, J. M., Sousa, N., Barthelemy, I., Fernandez, J. J., Fujisawa, H., Ibarra, B., and Carrascosa, J. L. (2000). Structural analysis of the bacteriophage T3 head-to-tail connector. *Journal of structural biology* **131:**146–155.

van der Poll, T., and Opal, S. M. (2009). Pathogenesis, treatment, and prevention of pneumococcal pneumonia. *Lancet* **374:**1543–1556.

Vander Byl, C., and Kropinski, A. M. (2000). Sequence of the genome of Salmonella bacteriophage P22. *J. Bacteriol.* **182:**6472–6481.
Vershon, A. K., Liao, S. M., McClure, W. R., and Sauer, R. T. (1987a). Bacteriophage P22 Mnt repressor: DNA binding and effects on transcription *in vitro*. *J. Mol. Biol.* **195:**311–322.
Vershon, A. K., Liao, S. M., McClure, W. R., and Sauer, R. T. (1987b). Interaction of the bacteriophage P22 Arc repressor with operator DNA. *J. Mol. Biol.* **195:**323–331.
Villafane, R., Zayas, M., Gilcrease, E. B., Kropinski, A. M., and Casjens, S. R. (2008). Genomic analysis of bacteriophage epsilon 34 of *Salmonella enterica* serovar Anatum (15+). *BMC Microbiol.* **8:**227.
Wang, J., Hofnung, M., and Charbit, A. (2000a). The C-terminal portion of the tail fiber protein of bacteriophage lambda is responsible for binding to LamB, its receptor at the surface of *Escherichia coli* K-12. *J. Bacteriol.* **182:**508–512.
Wang, S., Chang, J. R., and Dokland, T. (2006). Assembly of bacteriophage P2 and P4 procapsids with internal scaffolding protein. *Virology* **348:**133–140.
Wang, S., Palasingam, P., Nøkling, R. H., Lindqvist, B. H., and Dokland, T. (2000). *In vitro* assembly of bacteriophage P4 procapsids from purified capsid and scaffolding proteins. *Virology* **275:**133–144.
Wang, W. F., Cheng, X., and Molineux, I. J. (1999). Isolation and identification of fxsA, an *Escherichia coli* gene that can suppress F exclusion of bacteriophage T7. *J. Mol. Biol.* **292:**485–499.
Warren, D., Sam, M. D., Manley, K., Sarkar, D., Lee, S. Y., Abbani, M., Wojciak, J. M., Clubb, R. T., and Landy, A. (2003). Identification of the lambda integrase surface that interacts with Xis reveals a residue that is also critical for Int dimer formation. *Proc. Natl. Acad. Sci. USA* **100:**8176–8181.
Węgrzyn, A., Taylor, K., and Węgrzyn, G. (1996). The cbpA chaperone gene function compensates for dnaJ in lambda plasmid replication during amino acid starvation of Escherichia coli. *J. Bacteriol.* **178:**5847–5849.
Węgrzyn, A., and Węgrzyn, G. (2001). Inheritance of the replication complex: A unique or common phenomenon in the control of DNA replication? *Arch. Microbiol.* **175:**86–93.
Węgrzyn, G., and Węgrzyn, A. (2002). Stress responses and replication of plasmids in bacterial cells. *Microb. Cell Fact.* **1:**2.
Weigele, P. R., Sampson, L., Winn-Stapley, D., and Casjens, S. R. (2005). Molecular genetics of bacteriophage P22 scaffolding protein's functional domains. *J. Mol. Biol.* **348:**831–844.
Werts, C., Michel, V., Hofnung, M., and Charbit, A. (1994). Adsorption of bacteriophage lambda on the LamB protein of *Escherichia coli* K-12: point mutations in gene J of lambda responsible for extended host range. *J. Bacteriol.* **176:**941–947.
White, J. H., and Richardson, C. C. (1987). Gene 18 protein of bacteriophage T7: Overproduction, purification, and characterization. *J. Biol. Chem.* **262:**8845–8850.
Wickner, S. (1984). DNA-dependent ATPase activity associated with phage P22 gene 12 protein. *J. Biol. Chem.* **259:**14038–14043.
Wickner, S. H., and Zahn, K. (1986). Characterization of the DNA binding domain of bacteriophage lambda O protein. *J. Biol. Chem.* **261:**7537–7543.
Wood, L. F., Tszine, N. Y., and Christie, G. E. (1997). Activation of P2 late transcription by P2 Ogr protein requires a discrete contact site on the C terminus of the alpha subunit of *Escherichia coli* RNA polymerase. *J. Mol. Biol.* **274:**1–7.
Xiang, Y., Leiman, P. G., Li, L., Grimes, S., Anderson, D. L., and Rossmann, M. G. (2009). Crystallographic insights into the autocatalytic assembly mechanism of a bacteriophage tail spike. *Mol. Cell.* **34:**375–386.
Xiang, Y., Morais, M. C., Cohen, D. N., Bowman, V. D., Anderson, D. L., and Rossmann, M. G. (2008). Crystal and cryoEM structural studies of a cell wall degrading enzyme in the bacteriophage phi29 tail. *Proc. Natl. Acad. Sci. USA* **105:**9552–9557.

Xu, J., Hendrix, R. W., and Duda, R. L. (2004). Conserved translational frameshift in dsDNA bacteriophage tail assembly genes. *Mol. Cell.* **16:**11–21.

Yang, F., Forrer, P., Dauter, Z., Conway, J. F., Cheng, N., Cerritelli, M. E., Steven, A. C., Pluckthun, A., and Wlodawer, A. (2000). Novel fold and capsid-binding properties of the lambda-phage display platform protein gpD. *Nat. Struct. Biol.* **7:**230–237.

Yin, Y. W., and Steitz, T. A. (2002). Structural basis for the transition from initiation to elongation transcription in T7 RNA polymerase. *Science* **298:**1387–1395.

Yin, Y. W., and Steitz, T. A. (2004). The structural mechanism of translocation and helicase activity in T7 RNA polymerase. *Cell* **116:**393–404.

Yoshikawa, H., Elder, J. H., and Ito, J. (1986). Comparative studies on the small *Bacillus* bacteriophages. *J. Gen. Appl. Microbiol.* **155:**392–401.

Youderian, P., and Susskind, M. M. (1980). Identification of the products of bacteriophage P22 genes, including a new late gene. *Virology* **107:**258–269.

Yura, K., Yamaguchi, A., and Go, M. (2006). Coverage of whole proteome by structural genomics observed through protein homology modeling database. *J. Struct. Funct. Genom.* **7:**65–76.

Zahn, K., and Blattner, F. R. (1985). Binding and bending of the lambda replication origin by the phage O protein. *EMBO J.* **4:**3605–3616.

Zahn, K., and Blattner, F. R. (1987). Direct evidence for DNA bending at the lambda replication origin. *Science* **236:**416–422.

Zahn, K., and Landy, A. (1996). Modulation of lambda integrase synthesis by rare arginine tRNA. *Mol. Microbiol.* **21:**69–76.

Zhang, X., and Studier, F. W. (1995). Isolation of transcriptionally active mutants of T7 RNA polymerase that do not support phage growth. *J. Mol. Biol.* **250:**156–168.

Zhang, X., and Studier, F. W. (1997). Mechanism of inhibition of bacteriophage T7 RNA polymerase by T7 lysozyme. *J. Mol. Biol.* **269:**10–27.

Zhang, X., and Studier, F. W. (2004). Multiple roles of T7 RNA polymerase and T7 lysozyme during bacteriophage T7 infection. *J. Mol. Biol.* **340:**707–730.

Ziermann, R., and Calendar, R. (1990). Characterization of the cos sites of bacteriophages P2 and P4. *Gene* **96:**9–15.

Zillig, W., Fujiki, H., Blum, W., Janekovic, D., Schweiger, M., Rahmsdorf, H., Ponta, H., and Hirsch-Kauffmann, M. (1975). *In vivo* and *in vitro* phosphorylation of DNA-dependent RNA polymerase of *Escherichia coli* by bacteriophage-T7-induced protein kinase. *Proc. Natl. Acad. Sci. USA* **72:**2506–2510.

Zinder, N. D., and Lederberg, J. (1952). *Genetic exchange in Salmonella. J. Bacteriol.* **64:**679–699.

Żylicz, M., Ang, D., Liberek, K., and Georgopoulos, C. (1989). Initiation of lambda DNA replication with purified host- and bacteriophage-encoded proteins: The role of the dnaK, dnaJ and grpE heat shock proteins. *EMBO J.* **8:**1601–1608.

CHAPTER 7

Endolysins as Antimicrobials

Daniel C. Nelson,[*,†] Mathias Schmelcher,[‡] Lorena Rodriguez-Rubio,[§] Jochen Klumpp,[||] David G. Pritchard,[¶] Shengli Dong,[¶] and **David M. Donovan[‡,1]**

Contents

[*] Institute for Bioscience and Biotechnology Research, University of Maryland, Rockville, Maryland, USA
[†] Department of Veterinary Medicine, Virginia-Maryland Regional College of Veterinary Medicine, College Park, Maryland, USA
[‡] Animal Biosciences and Biotechnology Laboratory, ANRI, ARS, USDA, Beltsville, Maryland, USA
[§] Instituto de Productos Lácteos de Asturias (IPLA-CSIC), Villaviciosa, Asturias, Spain
[||] Institute of Food, Nutrition and Health, ETH Zurich, Zurich, Switzerland
[¶] Department of Biochemistry and Molecular Genetics, University of Alabama at Birmingham, Birmingham, Alabama, USA
[1] Corresponding author, E-mail: david.donovan@ars.usda.gov

Advances in Virus Research, Volume 83
ISSN 0065-3527, DOI: 10.1016/B978-0-12-394438-2.00007-4

Abstract Peptidoglycan (PG) is the major structural component of the bacterial cell wall. Bacteria have autolytic PG hydrolases that allow the cell to grow and divide. A well-studied group of PG hydrolase enzymes are the bacteriophage endolysins. Endolysins are PG-degrading proteins that allow the phage to escape from the bacterial cell during the phage lytic cycle. The endolysins, when purified and exposed to PG externally, can cause "lysis from without." Numerous publications have described how this phenomenon can be used therapeutically as an effective antimicrobial against certain pathogens. Endolysins have a characteristic modular structure, often with multiple lytic and/or cell wall-binding domains (CBDs). They degrade the PG with glycosidase, amidase, endopeptidase, or lytic transglycosylase activities and have been shown to be synergistic with fellow PG hydrolases or a range of other antimicrobials. Due to the coevolution of phage and host, it is thought they are much less likely to invoke resistance. Endolysin engineering has opened a range of new applications for these proteins from food safety to environmental decontamination to more effective antimicrobials that are believed refractory to resistance development. To put phage endolysin work in a broader context, this chapter includes relevant studies of other well-characterized PG hydrolase antimicrobials.

ABBREVIATIONS:

CBD	cell wall-binding domain;
CFU	colony-forming unit;
CHAP	cysteine, histidine-dependent amidohydrolase/peptidase;
CPP	cell-penetrating peptides;
CSF	cerebrospinal fluid;
GlcNAc	*N*-acetylglucosamine;
HIV	Human Immunodeficiency Virus
HPLC	high-pressure liquid chromatography

IV	intravenous
MBC	minimum bactericidal concentration;
mDAP	*meso*-diaminopimelic acid;
MIC	minimum inhibitory concentration;
MRSA	methicillin-resistant *Staphylococcus aureus*;
MS	mass spectrometry;
MurNAc	*N*-acetylmuramic acid;
OD	optical density (ΔOD; change in OD);
PG	peptidoglycan;
PTD	protein transduction domain;
SDS-PAGE	sodium dodecyl sulfate–polyacrylamide gel electrophoresis;
TAT	transactivator of transcription domain

I. INTRODUCTION

The bacterial peptidoglycan (PG) is a protective barrier as well as a structural component of the bacterial cell wall that defines its shape. Notably, the PG supports the internal turgor pressure essential for survival of the prokaryotic cell. PG hydrolase generically describes a wide range of lytic enzymes that act upon the bacterial PG and can be classified into several groups based on their origin. An "autolysin" is a PG hydrolase produced and regulated by the bacterial cell for growth, division, maintenance, and repair of the PG. In contrast, an "exolysin" is an enzyme secreted by a bacterial cell that functions to lyse the PG of a different strain or species occupying the same ecological niche. One of the most-studied bacterial exolysins is lysostaphin, a PG hydrolase secreted by *Staphylococcus simulans* that cleaves the *Staphylococcus aureus* PG, but does not harm the *S. simulans* PG (Schindler and Schuhardt, 1964). In addition to bacterial exolysins, eukaryotic cells can secrete their own exolysins. For example, lysozyme found in human saliva and tears is a eukaryotic exolysin that is part of the innate immune system providing protection against bacterial invasion.

Peptidoglycan hydrolases are also used extensively by bacteriophage (phage), for infection and/or release from a bacterial host. Particle-associated PG hydrolases can produce "lysis from without," a term used to describe bacterial lysis in the absence of the full lytic infection cycle, as first described by Delbrück (1940). Work by Moak and Molineux (2004) demonstrated that PG hydrolases were associated with numerous phage particles infecting either Gram-negative or Gram-positive bacteria. These lytic structural proteins, which are mostly tail associated, cause localized degradation of the cell wall to enable infection of the bacterial host. Alternatively, phages encode PG hydrolases that, along with holins, are part of the lytic cassette. Holins are produced during the late stages of a phage infection cycle to perforate the inner bacterial membrane, thus allowing the PG hydrolases that have accumulated in the cytoplasm to gain access to the PG. The result is bacterial lysis and release of progeny phage completing the infection cycle (Young, 1992). Because these PG hydrolases lyse "from within," they are referred to as "endolysins," or simply "lysins."

Significantly, exogenous addition of a phage endolysin or a bacterial exolysin to a susceptible host can be exploited to produce lysis from without due to the high osmotic pressure within the cell [≈5 atmospheres for Gram-negative organisms and up to 50 atmospheres for Gram-positive organisms (Seltman and Holst, 2001)]. The use of purified phage endolysins or other naturally occurring PG hydrolases as antimicrobial agents against Gram-positive pathogens is the theme of this chapter [for prior reviews, see Callewaert *et al.* (2010), Fischetti (2005), Fischetti *et al.* (2006), Hermoso *et al.* (2007), and Loessner (2005)]. Due to the presence of an outer membrane in Gram-negative bacteria, an exogenously added PG hydrolase will usually not gain access to the PG without surfactant or some other mechanism to translocate the protein across the outer membrane. Nonetheless, reports are beginning to emerge in the literature that describe fusions of Gram-negative endolysins that will lyse these pathogens from without, which is discussed at the end of this chapter.

II. PEPTIDOGLYCAN STRUCTURE

The peptidoglycan is a three-dimensional lattice of peptide and glycan moieties. A polymer of alternating *N*-acetylmuramic acid (MurNAc) and *N*-acetylglucosamine (GlcNAc) residues coupled by β(1→4) linkages comprises the "glycan" component of the PG (Fig. 1). This polymer displays little variation between bacterial species [for a review, see Schleifer and Kandler (1972)]. The glycan polymer is in turn linked covalently to a short stem peptide through an amide bond between MurNAc and an L-alanine, the first amino acid of the "peptide" component. The remainder of the stem peptide is composed of alternating L- and D-form amino acids that are fairly well conserved in Gram-negative organisms, but is variable in composition for Gram-positive organisms. For many Gram-positive organisms, the third residue of the stem peptide is L-lysine, which is cross-linked to an opposing stem peptide on a separate glycan polymer through an interpeptide bridge, the composition of which varies between species. For example, the interpeptide bridge of *S. aureus* is composed of pentaglycine (depicted in Fig. 1), whereas the interpeptide bridge of *Streptococcus pyogenes* is two L-alanines. In Gram-negative organisms and some genera of Gram-positive bacteria (i.e., *Bacillus* and *Listeria*), a *meso*-diaminopimelic acid (mDAP) residue is present at position number three of

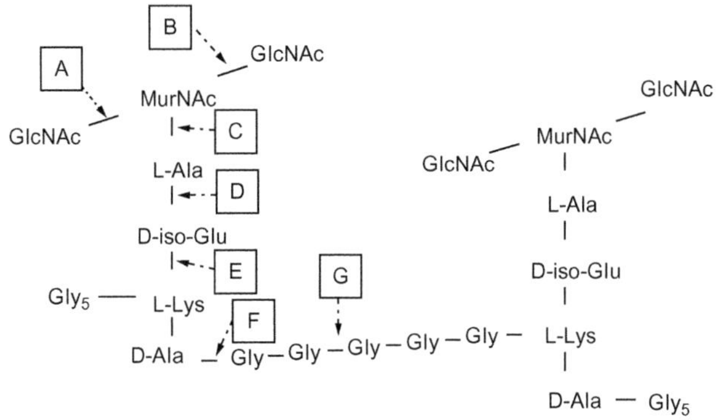

FIGURE 1 Structure of *Staphylococcus aureus* bacterial PG and cleavage sites by PG hydrolases. (A) An *N*-acetylglucosaminidase hydrolyzes the glycan component of the PG on the reducing side of GlcNAc. (B) In contrast, an *N*-acetylmuramidase (also known as "muramidase" or "lysozyme") hydrolyzes the glycan component of the PG on the reducing side of MurNAc. Likewise, lytic transglycosylases cleave the same bond, but form N-acetyl-1,6-anhydro-muramyl intermediates during cleavage. (C) An *N*-acetylmuramoyl-L-alanine amidase cleaves a critical amide bond between the glycan moiety (MurNAc) and the peptide moiety (L-alanine) of the cell wall. This activity is sometimes referred to generically as an "amidase." (D–G) An endopeptidase cleaves an amide bond between two amino acids. This type of activity may occur in the stem peptide of the PG, as in the case of the *Listeria* endolysins, Ply500 and Ply118 (D), or the streptococcal endolysin, λSa2 (E). Alternatively, an endopeptidase can cleave the interpeptide bridge as displayed by the staphylococcal endolysin Φ11 (F) or the staphylococcal bacteriocin, lysostaphin. (G) Note that the structure of the *S. aureus* PG is depicted for illustration purposes. Other bacterial species have interpeptide bridges composed of different amino acids or may lack an interpeptide bridge altogether. In these organisms, a mDAP replaces L-Lys and directly cross-links to the terminal D-Ala of the opposite peptide chain.

the stem peptide instead of L-lysine. In these organisms, mDAP cross-links directly to the terminal D-alanine of the opposite stem peptide (i.e., no interpeptide bridge). Whether an interpeptide bridge is present or not, a transpeptidation reaction joining opposing stem peptides gives rise to the three-dimensional lattice that is the hallmark of the bacterial peptidoglycan. Notably, several antibiotics target the transpeptidation reaction because the cross-linking is so critical to proper formation and integrity of the cell wall and survival of the organism.

III. ENDOLYSIN ACTIVITIES AND STRUCTURE

A. Enzymatic activities

Due to the moderately conserved overall structure of the PG, there are limited types of covalent bonds available for cleavage by endolysins and other PG hydrolases (Fig. 1). In general, there are four mechanistic classes associated with PG hydrolases: glycosidase, endopeptidase, a specific amidohydrolase, and lytic transglycosylase. One type of glycosidase, known as an *N*-acetylglucosaminidase, cleaves the glycan component of the PG on the reducing side of GlcNAc (Fig. 1A). This type of activity is found frequently in autolysins, such as AltA from *Enterococcus faecalis* (Mesnage *et al.*, 2008) or AcmA, AcmB, AcmC, and AcmD from *Lactococcus lactis* (Steen *et al.*, 2007). However, with the exception of the streptococcal LambdaSa2 endolysin (Pritchard *et al.*, 2007), this activity has not been associated with phage endolysins. A second type of glycosidic activity is an *N*-acetylmuramidase, which cleaves the glycan component of the PG on the reducing side of MurNAc (Fig. 1B). This activity is referred to commonly as a "muramidase" or "lysozyme" and is found frequently in autolysins, exolysins, and phage endolysins, including the pneumococcal Cpl-1 endolysin (Garcia *et al.*, 1987) and the streptococcal B30 endolysin (Pritchard *et al.*, 2004).

The second class of PG hydrolases is an *N*-acetylmuramoyl-L-alanine amidase, a specific amidohydrolase that cleaves a critical amide bond between the glycan moiety (MurNAc) and the peptide moiety (L-alanine) of the PG (Fig. 1 C) This activity is associated more often with bacteriophage endolysins than autolysins or exolysins. The reasons for this are not clear. However, because hydrolysis of this bond separates the glycan polymer from the stem peptide, such activity is speculated to be more destabilizing to the PG than hydrolysis of other bonds and may be favored evolutionarily by bacteriophages that require rapid lysis of host cells for the dissemination of progeny phage. This activity has been demonstrated for the amidase domain of the staphylococcal phage Φ11 endolysin (Navarre *et al.*, 1999), the phage K endolysin, LysK (Becker *et al.*, 2009a; Donovan *et al.*, 2009), and the *Listeria* phage endolysins Ply511 (Loessner *et al.*, 1995b) and PlyPSA (Korndorfer *et al.*, 2006).

The third class of PG hydrolases is that of an endopeptidase (i.e., protease), which cleaves peptide bonds between two amino acids. This cleavage may occur in the stem peptide, such as the listerial Ply500 and Ply118 L-alanyl-D-glutamate endolysins (Loessner *et al.*, 1995b), or in the interpeptide bridge, such as the staphylococcal Φ11 D-alanyl-glycyl endolysin (Navarre *et al.*, 1999) or the lysostaphin exolysin (Figs. 1D–1 G).

The fourth and final class of PG lytic enzymes is the lytic transglycosylase. By definition, these enzymes are not true "hydrolases" because they do not require water to catalyze PG cleavage. They are very similar to muramidases in that they cleave the $\beta(1\rightarrow4)$ linkages between *N*-acetylmuramyl and *N*-acetylglucosaminyl residues of the PG (Fig. 1B), but they form a N-acetyl-1,6-anhydro-muramyl moiety residue during glycosidic cleavage and thus belong to a different mechanistic class than the lysozymes (Holtje and Tomasz, 1975). The phage Lambda endolysin (Taylor and Gorazdowska, 1974) and the gp144 endolysin from the ΦKZ bacteriophage (Paradis-Bleau *et al.*, 2007) were both confirmed biochemically to be lytic transglycosylases.

B. Biochemical determination of endolysin specificity

Numerous studies have investigated the specificity of endolysins by assaying the cleavage sites on purified PG (Dhalluin *et al.*, 2005; Fukushima *et al.*, 2007, 2008; Loessner *et al.*, 1998; Navarre *et al.*, 1999; Pritchard *et al.*, 2004). Classic biochemical methods, such as the Park–Johnson method, can be used to measure an increase of reducing sugar moieties as an indication of glycosidase activity by a reduction of ferricyanide to ferrocyanide (Park and Johnson, 1949; Spiro, 1966). A variation of the method using sodium borohydride to reduce digested cell wall samples (Ward, 1973) has also been used frequently (Deutsch *et al.*, 2004; Dhalluin *et al.*, 2005; Scheurwater and Clarke, 2008; Vasala *et al.*, 1995).

Endopeptidase or L-alanine amidase activities can be observed by an increase of free amine groups as measured by a trinitrophenylation reaction described originally by Satake *et al.* (1960) and modified by Mokrasch (1967). N-terminal sequencing of digestion products (i.e., Edman degradation) can also reveal cleavage sites of a PG hydrolase possessing endopeptidase activity (Navarre *et al.*, 1999; Pritchard *et al.*, 2004). Alternatively, digestion products can be labeled with 1-fluoro-2,4-dinitrobenzene, followed by HCl hydrolysis and reverse-phase high-pressure liquid chromatography (HPLC) (Fukushima *et al.*, 2007). HPLC peaks can be analyzed by mass spectrometry (MS) and resulting fragment ions by MS–MS analysis (Fig. 2) (Becker *et al.*, 2009a; Fukushima *et al.*, 2008; Navarre *et al.*, 1999). Many of the techniques described earlier were used in an elegant series of experiments that showed that the streptococcal phage B30 endolysin contains both glycosidase and endopeptidase activity within the same protein (Baker *et al.*, 2006; Pritchard *et al.*, 2004).

C. Confusion over historical endolysin nomenclature

The assignment of nomenclature to endolysins has been less than ideal. Decades ago, endolysins were simply referred to as "lysozymes," a generic term often applied to PG hydrolases despite a lack of biochemical evidence characterizing their enzymatic activity. Unfortunately, many of these older designations persist to this day. The endolysin of the T7 bacteriophage continues to be called the "T7 lysozyme" in the literature despite experimental evidence dating back to 1973 showing that it is actually an *N*-acetylmuramoyl-L-alanine amidase rather than an *N*-acetylmuramidase (i.e., lysozyme) (Inouye *et al.*, 1973). Likewise, the λ endolysin was shown to be a lytic transglycosylase 35 years ago, but the "lysozyme" moniker continues in the current literature.

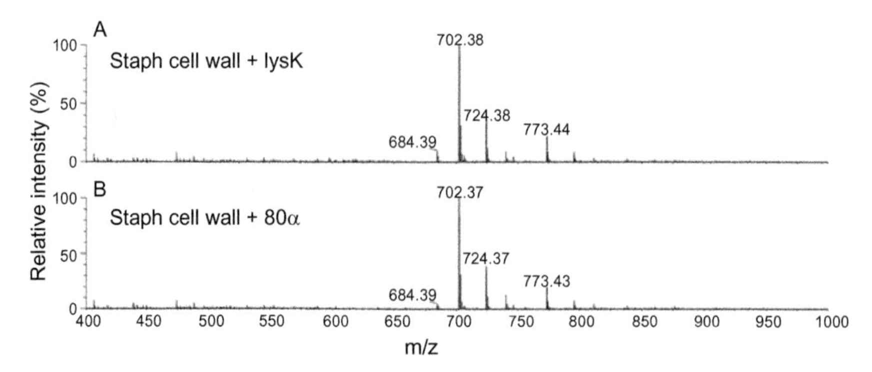

FIGURE 2 Electron spray ionization mass spectrometry determination of LysK and phi80α endolysin cut sites in *S. aureus* PG. Purified *S. aureus* PG was digested with LysK and phi80α endolysin under identical conditions as described in Becker *et al.* (2009). Digests were filtered through 5 K cutoff ultrafilters; these filtrates were processed further through disposable charcoal columns (CarboPak). The bound muropeptides were eluted with 50% acetonitrile and subjected to mass spectrometry. *m/z*, mass-to-charge ratio.

Another challenge is the generic classification of many endolysins simply as "amidases," which is used ubiquitously to describe both *N*-acetylmuramoyl-L-alanine amidases and endopeptidases, the latter being exclusive to hydrolysis of an amide bond between two amino acids. To complicate this issue further, a protein family called CHAP (cysteine, histidine-dependent amidohydrolase/peptidase) has emerged as a common domain found in bacteriophage endolysins (Bateman and Rawlings, 2003). Experimental evidence shows that the CHAP domain of the group B streptococcal B30 lysin is a D-alanyl-L-alanyl endopeptidase (Pritchard *et al.*, 2004), whereas the CHAP domain of the group A streptococcal PlyC lysin is an *N*-acetylmuramoyl-L-alanine amidase (Fischetti *et al.*, 1972; Nelson *et al.*, 2006). Finally, many endolysin catalytic domains are alleged to possess a particular activity based exclusively on limited homology to another endolysin domain with a putative function. When actual experiments are conducted to determine cleavage specificities, the results are often contrary to the function assigned by bioinformatic analysis. For example, *in silico* analysis suggests that the streptococcal endolysins λSa1 and λSa2 contain *N*-acetylmuramoyl-L-alanine amidase activities. However, utilizing electrospray ionization mass spectrometry, Pritchard *et al.* (2007) not only showed an absence of *N*-acetylmuramoyl-L-alanine amidase activity, but provided evidence that these enzymes function as D-glutaminyl-L-lysine endopeptidases. Clearly, more rigorous biochemical characterization of bacteriophage endolysins will help better define and predict the catalytic classes of these enzymes.

D. Endolysin modular structure

1. Gram-negative endolysin structure

The Gram-negative PG, which lies subjacent to the outer membrane in the periplasmic space, is relatively thin and undecorated by surface proteins or carbohydrates. Most lysins from phage that infect Gram-negative hosts are single domain globular proteins typically composed of only a single catalytic domain and have a mass of 15 to 20 kDa. However, two Gram-negative phage endolysins (*Pseudomonas* phage endolysins KZ144 and EL188) have been shown to harbor both a lytic domain and an N-terminal cell wall-binding domain (CBD) (Briers *et al.*, 2007). The first 83 amino acids of KZ144 have been shown to be sufficient for high-affinity binding to *Pseudomonas aeruginosa* cell walls (Briers *et al.*, 2009). Moreover, this domain was shown to bind to Gram-negative PG from all species on which it was tested (after chemical treatments to remove the outer membrane) (Briers *et al.*, 2007).

2. Gram-positive endolysin structure

In contrast to Gram-negative bacteria, Gram-positive organisms contain no protective outer membrane, but rather have a much thicker (up to 40 layers) PG layer that is highly cross-linked and decorated with surface carbohydrates and proteins. Endolysins from Gram-positive-infecting bacteriophage typically utilize a modular design (Diaz *et al.*,1990), having one or more catalytic domains and a CBD that recognizes epitopes on the surface of susceptible organisms, often giving rise to strain- or near-species-specific binding (Schmelcher *et al.*, 2010). Typically, a flexible interdomain linker sequence connects the catalytic domain(s) to the CBD (Korndorfer *et al.*, 2006).

Nearly all Gram-positive phage endolysins and autolysins are the products of single genes, although group I introns are often found within these genes and have been reported for *Streptococcus* (Foley *et al.*, 2000) and *Staphylococcus* (Becker *et al.*, 2009b; Kasparek *et al.*, 2007; O'Flaherty *et al.*, 2004). The gene encoding the streptococcal C1 phage endolysin, PlyC, was originally believed to contain an intron (Nelson *et al.*, 2003), but the C1 endolysin was later shown to be synthesized from two genes. This enzyme is composed of a gene product, PlyCA, that contains the catalytic domain and eight identical copies of a second gene product, PlyCB, which harbors the CBD (Nelson *et al.*, 2006). To date, no other multimeric lysin has been identified, and the implications for a multigene, heterononomer (nine subunit protein) are not abundantly clear. Nonetheless, nanogram quantities of PlyC can achieve $\approx$7 log killing of streptococcal cells within seconds, making PlyC several orders of magnitude more active than any other PG hydrolase ever described (Nelson *et al.*, 2001).

The three-dimensional crystal structure of known endolysin lytic domains was reviewed by Hermoso *et al.* (2007). A very complete discussion of the PG hydrolase endopeptidase activities and their active site structure was also presented by Bochtler and colleagues (Firczuk and Bochtler, 2007). Interdomain linker sequences between the catalytic and CBD domains can vary in size and can impart an inherent flexibility to these proteins, making crystallography of full-length endolysins challenging. Many attempts have yielded only the structures of individual catalytic domains or isolated CBDs (Korndorfer *et al.*, 2008; Low *et al.*, 2005; Porter *et al.*, 2007; Silva-Martin *et al.*, 2010). Only a few full-length structures have become available, including PlyPSA, a listerial *N*-acetylmuramoyl-L-alanine amidase (Korndorfer *et al.*, 2006), and Cpl-1, a pneumococcal *N*-acetylmuramidase (Hermoso *et al.*, 2003). Remarkably, both structures reveal extreme compartmentalization displayed by the individual domains (Bustamante *et al.*, 2010; Monterroso *et al.*, 2008).

3. Domain conservation of Gram-positive endolysins

Alignment of conserved PG hydrolase domain sequences is available in public data sets (e.g., Pfam; http://pfam.jouy.inra.fr/). Such comparisons have identified numerous conserved domains shared across many genera for both binding to the bacterial surface (CBDs) and PG digestion (lytic domains). Through a limited number of site-directed mutagenic studies, invariant amino acid residues conserved in domain sequences have been identified. Primarily, histidine residues have been identified that, when mutated, can destroy the hydrolytic activity of the M23 endopeptidase domain (Fujiwara *et al.*, 2005) or the cysteine, histidine-dependent amidohydrolases/peptidases domain (Bateman and Rawlings, 2003; Huard *et al.*, 2003; Nelson *et al.*, 2006; Pritchard *et al.*, 2004; Rigden *et al.*, 2003).

Using public data sets and PubMed, the authors have attempted to compile known PG hydrolase sequences for each of three genera—*Staphylococcus, Streptococcus,* and *Enterococcus*. These protein structures are collated in Figures 3–5. This summary sheds light on the degree of domain conservation and the range of lytic protein domain organization within and among these closely related genera. Within each genus, endolysins have been collated into groups based on protein architecture and sequence homology. Group members are listed in Tables I–III. Each group has mostly >90% within group identity at the amino acid residue level, and between group identities is mostly less than 50%. There are also stand-alone lysins with no apparent homologues yet reported. There has not been an attempt to assign a species to each of the endolysins within a genus due to the high frequency of mobile genetic elements and lateral gene transfer known to exist within each (Lindsay, 2008; Palmer *et al.*, 2010; Rossolini *et al.*, 2010). Each of the domains listed in Figures 3–5 can be found in public data sets describing conserved domains (PFAM: http://pfam.sanger.ac.uk/ or NCBI-conserved domain database: http://www.ncbi.nlm.nih.gov/Structure/cdd/cdd.shtml).

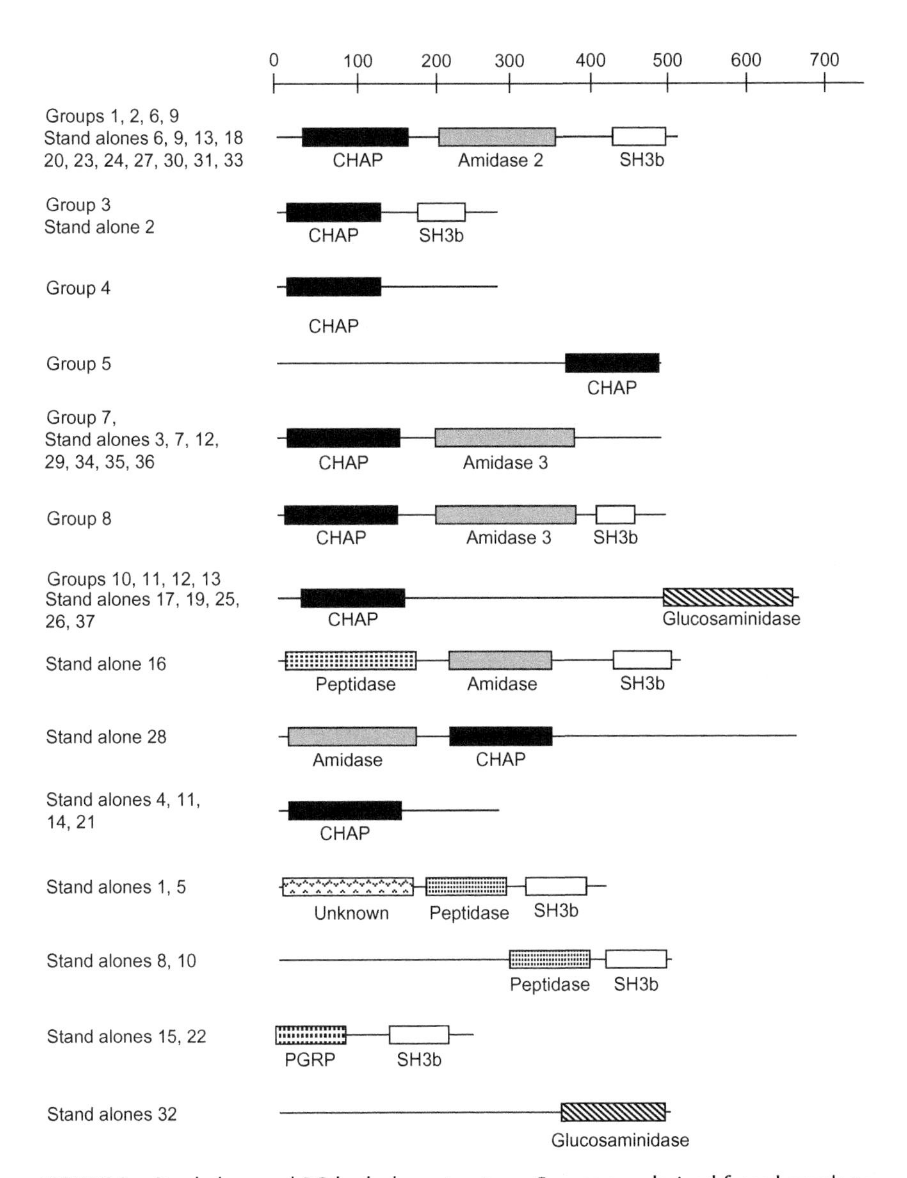

FIGURE 3 Staphylococcal PG hydrolase structure. Groups are derived from homology clustering performed in BLAST, NCBI of the proteins described in Table I. Scale bar represents number of amino acids. Domains are defined more clearly in the PFAM database http://www.sanger.ac.uk/resources/databases/pfam.html. White boxes represent CBDs. SH3b, bacterial Src homology 3 domain (Ponting *et al.*, 1999; Whisstock and Lesk, 1999); PGRP, peptidoglycan recognition protein (Dziarski and Gupta, 2006).

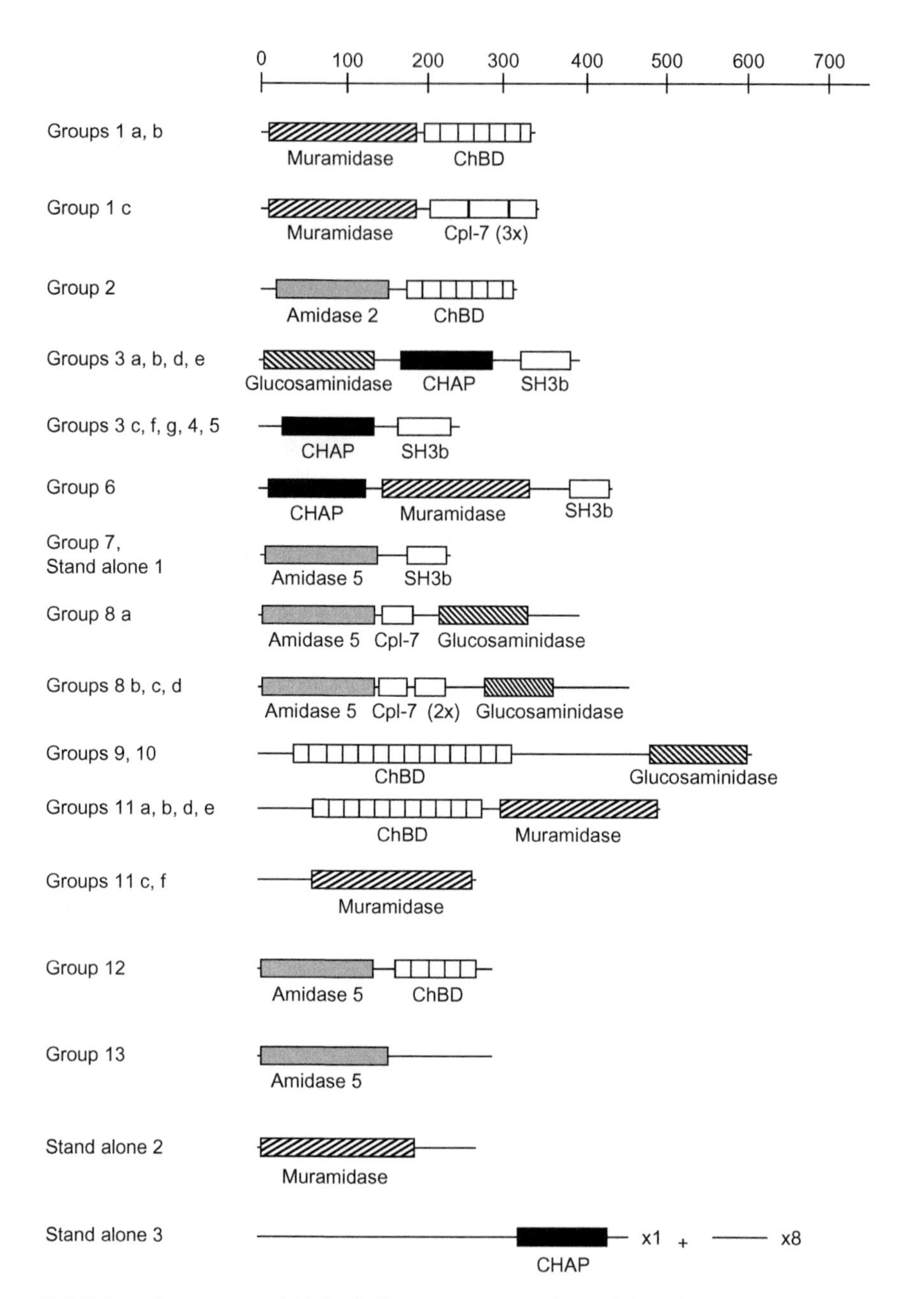

FIGURE 4 Streptococcal PG hydrolases. Groups are derived from homology clustering performed in BLAST, NCBI of the proteins described in Table II. Scale bar represents number of amino acids. Domains are defined more clearly in the PFAM database http://www.sanger.ac.uk/resources/databases/pfam.html. White boxes represent CBDs. ChBD, choline-binding domain (Hermoso *et al.*, 2003); Cpl-7, cell wall-binding domain (Garcia *et al.*, 1990); SH3b, bacterial Src homology 3 domain (Ponting *et al.*, 1999; Whisstock and Lesk, 1999). Stand alone protein 3 is a multimeric lysin consisting of 1 big subunit and 8 copies of a small subunit.

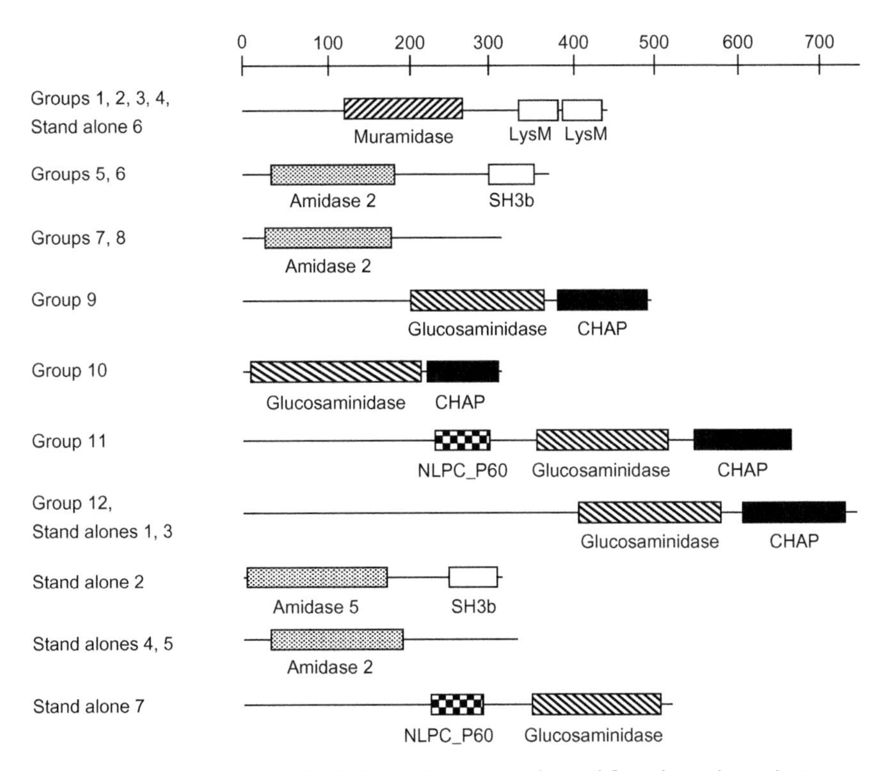

FIGURE 5 Enterococcal PG hydrolases. Groups are derived from homology clustering performed in BLAST, NCBI of the proteins described in Table III. Scale bar represents number of amino acids. Domains are defined more clearly in the PFAM database http://www.sanger.ac.uk/resources/databases/pfam.html. White boxes represent CBDs. LysM, (Bateman and Bycroft, 2000; Joris *et al*., 1992); SH3b, bacterial Src homology 3 domain (Ponting *et al*., 1999; Whisstock and Lesk, 1999); NLP_P60 (Anantharaman and Aravind, 2003).

4. Endolysins with multiple catalytic domains

Although it is well established that single domain endolysins can lyse the target pathogen (Sanz *et al.*, 1996), numerous endolysins harbor two short lytic domains (≈100–200 amino acids), each encoding a different catalytic activity. A few examples of dual domain endolysins for which the cut sites are known include: (1) the staphylococcal Φ11 endolysin has both *N*-acetylmuramoyl-L-alanine amidase and D-alanyl-glycyl endopeptidase catalytic activities (Navarre *et al.*, 1999), (2) the group B streptococcal lysin B30 was shown to have both *N*-acetylmuramidase and D-alanyl-L-alanyl endopeptidase catalytic activity on purified PG (Pritchard *et al.*, 2004), (3) the streptococcal λSa2 phage endolysin has N-terminal D-glutaminyl-L-lysine endopeptidase activity and an *N*-acetylglucosaminidase C-terminal domain (Pritchard *et al.*, 2007), and (4) LysK is the staphylolytic phage K

TABLE I Staphylococcal PG hydrolases

	AA	Accession #
Group 1		
Putative lysin [*Staphylococcus* phage K]	495	YP_024461
Endolysin [*Staphylococcus* phage 812]	494	ABL87139
Group 2		
N-Acetylmuramoyl-L-alanine amidase [*S. epidermidis* M23864:W2(grey)]	487	ZP_06612943
Autolysin (*N*-acetylmuramoyl-L-alanine amidase) [*S. caprae* C87]	487	ZP_07841306
Group 3		
Amidase [*Staphylococcus* phage 44AHJD]	250	NP_817310
ORF009 [*Staphylococcus* phage 66]	250	YP_239469
Amidase [*Staphylococcus* phage SAP-2]	249	YP_001491539
Group 4		
Lytic enzyme [*S. aureus* subsp. *aureus* N315]	251	NP_375054
Autolysin [*S. aureus* subsp. *aureus* MR1]	251	ZP_06859751
Lytic enzyme [*S. aureus* subsp. *aureus* MW2]	251	NP_646703
Autolysin [*S. aureus* subsp. *aureus* MSSA476]	251	YP_043983
Gametolysin [*S. aureus* subsp. *aureus* A017934/97]	251	ZP_06376153
N-Acetylmuramoyl-L-alanine amidase [*S. aureus* subsp. *aureus* H19]	251	ZP_06343995
Lytic enzyme (*N*-acetylmuramoyl-L-alanine amidase) [*Staphylococcus* prophage phiPV83]	251	NP_061648
ORF017 [*Staphylococcus* phage 42E]	251	YP_239884
Group 5		
Hypothetical protein 44AHJD_11 [*Staphylococcus* phage 44AHJD]	479	NP_817306
ORF004 [*Staphylococcus* phage 66]	487	YP_239474
Hypothetical protein SAP2_gp10 [*Staphylococcus* phage SAP-2][1]	478	YP_001491535
Group 6		
Amidase [*Staphylococcus* phage phi2958PVL]	484	YP_002268027
Amidase (peptidoglycan hydrolase) [*Staphylococcus* phage PVL]	484	NP_058463
Amidase [*Staphylococcus* phage tp310-1]	484	YP_001429893
Truncated amidase [*S. aureus* subsp. *aureus* MW2]	484	NP_646197
Amidase [*S. aureus* A6224]	484	ZP_05696927
ORF006 [*Staphylococcus* phage 96]	484	YP_240259
prophage amidase, putative [*S. aureus* subsp. aureus ED133]	484	ADI96879
putative amidase [*S. aureus* subsp. *aureus* ED98]	484	YP_003282866
amidase [*Staphylococcus* phage phiSLT]	484	NP_075522
amidase [*S. aureus* subsp. *aureus* ST398]	484	CAQ48834
77ORF005 [*Staphylococcus* phage 77]	484	NP_958622

TABLE I (*continued*)

	AA	Accession #
Amidase [*S. aureus* subsp. *aureus* MRSA252]	484	YP_040898
Prophage L54a, amidase, putative [*S. aureus* subsp. *aureus* COL]	484	YP_185281
Prophage L54a, amidase, putative [*S. aureus* subsp. *aureus* CGS03]	484	EFT84462
Amidase [*Staphylococcus* phage tp310-2]	484	YP_001429961
Amidase [*S. aureus* subsp. *aureus* MSSA476]	484	YP_043081
Putative endolysin [*Staphylococcus* phage phiSauS-IPLA35]	484	YP_002332423
N-Acetylmuramoyl-L-alanine amidase [*S. aureus* A10102]	484	ZP_06334988
Peptidoglycan hydrolase [*Staphylococcus* phage phi12]	484	NP_803355
N-Acetylmuramoyl-L-alanine amidase [ORF007 *Staphylococcus* phage 47]	484	%YP_240025
Peptidoglycan hydrolase, putative [*S. aureus* subsp. *aureus* 132]	484	ZP_06378887
Amidase [*S. aureus* A6300]	484	ZP_05693770
N-Acetylmuramoyl-L-alanine amidase [*S. aureus* A9765]	484	ZP_06329456
Amidase [*S. aureus* subsp. *aureus* 65-1322]	484	ZP_05604610
Group 7		
Amidase [*Staphylococcus* phage CNPH82]	460	YP_950628
Phage amidase [*Staphylococcus* phage PH15]	460	YP_950690
Bacteriophage amidase [*S. epidermidis* M23864: W1][2]	460	ZP_04819028
Group 8		
CHAP domain-containing protein [*S. aureus* subsp. *aureus* JH9]	470	YP_001246290
Bacteriophage amidase [*S. aureus* subsp. *aureus* USA300_TCH959]	473	ZP_04865682
Phage amidase [*S. aureus* subsp. *aureus* 132]	470	ZP_06378624
Phage amidase [*S. aureus* subsp. *aureus* MR1]	470	ZP_06859762
Phage amidase [*S. aureus* subsp. *aureus* ED98]	470	YP_003281797
CHAP domain-containing protein [*S. aureus* A6300]	470	ZP_05694219
Similar to phage phi PVL amidase [*Staphylococcus* phage phiETA]	470	NP_510959
Amidase [*Staphylococcus* phage phiETA2]	470	YP_001004328
Amidase [*Staphylococcus* phage phiETA3]	470	YP_001004396
Group 9		
Autolysin (*S. aureus*)[3]	481	LYTA_STAAU

(*continued*)

TABLE I *(continued)*

	AA	Accession #
Amidase [*Staphylococcus* phage 80alpha][4]	481	AAB39699
Phage amidase [*S. aureus* subsp. *aureus* str. Newman]	481	YP_001332073
Amidase [*S. aureus* A9719]	486	ZP_05684021
N-Acetylmuramoyl-L-alanine amidase [*S. aureus* subsp. *aureus* D139]	484	ZP_06324909
N-Acetylmuramoyl-L-alanine amidase [*S. aureus* A9765]	484	ZP_06327634
ORF007 [*Staphylococcus* phage 29]	481	YP_240560
Autolysin [*S. aureus* subsp. *aureus* NCTC 8325]	481	YP_500516
Autolysin, hypothetical phage protein [*S. aureus* subsp. *aureus* TW20]	481	CBI48272
Amidase [*S. aureus* subsp. *aureus* Mu50]	481	NP_371437
ORF006 [*Staphylococcus* phage 88]	481	YP_240699
Endolysin [*Staphylococcus* phage phiMR11]	481	YP_001604156
Putative cell wall hydrolase [*Staphylococcus* phage phiMR25]	481	YP_001949866
N-Acetylmuramoyl-L-alanine amidase [*S. aureus* subsp. *aureus* C427]	484	ZP_06327377
N-Acetylmuramoyl-L-alanine amidase [*S. aureus* subsp. *aureus* JH9]	481	YP_001246457
ORF007 [*Staphylococcus* phage 55]	481	YP_240484
N-Acetylmuramoyl-L-alanine amidase [*S. aureus* A6300]	486	ZP_05693156
ORF007 [*Staphylococcus* phage 69]	481	YP_239596
ORF007 [*Staphylococcus* phage 52A]	481	YP_240634
N-Acetylmuramoyl-L-alanine amidase [*S. aureus* subsp. *aureus* MN8]	481	ZP_06948777
ORF006 [*Staphylococcus* phage 92]	481	YP_240773
Autolysin [*S. aureus* subsp. *aureus* JKD6009]	481	ZP_03566881
Phage amidase [*S. aureus* A9635]	484	ZP_05687279
Phage-related amidase [*S. aureus* subsp. *aureus* CGS00]	481	EFU23738
Autolysin (*N*-acetylmuramoyl-L-alanine amidase) [*S. aureus* subsp. *aureus* ST398]	481	CAQ49916
Group 10		
Cell wall hydrolase [*Staphylococcus* phage 11]	632	NP_803302
ORF004 [*Staphylococcus* phage 69]	632	YP_239591
Cell wall hydrolase [*Staphylococcus* phage phiNM]	632	YP_874009
Cell wall hydrolase [*Staphylococcus* phage TEM126]	632	ADV76510
Autolysin [*S. aureus* A9765]	632	ZP_06327630

TABLE I (*continued*)

	AA	Accession #
Mannosyl-glycoprotein endo-β-*N*-acetylglucosaminidase [*S. aureus* subsp. *aureus* JH9]	632	YP_001246286
Mannosyl-glycoprotein endo-β-*N*-acetylglucosaminidase [*S. aureus* A8115]	632	ZP_05690673
Mannosyl-glycoprotein endo-β-*N*-acetylglucosaminidase [*S. aureus* subsp. *aureus* CGS03]	589	EFT84342
Phage *N*-acetylglucosamidase [*S. aureus* subsp. *aureus* CGS00]	632	EFU23742
ORF004 [*Staphylococcus* phage 85]	632	YP_239746
Phage *N*-acetylglucosamidase [*S. aureus* subsp. *aureus* str. Newman]	632	YP_001331343
Cell wall hydrolase [*S. aureus* subsp. *aureus* Mu50]	632	NP_371433
Cell wall hydrolase [*Staphylococcus* phage phiETA2]	632	YP_001004324
Cell wall hydrolase [*Staphylococcus* phage SAP-26]	632	YP_003857090
Putative tail-associated cell wall hydrolase [*Staphylococcus* phage phiMR25]	632	YP_001949862
Mannosyl-glycoprotein endo-β-*N*-acetylglucosaminidase [*S. aureus* subsp. *aureus* D139]	632	ZP_06324913
Mannosyl-glycoprotein endo-β-*N*-acetylglucosaminidase [*S. aureus* subsp. *aureus* C427]	632	ZP_06327381
Lyz [*Staphylococcus* phage 80alpha]	632	YP_001285381
ORF004 [*Staphylococcus* phage 53]	632	YP_239671
Phage-related cell wall hydrolase [*S. aureus* RF122][5]	634	YP_417168
Group 11		
ORF004 [*Staphylococcus* phage 71]	624	YP_240403
Similar to phage phi187 cell hydrolase Ply187 [*Staphylococcus* phage phiETA]	624	NP_510955
Mannosyl-glycoprotein endo-β-*N*-acetylglucosaminidase [*S. aureus* subsp. *aureus* 132]	624	ZP_06378620
Mannosyl-glycoprotein endo-β-*N*-acetylglucosaminidase [*S. aureus* subsp. *aureus* str. CF-Marseille]	624	ZP_04837774
Conserved hypothetical protein [*S. aureus* A9635]	624	ZP_05687283
ORF004 [*Staphylococcus* phage 55]	624	YP_240479

(*continued*)

TABLE I (*continued*)

	AA	Accession #
Cell wall hydrolase [*Staphylococcus* phage phiETA3]	624	YP_001004392
Tail tip protein [*Staphylococcus* phage phiMR11]	624	YP_001604152
ORF004 [*Staphylococcus* phage ROSA]	624	YP_240329
ORF004 [*Staphylococcus* phage 96]	624	YP_240255
ORF004 [*Staphylococcus* phage 88]	624	YP_240695
ORF004 [*Staphylococcus* phage 29]	624	YP_240556
ORF005 [*Staphylococcus* phage X2]	624	YP_240843
Mannosyl-glycoprotein endo-β-*N*-acetylglucosaminidase [*S. aureus* subsp. *aureus* JKD6009]	624	ZP_03566885
Hypothetical protein HMPREF0776_1895 [*S. aureus* subsp. *aureus* USA300_TCH959]	624	ZP_04865678
Group 12		
Hydrolase [*Staphylococcus* phage PH15][7]	633	YP_950686
Hydrolase [*S. epidermidis* BCM-HMP0060][8]	607	ZP_04824942
Amidase [*Staphylococcus* phage CNPH82]	633	YP_950623
N-Acetylmuramoyl-L-alanine amidase [*S. epidermidis* M23864:W2(grey)][9]	635	ZP_06614671
Group 13		
Bifunctional autolysin Atl/*N*-acetylmuramoyl-L-alanine amidase/endo-β-*N*-acetylglucosaminidase [*S. pseudintermedius* HKU10-03][10]	629	YP_004148762
ORF002 [*Staphylococcus* phage 187]	628	YP_239513
Cell wall hydrolase Ply187 [*Staphylococcus* phage 187]	628	CAA69022
Stand-alone proteins		
1 Lysostaphin [*S. simulans*]	389	AAA26655
2 Endolysin [*Staphylococcus* phage 812]	284	ABL87142
3 Lytic enzyme, amidase [*S. aureus*]	426	ACZ59017
4 Endolysin [*Staphylococcus* phageSA4]	267	ADR02788
5 Glycyl-glycine endopeptidase ALE1	362	ALE1-STACP
6 Lysine [bacteriophage phi WMY]	477	BAD83402
7 Phage amidase [*S. aureus* subsp. *aureus* TW20]	500	CBI50050
8 Lysostaphin	480	LSTP_STAST
9 Phage *N*-acetylmuramoyl-L-alanine amidase [*S. lugdunensis* HKU09-01]	488	YP_003472450
10 Lysostaphin [*S. simulans* bv. *staphylolyticus*]	452	YP_003505772
11 Autolysin [*S. pseudintermedius* HKU10-03]	251	YP_004148764
	463	YP_189215

TABLE I (*continued*)

	AA	Accession #
12 Prophage, amidase, putative [*S. epidermidis* RP62A]		
13 ORF015 [*Staphylococcus* phage Twort]	467	YP_238716
14 ORF021 [*Staphylococcus* phage 85]	213	YP_239752
15 ORF018 [*Staphylococcus* phage 85]	237	YP_239755
16 ORF007 [*Staphylococcus* phage 2638A]	486	YP_239818
17 ORF004 [*Staphylococcus* phage 37]	639	YP_240099
18 ORF006 [*Staphylococcus* phage 37]	481	YP_240103
19 ORF003 [*Staphylococcus* phage EW]	630	YP_240176
20 ORF007 [*Staphylococcus* phage EW]	482	YP_240182
21 ORF018 [*Staphylococcus* phage X2]	213	YP_240847
22 ORF019 [*Staphylococcus* phage X2]	210	YP_240849
23 Amidase (peptidoglycan hydrolase) [*S. haemolyticus* JCSC1435]	464	YP_253663
24 *N*-Acetylmuramoyl-L-alanine amidase [*S. haemolyticus* JCSC1435]	494	YP_254248
25 Hypothetical protein SH2336 [*S. haemolyticus* JCSC1435]	647	YP_254251
26 Mannosyl-glycoprotein endo-β-*N*-acetylglucosaminidase [*S. capitis* SK14]	626	ZP_03614366
27 Autolysin [*S. warneri* L37603]	477	ZP_04679079
28 Possible *N*-acetylmuramoyl-L-alanine amidase [*S. epidermidis* BCM-HMP0060]	574	ZP_04824947
29 Conserved hypothetical protein [*S. aureus* subsp. *aureus* E1410]	325	ZP_05610313
30 Peptidoglycan hydrolase [*S. aureus* A9299]	405	ZP_05688267
31 Amidase [*S. aureus* A9299]	405	ZP_05688584
32 Conserved hypothetical protein [*S. aureus* A6300]	494	ZP_05694215
33 Bacteriophage amidase [*S. epidermidis* M23864: W2(grey)]	467	ZP_06614678
34 *N*-Acetylmuramoyl-L-alanine amidase [*S. aureus* A8819]	394	ZP_06817547
35 Petidoglycan hydrolase, putative [*S. aureus* subsp. *aureus* MR1]	392	ZP_06859771
36 *N*-Acetylmuramoyl-L-alanine amidase [*S. aureus* A8796]	419	ZP_06930779
37 *N*-Acetylmuramoyl-L-alanine amidase [*S. aureus* subsp. *aureus* ATCC BAA-39]	564	ZP_07361756

Identities within groups are generally $\geq$90%.
Exceptions: [1]89%; [2]87%; [3]89%; [4]89%; [5]88%; [6]88%; [7]89%; [8]87%; [9]86%; [10]84%.

TABLE II Streptococcal PG hydrolases

	AA	Accession #
Group 1a		
Cpl-1 [*S. pneumoniae*]	339	NP_044837.1
Cpl-9 [*S. pneumoniae*]	339	P19386.1
Group 1b		
PH10 lysin [*S. oralis*]	334	YP_002925184.1
Group 1c		
Cpl-7 [*S. pneumoniae*]	342	P19385.1
Group 2a		
Autolysin [*S. pneumoniae* SP3-BS71]	318	ZP_01819152.1
Lytic amidase [*S. pneumoniae* SP195]	318	ZP_02714370.1
Autolysin [*S. pneumoniae* SP11-BS70]	318	ZP_01824138.1
Lytic amidase [*S. pneumoniae* CDC1873-00]	318	ZP_02708645.1
Autolysin [*S. pneumoniae* SP19-BS75]	318	ZP_01832999.1
Lytic amidase [*S. pneumoniae* 670-6B]	318	YP_003880285.1
Lytic amidase [*S. pneumoniae* Hungary19A-6]	318	YP_001693491.1
Autolysin [*S. pneumoniae* SP6-BS73]	318	ZP_01821560.1
Autolysin [*S. pneumoniae* AP200]	318	YP_003875665.1
MM1 lysin [*S. pneumoniae*]	318	NP_150182.1
Lytic amidase [*S. pneumoniae* SP195]	318	ZP_02712971.1
VO1 amidase [*S. pneumoniae*]	318	CAD35393.1
HB-3 amidase [*S. pneumoniae*]	318	P32762.1
Lytic amidase [*S. pneumoniae* CDC3059-06]	318	ZP_02718952.1
Lytic amidase [*S. pneumoniae* 70585]	318	YP_002739391.1
Lytic amidase [*S. pneumoniae* SP-BS293]	318	ZP_07345341.1
Lytic amidase [*S. pneumoniae* P1031]	318	YP_002737318.1
Autolysin [*S. pneumoniae* SP23-BS72]	318	ZP_01835850.1
Group 2b		
Autolysin [*S. pneumoniae*]	313	AAK29073.1
Autolysin [*S. pneumoniae* TIGR4]	318	NP_346365.1
Amidase [*S. pneumoniae* R6]	318	NP_359346.1
Putative amidase [*S. pneumoniae* INV104]	318	CBW37351.1
Autolysin [*S. pneumoniae* SP3-BS71]	318	ZP_01818711.1
VO1 amidase [*S. pneumoniae* 8249]	318	CAD35389.1
LytA amidase [*S. pneumoniae*]	318	CAJ34409.1
LytA amidase [*S. pneumoniae*]	318	CAJ34410.1
Autolysin [*S. pneumoniae* 670-6B]	318	YP_003880176.1

TABLE II (*continued*)

	AA	Accession #
Autolysin [*S. pneumoniae*]	313	AAK29074.1
Autolysin [*S. pneumoniae* CDC1087-00]	318	ZP_02711922.1
Autolysin [*S. pneumoniae*]	313	CBE65469.1
LytA autolysin [*S. pneumoniae*]	302	CAB53774.1
Autolysin [*S. pneumoniae* SP11-BS70]	318	ZP_01825916.1
LytA autolysin [*S. pneumoniae*]	302	CAB53770
Autolysin [*S. pneumoniae* 670-6B]	318	YP_003878279.1
Autolysin [*S. pneumoniae* SP14-BS69]	318	ZP_01828965.1
Autolysin [*S. pneumoniae* JJA]	318	YP_002736862.1
Group 2c		
LytA amidase [*S. pneumoniae*]	316	CAD12111.1
Amidase [*S. mitis* SK597]	316	ZP_07640915.1
LytA amidase [*S. pneumoniae*]	316	CAD12115.1
LytA amidase [*S. pneumoniae* sp. 1504]	316	CAJ34416.1
LytA amidase [*S. pneumoniae*]	316	CAD12112.1
LytA amidase [*S. pneumoniae*]	316	CAD12116.1
LytA amidase [*S. pneumoniae*]	316	CAD12106.1
LytA amidase [*S. pseudopneumoniae*]	316	CAJ34411.1
LytA amidase [*S. pneumoniae*]	316	CAD12108.1
LytA amidase [*S. pneumoniae* sp. 578]	316	CAJ34413.1
LytA amidase [*S. pneumoniae* sp. 3072]	316	CAJ34420.1
LytA amidase [*S. pneumoniae*]	316	CAD12113.1
LytA amidase [*S. pneumoniae*]	316	CAD12110.1
LytA amidase [*S. pneumoniae* sp. 2410]	316	CAJ34419.1
LytA101 [*S. pneumoniae*]	316	AAB23082.1
Autolysin [*S. mitis*]	300	CAB76388.1
Autolysin [*Streptococcus* sp.]	300	CAB76391.1
LytA amidase [*S. pneumoniae*]	316	CAD12114.1
Autolysin [*Streptococcus* sp.]	300	CAB76389.1
Autolysin [*Streptococcus* sp.]	300	CAB76392.1
LytA amidase [*S. pneumoniae* sp. 1237]	316	CAJ34414.1
Autolysin [*Streptococcus* sp.]	300	CAB76394.1
LytA amidase [*S. pneumoniae*]	316	CAD12109.1
LytA amidase [*S. pneumoniae*]	316	CAD12107.2
Autolysin [*Streptococcus* sp.]	300	CAB76390.1
Group 2d		
LytA amidase [*S. mitis* B6]	318	YP_003445618.1
LytA-like amidase [*S. mitis*]	318	CAF02035.1
EJ-1 lysin [*S. pneumoniae*]	316	NP_945312.1

(*continued*)

TABLE II *(continued)*

	AA	Accession #
Group 3a		
Putative lysin [*S. pyogenes* phage 315.2]	402	NP_664726.1
Putative amidase [*S. pyogenes* phage 315.1]	401	NP_664535.1
Phage-associated lysin [*S. pyogenes* NZ131]	402	YP_002286426.1
spyM18_0777 [*S. pyogenes* MGAS8232]	401	NP_606945.1
Phage-associated lysin [*Streptococcus* phage 9429.1]	404	YP_596324.1
spyM18_1750 [*S. pyogenes* MGAS8232]	401	NP_607778.1
Amidase [*S. pyogenes* MGAS10394]	401	YP_060660.1
Putative phage amidase [*S. pyogenes* str. Manfredo]	401	YP_001128106.1
Spy_1438 [*S. pyogenes* M1 GAS]	401	NP_269522.1
spyM18_1448 [*S. pyogenes* MGAS8232]	401	NP_607527.1
Amidase [*S. pyogenes* ATCC 10782]	401	ZP_07461342.1
Amidase [*S. pyogenes* ATCC10782]	401	ZP_07460525.1
Group 3b		
Phage-associated lysin [*S. pyogenes* MGAS10394]	400	YP_059383.1
370.1 lysin [*S. pyogenes*]	400	NP_268942.1
Amidase [*S. pyogenes* ATCC 10782]	400	ZP_07461599.1
Lysin [*S. dysgalactiae* subsp. *equisimilis* GGS_124]	400	YP_002996819.1
P9 lysin [*S. equi* phage P9]	400	YP_001469230.1
Group 3c		
315.6 lysin [*S. pyogenes* MGAS315]	244	NP_665215.1
SPs0453 [*S. pyogenes* SSI-1]	226	NP_801715.1
SPs1121 [*S. pyogenes* SSI-1]	226	NP_802383.1
Group 3d		
Phage-associated lysin [*S. equi* subsp. *equi* 4047]	404	YP_002745608.1
Phage amidase [*S. equi* subsp. *equi* 4047]	403	YP_002746965.1
Group 3e		
Phage-associated lysin [*S. pyogenes* MGAS5005]	398	YP_282779.1
Phage 2096.1 lysin [group A *Streptococcus*]	398	YP_600196.1
Phage amidase [*S. equi* subsp. *equi* 4047][1]	398	YP_002746181.1

TABLE II (*continued*)

	AA	Accession #
Group 3f		
spyM18_1242 [*S. pyogenes* MGAS8232]	161	NP_607353.1
Group 3g		
Phage-associated lysin [*S. pyogenes* MGAS10394]	213	YP_060304.1
Group 4		
Putative phage lysin [*S. pyogenes* phage 315.5]	254	NP_665110.1
SpyoM01000009 [*S. pyogenes* M49 591]	251	ZP_00366664.1
Phage-associated lysin [*S. pyogenes* MGAS5005]	254	YP_282364.1
Group 5a		
Phi3396 lysin [*S. dysgalactiae* subsp. *equisimilis*]	253	YP_001039943.1
Phage NZ131.2 lysin [*S. pyogenes*]	249	YP_002285797.1
Phage-associated lysin [*S. pyogenes* MGAS10394]	250	YP_060862.1
Group 5b		
Phage-associated lysin [*S. pyogenes* MGAS10394]	203	YP_060515.1
Group 6a		
Phage 9429.2 lysin [*S. pyogenes*]	373	YP_596581.1
Group 6b		
B30 lysin [*S. agalactiae*]	445	AAN28166.2
49.7 kDA protein [*S. equi*]	444	AAF72807.1
Putative lysin [*S. pyogenes* phage 370.3]	444	NP_269184.1
PlyGBS [*S. agalactiae* phage NCTC11261]	443	AAR99416.1
Phage-associated lysin [*S. pyogenes* MGAS6180]	444	YP_280438.1
Prophage LambdaSa03 endolysin [*S. agalactiae*]	443	YP_329285.1
49.7 kDa protein [*S. agalactiae* 18RS21]	447	ZP_00780878.1
Putative phage lysin [*S. pyogenes* strain Manfredo]	444	YP_001128574.1
Phage lysin [*S. equi* subsp. *equi* 4047]	444	YP_002747253.1
Group 7		
LambdaSa1 lysin [*S. agalactiae* 2603 V/R]	239	NP_687631.1
Endolysin [*S. agalactiae* H36B]	248	ZP_00782522.1

(*continued*)

TABLE II (*continued*)

	AA	Accession #
Group 8a		
Putative amidase [*S. pyogenes* phage 315.3]	404	NP_664900.1
Putative amidase [*S. pyogenes* MGAS8232]	405	NP_606641.1
Phage protein [*S. pyogenes* MGAS10750]	405	YP_602773.1
Putative phage lysin [*S. pyogenes* str. Manfredo]	402	YP_001128256.1
Group 8b		
LambdaSa2 lysin [*S. dysgalactiae* subsp. *equisimilis* GGS_124]	449	YP_002997317.1
Group 8c		
LambdaSa2 lysin [*S. agalactiae* 2603 V/R]	468	NP_688827.1
Group 8d		
SMP lysin [*S. suis*]	481	YP_950557.1
Group 9a		
Cell wall-binding repeat family protein [*S. mitis* SK321]	568	ZP_07643272.1
Cell wall-binding repeat family protein [*S. mitis* SK597]	570	ZP_07641594.1
Endo-β-*N*-acetylglucosaminidase [*S. mitis* NCTC 12261]	568	ZP_07645063.1
LytB [*S. mitis*]	568	ACO37163.1
LytB [*S. mitis* B6]	570	YP_003446078.1
Group 9b		
Endo-β-*N*-acetylglucosaminidase [*S. pneumoniae* 70585]	702	YP_002740268.1
Endo-β-*N*-acetylglucosaminidase [*S. pneumoniae* G54]	702	YP_002037600.1
Endo-β-*N*-acetylglucosaminidase [*S. pneumoniae* Hungary19A-6]	702	YP_001694410.1
Endo-β-*N*-acetylglucosaminidase [*S. pneumoniae* P1031]	702	YP_002738134.1
Endo-β-*N*-acetylglucosaminidase [*S. pneumoniae* Taiwan19F-14]	702	YP_002742657.1
Endo-β-*N*-acetylglucosaminidase [*S. pneumoniae* BS397]	702	ZP_07350631.1
Group 9c		
Endo-β-*N*-acetylglucosaminidase [*S. pneumoniae* SP-BS293]	614	ZP_07345852.1

TABLE II (*continued*)

	AA	Accession #
Endo-β-*N*-acetylglucosaminidase [*S. pneumoniae*]	614	AAK19156.1
Endo-β-*N*-acetylglucosaminidase [*S. pneumoniae* CDC1087-00]	614	ZP_02710425.1
Endo-β-*N*-acetylglucosaminidase [*S. pneumoniae* INV104]	614	CBW36509.1
LytB [*Spneumoniae* AP200]	614	YP_003876588.1
Group 9d		
Endo-β-*N*-acetylglucosaminidase [*S. pneumoniae* CGSP14]	677	YP_001835658.1
Group 9e		
Endo-β-*N*-acetylglucosaminidase [*S. pneumoniae* CCRI 1974]	658	ZP_04525138.1
Endo-β-*N*-acetylglucosaminidase [*S. pneumoniae* CDC0288-04]	658	ZP_02715197.1
Endo-β-*N*-acetylglucosaminidase [*S. pneumoniae* CDC3059-06]	658	ZP_02718537.1
Endo-β-*N*-acetylglucosaminidase [*S. pneumoniae* JJA]	658	YP_002735981.1
Endo-β-*N*-acetylglucosaminidase [*S. pneumoniae* SP23-BS72]	658	ZP_01834875.1
Endo-β-*N*-acetylglucosaminidase [*S. pneumoniae* MLV-016]	658	ZP_02721563.1
Endo-β-*N*-acetylglucosaminidase [*S. pneumoniae* TIGR4]	658	NP_345446.1
Endo-β-*N*-acetylglucosaminidase [*S. pneumoniae* SP3-BS71]	658	ZP_01817975.1
Group 9f		
Endo-β-*N*-acetylglucosaminidase [*S. pneumoniae* INV200]	721	CBW34519.1
Endo-β-*N*-acetylglucosaminidase [*S. pneumoniae* R6]	721	NP_358461.1
Endo-β-*N*-acetylglucosaminidase [*S. pneumoniae* TCH8431/19A]	721	YP_003724965.1
Group 10		
Endo-β-*N*-acetylglucosaminidase [*S. mitis* ATCC6249]	750	ZP_07462509.1
Endo-β-*N*-acetylglucosaminidase [*S. sanguinis* ATCC49296]	750	ZP_07887886.1

(*continued*)

TABLE II (*continued*)

	AA	Accession #
Endo-β-*N*-acetylglucosaminidase [*Streptococcus* sp. oral taxon str. 73H25AP]	750	ZP_07458768.1
Group 11a		
Lysozyme [*S. mitis* NCTC 12261]	525	ZP_07644807.1
LytC Cpb13 [*S. mitis* B6]	536	YP_003446665.1
Group 11b		
Cell wall-binding protein [*S. mitis* SK564]	504	ZP_07642782.1
Cell wall-binding protein [*S. mitis* SK597]	504	ZP_07641292.1
Cell wall-binding protein [*S. mitis* SK321]	493	ZP_07642984.1
GROUP 11c		
Lysozyme [*S. pneumoniae* SP3-BS71]	270	ZP_01818179.1
Group 11d		
1,4-β-*N*-Acetylmuramidase [*S. pneumoniae* CDC1873-00]	490	ZP_02708500.1
1,4-β-*N*-Acetylmuramidase [*S. pneumoniae* P1031]	490	YP_002738710.1
1,4-β-*N*-Acetylmuramidase [*S. pneumoniae* SP11-BS70]	490	ZP_01824964.1
1,4-β-*N*-Acetylmuramidase [*S. pneumoniae* SP9-BS68]	490	ZP_01822918.1
1,4-β-*N*-Acetylmuramidase [*S. pneumoniae* 70585]	490	YP_002740840.1
1,4-β-*N*-Acetylmuramidase [*S. pneumoniae* CDC1087-00]	490	ZP_02711346.1
1,4-β-*N*-Acetylmuramidase [*S. pneumoniae* TCH8431/19A]	501	YP_003725251.1
1,4-β-*N*-Acetylmuramidase [*S. pneumoniae* R6]	501	NP_359024.1
1,4-β-*N*-Acetylmuramidase [*S. pneumoniae*]	492	AAK19157.1
ATP-dependent protease [*S. pneumoniae* SP6-BS73]	490	ZP_01820060.1
Endo-β-*N*-acetylglucosaminidase [*S. pneumoniae* G54]	490	YP_002038205.1
Lysozyme [*S. pneumoniae* Taiwan 19 F-14]	493	YP_002742915.1
Lysozyme [*S. pneumoniae* BS455]	490	ZP_07341428.1
Lysozyme [*S. pneumoniae* CGSP14]	501	YP_001836276.1
LytC autolysin [*S. pneumoniae*]	501	CAA08765.1

TABLE II (*continued*)

	AA	Accession #
Putative choline-binding glycosyl hydrolase [*S. pneumoniae* INV104]	490	CBW37026.1
Putative choline-binding glycosyl hydrolase [*S. pneumoniae* ATCC700669]	490	YP_002511487.1
SpneCMD 07616 [*S. pneumoniae* str. Canada MDR 19 F]	490	ZP_06964203.1
SpneT 0200379 [*S. pneumoniae* TIGR4]	490	ZP_01409152.1
Group 11e		
1,4-β-*N*-Acetylmuramidase [*S. pneumoniae* SP14-BS69]	311	ZP_01828088.1
Group 11f		
Lysozyme [*S. pneumoniae* SP19-BS75]	227	ZP_01833670.1
Group 12a		
Pal [*S. pneumoniae* phage DP-1]	296	O03979.1
Group 12b		
gp56 [*Streptococcus* phage SM1]	295	NP_862895.1
Group 13a		
S3b lysin [*S. thermophilus*][2]	206 + 82 [5]	AAF24749.1
DT1 lysin [*S. thermophilus*]	200 + 75 [5]	NP_049413.1 + NP_049415.2
ALQ13.2 lysin [*S. thermophilus*]	200 + 75 [5]	YP_003344870.1 + YP_003344872.1
Orf28 [*S. thermophilus* phage 858]	200 + 75 [5]	YP_001686822.1+ YP_001686825.1
Phage 2972 lysin [*S. thermophilus*][3]	199 + 75 [5]	YP_238509.1 + YP_238512.1
Group 13b		
Putative phage PH15 endolysin [*S. gordonii*]	283	YP_001974380.1
Group 13c		
Abc2 lysin [*S. thermophilus*]	281	YP_003347431.1
ORF44 [*S. thermophilus* phage 7201]	281	NP_038345.1
Phage 5093 lysin [*S. thermophilus* CSK939]	281	YP_002925118.1
Phage O1205 p51 [*S. thermophilus* CNRZ1205][4]	281	NP_695129.1
Group 13d		
Sfi11 lysin [*S. thermophilus*]	288	NP_056699.1
Sfi18 lysin [*S. thermophilus*]	288	AAF63073.1

(*continued*)

TABLE II *(continued)*

	AA	Accession #
Sfi19 lysin [*S. thermophilus*]	288	NP_049942.1
Sfi21 lysin [*S. thermophilus*]	288	NP_049985.1
Group 13e		
STRINF 01560 [*S. infantarius* subsp. *infantarius* ATCC BAA-102]	281	ZP_02920679.1
Stand-alone proteins		
1 700P1 lysin [*S. uberis*]	236	ABB02702.1
2 Phage M102 gp19 [*S. mutans*]	273	YP_002995476.1
3 PlyC [Group A *Streptococcus* phage C1]	465 + 72[6]	NP_852017.2

Identities within groups are generally ≥90%.
Exceptions: [1]88%; [2]88%; [3]84%; [4]86%;
[5] encoded by two coding regions separated by an intron;
[6] multimeric protein consisting of two gene products.

endolysin featuring a CHAP endopeptidase and an amidase domain but shares less than 50% amino acid sequence identity with the Φ11 endolysin, despite cleaving identical bonds on purified staphylococcal PG (Becker *et al.*, 2009a).

The presence of two catalytic domains does not necessarily indicate that both are equally active when lysing from without. The streptococcal λSa2 phage endolysin D-glutaminyl-L-lysine endopeptidase activity domain was shown via deletion analysis to be responsible for almost all of the hydrolytic activity of this enzyme, whereas its *N*-acetylglucosaminidase domain was found to be almost devoid of activity (Donovan and Foster-Frey, 2008). The same dominant domain phenomenon was demonstrated with both deletion and site-directed mutational analysis for the streptococcal B30 phage endolysin [99% identical to PlyGBS (Cheng and Fischetti, 2007)]. The N-terminal D-alanyl-L-alanyl endopeptidase domain is responsible for virtually all *in vitro* streptolytic activity and the glycosidase domain is silent in these assays (Donovan *et al.*, 2006b), despite both domains showing catalytic activity on purified PG (Pritchard *et al.*, 2004). There is no current explanation for this recurrent pattern of a highly conserved lytic domain that is seemingly inactive (when applied externally) in these unrelated streptococcal proteins (λSa2 vs B30). These two proteins share little in the way of domain architecture (lytic-CBD-CBD-lytic vs lytic-lytic-CBD), there are virtually no conserved sequences between them, and each utilizes an unrelated CBD (Cpl-7-like vs SH3b).

This pattern is not limited to the streptococcal lysins. Interestingly, inactive lytic domains are also observed in staphylolytic endolysins. The staphylolytic Φ11 endolysin was shown to have a very active N-terminal D-alanyl-glycyl endopeptidase domain via deletion analysis (Donovan *et al.*, 2006c; Sass and Bierbaum, 2007) and a nearly silent *N*-acetylmuramoyl-L-

TABLE III Enterococcal PG hydrolases

	AA	Accession #
Group 1		
Endolysin, putative [*E. faecalis* V583]	433	NP_814147.1
Endolysin [*E. faecalis* ATCC 29200]	433	ZP_04437810.1
Lysin [*E. faecalis* DS5]	433	ZP_05562195.1
Lysin [*E. faecalis* T1]	433	ZP_05423767.1
Lysin [*E. faecalis* HIP11704]	433	ZP_05568662.1
Endolysin [phage phiFL4A]	433	YP_003347409.1
Endolysin [*E. faecalis* V583]	433	NP_816427.1
Lysin [*E. faecalis* AR01/DG]	433	ZP_05593964.1
Endolysin [*E. faecalis* X98]	433	ZP_05598729.1
Endolysin [phage phiFL1A]	433	YP_003347517.1
Endolysin [phage phiFL2A]	433	YP_003347352.1
Endolysin [phage phiFL1B]	433	ACZ63822.1
Endolysin [phage phiFL1C]	433	ACZ63895.1
Endolysin [phage phiFL2B]	433	ACZ64018.1
Endolysin [*E. faecalis* T8]	433	ZP_05558876.1
Lysin [*E. faecalis* JH1]	433	ZP_05573731.1
Group 2		
Lysin [*E. faecalis* Merz96]	419	ZP_05565596.1
Endolysin [*E. faecalis* R712]	419	ZP_06629599.1
Endolysin [*E. faecalis* S613]	419	ZP_06631635.1
Endolysin [phage phiEf11]	419	YP_003358816.1
Endolysin [*E. faecalis* X98]	419	ZP_05599066.1
Endolysin [*E. faecalis* CH188]	419	ZP_05585395.1
Endolysin [phage phiFL3A]	419	YP_003347625.1
Endolysin [phage phiFL3B]	419	ACZ64148.1
Lysin [*E. faecalis* JH1]	419	ZP_05572412.1
Lysin [*E. faecalis* D6]	419	ZP_05581557.1
Group 3		
Endolysin [*E. faecalis* ATCC 29200]	412	ZP_04438395.1
Phage lysin [*E. faecalis* T1]	412	ZP_05422953.1
Endolysin [*E. faecalis* V583]	413	NP_815667.1
Phage lysin [*E. faecalis* HIP11704]	413	ZP_05568908.1
Phage lysin [*E. faecalis* E1Sol]	413	ZP_05576004.1
Endolysin [*E. faecalis* TX1322]	413	ZP_04434151.1
Endolysin [*E. faecalis* CH188]	413	ZP_05584633.1
Phage lysin [*E. faecalis* ATCC 4200] [1]	413	ZP_05476312.1
Endolysin [*E. faecalis* TUSoD Ef11]	394	ZP_04647652.1
Endolysin [*E. faecalis* T8]	413	ZP_05559457.1

(continued)

TABLE III (*continued*)

	AA	Accession #
Group 4		
Endolysin [*E. faecium* E1039]	394	ZP_06675756.1
Endolysin [*E. faecium* E1039]	425	ZP_06674744.1
Group 5		
PlyP100 [*E. faecalis* HIP11704]	322	ZP_05566775.1
Endolysin [*E. faecalis* Merz96]	322	ZP_05564324.1
Endolysin [*E. faecalis* R712]	368	ZP_06628454.1
Endolysin [*E. faecalis* S613]	368	ZP_06632418.1
Endolysin [*E. faecalis* DS5]	322	ZP_05561234.1
Endolysin [*E. faecalis* T8]	351	ZP_05557995.1
Endolysin [*E. faecalis* V583]	368	NP_815207.1
Endolysin [*E. faecalis* R712]	368	ZP_06628239.1
Endolysin [*E. faecalis* S613]	368	ZP_06633896.1
Endolysin [*E. faecalis* Fly1]	341	ZP_05579618.1
Group 6		
Amidase [*E. faecalis* TX0104]	374	ZP_03948603.1
Amidase [*E. faecalis* HH22]	374	ZP_03983131.1
Amidase [*E. faecalis* TX1322]	374	ZP_04434756.1
Endolysin [*E. faecalis* R712]	374	ZP_06629056.1
Endolysin [*E. faecalis* S613]	374	ZP_06632253.1
Endolysin [*E. faecalis* V583]	365	NP_815016.1
Endolysin [*E. faecalis* ATCC 29200]	374	ZP_04438946.1
Endolysin [*E. faecalis* TUSoD Ef11]	365	ZP_04647840.1
Endolysin [*E. faecalis* X98]	365	ZP_05599811.1
Endolysin [*E. faecalis* T8]	361	ZP_05558304.1
Endolysin [*E. faecalis* ATCC 4200]	352	ZP_05475717.1
Endolysin [*E. faecalis* JH1]	350	ZP_05573170.1
Endolysin [*E. faecalis* HIP11704]	345	ZP_05569483.1
Endolysin [*E. faecalis* Fly1]	345	ZP_05579809.1
Endolysin [*E. faecalis* Merz96]	345	ZP_05566285.1
Endolysin [*E. faecalis* AR01/DG]	345	ZP_05592904.1
Endolysin [*E. faecalis* DS5]	345	ZP_05562950.1
Group 7		
Amidase [*E. faecium* 1,141,733]	338	ZP_05666679.1
Amidase [*E. faecium* Com15]	339	ZP_05677833.1
Amidase [*E. faecium* 1,231,501]	338	ZP_05664801.1
Amidase [*E. faecium* E980]	339	ZP_06681905.1
Amidase [*E. faecium* 1,230,933]	339	ZP_05659803.1
Amidase [*E. faecium* U0317]	339	ZP_06702043.1
Amidase [*E. faecium* 1,231,408]	339	ZP_05673558.1

TABLE III (*continued*)

	AA	Accession #
Amidase [*E. faecium* Com15]	338	ZP_05678707.1
Amidase [*E. faecium* 1,231,410]	339	ZP_05671179.1
Amidase [*E. faecium* E980]	336	ZP_06683607.1
Amidase [*E. faecium* E1071]	339	ZP_06680220.1
Amidase, family 2 [*E. faecium* C68]	320	ZP_05832333.1
Amidase [*E. faecium* 1,230,933]	336	ZP_05659231.1
Amidase [*E. faecium* 1,231,502]	336	ZP_05662248.1
Amidase [*E. faecium* U0317]	336	ZP_06700224.1
Amidase [*E. faecium* 1,231,501]	338	ZP_05663923.1
Amidase [*E. faecium* 1,231,410]	321	ZP_05671689.1
Amidase, family 2 [*E. faecium* TC 6]	323	ZP_05924003.1
Amidase, family 2 [*E. faecium* D344SRF]	323	ZP_06447215.1
Amidase [*E. faecium* 1,231,502]	306	ZP_05663252.1
Amidase [*E. faecium* E1636]	308	ZP_06695864.1
Group 8		
Amidase, family 2 [*E. faecium* DO]	341	ZP_00602919.1
Amidase [*E. faecium* E1162]	341	ZP_06676885.1
Amidase [*E. faecium* 1,231,408]	341	ZP_05673081.1
Amidase [*E. faecium* 1,231,410]	323	ZP_05671663.1
Amidase, family 2 [*E. faecium* C68]	322	ZP_05833245.1
Amidase [*E. faecium* E1636]	310	ZP_06694650.1
Amidase [*E. faecium* 1,231,502]	291	ZP_05661451.1
Group 9		
Amidase [*E. faecalis* V583]	503	NP_814047.1
Amidase [*E. faecalis* HH22]	503	ZP_03985946.1
Amidase [*E. faecalis* T11]	503	ZP_05595649.1
Amidase [*E. faecalis* Fly1]	503	ZP_05578550.1
Amidase [*E. faecalis* TX0104]	503	ZP_03950088.1
Amidase [*E. faecalis* AR01/DG]	503	ZP_05594613.1
Amidase [*E. faecalis* Merz96]	503	ZP_05564795.1
Amidase, family 4 [*E. faecalis* R712]	503	ZP_06628637.1
Amidase, family 4 [*E. faecalis* S613]	503	ZP_06632633.1
Amidase, family 4 [*E. faecalis* T8]	503	ZP_05560568.1
Amidase [*E. faecalis* HIP11704]	503	ZP_05568347.1
Amidase [*E. faecalis* ATCC 4200]	503	ZP_05475182.1
Amidase [*E. faecalis* TX1322]	503	ZP_04435643.1
Amidase [*E. faecalis* X98]	503	ZP_05598533.1
Amidase [*E. faecalis* ATCC 29200]	501	ZP_04439231.1
Amidase [*E. faecalis* DS5]	503	ZP_05560989.1
Amidase [*E. faecalis* E1Sol]	503	ZP_05575902.1

(*continued*)

TABLE III (*continued*)

	AA	Accession #
Amidase [*E. faecalis* JH1]	503	ZP_05572849.1
Amidase [*E. faecalis* TUSoD Ef11]	501	ZP_04648145.1
Group 10		
Amidase [*E. faecalis* TX0104]	309	ZP_03948310.1
Amidase, family 4 [*E. faecalis* R712]	309	ZP_06630528.1
Amidase, family 4 [*E. faecalis* S613]	309	ZP_06633335.1
Group 11		
Amidase [*E. faecalis* T1]	663	ZP_05423074.1
Amidase [*E. faecalis* T11]	649	ZP_05596538.1
Amidase [*E. faecalis* Fly1]	652	ZP_05579285.1
Amidase [*E. faecalis* E1Sol]	649	ZP_05576670.1
Amidase [*E. faecalis* V583]	652	NP_815520.1
Amidase [*E. faecalis* TX0104]	652	ZP_03949059.1
Amidase [*E. faecalis* HH22]	652	ZP_03983681.1
Amidase, family 4 [*E. faecalis* R712]	652	ZP_06629298.1
Amidase, family 4 [*E. faecalis* S613]	652	ZP_06633447.1
Group 12		
Amidase [*E. casseliflavus* EC20]	655	ZP_05655421.1
Amidase [*E. casseliflavus* EC30]	650	ZP_05645789.1
Amidase [*E. casseliflavus* EC10]	650	ZP_05652119.1
Stand-alone proteins		
1 Amidase [*E. gallinarum* EG2]	703	ZP_05649621.1
2 PlyV12 [phage phi1]	314	AAT01859.1
3 Amidase [*E. casseliflavus* EC20]	715	ZP_05656866.1
4 Amidase [phage phiEF24C]	289	YP_001504118.1
5 Amidase [phage EFAP-1]	328	YP_002727874.1
6 Endolysin [*E. faecalis* HH22]	270	ZP_03985506.1
7 Amidase [*E. faecalis* T3]	523	ZP_05503383.1

Identities within groups are generally $\geq$90%. Exception: [1]89%.

alanine amidase domain (Sass and Bierbaum, 2007). The staphylococcal phage endolysin LysK shares a high degree of domain architecture with the Φ11 endolysin and shows the same pattern of a highly active N-terminal CHAP endopeptidase domain (Becker *et al.*, 2009a; Horgan *et al.*, 2009) and a nearly silent second lytic (amidase) domain. This pattern also shows up in numerous (but not all) SH3b containing staphylococcal endolysins (D. M. Donovan, unpublished data). The fact that this pattern is occurring in seemingly unrelated proteins and in more than one genera begs the question of why would this be evolutionarily conserved. A discussion of potential explanations has been presented previously (Donovan and Foster-Frey, 2008) and thus will not be repeated here, but the most likely explanation lies in the potential (unidentified) differences between lysis from without (where these nearly silent domains have been identified) vs. lysis from within. What is needed are a series of experiments that test the effect of a mutant endolysin gene, with either the active or the silent domain ablated, in a wild-type phage lytic cycle.

E. Measuring endolysin activity

The catabolic activity of PG hydrolases has been studied and quantified for many years. The earliest assays did not focus on antimicrobial activity but rather used PG hydrolase enzymes to degrade PG in order to elicit PG structure (Schleifer and Kandler, 1972; Weidel and Pelzer, 1964). These early studies laid the ground work for identification of the enzymes as antimicrobials. It should be noted that although multiple assays have been used to quantify PG hydrolase activity, there can be quantitative discrepancies from assay to assay (Kusuma and Kokai-Kun, 2005). Similarly, measuring PG hydrolase enzymatic activity is not the same as measuring PG hydrolase antimicrobial activity (which by definition must assay live cells). Nonetheless, what follows is a list of both qualitative and quantitative assays that have been employed in the study of PG hydrolases.

Turbidity reduction assays: A decrease in light scattering (i.e., turbidity reduction) of a suspension of live cells, nonviable cells (heat killed or autoclaved), or cell wall preparation/extract can be used in a spectrophotometer to assay the activity of PG hydrolases. The reduction in optical density over time (minutes or hours) can be used to calculate a rate of hydrolysis (Fig. 6). Results are compared to a "no-enzyme added, buffer-only control" preparation treated identically for the same period of time. In this manner, a specific activity of the enzyme preparation can be reported as ΔOD/time/μg lysin protein. Critical to the interpretation of these assays are considerations for whether (1) the assay is performed in the linear range of enzyme activity with excess substrate always present; (2) the maintenance of a homogeneous substrate solution (to avoid the substrate settling out of solution); and (3) the requirement for an identically treated no-enzyme control sample, the OD of which must be subtracted from the experimental sample result. There are published results using spectrophotometric turbidity reduction assays to quantify enzyme activity (Filatova *et al.*, 2010) and even determine kinetic constants (Mitchell *et al.*, 2010). However, some caution should be used when interpreting the results because a loss of optical density is not always directly equated with antimicrobial activity (Fig. 6). Furthermore, variation in the assay between laboratories and arbitrary unit definitions often makes comparison of lytic activities difficult. Activities of phage-encoded

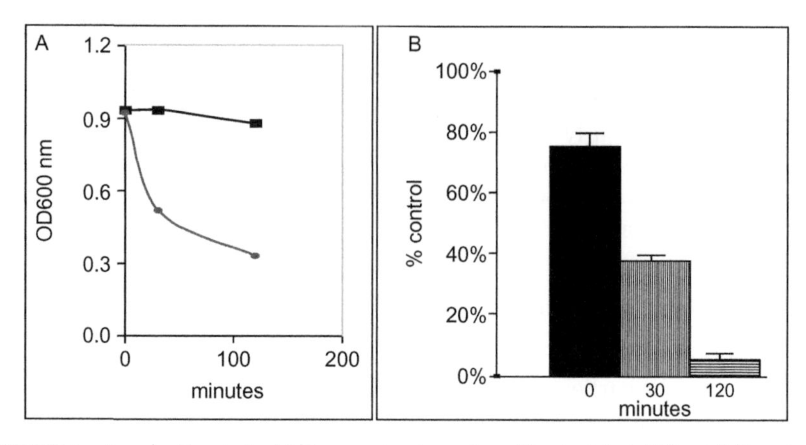

FIGURE 6 A reduction in turbidity equates to reduced bacterial viability. (A) Twenty-five micrograms of Φ11 endolysin [construct Φ11-194 (Donovan *et al.*, 2006)] protein (circles) and *S. aureus* cells alone (squares) were monitored for 120 min in a turbidity reduction assay. (B) Treated (Φ11-194) and nontreated (cells alone) turbidity assay samples were diluted serially and plated onto tryptic soy agar plates at 0, 30, and 120 min. Results shown reflect the CFU/ml of treated cells expressed as a percentage of the viable counts of the untreated control sample. Error bars: SEM.

and bacterial PG hydrolases reportedly range from 10^2 to 10^8 ''units'' per milligram protein (Fukushima *et al.*, 2007; Loeffler *et al.*, 2003; Loessner *et al.*, 1995a; Nelson *et al.*, 2001; Vasala *et al.*, 1995; Yoong *et al.*, 2006).

Zymogram assay: Zymograms are a simple way to follow PG hydrolase activity during purification. Briefly, endolysin preparations are electrophoresed in duplicate sodium dodecyl sulfate (SDS)–polyacrylamide gel electrophoresis gels. The gels are prepared either with or without the target cells or extracted PG embedded in the gel during polymerization. Following electrophoresis, the gel is soaked for 1 hr in a buffer compatible with the lytic enzyme to remove the SDS. Appearance of a cleared region in the opaque gel indicates that cells embedded in the gel were lysed at that location, most likely due to a lytic protein/agent in the gel. This too is not an antimicrobial assay per se as the bacterial cells are often heat treated before mixing them with the gel matrix and are obviously SDS treated. Nonetheless, a zymogram is particularly useful for identifying putative PG hydrolases and offers a higher sensitivity level than the turbidity reduction assays.

Minimum inhibitory concentration (MIC) and minimum bactericidal concentration (MBC): MIC and MBC are classical assays for quantifying the antimicrobial activity of a variety of drugs. The protocols are described in detail in bacteriological manuals (Jones *et al.*, 1985). Briefly, a 2× dilution series (100, 50, 25 μg, etc.) of the compound to be assayed (i.e., antibiotic or

PG hydrolase) is established in a defined volume (usually in a 96-well plate) of growth media to which a constant number of colony-forming units (CFUs) is added (i.e., 1×10^5) and incubated overnight at 37 °C. After 20 hr, wells are examined for growth or no growth (turbid or clear) (Becker *et al.*, 2009a). The lowest concentration of the compound that can inhibit overnight growth is the MIC (usually reported in μg/ml). For MBC, an aliquot of wells with no apparent growth (clear to the eye) is plated onto agar growth media, and the lowest concentration of the compound that results in no CFUs (no viable cells) is the MBC (μg/ml). All PG hydrolase enzymes are not amenable to the MIC assay for reasons unknown. For these enzymes, cleared wells are never obtained, despite highly active PG hydrolase activity in multiple other PG hydrolase assays (D. M. Donovan, unpublished data).

Plate lysis (spot on lawn): A log growth-phase culture of target bacteria is plated onto media agar plates (e.g., 0.6 ml of culture per 100-mm plate) and allowed to air dry (≈15 min) at room temperature. Ten microliter-aliquots of known concentration(s) of the PG hydrolase are spotted onto the lawn and allowed to air dry (≈10 min) at room temperature. Plates are incubated at optimal growth temperature, and plates are assayed after overnight growth. A cleared spot on an opaque lawn indicates lytic antimicrobial activity of the PG hydrolase. Relative activity levels can be obtained by spotting a dilution series on the plate.

The disk diffusion assay is a variation of the plate assay, but opposed to spotting a known concentration directly onto a recently plated lawn of bacteria, a disk of sterile filter paper with a known concentration of PG hydrolase embedded in the disk is placed on the surface of the lawn and a ring of growth inhibition or lysis is observed after overnight growth. This method is not only dependent on a lytic agent, but simultaneously requires that the compound does not stick to the filter and can diffuse through the agar growth media.

Soft agar overlay assay: For screening of expression libraries for clones producing PG hydrolases, a soft agar overlay assay can be performed (Loessner *et al.*, 1995b; Schuch *et al.*, 2009). Replica plates containing an inducer of protein expression (e.g., isopropyl-β-D-thiogalactoside) are created from original agar plates containing transformant colonies. The replica plates are incubated at 37 °C for up to 6 hr to allow protein production. Then, the colonies are exposed to saturated chloroform vapor for ≈5 min in order to disintegrate the cytoplasmic membrane and externalize the expressed proteins and are immediately overlaid with soft agar (0.4% agar in water or buffer) containing bacterial substrate cells at high concentration. After incubation at room temperature (30 min to 18 hr), lytic phenotypes can be identified by clear halos in the turbid soft agar layer. Subsequently, positive clones can be picked from original plates for plasmid isolation and genetic characterization.

Interestingly, although each of these assays can quantify the lytic activity of PG hydrolases, when a comparison of four different assays (i.e., turbidity, disk diffusion, MIC, and MBC) was utilized to quantify the antimicrobial activity of lysostaphin, results were not always directly comparable between assays (Kusuma and Kokai-Kun, 2005). A similar result indicating qualitative but not quantitative agreement between assays was demonstrated with zymogram, turbidity reduction, MIC, and plate lysis assays using constructs of LysK, the staphylococcal phage K endolysin (Becker *et al.*, 2009a). A reasonable explanation for this quandary was proposed by Kusuma and Kokai-Kun (2005), acknowledging that bacteria express different surface factors in liquid media than on solid media (culture media can affect capsular polysaccharide production in *S. aureus*). They also suggest that the MIC assay may not be the most appropriate assay for a rapidly acting lytic enzyme, as the MIC assay measures growth inhibition while PG hydrolases probably kill the initial inocula rapidly.

F. Cell wall-binding domains of Gram-positive endolysins

Numerous domains have been assigned CBD status (see Figs. 3–5). Very few of these have been demonstrated unequivocally to be true CBDs. However, their ability to confer altered species/cell wall specificity is highly suggestive and thus CBD status has been assigned. One of the first PG hydrolase-binding domains identified was the Cpl-7 domain of the pneumococcal amidase autolysin, which requires choline or ethanolamine to achieve full activation (Garcia *et al.*, 1990). Similar Cpl-7-like CBDs have been found in a group B streptococcal λSa2 phage endolysin (Pritchard *et al.*, 2007) that appear to be essential for lytic activity (Donovan and Foster-Frey, 2008).

Another of the most well-studied PG hydrolase CBDs is that of the M23 glycyl-glycine endopeptidase, lysostaphin, and its homologue ALE-1 that is 80% identical in both lytic domains and CBDs. The lysostaphin bacterial src homology 3 (SH3b) CBD binds to the pentaglycine interpeptide bridge of the *S. aureus* PG (Grundling and Schneewind, 2006). The regions and exact amino acid residues involved in this binding have been identified in the C-terminal domain via site-directed mutagenesis of ALE-1 (Lu *et al.*, 2006). It has been reported that both lysozyme and lysostaphin are more active when the C terminus of the target of RNAIII activating protein (TRAP) is present in the staphylococcal cell wall. Binding studies indicate that the binding of these two lytic enzymes to the staphylococcal cell surface is favored by the TRAP protein C terminus (Yang *et al.*, 2008). Additional (SH3b) domains are found on many phage endolysins and appear to bind to the cell wall in an as yet undetermined manner.

For some species, CBD recognition of an epitope is analogous to recognition of a cell surface receptor by a phage tail fiber. In fact, some evidence shows that these two disparate types of proteins have evolved to target identical epitopes. For example, the γ phage of *Bacillus anthracis* forms plaques on all tested *B. anthracis* strains as well as *Bacillus cereus* 4342, which is considered a *B. anthracis* transition state strain, but not other *B. cereus* strains (Schuch *et al.*, 2002). Significantly, the lytic range of γ-phage endolysin, PlyG, mirrors the host range of the phage. In a similar fashion to pneumococcal phage tail fibers (Lopez *et al.*, 1982), pneumococcal lysin CBDs are known to bind choline in the pneumococcal cell wall (Hermoso *et al.*, 2003; Lopez *et al.*, 1982, 1997). Some CBDs of *Listeria* phage endolysins are, in fact, not just species specific, but through binding to presumably teichoic acid moieties achieve serovar or even strain specificity (Kretzer *et al.*, 2007; Loessner *et al.*, 2002; Schmelcher *et al.*, 2010). However, these highly specific endolysins are exceptions rather than the rule. In most cases, the specificity of the phage is more restrictive than its encoded endolysin. The C1 bacteriophage only forms plaques on group C streptococci, yet its endolysin, PlyC, efficiently lyses groups A, C, and E streptococci (Krause, 1957), as well as *Streptococcus uberis* (D. C. Nelson, unpublished observation). An extreme example would be PlyV12, an endolysin derived from the enterococcal phage φ1. This enzyme not only lyses *E. faecalis* and *E. faecium*, but also lyses almost all streptococcal strains (groups A, B, C, E, F, G, L, and N streptococci, *S. uberis*, *S. gordonii*, *S. intermedius*, and *S. parasanguis*), as well as staphylococcal strains (*S. aureus* and *S. epidermidus*) (Yoong *et al.*, 2004). Similarly, the *Acinetobacter baumannii* phage ΦAB2 endolysin is reported to lyse both Gram-positive and Gram-negative bacteria (Lai *et al.*, 2011).

IV. GRAM-POSITIVE ENDOLYSINS AS ANTIMICROBIALS

A. *In vivo* activity

Phage endolysins have been studied extensively for half a century, particularly those endolysins from the T-even phage that infect Gram-negative hosts. However, it has only been since 2001 that scientists have begun evaluating the use of endolysins, specifically endolysins from phage that infect Gram-positive hosts, in animal infection models of human disease. Table IV shows a complete list to date of all *in vivo* therapeutic trials that utilize bacteriophage-encoded endolysins, which are summarized here.

Nelson *et al.* (2001) were the first to use a purified phage endolysin in an *in vivo* model. It was found that oral administration of an endolysin (250 U) from the streptococcal C1 bacteriophage provided protection from upper respiratory colonization in mice challenged with 10^7 *Streptococcus*

TABLE IV Summary of *in vivo* studies with phage endolysins as antimicrobials.

Bacteria	Phage	Endolysin	Reference
Streptococcus pneumoniae	Cp-1	Cpl-1	Loeffler *et al.*, 2001 Loeffler *et al.*, 2003 Loeffler & Fischetti, 2003 Jado *et al.*, 2003 Entenza *et al.*, 2005 McCullers *et al.*, 2007 Grandgirard *et al.*, 2008
Streptococcus pneumonia	Dp-1	PAL	Loeffler & Fischetti, 2003 Jado *et al.*, 2003
Streptococcus pyogenes	C1	C1*	Nelson *et al.*, 2001
Streptococcus agalactiae	NCTC 11361	PlyGBS	Cheng *et al.*, 2005
Bacillus anthracis	γ	PlyG	Schuch *et al.*, 2002
	N/A**	PlyPH	Yoong *et al.*, 2006
Staphylococcus aureus	MR11	MV-L	Rashel *et al.*, 2007
	N/A***	ClyS	Daniel *et al.*, 2010
	Bacteriophage K	CHAPk	Fenton *et al.*, 2010
	GH15	LysGH15	Gu *et al.*, 2011

* Renamed PlyC according to (Nelson *et al.*, 2006)
** This endolysin was amplified from a prophage of the *Bacillus anthracis* Ames strain
*** Chimeric construct from the bacteriophage Twort and PhiNM3 endolysins

pyogenes (i.e., group A streptococci) (28.5% infected for endolysin treatment vs 70.5% infected for phosphate-buffered treatment). Furthermore, when 500 U of this streptococcal endolysin, named PlyC in a later publication (Nelson *et al.*, 2006), was administered orally to nine heavily colonized mice, no detectable streptococci were observed via oral swabs 2 hr post-endolysin treatment (Nelson *et al.*, 2001). Based on these results, the authors coined the term "enzybiotic" to describe the therapeutic potential of not only the streptococcal endolysin, but all bacteriophage-derived endolysins.

PlyGBS is another phage endolysin that is active against group A streptococci as well as groups B, C, G, and L streptococci (Cheng *et al.*, 2005). This enzyme was tested in a murine vaginal model of *Streptococcus agalactiae* (i.e., group B streptococci) colonization as a potential therapeutic for pregnant women to prevent transmission of neonatal meningitis-causing streptococci to newborns. A single vaginal dose of 10 U was shown to decrease colonization of group B streptococci by ≈3 logs. Significantly, PlyGBS was found to have a pH optimum ≈5.0, which is similar to the range normally found within the human vaginal tract. Moreover, this enzyme did not possess bacteriolytic activity against common vaginal microflora such as *Lactobacillus acidophilus.*

The most extensively studied endolysins in animal models are Cpl-1, an *N*-acetylmuramidase from the Cp-1 pneumococcal phage, and PAL, an *N*-acetylmuramoyl-L-alanine amidase from the Dp-1 pneumococcal phage. PAL (100 U/ml) was shown to cause a ≈4 log drop in viability in 30 s of 15 different *Streptococcus pneumoniae* serotypes representing multidrug-resistant isolates and those that contain a heavy polysaccharide capsule (Loeffler *et al.*, 2001). In a mouse model of nasopharyngeal carriage, 1400 U of PAL was shown to eliminate all pneumococci and 700 U was shown to significantly reduce bacterial counts, suggesting a dose response. In another study, Cpl-1 was shown to be effective both in a mucosal colonization model and in blood via a pneumococcal bacteremia model (Loeffler *et al.*, 2003). Because the catalytic domains of PAL and Cpl-1 hydrolyze different bonds in the pneumococcal peptidoglycan, they were shown to be synergistic when used in combination *in vitro* (Loeffler and Fischetti, 2003), which was later confirmed *in vivo* in a murine intraperitoneal infection model (Jado *et al.*, 2003). In a study on the effectiveness of endolysins against *in vivo* biofilms, Cpl-1 was shown to work on established pneumococci in a rat endocarditis model (Entenza *et al.*, 2005) [although biofilms were not specifically described, it is widely accepted that biofilms play a major role in endocarditis (Simões, 2011)]. Infusion of 250 mg/kg was able to sterilize 10^5 colony-forming unit (CFU)/ml pneumococci in blood within 30 min and reduce bacterial titers on heart valve vegetations by >4 log CFU/g in 2 hr. In an infant rat model of pneumococcal meningitis, a single intracisternal injection (20 mg/kg) of Cpl-1 resulted in a 3 log decrease of pneumococci in the cerebrospinal fluid (CSF), and an intraperitoneal injection (200 mg/kg) led to a decrease of 2 orders of magnitude in the CSF (Grandgirard *et al.*, 2008). Finally, because pneumococci are often early colonizers to which additional pathogens and viruses adhere, Cpl-1 treatment of mice colonized with *S. pneumoniae* in an otitis media model was shown to significantly reduce secondary colonization by challenge with influenza virus (McCullers *et al.*, 2007).

Several phage endolysins have also been used against vegetative cells and germinating spores of *Bacillus* species. Fifty units of PlyG, an endolysin isolated from the *B. anthracis* γ phage, was shown to rescue 13 out of 19 mice in an intraperitoneal mouse model of infection (Schuch *et al.*, 2002). Furthermore, of the mice that died, fatal infection took between 6 and 14 hr to develop, whereas all non-PlyG-treated mice died within 5 hr. Significantly, this enzyme displayed a favorable thermostability profile and was able to remain fully active after heating to 60 °C for an hour. Moreover, the extreme lytic specificity of this enzyme toward *B. anthracis* and not other *Bacillus* species was exploited for diagnostic purposes in a luminescent-based ATP assay of *B. anthracis* cell lysis. A second *Bacillus* lysin, PlyPH, is unique in that it has a relatively high activity over a broad pH range, from pH 4.0 to 10.5. This enzyme also protected 40% of mice in an intraperitoneal *Bacillus* infection model compared to 100% death in control mice (Yoong *et al.*, 2006). Taken together, the robust and specific properties of *Bacillus* endolysins make them amenable to therapeutic treatment and diagnostics of *B. anthracis*.

The prevalence of methicillin-resistant *S. aureus* (MRSA) as a primary source of nosocomial infection and community-acquired MRSA as an emerging public health threat has generated a considerable amount of interest in identifying and evaluating highly active staphylococcal endolysins. The first anti-staphylococcal endolysin investigated *in vivo* was MV-L, which was cloned from the ΦMR11 bacteriophage (Rashel *et al.*, 2007). This enzyme lysed all tested staphylococcal strains rapidly, including MRSA and vancomycin-resistant clones. *In vivo*, 310 U of this enzyme reduced MRSA nasal colonization ≈3 logs and 500 U provided complete protection in an intraperitoneal model of staphylococcal infection when administered 30 min postinfection. At 60 min postinfection, the same amount of enzyme provided protection in 60% of mice vs controls. Another staphylococcal endolysin, ClyS, is a chimera between the N-terminal catalytic domain of the Twort phage endolysin (Loessner *et al.*, 1998) and the C-terminal CBD of the ΦNM3 phage endolysin (Daniel *et al.*, 2010). Like MV-L, this enzyme displayed potent bacteriolytic properties against multidrug-resistant staphylococci *in vitro*. In a mouse MRSA colonization model, 2-log reductions in viability were observed 1 hr following a single treatment of 960 μg ClyS. Similarly, a single dose (1 mg) of ClyS provided protection when administered 3 hr post-staphylococcal challenge in an intraperitoneal septicemia model. Notably, ClyS showed synergy *in vivo* with oxacillin at doses that were not protective individually against MRSA in this infection model. Most recently, 50 μg of an endolysin from the GH15 phage, LysGH15, showed 100% protection in a mouse intraperitoneal model of septicemia (Gu *et al.*, 2011) and 925 μg of CHAPk, a truncated version of LysK, caused a 2 log drop in nasal colonization of mice 1 hr post-treatment (Fenton *et al.*, 2010a,b).

In addition to phage-encoded endolysins, a large body of *in vivo* work devoted to lysostaphin, a bacterial-derived exolysin, should not be overlooked. Lysostaphin was first identified in 1964 (Schindler and Schuhardt, 1964), and the therapeutic potential of this enzyme has been studied intensely for almost 50 years. To name but a few *in vivo* experiments, this enzyme has been investigated in animal models of burn infections

(Cui *et al.*, 2011), ocular infections (Dajcs *et al.*, 2001, 2002), systemic infections (Kokai-Kun *et al.*, 2007), keratitis models (Dajcs *et al.*, 2000), nasal colonization (Kokai-Kun *et al.*, 2003), and aortic valve endocarditis (Climo *et al.*, 1998; Patron *et al.*, 1999). In addition to human disease, *S. aureus* is the major cause of acute bovine mastitis in milking cows. As such, lysostaphin has been evaluated for therapeutic use in mouse mammary models (Bramley and Foster, 1990) and bovine mastitis models (Oldham and Daley, 1991). Transgenic mice and cows expressing mammary lysostaphin have even been produced and studied for anti-mastitic phenotypes (Kerr *et al.*, 2001; Wall *et al.*, 2005). The transgenic cows produced lysostaphin at concentrations ranging from 0.9 to 14 μg/ml in their milk. Protection appeared to be dose dependent, with a minimum concentration of 3 μg/ml in milk required for complete protection.

B. Immune responses

Due to the proteinaceous nature of PG hydrolases and their potential use as human and animal therapeutics, potential adverse immune responses must be considered, including the generation of antibodies, to these enzymes. It is envisioned that PG hydrolases might be applied topically to mucous membranes (oral, nasal, or vaginal cavities), intravenous, or even intramammary in the case of bovine mastitis.

To address these questions, serum antibodies were raised to phage endolysins specific to *B. anthracis* (PlyG), *S. pyogenes* (PlyC), or *S. pneumoniae* (Pal). When high titers of these antibodies were mixed *in vitro* with endolysins, killing of the target microbe was slowed, but not stopped (Fischetti, 2005; Loeffler *et al.*, 2003). In another study, Cpl-1, a pneumococcal endolysin, was injected intravenously (IV) three times per week into mice for 4 weeks, resulting in positive IgG antibodies against Cpl-1 in five of six mice. Vaccinated and naive control mice were then challenged IV with pneumococci and the mice were treated IV with 200 μg Cpl-1 after 10 hr. Bacteremic titers were reduced within 1 min to the same level in both groups of mice (Loeffler *et al.*, 2003). Furthermore, Western blot analysis revealed that both of the phage lytic enzymes Cpl-1 and Pal elicited antibodies 10 days after a 200-μg injection in mice, but the second injection (at 20 days) also reduced the bacteremia profile 2–3 log units, indicating that the antibodies were not neutralizing *in vivo*. All mice recovered fully with no apparent adverse side effects or anaphylaxis noted (Jado *et al.*, 2003). Taken together, these studies suggest that while antibodies can be readily raised to endolysins, they do not neutralize their hydrolytic activity *in vitro* or *in vivo*.

In studies performed with a catheter-induced *S. aureus* endocarditis model, lysostaphin was tolerated following administration by the systemic route with minimal adverse effects (Climo *et al.*, 1998). Rabbits injected weekly with lysostaphin (15 mg/kg) for 9 weeks by the IV route produced serum antibodies to lysostaphin that resulted in an eightfold reduction in its lytic activity, consistent with earlier work (Schaffner *et al.*, 1967), but no adverse immune response. It is believed that high purity and the absence of Gram-negative lipopolysaccharide are essential for guaranteeing a minimal host immune response.

C. Resistance development

The near-species specificity of phage endolysins avoids many pitfalls associated with broad-range antimicrobial treatments. For example, broad-range antimicrobials, when used alone, lead to selection for resistant strains, not just in the target pathogen, but also in coresident commensal bacteria exposed to the drug. The acquisition of antibiotic resistance is often accomplished by the transfer of DNA sequences from a resistant strain to a susceptible strain (Johnsborg and Håvarstein, 2009). This transfer is not necessarily species or genus limited and can lead to commensal bacteria that are both antibiotic resistant and can serve as carriers of these DNA elements for propagation to neighboring bacteria. Those neighboring strains (i.e., potential pathogens) with newly acquired resistance elements can emerge as antibiotic-resistant strains during future treatment episodes and be distributed further in the bacterial community. Thus, in order to reduce the spread of antibiotic resistance, it is recommended to avoid subjecting commensal bacterial communities to broad-range antibiotics.

To date, there are no reports of strains resistant to phage endolysins. Two reports have attempted to identify resistant strains [summarized in Fischetti (2005)]. In brief, three species, *S. pneumonia*, *S. pyogenes*, and *B. anthracis*, were tested with repeated exposure to sublethal doses of phage endolysins specific to each species. Surviving bacteria were then challenged with a lethal dose and there was no notable change in susceptibility. In another study, *Bacillus* species were exposed to chemical mutagens that increased the frequency of antibiotic resistance several orders of magnitude. In contrast, these organisms remained fully sensitive to PlyG, a *B. anthracis*-specific endolysin (Schuch *et al.*, 2002). A likely explanation for the lack of observed resistance in endolysins as put forth by Fischetti (2005) is that the bacterial host and phage have coevolved such that the phage might have evolved endolysins to target immutable bonds in order to ensure its survival and release from the host. Thus, resistance to phage endolysins is expected to be a very rare event.

Despite the lack of observed resistance in phage endolysins, there are reports of resistance to other types of PG hydrolases, specifically exolysins. Lysozyme is a human exolysin with catalytic (muramidase) and cationic antimicrobial peptide activities. It is secreted by epithelial cells and is present on mucous membranes and in the granules of phagocytes. Degradation of the bacterial peptidoglycan by lysozyme yields peptidoglycan fragments that can elicit a strong host immune response and recruitment of immune cells. Bacterial resistance to lysozyme has been accomplished through a variety of modifications that the bacteria can incorporate into the peptidoglycan backbone [for reviews, see Davis and Weiser (2011) and Vollmer (2008)].

Similarly, at least two genes can confer resistance to the lysostaphin exolysin, which targets the bonds of the staphylococcal PG interpeptide bridge. *S. simulans* produces lysostaphin and avoids its lytic action by the product of the lysostaphin immunity factor (*lif*) gene [same as endopeptidase resistance gene (*epr*) (DeHart *et al.*, 1995)] that resides on a native plasmid (pACK1) (Thumm and Gotz, 1997). The *lif* gene product functions by inserting serine residues into the PG cross bridge, thus interfering with the ability of the glycyl-glycine endopeptidase to recognize and cleave this structure. Mutations in the *S. aureus femA* gene (factor essential for methicillin resistance) (Sugai *et al.*, 1997) result in a change in the muropeptide interpeptide cross bridge from pentaglycine to a single glycine, rendering *S. aureus* resistant to the lytic action of lysostaphin. MRSA have been shown to mutate *femA* when exposed *in vitro* or *in vivo* to subinhibitory doses of lysostaphin (Climo *et al.*, 2001). Interestingly, in one report, MRSA strains that did develop resistance to lysostaphin via the *femA* gene showed a reduced fitness compared to their parental counter parts, were fivefold less virulent in a rodent kidney infection model, and were treated easily with β-lactam antibiotics (Kusuma *et al.*, 2007).

Grundling *et al.* (2006) identified the *lyrA* gene (lysostaphin resistance A) that, when mutated by a transposon gene insertion, reduced staphylococcal susceptibility to lysostaphin. Although some structural changes in PG were noted in the *lyrA* mutant, PG purified from the *lyrA* mutant was susceptible to lysostaphin and the Φ11 endolysin, suggesting that additional unidentified alterations in the *S. aureus* cell wall envelope might mediate resistance in the *lyrA* mutant.

D. Synergy

Antimicrobial synergy has been demonstrated for multiple PG hydrolases in combination with other PG hydrolases, as well as numerous other classes of antimicrobials. Synergy between two PG hydrolases was shown with LysK and lysostaphin via the checkerboard assay (Becker *et al.*, 2008, 2009a). This is consistent with the two enzymes having unique cut sites. Lysostaphin has also been shown to be synergistic in the checkerboard assay with the cationic peptide antimicrobial ranalexin (Graham and Coote, 2007); this combination has been demonstrated to be an effective surface disinfectant (Desbois *et al.*, 2010). Lysostaphin was also shown to be synergistic with β-lactams against MRSA. This combination is uniquely promising in that when lysostaphin-resistant staphylococci are generated by modifying the pentaglycine bridge of the PG; these cell wall-altered strains are often hypersusceptible to β-lactams (Kiri *et al.*, 2002). The pneumococcal Cpl-1 endolysin is synergistic with either penicillin or gentamicin (Djurkovic *et al.*, 2005) and with the Pal amidase (Jado *et al.*, 2003; Loeffler and Fischetti, 2003). The phage endolysin LysH5, which has been shown to eradicate *S. aureus* in milk (Obeso *et al.*, 2008), is synergistic with nisin (Garcia *et al.*, 2010a). Nisin was also shown to be synergistic with lysozyme against lactic acid bacteria (Chun and Hancock, 2000). Finally, ClyS, a fusion lysin described earlier, has been shown to be better than mupirocin at eradicating staphylococcal skin infections (Pastagia *et al.*, 2011) and is synergistic with oxacillin (Daniel *et al.*, 2010).

E. Biofilms

A high level of antimicrobial resistance is achieved by many pathogens through the multifaceted changes that accompany growth in a biofilm. Biofilms are sessile forms of bacterial colonies that attach to a mechanical or prosthetic device or a layer of mammalian cells and have an extensive extracellular matrix. The National Institutes of Health (NIH) estimate that 80% of human bacterial infections involve biofilms (http://grants.nih.gov/grants/guide/pa-files/PA-06-537.html) (Sawhney and Berry, 2009). Bacteria in biofilms can be orders of magnitude more resistant to antibiotic treatment than their planktonic (liquid culture) counterparts (Amorena *et al.*, 1999).

Several mechanisms are thought to contribute to the antimicrobial resistance associated with biofilms: (1) delayed or restricted penetration of antimicrobial agents through the biofilm exopolysaccharide matrix; (2) decreased metabolism and growth rate of biofilm organisms that resist killing by compounds that only attack actively growing cells; (3) increased accumulation of antimicrobial-degrading enzymes; (4) enhanced exchange rates of drug resistance genes; and (5) increased antibiotic tolerance (as opposed to resistance) through expression of stress response genes, phase variation, and biofilm-specific phenotype development (Emori and Gaynes, 1993; Fux *et al.*, 2003; Høiby *et al.*, 2010; Keren *et al.*, 2004; Lewis, 2001).

Little work has been done to specifically test phage endolysins for their antibiofilm activity. Φ11 endolysin (Sass and Bierbaum, 2007) and lysostaphin have been shown to eliminate static staphylococcal biofilms (Walencka *et al.*, 2005; Wu *et al.*, 2003), as has LysK (O'Flaherty *et al.*, 2005). Lysostaphin was also shown to eliminate staphylococcal biofilms in jugular vein-catheterized mice (Kokai-Kun *et al.*, 2009). The *S. aureus* SAP-2 phage endolysin SAL-2, which is nearly identical to the phage P68 endolysin, was also reported to eliminate *S. aureus* biofilms (Son *et al.*, 2010). Alternative strategies for eradicating biofilms are necessary, including catalytic enzymes to destroy the matrix. Bacteriophage and phage lytic enzymes are a potential new source of antibiofilm therapy (Donlan, 2008).

F. Disinfectant use

Decontamination of environmental pathogens is another area where PG hydrolases may find a niche in the marketplace. Although most disinfectants have broad-spectrum efficacy, one can envision environments where targeted decontamination of a pathogen by a narrow-spectrum endolysin would be sufficient. For example, endolysins targeting MRSA may have utility in nursing homes, surgical suites, or athletic locker rooms; endolysins effective against *B. anthracis* may be important for decontamination of suspected exposures; those against *Listeria monocytogenes* would have applications in meat-packing or food-processing facilities; and enzymes against group A streptococci could be used to reduce bacterial loads in child care settings.

Endolysins avoid several problems associated with chemical disinfectants. By their enzymatic nature, endolysins do not rely on potentially toxic reactive groups utilized by chemical disinfectants. As proteins, they are inherently biodegradable and noncorrosive (i.e., a "green" disinfectant). Finally, due to the high affinity of their binding domains for the bacterial peptidoglycan and their ability to concentrate on the cell surface, endolysins may not be as susceptible to dilution factors as chemical disinfectants.

To date, the literature is sparse with examples of PG hydrolases used for disinfecting purposes. Nonetheless, lysostaphin and the cationic peptide antimicrobial ranalexin have been shown to be synergistic at killing MRSA on solid surfaces (Graham and Coote, 2007). Similarly, the same combination was found to kill MRSA on human skin within 5 min using an *ex vivo* assay (Desbois *et al.*, 2010). In one unique application, lysostaphin attached to nanotubes and mixed with latex paint was shown to retain anti-staphylococcal properties on painted surfaces (Pangule *et al.*, 2010).

For endolysins, only PlyC has been tested specifically as an environmental disinfectant (Hoopes *et al.*, 2009). PlyC lyses several streptococcal species, including *S. equi*, the causative organism of equine strangles disease. This highly contagious disease of horses is transmitted through shedding of live bacteria from nasal secretions and abscess drainage onto common surfaces in a stall or barn. Chemical disinfectants can be effective against *S. equi*, but inactivation by environmental factors, damage to equipment, and toxicity are of concern. PlyC was found to be 1000 times more active on a per weight basis ($\approx$150,000 times more active on a molar basis) than a commercially available oxidizing disinfectant. Significantly, 1 μg of PlyC was able to sterilize 10^8 CFU/ml of *S. equi* in 30 min. Based on these findings, the authors performed a standard battery of tests approved by the Association of Official Analytical Chemists, including the use dilution method for testing disinfectants and germicidal spray products tests. PlyC passed the use dilution method, which validates disinfectant claims, and was shown to eradicate or significantly reduce the *S. equi* load on the equipment of various porosities found commonly in horse stables. Finally, PlyC was shown to retain effectiveness when tested in the presence of nonionic detergents, hard water, and organic material.

G. Food safety

The use of phage and phage products for use in food safety has been reviewed (Hagens and Loessner, 2010; Hermoso *et al.*, 2007; O'Flaherty *et al.*, 2009). ListShield and Ecoshield from Intralytix and LISTEX™ from MICREOS Food Safety are phage preparations designed to protect food from *L. monocytogenes* or *Escherichia coli.* One regulatory distinction between phages and endolysins is that phages are considered a natural product and most endolysins are purified from a recombinant expression system, thus increasing the hurdles in the approval process.

The specific use of PG hydrolases to protect food from bacterial pathogens has also been reviewed (Callewaert *et al.*, 2010; Garcia *et al.*, 2010b; Loessner, 2005; Stark *et al.*, 2010). Despite extensive exploration in this area, at this writing, there are no approved enzybiotics (endolysins) for use in/on foods for human consumption. However, approval is anticipated eventually in light of the acceptance in 2006 by the U.S. Food and Drug Administration for the use of *Listeria* bacteriophage on sliced meat products (http://edocket.access.gpo.gov/2006/pdf/E6-13621.pdf).

Peptidoglycan hydrolases are effective antimicrobials when introduced into foodstuffs via transgene expression, but the safety of consumption of transgenic food products is still a highly debated topic worldwide. Transgenic goat milk containing human lysozyme could protect from mastitis *in vitro* and showed benefits in animal health for goats drinking transgenic milk (Maga *et al.*, 2006a,b). Similarly, pigs (Tong *et al.*, 2010) and cattle (Yang *et al.*, 2011) expressing lysozyme in the mammary gland have been created. Lysostaphin transgenic cattle were also protected from an intramammary *S. aureus* challenge (Wall *et al.*, 2005). A human lysozyme-expressing vector for injection into cattle mammary glands has also been created and reported to reduce mastitis symptoms within days (Sun *et al.*, 2006).

Expression of PG hydrolases in plants might serve multiple purposes: as a final stage to protect food products from food pathogens or a method to protect crop production from plant pests, and plant systems might be a better source of the PG hydrolase in quantities needed for commercialization as opposed to fermentation-derived recombinant proteins. Potatoes can be protected from the phytopathogen *Erwinia amylovora* by transgenic expression of the T4 lysozyme (During *et al.*, 1993). Transgenic rice expressing human lysozyme has also been created [reviewed in Boothe *et al.* (2010)], as have transgenic plants expressing a group B streptococcal endolysin, which was highly expressed in the chloroplasts (Oey *et al.*, 2009).

Nontransgenic uses of PG hydrolases in food applications are limited. Surface application of the phiEa1h (T4 lysozyme) endolysin on pears reduced the effects of an *Erwinia* challenge (Kim *et al.*, 2004). The staphylococcal phage endolysin LysH5 killed *S. aureus* in pasteurized milk *in vitro* (Obeso *et al.*, 2008) and was shown to be synergistic with nisin, a lactococcal bacteriocin that has achieved generally recognized as safe status (Garcia *et al.*, 2010a). Fusion of a streptococcal B30 endolysin and lysostaphin was also able to kill both streptococci and staphylococci in milk products (Donovan *et al.*, 2006a). An endolysin from *Clostridium tyrobutyricum* (Mayer *et al.*, 2010), which produces cheese spoilage, is also active in milk. Other clostridial endolysins that kill food pathogens have been reported (Simmons *et al.*, 2010; Zimmer *et al.*, 2002). Lactic acid bacteria engineered to secrete lysostaphin and a *Listeria* endolysin (Tan *et al.*, 2008; Turner *et al.*, 2007) or *Listeria* endolysin alone (Gaeng *et al.*, 2000; Stentz *et al.*, 2010) or *Clostridium* endolysin (Mayer *et al.*, 2008) have been produced, but the ability to protect foodstuffs from these pathogens has not yet been reported.

A very relevant role that endolysins play in food safety is based on the high specificity of their CBDs. These recognition domains have been used to develop rapid and sensitive identification, detection, and differentiation systems (Fujinami *et al.*, 2007; Schmelcher *et al.*, 2010). Magnetic beads coated with recombinant CBDs enabled immobilization and recovery of more than 90% of *L. monocytogenes* cells from food samples (Kretzer *et al.*, 2007; Walcher *et al.*, 2010).

V. ENGINEERING ENDOLYSINS

A. Swapping and/or combining endolysin domains

There are numerous examples in the literature of engineered PG hydrolases that range from site-directed mutant constructs used to identify essential amino acids in catalytic or CBD domains to novel fusion constructs for the purpose of making a better antimicrobial. Some of the earliest fusions were created by the exchange of CBDs of pneumococcal autolysins and phage endolysins (Diaz *et al.*, 1991; Garcia *et al.*, 1990). Fusion of clostridial or lactococcal *N*-acetylmuramidase catalytic domains to choline-binding domains from pneumococcal endolysin CBDs resulted in choline dependence of the chimeric enzyme (Croux *et al.*, 1993a,b; Lopez *et al.*, 1997). In a reverse approach, a clostridial CBD was fused C-terminally to a catalytic domain of the pneumococcal autolysin LytA, increasing its activity against clostridial cell walls considerably (Croux *et al.*, 1993a). In another study, the catalytic domain of the lactococcal phage Tuc2009 gained activity against choline-containing pneumococcal cell walls by fusion to the CBD of LytA (Sheehan *et al.*, 1996). The ability to swap catalytic and CBDs is not limited to choline-binding domains. The exchange of *Listeria* phage endolysin CBDs of different serovar specificity resulted in swapped lytic properties of the chimeras and enhanced lytic activity against certain strains (Schmelcher *et al.*, 2011). In the same study, heterologous tandem CBD constructs were shown to combine the binding properties of both individual CBDs, providing them with extended recognition properties. Furthermore, a duplication of a CBD resulted in a 50-fold increase in affinity to the listerial cell wall, making this protein a useful tool for bacterial detection. Combined with an enzymatically active catalytic domain, this increased affinity resulted in enhanced lytic activity at high ionic strength. Another chimeric endolysin (P16-17) was constructed with the N-terminal predicted D-alanyl-glycyl endopeptidase domain and the C-terminal CBD of the *S. aureus* phage P16 endolysin and the P17 minor coat protein, respectively. This approach was also a domain swap, which improved the solubility of the fusion over the parental hydrolases greatly, allowing purification and experiments to demonstrate strong antimicrobial activity toward *S. aureus* (Manoharadas *et al.*, 2009).

A series of intergeneric PG hydrolase fusions between the streptococcal B30 endolysin and the staphylolytic lysostaphin demonstrate activity against both pathogens (Donovan *et al.*, 2006a). These constructs relied on the streptococcal and staphylococcal lytic domains maintaining their parental specificities, with just the lysostaphin SH3b CBD. This dual lytic specificity challenges the dogma wherein the SH3b domain was believed to be essential for endolysin specificity (Baba and Schneewind, 1996). More recently, this theme has been expanded to include the streptococcal phage λSa2 endolysin CHAP endopeptidase domain fused to the ≈92 amino acid staphylococcal SH3b CBDs from either lysostaphin or LysK. These constructs show full activity against both streptococcal and staphylococcal pathogens in numerous *in vitro* assays (Becker *et al.*, 2009b), presumably due to the conserved bonds that this lytic domain recognizes and cleaves (γ-D-glutaminyl-L-lysine) in both streptococcal and staphylococcal PG. Again, the staphylococcal SH3b CBDs enhanced lytic activity on the cell walls of both genera. This dual activity argues against genera- or species-specific binding of the lysostaphin SH3b domain as has been reported (Grundling and Schneewind, 2006; Lu *et al.*, 2006).

A more recent fusion, ClyS, described earlier is reported to be effective at curing murine topical infections of *S. aureus* (Pastagia *et al.*, 2011) and is effective in combination with classical antibiotics at eradicating multidrug-resistant strains of *S. aureus* in a mouse model of nasal colonization (Daniel *et al.*, 2010).

Other more trivial modifications of PG hydrolases have also been reported, such as the addition of a His tag for ease of purification. Although such tags are considered a minor modification, rarely has the effect of such a modification been examined on lytic activity. One study examined the effect of an N- or C-terminal His tag on lysostaphin with the resultant activities being 80 and 20% of the nontagged version, respectively (Becker *et al.*, 2011). That same publication also looked at microdeletions (6 amino acid increments) in the N terminus of lysostaphin. Deletion of the first 3 or 6 residues has no significant effect on minimum inhibitory concentration, whereas deletion to residue 11 reduces the MIC to ≈40% of wild type with decreasing MICs for larger deletions. The lack of reproducibility of quantitative results between PG hydrolase assays for lysostaphin was first described by Kusuma and Kokai-Kun (2005); that finding was confirmed with turbidity reduction and plate lysis assays where N-terminal microdeletions of lysostaphin did not show significant reduction in lytic activity until 21 residues were deleted, resulting in only 17% of wild-type activity (Becker *et al.*, 2011).

Other minimally altered constructs are those where single amino acids are purposefully altered to examine the effect on lytic activity. Pritchard *et al.* (2004) altered conserved amino acids in the streptococcal B30 endolysin CHAP and lysozyme domains, which resulted in a sequential loss of activity from each domain. When analyzed on live bacteria, it was made clear that the B30 endolysin CHAP domain was the primary source of lytic activity from this dual domain endolysin when lysing "from without" (Donovan *et al.*, 2006b). Site-directed mutagenesis and deletion analysis of the *B. anthracis* phage lysin PlyG were essential in defining the binding domain and active site residues (Kikkawa *et al.*, 2007, 2008), as for PlyC that was also examined in this way (Nelson *et al.*, 2006). Similarly, site-directed mutations altering histidine codons in the staphylococcal glycyl-glycine PG hydrolase ALE-1 have been used to define essential amino acids in the M23 endopeptidase domain (Fujiwara *et al.*, 2005). Mutations of the ALE-1 CBD, when fused to GFP, were used to define those amino acids essential for cell wall binding (Lu *et al.*, 2006).

Further site-directed mutations of lysostaphin were examined when a lysostaphin transgene was expressed in the mammary gland of both mice (Kerr *et al.*, 2001) and dairy cattle (Wall *et al.*, 2005). Transgenic lysostaphin showed reduced activity due to N-linked glycosylation (Kerr *et al.*, 2001). Subsequently, two Asn codons (residues 125 and 232) were modified to encode Glu in order to ablate the N-linked glycosylation. The result was a secreted functional lysostaphin, however, with a 5- to 10-fold reduction in lytic activity compared to wild-type lysostaphin (Kerr *et al.*, 2001). By separating the two altered residues on separate constructs, it was shown that the N125Q modification alone was primarily responsible for this reduction in activity (Becker *et al.*, 2011). By homology to the well-characterized LytM (a closely related LAS metalloprotease) (Firczuk *et al.*, 2005), residue

125 is likely to reside in the catalytic domain of lysostaphin and thus may alter the enzymes ability to bind the substrate. When mapped to the crystal structure of LytM (Firczuk *et al.*, 2005) in the presence of a substrate analogue bound to a glycine-rich loop in the active site cleft, mutation of the equivalent residue (LytM N303Q) added an additional carbon into the side chain in the predicted active site. It is predicted that this might crowd the substrate analog and therefore interfere with substrate binding in the active site cleft (Firczuk *et al.*, 2005; Becker *et al.*, 2011).

Numerous engineered truncations of PG hydrolases have been described in the literature that were created primarily for defining active residues in lytic domains. A partial list includes the Twort endolysin (Loessner *et al.*, 1998), B30 endolysin (Donovan *et al.*, 2006b), λSa2 endolysin (Donovan and Foster-Frey, 2008), Φ11 endolysin (Donovan *et al.*, 2006c; Sass and Bierbaum, 2007), and the *Bacillus amyloliquifaciens* endolysin (Morita *et al.*, 2001). Some of these efforts have yielded truncations with a greater lytic specific activity than the full-length PG hydrolase, for example, the staphylococcal LysK (Horgan *et al.*, 2009). One such hyperactive truncation construct was the result of a random mutagenesis experiment, which also resulted in the incorporation of unpredicted sequences at the C terminus of the streptococcal PlyGBS endolysin (Cheng and Fischetti, 2007). The authors suggest that this enhanced activity may be potentially due to both a reduced size and the lack of full-length CBD, allowing the enzyme to move more quickly between substrate-binding sites and thus lyse more cells. Other studies suggest that the presence of a CBD increases lytic activity of an endolysin, presumably by bringing the catalytic domain in proximity of its substrate (Korndorfer *et al.*, 2006). However, duplication of a CBD, which results in a significant increase in binding affinity, was shown to reduce activity at a physiological salt concentration, which again may be explained by a loss of surface mobility (Schmelcher *et al.*, 2011).

Numerous works with fusion constructs further verify that PG hydrolases have evolved a modular design, with both lytic and CBD domains as first proposed by Diaz *et al.* (1990). When fused, these lytic domains can maintain their parental specificities for the PG bond cleaved and the species of cell wall recognized. These enzymes are candidate antimicrobials for the reasons outlined earlier, but most importantly, despite repeated attempts to identify them, no strains of host bacteria have been reported that can resist the lytic activities of their bacteriophage endolysins (Fischetti, 2005). In addition, numerous phage endolysins harbor dual lytic domains (see Figs. 3–5). Dual domain endolysins are predicted to be more refractory to resistant strain development (Fischetti, 2005). The Donovan laboratory has taken this one step further and reasoned that three lytic domains might create an antimicrobial that would be even more refractory to resistance development.

In theory, it is very rare that a bacterium can evade three, unique, simultaneous antimicrobial activities.

The authors have created several triple-lytic-domain anti-staphylococcal fusion constructs using the synergistic enzymes LysK and lysostaphin. Lysostaphin and LysK collectively harbor three cleavage domains that cleave at unique sites (described earlier). LysK and lysostaphin are also known to be active against multiple MRSA strains. The LysK-Lyso triple lytic domain construct described previously (Becker *et al.*, 2009b) is highly active against *S. aureus*, MRSA, and numerous coagulase negative staphylococci (unpublished data). Most importantly, all three lytic domains are active in the fusion construct, as demonstrated by electron spray ionization mass spectrometry of PG digestion products (Donovan *et al.*, 2009). Studies are underway to determine the efficacy of these and other triple-lytic-domain fusion endolysins in animal models of staphylococcal infection and to test for resistant strain development both *in vitro* and among the staphylococci retrieved from *in vivo* models.

B. Fusion of endolysins to protein transduction domains

It is apparent that the high antimicrobial resistance of some persistent pathogens is due to their ability to invade and reside intracellularly within eukaryotic cells. Some examples of bacteria that utilize this niche are *Legionella pneumophila*, *Mycobacteria turberculosis*, *Listeria monocytogenes* (Vazquez-Boland *et al.*, 2001), and *S. aureus*. There are numerous strategies that these intracellular residents have devised, including the creation of specialized vacuoles that block phagosome maturation into a phagolysosome and inhibition of phagosome acidification, to name a few (Garcia-del Portillo and Finlay, 1995). Alternative drug treatment systems for the delivery of antimicrobials to intracellular pathogens have been described (Imbuluzqueta *et al.*, 2010).

One proposed method involves fusing cell-penetrating peptides (CPPs) or protein transduction domains (PTDs) to PG hydrolases to enable these lytic enzymes access to intracellular bacteria (Borysowski and Gorski, 2010). CPPs or PTDs are usually highly positively charged regions that exist in naturally occurring proteins and are essential for the uptake of these proteins into target cells. The uptake mechanisms are likely cell type and peptide specific with some CPPs and their cargo traversing the membrane without involving pinocytosis, whereas others require pinocytotic uptake (Duchardt *et al.*, 2007; Joliot and Prochiantz, 2004). There are reports of noncharged peptide fragments that can also enhance transduction across the eukaryotic membrane, and some antimicrobial peptides can serve as CPPs and vice versa (Splith and Neundorf, 2011).

There are numerous reports on the use of CPPs to deliver bioactive molecules to a variety of cell types. Although no formal report exists in the literature for a PG hydrolase fused to a PTD for killing intracellular pathogens, there has been one patent application filed in 2009 wherein lysostaphin was fused to the HIV transactivator of transcription (TAT) protein transduction domain, Lyso-TAT (http://www.pat2pdf.org/patents/pat20110027249.pdf). In this application, the Lyso-TAT construct is reported to eradicate *S. aureus ex vivo* in cultured MAC-T mammary epithelial cells, bovine brain epithelia, human keratinocytes, and murine osteoblasts.

VI. GRAM-NEGATIVE ENDOLYSINS AS ANTIMICROBIALS

A. Background

The use of bacteriophage-encoded endolysins, or any type of PG hydrolase, to control Gram-negative pathogens has been very limited. Their effectiveness when added exogenously is hindered by the presence of the Gram-negative outer membrane, which is highly effective at excluding large molecules and is not present on Gram-positive cells. The endolysin-susceptible PG layer resides between an inner and outer membrane in Gram-negative organisms and, as such, is not exposed directly to the extracellular environment. An effective strategy to allow endolysins to translocate the outer membrane is vital for their use against Gram-negative pathogens.

There are numerous studies on the use of peptides, detergents, and chelators that can be used to permeabilize the Gram-negative outer membrane in combination with PG hydrolases (Vaara, 1992). As an example, 10 m*M* EDTA, used in combination with 50 μg/ml of the *Pseudomonas* endolysin EL188, decreased viable *P. aeruginosa* cells by 3 or 4 logs in 30 min depending on the strain tested (Briers *et al.*, 2011). Additionally, there have been studies in which various chemical moieties have been conjugated to PG hydrolases or hydrophobic peptides have been fused to them genetically in order to alter membrane permeability to these enzymes (Ito *et al.*, 1997; Masschalck and Michiels, 2003). All of these strategies can be applied to bacteriophage-derived endolysins and several specific examples are provided in the next section. However, each strategy also poses questions regarding their efficacy, practicality, and toxicity that must be determined empirically. Appreciably, agents that destabilize the Gram-negative outer membrane often destabilize eukaryotic cell membranes, both of which are similar lipid bilayers.

B. Nonenzymatic domains and recent successes

Some PG hydrolases and endolysins can kill pathogens via a mechanism completely separate from their ability to cleave the PG enzymatically. For example, the heat-denatured bacteriophage T4 lysozyme was found to retain 50% of its microbicidal activity despite a complete absence of muramidase activity (During *et al.*, 1993). The authors further identified three positively charged, amphipathic helices and showed that one of them, A4, exhibits 2.5 times more killing of *E. coli* than intact T4 lysozyme. A4 is proposed to act by membrane disruption due to its cationic nature. This action may be similar to that of other positively charged, amphipathic helices referred to collectively as host-defense peptides (Sahl and Bierbaum, 2008).

Similar to the T4 lysozyme, several additional endolysins have been identified that contain amphipathic or highly cationic regions in addition to their catalytic domains. Preliminary studies suggest that these endolysins are capable of producing lysis from without in a variety of Gram-positive and Gram-negative species. For example, LysAB2, the endolysin from the ΦAB2 *A. baumannii* phage, was found to degrade isolated cell walls of *A. baumannii* and *S. aureus* in a zymogram (Lai *et al.*, 2011). On live, viable cells, this enzyme was shown to be antibacterial toward several Gram-negative (*A. baumannii, E. coli, Salmonella enterica*) and Gram-positive (*Streptococcus sanguis, S. aureus, Bacillus subtilis*) strains. Significantly, LysAB2 contains a C-terminal amphipathic region that was shown by deletion analysis to be necessary for the observed antibacterial activity. A second example is the lys1521 endolysin from a *Bacillus amyloliquefaciens* phage, which possesses two cationic C-terminal regions. Using either a synthesized peptides of these regions or catalytically inactive mutants of the endolysin, the cationic regions alone were shown to be able to permeabilize the outer membrane of *P. aeruginosa*, a Gram-negative pathogen (Muyombwe *et al.*, 1999). The wild-type enzyme, containing an N-terminal catalytic domain and the two C-terminal cationic domains, displayed antibacterial activity against live *P. aeruginosa* (Orito *et al.*, 2004).

These successes have inspired renewed interest in the use of endolysins against Gram-negative bacteria, an idea once considered a nonstarter. Indeed, several new patents have been issued, which provide forward-looking insight into where the field is headed (see patents WO/2010/149792 and WO/2011/023702). It is expected that research focused on fusing endolysin catalytic domains with cationic peptides, polycationic peptides, amphipathic peptides, sushi peptides, hydrophobic peptides, defensins, and other antimicrobial peptides with the goal to improve endolysin-based therapy to Gram-negative pathogens will expand greatly in the coming years.

C. High-pressure treatment

In another approach, the use of high hydrostatic pressure (HHP) can dramatically increase the access of phage endolysins to the Gram-negative PG. While this may not have direct human applications, it does have potential applications for decontamination and food processing. HHP has several advantages: it can be bactericidal alone (Briers *et al.*, 2008; Hauben *et al.*, 1996; Masschalck *et al.*, 2000, 2001;Nakimbugwe *et al.*, 2006), it does not use heat so it will not compromise the quality of foodstuffs, and, most importantly, it is not considered to be a food additive. However, generating the required high pressures (200 to 500 MPa) can pose a cost hurdle. HHP has been used with a variety of antibacterials, including nisin, lactoferrin, and several PG (Briers *et al.*, 2008; Hauben *et al.*, 1996; Masschalck *et al.*, 2000, 2001; Nakimbugwe *et al.*, 2006).

Nakimbugwe *et al.* (2006) tested HHP in conjunction with six individual PG hydrolases, including phage endolysins from λ and T4, on 10 different bacterial strains (five each of Gram negative and positive). Both phage endolysins were active on four out of five of the Gram-negative bacteria and *Bacillus subtilis*, although the λ-derived endolysin showed greater activity on most of the strains. In a separate study, the efficacy of hen egg white lysozyme, a PG hydrolase, and the λ lysozyme, an endolysin with lytic transglycosylase activity, were tested in conjunction with HHP on skim milk (pH 6.8) and banana juice (pH 3.8) with four Gram-negative bacteria: *E. coli O157:H7*, *Shigella flexneri*, *Yersinia enterocolitica*, and *Salmonella typhimurium* (Nakimbugwe *et al.*, 2006). The λ lysozyme outperformed the PG hydrolase in a bacterial inactivation assay by almost 2 and 5 logs in skim milk and banana juice, respectively.

VII. CONCLUDING REMARKS

Multidrug-resistant superbugs have "raised the bar" in establishing a higher set of requirements for new antimicrobials. New antimicrobial agents should ideally eradicate multidrug-resistant pathogens, including those in biofilms, and successfully prevent further resistance development. PG hydrolases and their fusions have unique properties that make them ideal candidates for this much needed new class of therapeutics. PG hydrolases usually target a narrow range of closely related pathogens, avoiding selective pressures on unrelated commensal bacteria. They also target the cell surface and thus avoid the many resistance mechanisms that operate within the cell (e.g., modification of target, modification of agent, pumps to extrude the agent). PG hydrolases are effective against growing cells but can also target nondividing or slowly growing cells, for example, biofilms, which most antibiotics cannot. The modular nature of the phage endolysins and other PG hydrolases allow for naturally occurring and engineered lysins with two or more simultaneous lytic activities. It is expected to be a rare event that any pathogen can evade three simultaneous lytic activities. It is also worth noting that the ability to confer intracellular killing via PG hydrolase fusions to PTDs is nontrivial in light of the toxic levels required for most drugs to eradicate pathogens residing intracellularly. Similarly, PG hydrolases are synergistic with many classes of classical antimicrobials, thus potentially extending the clinical half-life of overused antibiotics. Although there are many advantages conferred by killing a drug-resistant pathogen via a lytic enzyme that lyses from without, the reality of increased antigen release that accompanies lysis of a systemic pathogen cannot be ignored. Similarly, the inherent hurdles of production costs and antigenicity of a protein antimicrobial are still awaiting full debate in the commercialization arena. However, despite these concerns, it is clear that biofilms are the major threat in human infectious disease, with NIH estimating that 80% and Centers for Disease Control and Prevention estimating that 65% of human infections are in the form of biofilms. It is also clear that conventional antimicrobials are poor eradicators of biofilms and that catalytic enzymes of some sort are going to be required to dissolve and eradicate persistent biofilms. Thus, the antigenicity of both the digestive enzyme used to treat the biofilm and the surge of bacterial antigens released upon cell lysis or biofilm degradation are hurdles that will need to be overcome in the unavoidable assault on bacterial biofilms. The authors believe that PG hydrolases are an ideal candidate class of novel antimicrobials with which to address these inevitable concerns.

ACKNOWLEDGMENTS

The U.S. Department of Agriculture (USDA) prohibits discrimination in all its programs and activities on the basis of race, color, national origin, age, disability, and, where applicable, sex, marital status, familial status, parental status, religion, sexual orientation, genetic information, political beliefs, reprisal, or because all or part of an individual's income is derived from any public assistance program. (Not all prohibited bases apply to all programs.) Persons with disabilities who require alternative means for communication of program information (Braille, large print, audiotape, etc.) should contact USDA's TARGET Center at (202) 720-2600 (voice and TDD). To file a complaint of discrimination, write to USDA, Director, Office of Civil Rights, 1400 Independence Avenue, S.W., Washington, DC 20250-9410 or call (800) 795-3272 (voice) or (202) 720-6382 (TDD). USDA is an equal opportunity provider and employer. This work was supported in part by NIH Grant 1RO1AI075077-01A1; NRI Grant 2007-35204-18395, and U.S. state department funds, all awards to DMD. Mention of trade names or commercial products in this article is solely for the purpose of providing specific information and does not imply recommendation or endorsement by the U.S. Department of Agriculture. Jochen Klumpp was funded via a fellowship under the OECD Co-operative Research Program: Biological Resource Management for Sustainable Agricultural Systems.

REFERENCES

Amorena, B., Gracia, E., Monzon, M., Leiva, J., Oteiza, C., Perez, M., Alabart, J. L., and Hernandez-Yago, J. (1999). Antibiotic susceptibility assay for *Staphylococcus aureus* in biofilms developed *in vitro*. *J. Antimicrob. Chemother.* **44**:43–55.

Anantharaman, V., and Aravind, L. (2003). Evolutionary history, structural features and biochemical diversity of the NlpC/P60 superfamily of enzymes. *Genome Biol.* **4:**R11.

Baba, T., and Schneewind, O. (1996). Target cell specificity of a bacteriocin molecule: a C-terminal signal directs lysostaphin to the cell wall of *Staphylococcus aureus*. *EMBO J.* **15:**4789–4797.

Baker, J. R., Liu, C., Dong, S., and Pritchard, D. G. (2006). Endopeptidase and glycosidase activities of the bacteriophage B30 lysin. *Appl. Environ. Microbiol.* **72:**6825–6828.

Bateman, A., and Bycroft, M. (2000). The structure of a LysM domain from *E. coli* membrane-bound lytic murein transglycosylase D (MltD). *J. Mol. Biol.* **299:**1113–1119.

Bateman, A., and Rawlings, N. D. (2003). The CHAP domain: A large family of amidases including GSP amidase and peptidoglycan hydrolases. *Trends Biochem. Sci.* **28:** 234–237.

Becker, S. C., Dong, S., Baker, J. R., Foster-Frey, J., Pritchard, D. G., and Donovan, D. M. (2009a). LysK CHAP endopeptidase domain is required for lysis of live staphylococcal cells. *FEMS Microbiol. Lett.* **294:**52–60.

Becker, S. C., Foster-Frey, J., and Donovan, D. M. (2008). The phage K lytic enzyme LysK and lysostaphin act synergistically to kill MRSA. *FEMS Microbiol. Lett.* **287:**185–191.

Becker, S. C., Foster-Frey, J., Powell, A., Kerr, D., and Donovan, D. M. (2011). Lysostaphin: Molecular changes that preserve staphylolytic activity. *In* "2010 International Conference on Antimicrobial Research" (A. Mendez-Vilas, ed.), pp. 18–22. World Scientific Publishing Co., Singapore.

Becker, S. C., Foster-Frey, J., Stodola, A. J., Anacker, D., and Donovan, D. M. (2009b). Differentially conserved staphylococcal SH3b_5 cell wall binding domains confer increased staphylolytic and streptolytic activity to a streptococcal prophage endolysin domain. *Gene* **443:**32–41.

Boothe, J., Nykiforuk, C., Shen, Y., Zaplachinski, S., Szarka, S., Kuhlman, P., Murray, E., Morck, D., and Moloney, M. M. (2010). Seed-based expression systems for plant molecular farming. *Plant Biotechnol. J.* **8:**588–606.

Borysowski, J., and Gorski, A. (2010). Fusion to cell-penetrating peptides will enable lytic enzymes to kill intracellular bacteria. *Med. Hypotheses* **74:**164–166.

Bramley, A. J., and Foster, R. (1990). Effects of lysostaphin on *Staphylococcus aureus* infections of the mouse mammary gland. *Res. Vet. Sci.* **49:**120–121.

Briers, Y., Cornelissen, A., Aertsen, A., Hertveldt, K., Michiels, C. W., Volckaert, G., and Lavigne, R. (2008). Analysis of outer membrane permeability of *Pseudomonas aeruginosa* and bactericidal activity of endolysins KZ144 and EL188 under high hydrostatic pressure. *FEMS Microbiol. Lett.* **280:**113–119.

Briers, Y., Schmelcher, M., Loessner, M. J., Hendrix, J., Engelborghs, Y., Volckaert, G., and Lavigne, R. (2009). The high-affinity peptidoglycan binding domain of *Pseudomonas* phage endolysin KZ144. *Biochem. Biophys. Res. Commun.* **383:**187–191.

Briers, Y., Volckaert, G., Cornelissen, A., Lagaert, S., Michiels, C. W., Hertveldt, K., and Lavigne, R. (2007). Muralytic activity and modular structure of the endolysins of *Pseudomonas aeruginosa* bacteriophages phiKZ and EL. *Mol. Microbiol.* **65:**1334–1344.

Briers, Y., Walmagh, M., and Lavigne, R. (2011). Use of bacteriophage endolysin EL188 and outer membrane permeabilizers against *Pseudomonas aeruginosa*. *J. Appl. Microbiol.* **110** (3)**:**778–785.

Bustamante, N., Campillo, N. E., Garcia, E., Gallego, C., Pera, B., Diakun, G. P., Saiz, J. L., Garcia, P., Diaz, J. F., and Menendez, M. (2010). Cpl-7, a lysozyme encoded by a pneumococcal bacteriophage with a novel cell wall-binding motif. *J Biol. Chem.* **285:**33184–33196.

Callewaert, L., Walmagh, M., Michiels, C. W., and Lavigne, R. (2010). Food applications of bacterial cell wall hydrolases. *Curr. Opin. Biotechnol.* **22:**164–171.

Cheng, Q., and Fischetti, V. A. (2007). Mutagenesis of a bacteriophage lytic enzyme PlyGBS significantly increases its antibacterial activity against group B streptococci. *Appl. Microbiol. Biotechnol.* **74:**1284–1291.

Cheng, Q., Nelson, D., Zhu, S., and Fischetti, V. A. (2005). Removal of group B streptococci colonizing the vagina and oropharynx of mice with a bacteriophage lytic enzyme. *Antimicrob. Agents Chemother.* **49:**111–117.

Chun, W., and Hancock, R. E. (2000). Action of lysozyme and nisin mixtures against lactic acid bacteria. *Int. J. Food Microbiol.* **60:**25–32.

Climo, M. W., Ehlert, K., and Archer, G. L. (2001). Mechanism and suppression of lysostaphin resistance in oxacillin-resistant *Staphylococcus aureus. Antimicrob. Agents Chemother.* **45:**1431–1437.

Climo, M. W., Patron, R. L., Goldstein, B. P., and Archer, G. L. (1998). Lysostaphin treatment of experimental methicillin-resistant *Staphylococcus aureus* aortic valve endocarditis. *Antimicrob. Agents Chemother.* **42:**1355–1360.

Croux, C., Ronda, C., Lopez, R., and Garcia, J. L. (1993a). Interchange of functional domains switches enzyme specificity: Construction of a chimeric pneumococcal-clostridial cell wall lytic enzyme. *Mol. Microbiol.* **9:**1019–1025.

Croux, C., Ronda, C., Lopez, R., and Garcia, J. L. (1993b). Role of the C-terminal domain of the lysozyme of *Clostridium acetobutylicum* ATCC 824 in a chimeric pneumococcal-clostridial cell wall lytic enzyme. *FEBS Lett.* **336:**111–114.

Cui, F., Li, G., Huang, J., Zhang, J., Lu, M., Lu, W., Huan, J., and Huang, Q. (2011). Development of chitosan-collagen hydrogel incorporated with lysostaphin (CCHL) burn dressing with anti-methicillin-resistant *Staphylococcus aureus* and promotion wound healing properties. *Drug Deliv.* **18:**173–180.

Dajcs, J. J., Austin, M. S., Sloop, G. D., Moreau, J. M., Hume, E. B., Thompson, H. W., McAleese, F. M., Foster, T. J., and O'Callaghan, R. J. (2002). Corneal pathogenesis of *Staphylococcus aureus* strain Newman. *Invest Ophthalmol. Vis. Sci.* **43:**1109–1115.

Dajcs, J. J., Hume, E. B., Moreau, J. M., Caballero, A. R., Cannon, B. M., and O'Callaghan, R. J. (2000). Lysostaphin treatment of methicillin-resistant *Staphylococcus aureus* keratitis in the rabbit(1). *Am. J. Ophthalmol.* **130:**544.

Dajcs, J. J., Thibodeaux, B. A., Hume, E. B., Zheng, X., Sloop, G. D., and O'Callaghan, R. J. (2001). Lysostaphin is effective in treating methicillin-resistant *Staphylococcus aureus* endophthalmitis in the rabbit. *Curr. Eye Res.* **22:**451–457.

Daniel, A., Euler, C., Collin, M., Chahales, P., Gorelick, K. J., and Fischetti, V. A. (2010). Synergism between a novel chimeric lysin and oxacillin protects against infection by methicillin-resistant *Staphylococcus aureus. Antimicrob. Agents Chemother.* **54:**1603–1612.

Davis, K. M., and Weiser, J. N. (2011). Modifications to the peptidoglycan backbone help bacteria to establish infection. *Infect. Immun.* **79:**562–570.

DeHart, H. P., Heath, H. E., Heath, L. S., LeBlanc, P. A., and Sloan, G. L. (1995). The lysostaphin endopeptidase resistance gene (epr) specifies modification of peptidoglycan cross bridges in *Staphylococcus simulans* and *Staphylococcus aureus. Appl. Environ. Microbiol.* **61:**1475–1479.

Delbruck, M. (1940). The growth of bacteriophage and lysis of the host. *J. Gen. Physiol* **23:**643–660.

Desbois, A. P., Lang, S., Gemmell, C. G., and Coote, P. J. (2010). Surface disinfection properties of the combination of an antimicrobial peptide, ranalexin, with an endopeptidase, lysostaphin, against methicillin-resistant *Staphylococcus aureus* (MRSA). *J. Appl. Microbiol.* **108:**723–730.

Deutsch, S. M., Guezenec, S., Piot, M., Foster, S., and Lortal, S. (2004). Mur-LH, the broad-spectrum endolysin of *Lactobacillus helveticus* temperate bacteriophage phi-0303. *Appl. Environ. Microbiol.* **70:**96–103.

Dhalluin, A., Bourgeois, I., Pestel-Caron, M., Camiade, E., Raux, G., Courtin, P., Chapot-Chartier, M. P., and Pons, J. L. (2005). Acd, a peptidoglycan hydrolase of *Clostridium difficile* with N-acetylglucosaminidase activity. *Microbiology* **151:**2343–2351.

Diaz, E., Lopez, R., and Garcia, J. L. (1991). Chimeric pneumococcal cell wall lytic enzymes reveal important physiological and evolutionary traits. *J. Biol. Chem.* **266:**5464–5471.

Diaz, E., Lopez, R., and Garcia, J. L. (1990). Chimeric phage-bacterial enzymes: A clue to the modular evolution of genes. *Proc. Natl. Acad. Sci. USA* **87:**8125–8129.

Djurkovic, S., Loeffler, J. M., and Fischetti, V. A. (2005). Synergistic killing of *Streptococcus pneumoniae* with the bacteriophage lytic enzyme Cpl-1 and penicillin or gentamicin depends on the level of penicillin resistance. *Antimicrob. Agents Chemother.* **49:**1225–1228.

Donlan, R. M. (2008). Biofilms on central venous catheters: Is eradication possible? *Curr. Top. Microbiol. Immunol.* **322:**133–161.

Donovan, D. M., and Foster-Frey, J. (2008). LambdaSa2 prophage endolysin requires Cpl-7-binding domains and amidase-5 domain for antimicrobial lysis of streptococci. *FEMS Microbiol. Lett.* **287:**22–33.

Donovan, D. M., Becker, S. C., Dong, S., Baker, J. R., Foster-Frey, J., and Pritchard, D. G. (2009). Peptidoglycan hydrolase enzyme fusions for treating multi-drug resistant pathogens. *Biotech. Intl.* **21:**6–10.

Donovan, D. M., Dong, S., Garrett, W., Rousseau, G. M., Moineau, S., and Pritchard, D. G. (2006a). Peptidoglycan hydrolase fusions maintain their parental specificities. *Appl. Environ. Microbiol.* **72:**2988–2996.

Donovan, D. M., Foster-Frey, J., Dong, S., Rousseau, G. M., Moineau, S., and Pritchard, D. G. (2006b). The cell lysis activity of the *Streptococcus agalactiae* bacteriophage B30 endolysin relies on the cysteine, histidine-dependent amidohydrolase/peptidase domain. *Appl. Environ. Microbiol.* **72:**5108–5112.

Donovan, D. M., Lardeo, M., and Foster-Frey, J. (2006c). Lysis of staphylococcal mastitis pathogens by bacteriophage phi11 endolysin. *FEMS Microbiol. Lett.* **265:**133–139.

Duchardt, F., Fotin-Mleczek, M., Schwarz, H., Fischer, R., and Brock, R. (2007). A comprehensive model for the cellular uptake of cationic cell-penetrating peptides. *Traffic* **8:**848–866.

During, K., Porsch, P., Fladung, M., and Lorz, H. (1993). Transgenic potato plants resistant to the phytopathogenic bacterium *Erwinia carotovora*. *Plant J.* **3:**587–598.

Dziarski, R., and Gupta, D. (2006). The peptidoglycan recognition proteins (PGRPs). *Genome Biol.* **7:**232.

Emori, T. G., and Gaynes, R. P. (1993). An overview of nosocomial infections, including the role of the microbiology laboratory. *Clin. Microbiol. Rev.* **6:**428–442.

Entenza, J. M., Loeffler, J. M., Grandgirard, D., Fischetti, V. A., and Moreillon, P. (2005). Therapeutic effects of bacteriophage Cpl-1 lysin against *Streptococcus pneumoniae* endocarditis in rats. *Antimicrob. Agents Chemother.* **49:**4789–4792.

Fenton, M., Casey, P. G., Hill, C., Gahan, C. G., Ross, R. P., McAuliffe, O., O'Mahony, J., Maher, F., and Coffey, A. (2010a). The truncated phage lysin CHAP(k) eliminates *Staphylococcus aureus* in the nares of mice. *Bioeng. Bugs.* **1:**404–407.

Fenton, M., Ross, P., McAuliffe, O., O'Mahony, J., and Coffey, A. (2010b). Recombinant bacteriophage lysins as antibacterials. *Bioeng. Bugs.* **1:**9–16.

Filatova, L. Y., Becker, S. C., Donovan, D. M., Gladilin, A. K., and Klyachko, N. L. (2010). LysK, the enzyme lysing *Staphylococcus aureus* cells: Specific kinetic features and approaches towards stabilization. *Biochimie* **92:**507–513.

Firczuk, M., and Bochtler, M. (2007). Folds and activities of peptidoglycan amidases. *FEMS Microbiol. Rev.* **31:**676–691.

Firczuk, M., Mucha, A., and Bochtler, M. (2005). Crystal structures of active LytM. *J. Mol. Biol.* **354:**578–590.

Fischetti, V. A. (2005). Bacteriophage lytic enzymes: Novel anti-infectives. *Trends Microbiol.* **13:**491–496.

Fischetti, V. A., Nelson, D., and Schuch, R. (2006). Reinventing phage therapy: Are the parts greater than the sum? *Nat. Biotechnol.* **24:**1508–1511.

Fischetti, V. A., Zabriskie, J. B., and Gotschlich, E. C. (1972). Physical, chemical, and biological properties of type 6 M-protein extracted with purified streptococcal phage-associated lysin. *In* "Fifth International Symposium on *Streptococcus pyogenes*" (M. J. Haverkorn, ed.), p. 26. Excerpta Medica, Amsterdam.

Foley, S., Bruttin, A., and Brussow, H. (2000). Widespread distribution of a group I intron and its three deletion derivatives in the lysin gene of *Streptococcus thermophilus* bacteriophages. *J. Virol.* **74:**611–618.

Fujinami, Y., Hirai, Y., Sakai, I., Yoshino, M., and Yasuda, J. (2007). Sensitive detection of *Bacillus anthracis* using a binding protein originating from gamma-phage. *Microbiol. Immunol.* **51:**163–169.

Fujiwara, T., Aoki, S., Komatsuzawa, H., Nishida, T., Ohara, M., Suginaka, H., and Sugai, M. (2005). Mutation analysis of the histidine residues in the glycylglycine endopeptidase ALE-1. *J. Bacteriol.* **187:**480–487.

Fukushima, T., Kitajima, T., Yamaguchi, H., Ouyang, Q., Furuhata, K., Yamamoto, H., Shida, T., and Sekiguchi, J. (2008). Identification and characterization of novel cell wall hydrolase CwlT: A two-domain autolysin exhibiting N-acetylmuramidase and DL-endopeptidase activities. *J. Biol. Chem.* **283:**11117–11125.

Fukushima, T., Yao, Y., Kitajima, T., Yamamoto, H., and Sekiguchi, J. (2007). Characterization of new L, D-endopeptidase gene product CwlK (previous YcdD) that hydrolyzes peptidoglycan in *Bacillus subtilis*. *Mol. Genet. Genomics* **278:**371–383.

Fux, C. A., Stooodley, P., Hall-Stoodley, L., and Costerton, J. W. (2003). Bacterial biofilms: A diagnostic and therapeutic challenge. *Expert Rev. Anti. Infect. Ther.* **1:**667–683.

Gaeng, S., Scherer, S., Neve, H., and Loessner, M. J. (2000). Gene cloning and expression and secretion of *Listeria monocytogenes* bacteriophage-lytic enzymes in *Lactococcus lactis*. *Appl. Environ. Microbiol.* **66:**2951–2958.

Garcia, J. L., Garcia, E., Arraras, A., Garcia, P., Ronda, C., and Lopez, R. (1987). Cloning, purification, and biochemical characterization of the pneumococcal bacteriophage Cp-1 lysin. *J. Virol.* **61:**2573–2580.

Garcia, P., Garcia, J. L., Garcia, E., Sanchez-Puelles, J. M., and Lopez, R. (1990). Modular organization of the lytic enzymes of *Streptococcus pneumoniae* and its bacteriophages. *Gene* **86:**81–88.

Garcia, P., Martinez, B., Rodriguez, L., and Rodriguez, A. (2010a). Synergy between the phage endolysin LysH5 and nisin to kill *Staphylococcus aureus* in pasteurized milk. *Int. J. Food Microbiol.* **141:**151–155.

Garcia, P., Rodriguez, L., Rodriguez, A., and Martinez, B. (2010b). Food biopreservation: Promising strategies using bacteriocins, bacteriophages and endolysins. *Trends Food Sci. Technol.* **21:**373–382.

Garcia-del Portillo, F., and Finlay, B. B. (1995). The varied lifestyles of intracellular pathogens within eukaryotic vacuolar compartments. *Trends Microbiol.* **3:**373–380.

Graham, S., and Coote, P. J. (2007). Potent, synergistic inhibition of *Staphylococcus aureus* upon exposure to a combination of the endopeptidase lysostaphin and the cationic peptide ranalexin. *J. Antimicrob. Chemother.* **59:**759–762.

Grandgirard, D., Loeffler, J. M., Fischetti, V. A., and Leib, S. L. (2008). Phage lytic enzyme Cpl-1 for antibacterial therapy in experimental pneumococcal meningitis. *J. Infect. Dis.* **197:**1519–1522.

Grundling, A., Missiakas, D. M., and Schneewind, O. (2006). *Staphylococcus aureus* mutants with increased lysostaphin resistance. *J. Bacteriol.* **188:**6286–6297.

Grundling, A., and Schneewind, O. (2006). Cross-linked peptidoglycan mediates lysostaphin binding to the cell wall envelope of *Staphylococcus aureus*. *J. Bacteriol.* **188:**2463–2472.

Gu, J., Xu, W., Lei, L., Huang, J., Feng, X., Sun, C., Du, C., Zuo, J., Li, Y., Du, T., Li, L., and Han, W. (2011). LysGH15, a novel bacteriophage lysin, protects a murine bacteremia model efficiently against lethal methicillin-resistant *Staphylococcus aureus* infection. *J. Clin. Microbiol.* **49:**111–117.

Hagens, S., and Loessner, M. J. (2010). Bacteriophage for biocontrol of foodborne pathogens: Calculations and considerations. *Curr. Pharm. Biotechnol.* **11:**58–68.

Hauben, K. J. L., Wuytack, E. Y., Soontjens, C. C. F., and Michiels, C. W. (1996). High-pressure transient sensitization of *Escherichia coli* to lysozyme and nisin by disruption of outer-membrane permeability. *J. Food Protection* **59:**350–355.

Hermoso, J. A., Garcia, J. L., and Garcia, P. (2007). Taking aim on bacterial pathogens: From phage therapy to enzybiotics. *Curr. Opin. Microbiol.* **10:**461–472.

Hermoso, J. A., Monterroso, B., Albert, A., Galan, B., Ahrazem, O., Garcia, P., Martinez-Ripoll, M., Garcia, J. L., and Menendez, M. (2003). Structural basis for selective recognition of pneumococcal cell wall by modular endolysin from phage Cp-1. *Structure* **11:**1239–1249.

Høiby, N., Bjarnsholt, T., Givskov, M., Molin, S., and Ciofu, O. (2010). Antibiotic resistance of bacterial biofilms. *Int. J. Antimicrob. Agents* **35**(4)**:**322–332.

Holtje, J. V., and Tomasz, A. (1975). Lipoteichoic acid: A specific inhibitor of autolysin activity in *Pneumococcus*. *Proc. Natl. Acad. Sci. USA* **72:**1690–1694.

Hoopes, J. T., Stark, C. J., Kim, H. A., Sussman, D. J., Donovan, D. M., and Nelson, D. C. (2009). Use of a bacteriophage lysin, PlyC, as an enzyme disinfectant against *Streptococcus equi*. *Appl. Environ. Microbiol.* **75:**1388–1394.

Horgan, M., O'Flynn, G., Garry, J., Cooney, J., Coffey, A., Fitzgerald, J. F., Ross, R. P., and McAuliffe, O. (2009). Phage lysin LysK can be truncated to its CHAP domain and retain lytic activity against live antibiotic-resistant staphylococci. *Appl. Environ. Microbiol.* **75:**872–874.

Huard, C., Miranda, G., Wessner, F., Bolotin, A., Hansen, J., Foster, S. J., and Chapot-Chartier, M. P. (2003). Characterization of AcmB, an N-acetylglucosaminidase autolysin from *Lactococcus lactis*. *Microbiology* **149:**695–705.

Imbuluzqueta, E., Gamazo, C., Ariza, J., and Blanco-Prieto, M. J. (2010). Drug delivery systems for potential treatment of intracellular bacterial infections. *Front. Biosci.* **15:**397–417.

Inouye, M., Arnheim, N., and Sternglanz, R. (1973). Bacteriophage T7 lysozyme is an N-acetylmuramyl-L-alanine amidase. *J Biol. Chem.* **248:**7247–7252.

Ito, Y., Kwon, O. H., Ueda, M., Tanaka, A., and Imanishi, Y. (1997). Bactericidal activity of human lysozymes carrying various lengths of polyproline chain at the C-terminus. *FEBS Lett.* **415:**285–288.

Jado, I., Lopez, R., Garcia, E., Fenoll, A., Casal, J., and Garcia, P. (2003). Phage lytic enzymes as therapy for antibiotic-resistant *Streptococcus pneumoniae* infection in a murine sepsis model. *J. Antimicrob. Chemother.* **52:**967–973.

Johnsborg, O., and Håvarstein, L. S. (2009). Regulation of natural genetic transformation and acquisition of transforming DNA in *Streptococcus pneumoniae*. *FEMS Microbiol. Rev.* **33:**627–642.

Joliot, A., and Prochiantz, A. (2004). Transduction peptides: From technology to physiology. *Nat. Cell Biol.* **6:**189–196.

Jones, R. N., Barry, A. L., Gavan, T. L., and Washington, J. A., II (1985). Susceptibility tests: Microdilution and macrodilution broth procedures. *In* "Manual of Clinical Microbiology" (A. Balows, J. W. J. Hausler, and H. J. Shadomy, eds.), pp. 972–977. American Society for Microbiology, Washington, DC.

Joris, B., Englebert, S., Chu, C. P., Kariyama, R., Neo-Moore, L., Shockman, G. D., and Ghuysen, J. M. (1992). Modular design of the *Enterococcus hirae* muramidase-2 and *Streptococcus faecalis* autolysin. *FEMS Microbiol. Lett.* **70:**257–264.

Kasparek, P., Pantucek, R., Kahankova, J., Ruzickova, V., and Doskar, J. (2007). Genome rearrangements in host-range mutants of the polyvalent staphylococcal bacteriophage 812. *Folia Microbiol. (Praha)* **52:**331–338.

Keren, I., Kaldalu, N., Spoering, A., Wang, Y., and Lewis, K. (2004). Persister cells and tolerance to antimicrobials. *FEMS Microbiol. Lett.* **230:**13–18.

Kerr, D. E., Plaut, K., Bramley, A. J., Williamson, C. M., Lax, A. J., Moore, K., Wells, K. D., and Wall, R. J. (2001). Lysostaphin expression in mammary glands confers protection against staphylococcal infection in transgenic mice. *Nat. Biotechnol.* **19:**66–70.

Kikkawa, H., Fujinami, Y., Suzuki, S., and Yasuda, J. (2007). Identification of the amino acid residues critical for specific binding of the bacteriolytic enzyme of gamma-phage, PlyG, to *Bacillus anthracis. Biochem. Biophys. Res. Commun.* **363:**531–535.

Kikkawa, H. S., Ueda, T., Suzuki, S., and Yasuda, J. (2008). Characterization of the catalytic activity of the gamma-phage lysin, PlyG, specific for *Bacillus anthracis. FEMS Microbiol. Lett.* **286:**236–240.

Kim, W. S., Salm, H., and Geider, K. (2004). Expression of bacteriophage phiEa1h lysozyme in *Escherichia coli* and its activity in growth inhibition of *Erwinia amylovora. Microbiology* **150:**2707–2714.

Kiri, N., Archer, G., and Climo, M. W. (2002). Combinations of lysostaphin with beta-lactams are synergistic against oxacillin-resistant *Staphylococcus epidermidis. Antimicrob. Agents Chemother.* **46:**2017–2020.

Kokai-Kun, J. F., Chanturiya, T., and Mond, J. J. (2007). Lysostaphin as a treatment for systemic *Staphylococcus aureus* infection in a mouse model. *J. Antimicrob. Chemother.* **60:**1051–1059.

Kokai-Kun, J. F., Chanturiya, T., and Mond, J. J. (2009). Lysostaphin eradicates established *Staphylococcus aureus* biofilms in jugular vein catheterized mice. *J. Antimicrob. Chemother.* **64:**94–100.

Kokai-Kun, J. F., Walsh, S. M., Chanturiya, T., and Mond, J. J. (2003). Lysostaphin cream eradicates *Staphylococcus aureus* nasal colonization in a cotton rat model. *Antimicrob. Agents Chemother.* **47:**1589–1597.

Korndorfer, I. P., Danzer, J., Schmelcher, M., Zimmer, M., Skerra, A., and Loessner, M. J. (2006). The crystal structure of the bacteriophage PSA endolysin reveals a unique fold responsible for specific recognition of *Listeria* cell walls. *J. Mol. Biol.* **364:**678–689.

Korndorfer, I. P., Kanitz, A., Danzer, J., Zimmer, M., Loessner, M. J., and Skerra, A. (2008). Structural analysis of the L-alanoyl-D-glutamate endopeptidase domain of *Listeria* bacteriophage endolysin Ply500 reveals a new member of the LAS peptidase family. *Acta Crystallogr. D Biol. Crystallogr.* **64:**644–650.

Krause, R. M. (1957). Studies on bacteriophages of hemolytic streptococci. *I. Factors influencing the interaction of phage and susceptible host cell. J Exp. Med.* **106:**365–384.

Kretzer, J. W., Lehmann, R., Schmelcher, M., Banz, M., Kim, K. P., Korn, C., and Loessner, M. J. (2007). Use of high-affinity cell wall-binding domains of bacteriophage endolysins for immobilization and separation of bacterial cells. *Appl. Environ. Microbiol.* **73:**1992–2000.

Kusuma, C., Jadanova, A., Chanturiya, T., and Kokai-Kun, J. F. (2007). Lysostaphin-resistant variants of *Staphylococcus aureus* demonstrate reduced fitness *in vitro* and *in vivo. Antimicrob. Agents Chemother.* **51:**475–482.

Kusuma, C., and Kokai-Kun, J. (2005). Comparison of four methods for determining lysostaphin susceptibility of various strains of *Staphylococcus aureus. Antimicrob. Agents Chemother.* **49:**3256–3263.

Lai, M. J., Lin, N. T., Hu, A., Soo, P. C., Chen, L. K., Chen, L. H., and Chang, K. C. (2011). Antibacterial activity of *Acinetobacter baumannii* phage ΦAB2 endolysin (LysAB2) against both Gram-positive and Gram-negative bacteria. *Appl. Microbiol. Biotechnol.* **90** (2):529–539.

Lewis, K. (2001). Riddle of biofilm resistance. *Antimicrob Agents Chemother.* **45:**999–1007.
Lindsay, J. A. (2008). *Staphylococcus* Molecular Genetics. Caister Academic Press, Norfolk, UK.
Loeffler, J. M., Djurkovic, S., and Fischetti, V. A. (2003). Phage lytic enzyme Cpl-1 as a novel antimicrobial for pneumococcal bacteremia. *Infect. Immun.* **71:**6199–6204.
Loeffler, J. M., and Fischetti, V. A. (2003). Synergistic lethal effect of a combination of phage lytic enzymes with different activities on penicillin-sensitive and -resistant *Streptococcus pneumoniae* strains. *Antimicrob. Agents Chemother.* **47:**375–377.
Loeffler, J. M., Nelson, D., and Fischetti, V. A. (2001). Rapid killing of *Streptococcus pneumoniae* with a bacteriophage cell wall hydrolase. *Science* **294:**2170–2172.
Loessner, M. J. (2005). Bacteriophage endolysins: Current state of research and applications. *Curr. Opin. Microbiol.* **8:**480–487.
Loessner, M. J., Gaeng, S., Wendlinger, G., Maier, S. K., and Scherer, S. (1998). The two-component lysis system of *Staphylococcus aureus* bacteriophage Twort: A large TTG-start holin and an associated amidase endolysin. *FEMS Microbiol. Lett.* **162:**265–274.
Loessner, M. J., Kramer, K., Ebel, F., and Scherer, S. (2002). C-terminal domains of *Listeria monocytogenes* bacteriophage murein hydrolases determine specific recognition and high-affinity binding to bacterial cell wall carbohydrates. *Mol. Microbiol.* **44:**335–349.
Loessner, M. J., Schneider, A., and Scherer, S. (1995a). A new procedure for efficient recovery of DNA, RNA, and proteins from *Listeria* cells by rapid lysis with a recombinant bacteriophage endolysin. *Appl. Environ. Microbiol.* **61:**1150–1152.
Loessner, M. J., Wendlinger, G., and Scherer, S. (1995b). Heterogeneous endolysins in *Listeria monocytogenes* bacteriophages: A new class of enzymes and evidence for conserved holin genes within the siphoviral lysis cassettes. *Mol. Microbiol* **16:**1231–1241.
Lopez, R., Garcia, E., Garcia, P., and Garcia, J. L. (1997). The pneumococcal cell wall degrading enzymes: a modular design to create new lysins? *Microb. Drug. Resist.* **3:**199–211.
Lopez, R., Garcia, E., Garcia, P., Ronda, C., and Tomasz, A. (1982). Choline-containing bacteriophage receptors in *Streptococcus pneumoniae. J. Bacteriol.* **151:**1581–1590.
Low, L. Y., Yang, C., Perego, M., Osterman, A., and Liddington, R. C. (2005). Structure and lytic activity of a *Bacillus anthracis* prophage endolysin. *J. Biol. Chem.* **280:**35433–35439.
Lu, J. Z., Fujiwara, T., Komatsuzawa, H., Sugai, M., and Sakon, J. (2006). Cell wall-targeting domain of glycyl-glycine endopeptidase distinguishes among peptidoglycan cross-bridges. *J. Biol. Chem.* **281:**549–558.
Maga, E. A., Cullor, J. S., Smith, W., Anderson, G. B., and Murray, J. D. (2006a). Human lysozyme expressed in the mammary gland of transgenic dairy goats can inhibit the growth of bacteria that cause mastitis and the cold-spoilage of milk. *Foodborne. Pathog. Dis.* **3:**384–392.
Maga, E. A., Walker, R. K., Anderson, G. B., and Murray, J. D. (2006b). Consumption of milk from transgenic goats expressing human lysozyme in the mammary gland results in the modulation of intestinal microflora. *Transgenic Res.* **15:**515–519.
Manoharadas, S., Witte, A., and Blasi, U. (2009). Antimicrobial activity of a chimeric enzybiotic towards *Staphylococcus aureus. J. Biotechnol.* **139:**118–123.
Masschalck, B., Garcia-Graells, C., Van Haver, E., and Michiels, C. W. (2000). Inactivation of high pressure resistant *Escherichia coli* by nisin and lysozyme under high pressure. *Innov. Food Sci. Emerg. Technol.* **1:**39.
Masschalck, B., and Michiels, C. W. (2003). Antimicrobial properties of lysozyme in relation to foodborne vegetative bacteria. *Crit. Rev. Microbiol.* **29:**191–214.
Masschalck, B., Van, H. R., and Michiels, C. W. (2001). High pressure increases bactericidal activity and spectrum of lactoferrin, lactoferricin and nisin. *Int. J. Food Microbiol.* **64:**325–332.
Mayer, M. J., Narbad, A., and Gasson, M. J. (2008). Molecular characterization of a *Clostridium difficile* bacteriophage and its cloned biologically active endolysin. *J. Bacteriol.* **190:**6734–6740.

Mayer, M. J., Payne, J., Gasson, M. J., and Narbad, A. (2010). Genomic sequence and characterization of the virulent bacteriophage phiCTP1 from *Clostridium tyrobutyricum* and heterologous expression of its endolysin. *Appl. Environ. Microbiol.* **76:**5415–5422.

McCullers, J. A., Karlstrom, A., Iverson, A. R., Loeffler, J. M., and Fischetti, V. A. (2007). Novel strategy to prevent otitis media caused by colonizing *Streptococcus pneumoniae*. *PLoS Pathog.* **3:**e28.

Mesnage, S., Chau, F., Dubost, L., and Arthur, M. (2008). Role of N-acetylglucosaminidase and N-acetylmuramidase activities in *Enterococcus faecalis* peptidoglycan metabolism. *J. Biol. Chem.* **283:**19845–19853.

Mitchell, G. J., Nelson, D. C., and Weitz, J. S. (2010). Quantifying enzymatic lysis: Estimating the combined effects of chemistry, physiology and physics. *Phys. Biol.* **7:**046002.

Moak, M., and Molineux, I. J. (2004). Peptidoglycan hydrolytic activities associated with bacteriophage virions. *Mol. Microbiol.* **51:**1169–1183.

Mokrasch, L. C. (1967). Use of 2,4,6-trinitrobenzenesulfonic acid for the coestimation of amines, amino acids and proteins in mixtures. *Anal. Biochem* **18:**64–71.

Monterroso, B., Saiz, J. L., Garcia, P., Garcia, J. L., and Menendez, M. (2008). Insights into the structure-function relationships of pneumococcal cell wall lysozymes, LytC and Cpl-1. *J. Biol. Chem.* **283:**28618–28628.

Morita, M., Tanji, Y., Orito, Y., Mizoguchi, K., Soejima, A., and Unno, H. (2001). Functional analysis of antibacterial activity of *Bacillus amyloliquefaciens* phage endolysin against Gram-negative bacteria. *FEBS Lett.* **500:**56–59.

Muyombwe, A., Tanji, Y., and Unno, H. (1999). Cloning and expression of a gene encoding the lytic functions of *Bacillus amyloliquefaciens* phage: Evidence of an auxiliary lysis system. *J. Biosci. Bioeng.* **88:**221–225.

Nakimbugwe, D., Masschalck, B., Deckers, D., Callewaert, L., Aertsen, A., and Michiels, C. W. (2006). Cell wall substrate specificity of six different lysozymes and lysozyme inhibitory activity of bacterial extracts. *FEMS Microbiol. Lett.* **259:**41–46.

Navarre, W. W., Ton-That, H., Faull, K. F., and Schneewind, O. (1999). Multiple enzymatic activities of the murein hydrolase from staphylococcal phage phi11: Identification of a D-alanyl-glycine endopeptidase activity. *J. Biol. Chem.* **274:**15847–15856.

Nelson, D., Loomis, L., and Fischetti, V. A. (2001). Prevention and elimination of upper respiratory colonization of mice by group A streptococci by using a bacteriophage lytic enzyme. *Proc. Natl. Acad. Sci. USA* **98:**4107–4112.

Nelson, D., Schuch, R., Chahales, P., Zhu, S., and Fischetti, V. A. (2006). PlyC: A multimeric bacteriophage lysin. *Proc. Natl. Acad. Sci. USA* **103:**10765–10770.

Nelson, D., Schuch, R., Zhu, S., Tscherne, D. M., and Fischetti, V. A. (2003). Genomic sequence of C1, the first streptococcal phage. *J Bacteriol.* **185:**3325–3332.

Obeso, J. M., Martinez, B., Rodriguez, A., and Garcia, P. (2008). Lytic activity of the recombinant staphylococcal bacteriophage PhiH5 endolysin active against *Staphylococcus aureus* in milk. *Int. J. Food Microbiol.* **128:**212–218.

Oey, M., Lohse, M., Kreikemeyer, B., and Bock, R. (2009). Exhaustion of the chloroplast protein synthesis capacity by massive expression of a highly stable protein antibiotic. *Plant J.* **57:**436–445.

O'Flaherty, S., Coffey, A., Edwards, R., Meaney, W., Fitzgerald, G. F., and Ross, R. P. (2004). Genome of staphylococcal phage K: A new lineage of *Myoviridae* infecting gram-positive bacteria with a low G+C content. *J. Bacteriol.* **186:**2862–2871.

O'Flaherty, S., Coffey, A., Meaney, W., Fitzgerald, G. F., and Ross, R. P. (2005). The recombinant phage lysin LysK has a broad spectrum of lytic activity against clinically relevant staphylococci, including methicillin-resistant *Staphylococcus aureus*. *J. Bacteriol.* **187:**7161–7164.

O'Flaherty, S., Ross, R. P., and Coffey, A. (2009). Bacteriophage and their lysins for elimination of infectious bacteria. *FEMS Microbiol. Rev.* **33:**801–819.

Oldham, E. R., and Daley, M. J. (1991). Lysostaphin: Use of a recombinant bactericidal enzyme as a mastitis therapeutic. *J. Dairy Sci.* **74:**4175–4182.

Orito, Y., Morita, M., Hori, K., Unno, H., and Tanji, Y. (2004). *Bacillus amyloliquefaciens* phage endolysin can enhance permeability of *Pseudomonas aeruginosa* outer membrane and induce cell lysis. *Appl. Microbiol. Biotechnol.* **65:**105–109.

Palmer, K. L., Kos, V. N., and Gilmore, M. S. (2010). Horizontal gene transfer and the genomics of enterococcal antibiotic resistance. *Curr. Opin. Microbiol.* **13:**632–639.

Pangule, R. C., Brooks, S. J., Dinu, C. Z., Bale, S. S., Salmon, S. L., Zhu, G., Metzger, D. W., Kane, R. S., and Dordick, J. S. (2010). Antistaphylococcal nanocomposite films based on enzyme-nanotube conjugates. *ACS Nano.* **4:**3993–4000.

Paradis-Bleau, C., Cloutier, I., Lemieux, L., Sanschagrin, F., Laroche, J., Auger, M., Garnier, A., and Levesque, R. C. (2007). Peptidoglycan lytic activity of the *Pseudomonas aeruginosa* phage phiKZ gp144 lytic transglycosylase. *FEMS Microbiol. Lett.* **266:**201–209.

Park, J. T., and Johnson, M. J. (1949). A submicrodetermination of glucose. *J. Biol. Chem.* **181:**149–151.

Pastagia, M., Euler, C., Chahales, P., Fuentes-Duculan, J., Krueger, J. G., and Fischetti, V. A. (2011). A novel chimeric lysin shows superiority to mupirocin for skin decolonization of methicillin-resistant and -sensitive *Staphylococcus aureus* strains. *Antimicrob. Agents Chemother.* **55:**738–744.

Patron, R. L., Climo, M. W., Goldstein, B. P., and Archer, G. L. (1999). Lysostaphin treatment of experimental aortic valve endocarditis caused by a *Staphylococcus aureus* isolate with reduced susceptibility to vancomycin. *Antimicrob. Agents Chemother.* **43:**1754–1755.

Ponting, C. P., Aravind, L., Schultz, J., Bork, P., and Koonin, E. V. (1999). Eukaryotic signalling domain homologues in archaea and bacteria: Ancient ancestry and horizontal gene transfer. *J. Mol. Biol.* **289:**729–745.

Porter, C. J., Schuch, R., Pelzek, A. J., Buckle, A. M., McGowan, S., Wilce, M. C., Rossjohn, J., Russell, R., Nelson, D., Fischetti, V. A., and Whisstock, J. C. (2007). The 1.6 A crystal structure of the catalytic domain of PlyB, a bacteriophage lysin active against *Bacillus anthracis*. *J. Mol. Biol.* **366:**540–550.

Pritchard, D. G., Dong, S., Baker, J. R., and Engler, J. A. (2004). The bifunctional peptidoglycan lysin of *Streptococcus agalactieae* bacteriophage B30. *Microbiology* **150:**2079–2087.

Pritchard, D. G., Dong, S., Kirk, M. C., Cartee, R. T., and Baker, J. R. (2007). LambdaSa1 and LambdaSa2 prophage lysins of *Streptococcus agalactiae*. *Appl. Environ. Microbiol.* **73:**7150–7154.

Rashel, M., Uchiyama, J., Ujihara, T., Uehara, Y., Kuramoto, S., Sugihara, S., Yagyu, K., Muraoka, A., Sugai, M., Hiramatsu, K., Honke, K., and Matsuzaki, S. (2007). Efficient elimination of multidrug-resistant *Staphylococcus aureus* by cloned lysin derived from bacteriophage phi MR11. *J. Infect. Dis.* **196:**1237–1247.

Rigden, D. J., Jedrzejas, M. J., and Galperin, M. Y. (2003). Amidase domains from bacterial and phage autolysins define a family of gamma-D,L-glutamate-specific amidohydrolases. *Trends Biochem. Sci.* **28:**230–234.

Rossolini, G. M., Mantengoli, E., Montagnani, F., and Pollini, S. (2010). Epidemiology and clinical relevance of microbial resistance determinants versus anti-Gram-positive agents. *Curr. Opin. Microbiol.* **13:**582–588.

Sahl, H. G., and Bierbaum, G. (2008). Multiple activities in natural antimicrobials. *Microbe* **3:**467–473.

Sanz, J. M., Garcia, P., and Garcia, J. L. (1996). Construction of a multifunctional pneumococcal murein hydrolase by module assembly. *Eur. J. Biochem.* **235:**601–605.

Sass, P., and Bierbaum, G. (2007). Lytic activity of recombinant bacteriophage phi11 and phi12 endolysins on whole cells and biofilms of *Staphylococcus aureus*. *Appl. Environ. Microbiol.* **73:**347–352.

Satake, K., Okuyama, T., Ohashi, M., and Shinoda, T. (1960). The spectrophotometric determination of amine, amino acid and peptide with 2,4,6,-trinitrobenzene 1-sulfonic acid. *J. Biochem.* **47:**654–660.

Sawhney, R., and Berry, V. (2009). Bacterial biofilm formation, pathogenicity, diagnostics and control: An overview. *Indian J. Med. Sci.* **63:**313–321.

Schaffner, W., Melly, M. A., and Koenig, M. G. (1967). Lysostaphin: An enzymatic approach to staphylococcal disease. *II. In vivo studies. Yale J. Biol. Med.* **39:**230–244.

Scheurwater, E. M., and Clarke, A. J. (2008). The C-terminal domain of *Escherichia coli* YfhD functions as a lytic transglycosylase. *J. Biol. Chem.* **283:**8363–8373.

Schindler, C. A., and Schuhardt, V. T. (1964). Lysostaphin: A new bacteriolytic agent for the *Staphylococcus. Proc. Natl. Acad. Sci. USA* **51:**414–421.

Schleifer, K. H., and Kandler, O. (1972). Peptidoglycan types of bacterial cell walls and their taxonomic implications. *Bacteriol. Rev.* **36:**407–477.

Schmelcher, M., Shabarova, T., Eugster, M. R., Eichenseher, F., Tchang, V. S., Banz, M., and Loessner, M. J. (2010). Rapid multiplex detection and differentiation of *Listeria* cells by use of fluorescent phage endolysin cell wall binding domains. *Appl. Environ. Microbiol.* **76:**5745–5756.

Schmelcher, M., Tchang, V. S., and Loessner, M. J. (2011). Domain shuffling and module engineering of *Listeria* phage endolysins for enhanced lytic activity and binding affinity. *Microb. Biotechnol.* **4:**651–652.

Schuch, R., Fischetti, V. A., and Nelson, D. C. (2009). A genetic screen to identify bacteriophage lysins. *Methods Mol. Biol.* **502:**307–319.

Schuch, R., Nelson, D., and Fischetti, V. A. (2002). A bacteriolytic agent that detects and kills *Bacillus anthracis. Nature* **418:**884–889.

Seltman, G., and Holst, O. (2001). "The Bacterial Cell Wall". Springer Verlag, Berlin.

Sheehan, M. M., Garcia, J. L., Lopez, R., and Garcia, P. (1996). Analysis of the catalytic domain of the lysin of the lactococcal bacteriophage Tuc 2009 by chimeric gene assembling. *FEMS Microbiol. Lett.* **140:**23–28.

Silva-Martin, N., Molina, R., Angulo, I., Mancheno, J. M., Garcia, P., and Hermoso, J. A. (2010). Crystallization and preliminary crystallographic analysis of the catalytic module of endolysin from Cp-7, a phage infecting *Streptococcus pneumoniae. Acta Crystallogr. F Struct. Biol. Cryst. Commun.* **66:**670–673.

Simmons, M., Donovan, D. M., Siragusa, G. R., and Seal, B. S. (2010). Recombinant expression of two bacteriophage proteins that lyse *Clostridium perfringens* and share identical sequences in the C-terminal cell wall binding domain of the molecules but are dissimilar in their N-terminal active domains. *J. Agric. Food Chem.* **58:**10330–10337.

Simões, M. (2011). Antimicrobial strategies effective against infectious bacterial biofilms. *Curr. Med. Chem.* **18:**2129–2145.

Son, J. S., Lee, S. J., Jun, S. Y., Yoon, S. J., Kang, S. H., Paik, H. R., Kang, J. O., and Choi, Y. J. (2010). Antibacterial and biofilm removal activity of a podoviridae *Staphylococcus aureus* bacteriophage SAP-2 and a derived recombinant cell-wall-degrading enzyme. *Appl. Microbiol. Biotechnol.* **86:**1439–1449.

Spiro, R. G. (1966). Analysis of sugars found in glycoproteins. *Methods Enzymol.* **8:**3–26.

Splith, K., and Neundorf, I. (2011). Antimicrobial peptides with cell-penetrating peptide properties and vice versa. *Eur. Biophys. J.* **40**(4)**:**387–397.

Stark, C. J., Hoopes, J. T., Bonocoroa, R. P., and Nelson, D. C. (2010). Bacteriophage lytic enzymes as antimicrobials. *In* "Bacteriophage in the Detection and Control of Foodborne Pathogens" (P. V. Sabour and M. W. Griffith, eds.), pp. 137–156. ASM Press, Washington, DC.

Steen, A., van Schalkwijk, S., Buist, G., Twigt, M., Szeliga, W., Meijer, W., Kuipers, O. P., Kok, J., and Hugenholz, J. (2007). Lytr, a phage-derived amidase is most effective in induced lysis of *Lactococcus lactis* compared with other lactococcal amidases and glucosaminidases. *Int. Dairy J.* **17:**926–936.

Stentz, C. R., Bongaerts, R. J., Gunning, A. P., Gasson, M., and Shearman, C. (2010). Controlled release of protein from viable *Lactococcus lactis* cells. *Appl. Environ. Microbiol.* **76:**3026–3031.

Sugai, M., Fujiwara, T., Ohta, K., Komatsuzawa, H., Ohara, M., and Suginaka, H. (1997). epr, which encodes glycylglycine endopeptidase resistance, is homologous to *femAB* and affects serine content of peptidoglycan cross bridges in *Staphylococcus capitis* and *Staphylococcus aureus*. *J. Bacteriol.* **179:**4311–4318.

Sun, H. C., Xue, F. M., Qian, K., Fang, X. H., Qiu, H. L., Zhang, X. Y., and Yin, Z. H. (2006). Intramammary expression and therapeutic effect of a human lysozyme-expressing vector for treating bovine mastitis. *J. Zhejiang. Univ. Sci. B* **7:**324–330.

Tan, Y. P., Giffard, P. M., Barry, D. G., Huston, W. M., and Turner, M. S. (2008). Random mutagenesis identifies novel genes involved in the secretion of antimicrobial, cell wall-lytic enzymes by *Lactococcus lactis*. *Appl. Environ. Microbiol.* **74:**7490–7496.

Taylor, A., and Gorazdowska, M. (1974). Conversion of murein to non-reducing fragments by enzymes from phage lambda and Vi II lysates. *Biochim. Biophys. Acta* **342:**133–136.

Thumm, G., and Gotz, F. (1997). Studies on prolysostaphin processing and characterization of the lysostaphin immunity factor (Lif) of *Staphylococcus simulans* biovar *staphylolyticus*. *Mol. Microbiol.* **23:**1251–1265.

Tong, J., Wei, H., Liu, X., Hu, W., Bi, M., Wang, Y., Li, Q., and Li, N. (2010). Production of recombinant human lysozyme in the milk of transgenic pigs. *Transgenic Res.* **20** (2):417–419.

Turner, M. S., Waldherr, F., Loessner, M. J., and Giffard, P. M. (2007). Antimicrobial activity of lysostaphin and a *Listeria monocytogenes* bacteriophage endolysin produced and secreted by lactic acid bacteria. *Syst. Appl. Microbiol.* **30:**58–67.

Vaara, M. (1992). Agents that increase the permeability of the outer membrane. *Microbiol. Rev.* **56:**395–411.

Vasala, A., Valkkila, M., Caldentey, J., and Alatossava, T. (1995). Genetic and biochemical characterization of the *Lactobacillus delbrueckii* subsp. *lactis* bacteriophage LL-H lysin. *Appl. Environ. Microbiol.* **61:**4004–4011.

Vazquez-Boland, J. A., Kuhn, M., Berche, P., Chakraborty, T., Dominguez-Bernal, G., Goebel, W., Gonzalez-Zorn, B., Wehland, J., and Kreft, J. (2001). *Listeria* pathogenesis and molecular virulence determinants. *Clin. Microbiol. Rev.* **14:**584–640.

Vollmer, W. (2008). Structural variation in the glycan strands of bacterial peptidoglycan. *FEMS Microbiol. Rev.* **32:**287–306.

Walcher, G., Stessl, B., Wagner, M., Eichenseher, F., Loessner, M. J., and Hein, I. (2010). Evaluation of paramagnetic beads coated with recombinant Listeria phage endolysin-derived cell-wall-binding domain proteins for separation of *Listeria monocytogenes* from raw milk in combination with culture-based and real-time polymerase chain reaction-based quantification. *Foodborne Pathog. Dis.* **7:**1019–1024.

Walencka, E., Sadowska, B., Rozalska, S., Hryniewicz, W., and Rozalska, B. (2005). Lysostaphin as a potential therapeutic agent for staphylococcal biofilm eradication. *Pol. J. Microbiol.* **54:**191–200.

Wall, R. J., Powell, A., Paape, M. J., Kerr, D. E., Bannerman, D. D., Pursel, V. G., Wells, K. D., Talbot, N., and Hawk, H. W. (2005). Genetically enhanced cows resist intramammary *Staphylococcus aureus* infection. *Nat. Biotechnol.* **23:**445–451.

Ward, J. B. (1973). The chain length of the glycans in bacterial cell walls. *Biochem. J.* **133:**395–398.

Weidel, W., and Pelzer, H. (1964). Bagshaped macromolecules: A new outlook on bacterial cell walls. *Adv. Enzymol. Relat. Areas Mol. Biol.* **26:**193–232.

Whisstock, J. C., and Lesk, A. M. (1999). SH3 domains in prokaryotes. *Trends Biochem. Sci.* **24:**132–133.

Wu, J. A., Kusuma, C., Mond, J. J., and Kokai-Kun, J. F. (2003). Lysostaphin disrupts *Staphylococcus aureus* and *Staphylococcus epidermidis* biofilms on artificial surfaces. *Antimicrob. Agents Chemother.* **47:**3407–3414.

Yang, B., Wang, J., Tang, B., Liu, Y., Guo, C., Yang, P., Yu, T., Li, R., Zhao, J., Zhang, L., Dai, Y., and Li, N. (2011). Characterization of bioactive recombinant human lysozyme expressed in milk of cloned transgenic cattle. *PLoS One* **6:**e17593.

Yang, G., Gao, Y., Feng, J., Huang, Y., Li, S., Liu, Y., Liu, C., Fan, M., Shen, B., and Shao, N. (2008). C-terminus of TRAP in *Staphylococcus* can enhance the activity of lysozyme and lysostaphin. *Acta Biochim. Biophys. Sin. (Shanghai)* **40:**452–458.

Yoong, P., Schuch, R., Nelson, D., and Fischetti, V. A. (2004). Identification of a broadly active phage lytic enzyme with lethal activity against antibiotic-resistant *Enterococcus faecalis* and *Enterococcus faecium. J. Bacteriol.* **186:**4808–4812.

Yoong, P., Schuch, R., Nelson, D., and Fischetti, V. A. (2006). PlyPH, a bacteriolytic enzyme with a broad pH range of activity and lytic action against *Bacillus anthracis. J. Bacteriol.* **188:**2711–2714.

Young, R. (1992). Bacteriophage lysis: Mechanism and regulation. *Microbiol Rev.* **56:**430–481.

Zimmer, M., Vukov, N., Scherer, S., and Loessner, M. J. (2002). The murein hydrolase of the bacteriophage phi3626 dual lysis system is active against all tested *Clostridium perfringens* strains. *Appl. Environ. Microbiol.* **68:**5311–5317.

CHAPTER 8

Phage Recombinases and Their Applications

Kenan C. Murphy[1]

Contents

Department of Microbiology and Physiological Systems, University of Massachusetts Medical School, Worcester, Massachusetts, USA
[1] Corresponding author, E-mail: kenan.murphy@umassmed.edu

Advances in Virus Research, Volume 83
ISSN 0065-3527, DOI: 10.1016/B978-0-12-394438-2.00008-6

Abstract

The homologous recombination systems of linear double-stranded (ds)DNA bacteriophages are required for the generation of genetic diversity, the repair of dsDNA breaks, and the formation of concatemeric chromosomes, the immediate precursor to packaging. These systems have been studied for decades as a means to understand the basic principles of homologous recombination. From the beginning, it was recognized that these recombinases are linked intimately to the mechanisms of phage DNA replication. In the last decade, however, investigators have exploited these recombination systems as tools for genetic engineering of bacterial chromosomes, bacterial artificial chromosomes, and plasmids. This recombinational engineering technology has been termed "recombineering" and offers a new paradigm for the genetic manipulation of bacterial chromosomes, which is far more efficient than the classical use of nonreplicating integration vectors for gene replacement. The phage λ Red recombination system, in particular, has been used to construct gene replacements, deletions, insertions, inversions, duplications, and single base pair changes in the *Escherichia coli* chromosome. This chapter discusses the components of the recombination systems of λ, *rac* prophage, and phage P22 and properties of single-stranded DNA annealing proteins from these and other phage that have been instrumental for the development of this technology. The types of genetic manipulations that can be made are described, along with proposed mechanisms for both double-stranded DNA- and oligonucleotide-mediated recombineering events. Finally, the impact of this technology to such diverse fields as bacterial pathogenesis, metabolic engineering, and mouse genomics is discussed.

I. INTRODUCTION

Homologous recombination involves the exchange (or conversion) of DNA sequences between molecules that share significant base pair homology with one another. It is a universal process that is shared by most life forms that allows for the generation of diversity in a population, the alignment of chromosomes during meiosis, and the repair of damaged chromosomes. Proteins that cut, digest, unwind, align, and/or resolve DNA (and intermediate forms of DNA generated by recombination) are generally referred to as recombinases and include proteins with activities such as exonucleases, helicases, and single-stranded (ss)DNA-annealing functions. A set of defined functions within a cell that operate to promote the formation of recombination intermediates is often referred to as a recombination system. It may include functions that resolve recombinational intermediates into recombinant products, but often the resolution of these intermediates can be promoted by DNA replication. In general, a recurring theme in defining recombination pathways is that the processes of DNA recombination and replication are intimately linked, often in ways that are not clearly understood.

Once it was demonstrated that bacteriophage λ could promote homologous recombination (Jacob and Wollman, 1954; Kaiser, 1955), its recombination system became an intense area of study. The ability to find recombinant progeny from infected cells allowed investigators to easily measure recombination rates. The use of sucrose density gradients to separate labeled parental phage particles from their lighter recombinant progeny (Kellenberger *et al.*, 1961; Meselson, 1964; Meselson and Weigle, 1961) was instrumental in the study of recombination mechanisms. Later, the use of DNA replication blocks was employed to measure "pure" recombinants. Recombinant progeny phages were found in the unreplicated (heavy) peak on cesium-formate density gradients and thus were generated directly by recombination events, with little or no DNA synthesis involved (McMilin and Russo, 1972; Stahl and Stahl, 1971). These studies gave rise to mechanistic models of recombination for both phage λ and its *Escherichia coli* host that laid the foundation for studying recombination in other systems today, including higher eukaryotes. Efforts to study phage recombination systems were driven by the thought that they would prove simpler and more tractable compared to those of its host, *E. coli*. However, as phage systems are so mechanistically integrated with (and either stimulate or are dependent on) replication of the phage chromosome, the distinction between the processes of phage recombination and replication becomes obscured. For instance, in the absence of replication, recombination by λ Red is reduced drastically and becomes dependent on functions of the host recombination system (Enquist and Skalka, 1973; Stahl *et al.*, 1972a). Another example of the interplay between these two processes is seen in phage T4 where recombination is a key step for the initiation of replication of the T4 linear chromosome (for reviews, see (Kreuzer and Brister, 2010; Mosig, 1998).

The recombination systems of linear double-stranded (ds)DNA bacteriophage, examples of which are described later, play important roles for their respective phages. In lytic infections, recombination promotes at least one pathway for the generation of concatemeric DNA, the immediate precursor to packaging. In the population at large, recombination offers a means to generate genetic diversity, allowing phages to adapt to different environments as part of an evolutionary program. Recombination also can serve as a dsDNA repair pathway should the phage be subjected to a break in its chromosome. Since the late 1990s, however, phage recombination systems have been used for another purpose. Appropriated by genetic engineers and expressed in the absence of other phage functions, these recombination systems have been utilized as genetic tools for the manipulation of bacterial chromosomes, bacterial artificial chromosomes (BACs), and plasmids, a process that has been named "recombineering." Precisely why some phage recombinases have been so useful for this purpose is not clear. The answer is obviously rooted in the mechanistic details of how these recombination systems promote genetic exchange normally and how they are co-opted for use with small linear dsDNA polymerase chain reaction (PCR) products and oligonucleotides. What follows are descriptions of three recombination systems, two of which (λ Red and *rac* RecET) have been instrumental for the development of recombinational engineering technologies; the P22 system is included for comparative purposes, although it too has the capability to promote transformation with linear DNA substrates (Murphy, 1998). Both RecET and P22 recombination systems can substitute for λ Red in phage crosses (Gillen and Clark, 1974; Poteete and Fenton, 1984). The well-characterized phage T4 recombination system is not discussed here for two reasons: it has been the subject of extensive reviews previously (Kreuzer and Brister, 2010; Liu and Morrical, 2010; Mosig, 1998) and has not been reported to promote the types of recombineering events demonstrated by λ Red and RecET.

II. PHAGE RECOMBINATION SYSTEMS

A. Bacteriophage λ Red recombination system

Mostly known today for its ability to engineer bacterial chromosomes and plasmids, the bacteriophage λ Red recombination system has been studied in detail since its discovery in the mid-1960s. At that time, it was shown that phage λ could recombine efficiently in *E. coli recA* hosts, revealing that it encoded its own recombination system (Brooks and Clark, 1967; Takano, 1966; van de Putte *et al.*, 1966). Soon thereafter, a region central to the linear chromosome (near the cIII gene) was found to contain point mutants that abolished λ recombination in *recA* hosts (Echols and Gingery, 1968; Signer and Weil, 1968). These mutations became known as *red* (recombination defective) and were studied intensively in the wake of the discovery and characterization of the *E. coli* host recombination functions (*recA*, *recB*, and *recC*) (Stahl, 1998).

The *red* genes are located in the p_L-controlled operon of phage λ and are expressed as part of the delayed-early program in phage lytic development. The λ Red system consists of two functions that were originally designated as *red*β and *red*α genes, but are now known more commonly as *bet* and *exo* (Echols and Gingery, 1968; Signer and Weil, 1968). The *bet* gene encodes for the single-strand annealing Beta protein, and *exo* encodes the 5′-3′ double-stranded DNA λ exonuclease (λ Exo) function. Sequence analyses of phage chromosomes show that the presence of a *bet*-like function is often accompanied by the existence of genes showing homology to known phage exonucleases (Iyer *et al.*, 2002; Vellani and Myers, 2003). This gene pair of a synaptase and an exonuclease has been referred to as the Syn-Exo family of recombination systems that exists in many other phage chromosomes, such as *Bacillus subtilis* bacteriophage SPP1 (Ayora *et al.*, 2002; Vellani and Myers, 2003), and even in higher eukaryotic viruses, such as the UL12/ UL29 gene pair encoded by the herpes simplex type I virus (Balasubramanian *et al.*, 2010; Reuven *et al.*, 2003).

The Beta protein was first described as an interacting partner in early purification schemes of λ Exo (Radding and Shreffler, 1966; Radding *et al.*, 1971). Although the protein was purified and its gene had been implicated in recombination (Radding, 1970; Radding, 1971), the biological function of Beta was unclear until Kmiec and Holloman (1981) discovered that it could promote the annealing of complementary single-stranded DNA *in vitro*. The function of Beta is the key "recombinase" of the λ Red system, as Beta has been proposed to anneal ssDNA tails of the λ chromosome to ssDNA regions of the replication fork during a λ phage infection (Court *et al.*, 2002; Poteete, 2008), similar to the proposed role of Beta in recombineering events (Ellis *et al.*, 2001).

The λ Beta protein has been studied extensively *in vitro*. One of the most relevant observations from these studies is that a stable complex is formed between Beta bound to an oligonucleotide only when the complementary oligonucleotide is also present; a noncomplementary oligonucleotide does not produce a stable complex in this assay (Karakousis *et al.*, 1998). Also, if the oligonucleotides are annealed beforehand, a stable complex is not formed, consistent with the inability of Beta to bind to dsDNA *in vitro*. The exact structure of this complex, and to what degree oligonucleotides are annealed within the complex, is not currently known. However, one imagines that this complex likely represents a key intermediate in the Beta-promoted ssDNA-annealing event. The Beta protein has also been shown to direct strand exchange of an oligo bound to a ssDNA circle, and promote strand invasion into AT rich regions of a supercoiled plasmid (Li *et al.*, 1998; Rybalchenko and Radding, 2004).

The most dramatic structural feature of the Beta protein is the multimeric ring-like structure seen when Beta is examined under an electron microscope (Passy *et al.*, 1999). There are different forms of this structure depending on the presence or absence of ssDNA. In the absence of ssDNA, Beta forms small rings of around 12 subunits, revealing a donut-like structure with a central hole of diameter ≈35 Å. There are 12 projections extending out from each ring. Upon addition of ssDNA, larger rings are observed containing 16 subunits, which are ≈185 Å in diameter with a central hole of about 75 Å. These larger rings are not observed in the absence of ssDNA (Passy *et al.*, 1999). With the addition of complementary ssDNA, or dsDNA with ssDNA overhangs, left-handed helical filaments are observed that contain a variable pitch and have diameters similar to those seen with large rings (≈200 Å). The authors interpret these findings to suggest that the rings and filaments are two versions of a high-ordered structure that depends on the type of DNA bound. Rings are formed on ssDNA and act as initiators of an annealing event, whereas filaments are structures formed by the presence of annealed (duplex) dsDNA.

This ring-like structure of λ Beta is identical to that seen in earlier studies with the Erf protein, an analogous recombinase from phage P22 (Murphy *et al.*, 1987a; Poteete *et al.*, 1983) (see Section II.C). In these studies, projections emanating out of the rings were identified as C-terminal domains of the Erf subunits. A separate N-terminal domain was responsible for formation of the ring. A more recent analysis of the Beta protein has revealed a similar division of functional domains (Wu *et al.*, 2006), although the two proteins share no homology and are thought to have evolved separately (Iyer *et al.*, 2002). Proteolytic digestion of the Beta protein revealed a stable fragment consisting of residues 1–130 and a region from 131 to 177 that became resistant to digestion upon binding of ssDNA (Wu *et al.*, 2006). This description of an N-terminal region of Beta involved in DNA binding is consistent with an earlier study that showed that a 36 nucleotide oligonucleotide could be cross-linked by photoactivation to an N-terminal 20-kDa fragment of Beta (representing residues 1–184) (Mythili and Muniyappa, 1993). Oligonucleotides of 27 and 17 nucleotides were not cross-linked in these studies, suggesting a minimal size of 36 nucleotides for the binding of ssDNA to the Beta protein. The study by Wu *et al.* (2006) also revealed that the N-terminal 30 amino acid residues of Beta become susceptible to digestion following the binding of ssDNA, suggesting a conformational change in this region of the protein upon binding to substrate.

A study using atomic force microscopy has suggested that Beta rings without ssDNA bound are actually gapped ellipses, which show a right-handed helical pitch with 11 monomers of Beta per turn of the helix (Erler *et al.*, 2009). Upon binding of complementary oligonucleotides, the helix exhibited a left-handed curvature with 14 monomers of Beta per helical turn. Thus, the annealing event alters both the curvature and the handedness of the filament. In the model of annealing proposed by the authors, the binding of ssDNA to Beta disrupts the right-handed helix, resulting in a structure that promotes interactions between ssDNA molecules. Upon initiation of an annealing event, a second monomer of Beta is bound that promotes a stabilized association of the ssDNAs with continued elongation of a left-handed helix. Because dsDNA does not bind to Beta *in vitro*, one imagines that the two oligonucleotides are not completely annealed within the filament. In this scheme, Beta is envisioned to promote both initiation and elongation of the annealing reaction.

The interacting partner of Beta in the λ Red recombination system is λ exonuclease (or λ Exo) and has been characterized extensively (Carter and Radding, 1971; Cassuto and Radding, 1971; Little, 1967; Sriprakash *et al.*, 1975; Subramanian *et al.*, 2003). The enzyme is a 5′-3′ dsDNA exonuclease that requires magnesium, has a pH optimum of 9.5, and is highly processive ($\approx$3000-bp/binding event). λ Exo also requires a 5′ phosphate on the DNA end for optimal activity. The enzyme can bind to nicked DNA, but cannot initiate digestion from the nick, suggesting a role for the enzyme in strand assimilation (i.e., trimming of branched 5′ ssDNA), although this role has not been examined extensively (Cassuto and Radding, 1971). A study on the *Autographa californica* multiple nucleocapsid nucleopolyhedrovirus showed that its alkaline exonuclease (which belongs to the same family of nucleases as λ Exo) was important for nuclocapsid formation, perhaps as a result of alkaline exonuclease-promoted processing of branched-chain replication intermediates (Okano *et al.*, 2007).

The crystal structure of λ Exo has been solved (Kovall and Matthews, 1997), demonstrating that the enzyme exists as a trimer, consistent with biochemical studies of the purified protein. The trimer has a toroidal shape, with one end of a central channel within the trimer being twice as large as the opposite end. Duplex DNA is thought to enter the wide end of the channel. Following digestion of the 5′ strand by one of the active sites within the trimer, the 3′-ended strand has been proposed to exit the other end of the channel. The trimer encircles the dsDNA substrate, ensuring high processivity of the exonuclease function, similar to many other proteins bound to DNA via a sliding clamp model [e.g., the β sliding clamp of DNA polymerase III (Onrust *et al.*, 1995)]. A crystal structure of λ Exo with a 12 bp duplex DNA containing a 2 nucleotide extension on the 5′ end has also been solved, and reveals that dsDNA is unwound by 2 bp in the central channel, a hydrophobic wedge formed by several apolar residues facilitates unwinding at the ssDNA-dsDNA junction, and that the 5′ ended strand interacts with only one of the three active sites in the trimer (Zhang, 2011). A positively charged pocket at the end of the active site, which includes Arg28, binds the phosphate on the 5′ ended strand, which the authors propose to be important for an electrostatic "ratchet type" mechanism for λ Exo processivity. This is in agreement with a previous study showing the importance of this residue for processivity of λ Exo *in vitro* (Subramanian *et al.*, 2003).

Beta and Exo form a 1:1 complex *in vitro*, although the physiological significance of this interaction is not known (Carter and Radding, 1971). Because both proteins form multimers individually, the 1:1 association of this complex *in vitro* suggests that an even more highly ordered structure might occur *in vivo*. Alternatively, the 1:1 complex may represent a storage form of the proteins prior to the assembly of Exo trimers and Beta rings *in vivo*. From extensive biological characterization of Beta and Exo proteins (Carter and Radding, 1971; Cassuto and Radding, 1971; Kmiec and Holloman, 1981; Little, 1967; Muniyappa and Radding, 1986; Radding,

1971; Sriprakash *et al.*, 1975), the long-standing hypothesis for λ Red action on DNA is that Exo binds to dsDNA ends and promotes digestion of the 5′ strand in a processive fashion, liberating a recombinogenic 3′ ssDNA tail. Beta then binds to the ssDNA generated or is loaded actively by λ Exo. *In vitro* assays where Beta binding is followed by gel mobility shift assays of 1- to 2-kb DNA substrates suggest that Exo promotes more efficient loading of Beta onto ssDNA generated by Exo than is seen by Beta binding to preformed ssDNA (K. Murphy, unpublished observations). This effect is not observed with smaller dsDNA substrates generated by the annealing of two 60-base oligonucleotides. A similar role for Beta has been reported by Tolun (2007), who also revealed that Beta can limit the digestion of λ Exo, possibly by binding to partially processed substrates and preventing reinitiation of λ Exo digestion.

Another λ gene important for growth of λ *red* mutants in *recA* hosts was called the γ gene (Zissler *et al.*, 1971), but more recognized today as the λ *gam* gene. The *gam* gene was found to be critical for the development of the "late-mode" of phage DNA replication, leading to concatemeric phage chromosomes, the optimal substrate for DNA packaging into phage heads near the end of the phage lytic cycle (Enquist and Skalka, 1973). Host mutations in the *recB* gene, encoding for one of the subunits of the *E. coli* RecBCD dsDNA exonuclease/helicase (ExoV), masked the defect of a *gam* mutant, allowing growth of λ *red gam* mutants in a *recA* background (Zissler *et al.*, 1971). This observation led to the proposal that the Gam function targets and inhibits the RecBCD protein, an enzyme with potent exonuclease activity, which itself is an inhibitor of the rolling-circle mode of λ DNA replication. These suppositions were later verified by biochemical analyses of the interactions of the λ Gam protein with RecBCD *in vitro* and *in vivo* (Karu *et al.*, 1975; Murphy, 1991, 2007; Sakaki *et al.*, 1973). Gam is known to bind directly to RecBCD (through interactions with the RecB subunit) and prevents binding of the enzyme to dsDNA ends (Marsic *et al.*, 1993; Murphy, 2007). The structure of the Gam protein derived by X-ray crystallography suggests that it acts as a DNA mimic, occupying the same sites in the RecBCD structure where the separated strands of a dsDNA end are thought to bind (Court *et al.*, 2007).

What is the role of Gam in λ biology? It is known that the generation of multimers of the phage λ chromosome is necessary for λ phage growth, as monomeric chromosomes are not substrates for packaging (Dawson *et al.*, 1976; Enquist and Skalka, 1973; Stahl *et al.*, 1972b). The generation of multimers of the λ chromosome can occur via homologous recombination (using either λ Red or RecA) or by induction of the Gam-promoted (i.e., RecBCD-inhibited) rolling-circle mode of DNA replication. λ Red functions should also promote the rolling-circle mode of replication (to some degree), as the enzymatic action of Red at the tips of rolling circles should produce long ssDNA tails bound by Beta, a structure predicted to prevent binding of RecBCD (Prell and Wackernagel, 1980; Taylor and Smith, 1985). Thus, Gam plays an indirect role in phage λ Red recombination as an inhibitor of host protein complex (RecBCD) that would otherwise compete with the Red functions for access to dsDNA ends. This observation is also true for the efficient use of the λ Red system for genetic engineering (see below), where expression of Gam is required for proficient recombination of linear PCR-generated substrates into the *E. coli* chromosome.

Two other functions of phage λ, *orf* and *rap*, are involved qualitatively in λ recombination in wild-type *E. coli* hosts, but are not absolutely required for recombination. The *orf* gene encodes a small 15-kDa protein required for focusing phage recombination events near the site of a dsDNA break generated at *cos*, but only when phage DNA replication is blocked (Sawitzke and Stahl, 1997; Tarkowski *et al.*, 2002). It has been proposed that the Orf protein interferes with the *E. coli* single-stranded DNA binding protein (SSB), allowing access of the host RecA protein to λ Red-promoted recombination events (Sawitzke and Stahl, 1997). It also may protect ssDNA from digestion by host exonucleases. The Rap protein is analogous to the RusA resolvase, which can suppress the ultraviolet sensitivities of *ruvAB* or *ruvC* mutants (Mahdi *et al.*, 1996). However, the Rap protein is not required for resolving recombination events in a wild-type *E. coli* but, like Orf, helps focus recombination events near the *cos* sites in replication-blocked crosses (Tarkowski *et al.*, 2002). Under these conditions, the authors suggest that Rap may resolve recombination events before they are subjected to RecA-promoted branch migration toward central regions of the λ chromosome.

B. The Rac prophage RecET recombination system

The RecET recombination system was discovered as a mutation in *E. coli recB* and *recC* strains that suppressed the host recombination defect of these mutants (Barbour *et al.*, 1970; Templin *et al.*, 1972). This mutation, known as *sbcA* (suppression of *recBC*), upregulated the expression of a recombination system from an endogenous cryptic prophage of *E. coli* called Rac (Gillen *et al.*, 1981; Kaiser and Murray, 1979; Willis *et al.*, 1985). When used to promote conjugational recombination in *E. coli*, this pathway is known as the RecE pathway and requires RecA and many of the components of the RecF pathway (Gillen and Clark, 1974; Gillen *et al.*, 1981). For plasmid recombination in *E. coli*, the RecE pathway is independent of RecA (Luisi-DeLuca *et al.*, 1989). The Rac phage recombination system is encoded by two genes, *recE* and *recT*, which are functionally analogous to λ *exo* and λ *bet*, respectively. Studies have shown that the RecET system can substitute for λ Red for growth and recombination in *recA*⁻ *E. coli* hosts (Gillen and Clark, 1974; Gottesman *et al.*, 1974). Finally, an anti-RecBCD-like function, similar to the λ Gam protein, has not been described for the Rac prophage.

RecT is an ssDNA-annealing protein (Clark *et al.*, 1993; Hall *et al.*, 1993) and is similar in structure and function to the λ Beta protein. Thresher *et al.* (1995) used electron microscopy to visualize RecT alone and complexed to DNA. Like Beta, RecT forms oligomeric rings and C-shaped structures. However, unlike Beta, RecT formed filaments in the presence of ssDNA and, when combined with RecE exonuclease and dsDNA, was found to bind to ssDNA regions on the ends of dsDNA molecules, frequently arranging the DNA in a circular fashion (Thresher *et al*, 1995). The circles were presumably generated by interactions between RecT structures bound to the ends of the DNA. RecT has been shown to promote joint molecule formation by pairing circular ssDNA with homologous dsDNA linears and to promote strand transfer from the linear duplexes onto the circular ssDNA (Hall and Kolodner, 1994). While reminiscent of RecA-promoted strand transfer events, RecT-promoted strand transfer is independent of ATP and requires an end on the dsDNA substrate, properties not shared by RecA.

The RecE protein (aka ExoVIII) is a 5′-3′ dsDNA exonuclease that uses a processive mode of digestion, and thus is very similar to λ Exo. It requires magnesium and has a broad pH range of activity, peaking at pH 8.5 (Joseph and Kolodner, 1983a). No activity is observed on dsDNA circular substrates containing nicks or gaps. One interesting difference is that λ Exo has a marked preference for dsDNA substrates containing a 5′ phosphate, whereas RecE degrades dsDNA containing 5′ P and 5′–OH groups at nearly the same rates (Joseph and Kolodner, 1983b). The structure of the RecE exonuclease has been solved by X-ray crystallography (Zhang *et al.*, 2009). Like λ Exo, RecE has a toroidal-shaped structure with an internal central channel containing both wide and narrow openings. Both enzymes form oligomers, although RecE forms tetramers, whereas λ Exo is a trimer. Both exonucleases belong to a superfamily of endonuclease-like enzymes (Aravind *et al.*, 1999), but share no sequence homology. A comparison of the structures of the two exonucleases shows an interesting difference though. While they both form oligomeric structures and position their active sites on DNA in a similar way, if one aligns the monomers, the structures of their central channels face opposite directions, suggesting that these proteins are likely a product of independent convergent evolution (Zhang *et al.*, 2009).

The action of RecET proteins on dsDNA ends is thought to be conceptually similar to that seem by the Red proteins, that is, that RecE generates 3′ ssDNA tails, which are subsequently bound by RecT single-stranded DNA-annealing proteins (SSAP). It then, presumably, could promote recombination by any of the mechanisms discussed below for the λ Red system.

C. The bacteriophage P22 recombination system

Experiments by Norton Zinder in the Lederberg laboratory in early 1950s were designed to discover if *Salmonella typhimurium* could promote bacterial conjugation, as demonstrated previously in *E. coli* (Zinder and Lederberg, 1952). The mixing of different autotrophic mutants in genes involved in amino acid metabolism, followed by plating for prototrophs, led to the identification of one particular strain (#22 in the collection) that generated wild-type phenotypes at high frequency. Unexpectantly, it led to the discovery of a bacteriophage (now called P22) that could carry out cell-to-cell transfer of a gene, a process known today as viral transduction. P22 is a temperate bacteriophage of *S. typhimurium* that possesses a circularly permuted and terminally redundant chromosome (Levine, 1972; Poteete, 1988; Susskind and Botstein, 1978). One of the key steps in the life cycle of phage P22 is circularization of the infecting linear chromosome, which is promoted by the phage recombination system acting on the terminally repetitious ends of the phage DNA (Weaver and Levine, 1977). This mechanism of circularization is different from that seen with phage λ, where circularization proceeds by annealing of overlapping single-stranded *cos* (cohesive ends) sites. This distinction makes recombination an essential feature of P22 biology, which is not the case for phage λ. Similar to λ though, phage P22 generates concatemeric DNA for packaging (Botstein, 1968). Like λ, phage P22 generates chromosome multimers in two ways: a phage-promoted recombination event between two monomeric circles and the rolling-circle mode of DNA replication (Gilbert and Dressler, 1968; Weaver and Levine, 1977).

The recombination system of P22 consists of four proteins: Erf, Abc1, Abc2, and Arf. The ssDNA-annealing function is provided by the Erf protein (essential recombination function) (Botstein and Matz, 1970; Poteete, 1982; Weaver and Levine, 1977). The Erf protein was shown to bind to ssDNA (Poteete and Fenton, 1983). Given its ability to be replaced by the λ Beta protein in P22 phage crosses, Erf most likely possesses an ssDNA-annealing function similar to Beta. Erf was the first phage recombinase to show the ring-like quaternary structure under electron micrographs (Poteete *et al.*, 1983). As described earlier, Erf showed projections emanating from its rings, which were identified as the C-terminal domains of the individual subunits (Murphy *et al.*, 1987a). Erf fragments missing the C-terminal domain could still form rings, showing that this region of the protein was not required for either formation or stability of the rings (Murphy *et al.*, 1987a; Poteete *et al.*, 1983). However, the C-terminal domain is important for recombinase function *in vivo*, as P22 *erf* amber mutants expressing only the N-terminal ring-forming domain were deficient for growth in a *recA* host strain. Further analysis of purified Erf protein fragments revealed that a region important for binding ssDNA was located between the N-terminal ring-forming domain and the C-terminal domain of unknown function (Murphy *et al.*, 1987a).

Unlike the λ Beta protein, Erf does not have an associated 5′-3′ exonuclease partner, nor does it have an accompanying protein such as λ Gam, which binds to and inhibits the RecBCD enzyme. Amazingly though, these two functions are provided by the Abc2 (anti-recBCD) protein that binds to the host RecBCD enzyme and modifies its exonuclease function so that it works cooperatively with Erf to promote recombination *in vivo* (Murphy, 1994; Murphy and Lewis, 1993; Murphy *et al.*, 1987b). Effectively, the RecBCD protein is hijacked by Abc2 and made to work as part of the P22 recombination system. How this is accomplished is not clear. It is known that Abc2 binds to the RecC subunit of the RecBCD complex. In assays designed to measure *in vitro* nicking of RecBCD at its hot spot sequence Chi (3′GCTGGTGG5′), the Abc2–RecBCD complex was incapable of recognizing or responding to Chi (Murphy, 2000). (This inhibition of Chi activity is not relevant to P22 recombination, as P22-promoted recombination is independent of Chi.) More importantly, Abc2-modifed RecBCD still possesses dsDNA exonuclease activity, although it is qualitatively modified (Murphy, 2000). Digestion of the 5′ strand is upregulated by Abc2, not unlike a native RecBCD species after its encounter with Chi (Dillingham and Kowalczykowski, 2008). Although not yet tested, one possibility is that Abc2 prevents loading of RecA at Chi sites, thus explaining how the expression of Abc2 alone inhibits host RecABCD-promoted conjugational recombination events. It may be that Abc2 interferes with the interaction between RecA and RecBCD and, at the same time, allows the modified enzyme to cooperate with the P22 Erf protein to promote recombination, and likely other SSAPs as well. This latter supposition comes from the observation that coexpression of Abc2 with P22 Erf, λ Beta, and even the *Pseudomonas* phage D3 SSAP Orf-52 promotes the growth of λ *red gam* mutants in *E. coli recA* hosts (K. Murphy, unpublished observations). This hypothesis explains why Beta can substitute for Erf in a P22 infection, although Erf cannot substitute for Beta in a λ infection (Poteete and Fenton, 1984). Such a complementation would require an interaction between P22 Erf and λ Exo. It is known that the interacting partners of the λ Red and the *rac* prophage RecET system cannot substitute for one another (i.e., λ Beta cannot work with RecE and λ Exo cannot work with RecT to promote recombination) (Muyrers *et al.*, 2000). If exogenous phage SSAPs interact directly with Abc2-modified RecBCD, it would suggest that the modified enzyme interacts with a motif common to a variety of phage SSAPs. Alternatively, no direct interaction between an Abc2-modified RecBCD and an accompanying SSAP is required.

Other components of the P22 system play a less defined role, but are required for full growth and recombination of a P22 phage deleted of its recombination region in *Salmonella recA* hosts. The Abc1 protein was defined as assisting Abc2 for the full restoration of growth and recombination in P22 mutants deficient for recombination (Murphy and Lewis, 1993; Murphy *et al.*, 1987b). However, the ability of Abc1 to interact with RecBCD has not been studied. The Arf (accessory recombination function) protein is encoded by a small open reading frame upstream of the *erf* gene. It was originally not expected to be involved in recombination due to its small size (47 amino acids), the unusually acidic nature of the predicted protein (pI of 3.5), and the suspicion that the modest effects of a deletion identifying this open reading frame might have simply altered the expression level of the nearby *erf* gene (Semerjian *et al.*, 1989). Nonetheless, P22 phage containing either an in-frame deletion or an amber mutation in the *arf* gene showed a 4- to 5-fold decrease in the rate of recombination in P22 phage crosses, a defect that could be complemented by plasmids expressing the *arf* gene *in trans* (Poteete *et al.*, 1991). Given the highly acidic nature of the Arf protein, it was suggested that it might play a role as a DNA mimic, perhaps helping displace Erf from ssDNA during the annealing reaction. In this role, the Arf protein might not be specific for the phage recombination system, as its overexpression from a plasmid also stimulated a small (2.5-fold) effect in λ *red gam* phage recombination promoted by the host recombination system (Poteete *et al.*, 1991).

The P22 recombination system could promote recombination of its chromosome by any of the mechanisms discussed below for λ Red. This view is supported by the fact that the two recombination systems can substitute for one another (Poteete and Fenton, 1984). However, P22 has the added feature that recombination is essential for its life cycle, as circularization of its chromosome following infection is a prerequisite for growth (presumably via an ssDNA annealing mechanism between it terminally repetitious ends). It is interesting to speculate that this feature of the P22 life cycle might have dictated the greater complexity of its recombination system relative to phage λ (four P22 genes relative to two λ genes), providing a level of regulation that ensures recombination will take place consistently soon after infection. A summary of the recombination functions for phages lambda, rac and P22 is listed in Table I.

TABLE I Summary of phage recombination functions

Phage	Annealing function	Exonuclease	Anti-RecBCD	Accessory functions
λ	Beta	λ Exo	Gam	Rap, Orf
rac	RecT	RecE	None identified	None identified
P22	Erf	Abc2-modified RecBCD	Abc2-modified RecBCD	Arf, Abc1

III. SINGLE-STRANDED DNA-ANNEALING PROTEINS

The λ Beta, RecT, and Erf proteins discussed earlier all belong to a class of proteins known as single-stranded DNA-annealing proteins (Iyer *et al.*, 2002). They share many common structural and functional properties that include many of the characteristics described previously. Among them are a filament or ring-like quaternary structure, the ability to bind to and anneal complementary ssDNA in an ATP-independent fashion, and a proficiency for promoting RecA-dependent or -independent recombination events. In a comprehensive sequence comparison and evolutionary classification of SSAP-like proteins (Iyer *et al.*, 2002), three distinct superfamilies of annealing proteins were found, classified within families represented by the λ Beta protein, the P22 Erf protein, and the yeast Rad52 protein. There are no sequence homologies between superfamilies, suggesting that their similar functions and quaternary structures arose as a result of independent, yet converging evolutionary pathways. The Rad52 protein is a key protein in various pathways of recombination in *Sacchromyces cerevisea* and humans (Bi *et al.*, 2004; Kumar and Gupta, 2004). It, like other SSAPs, forms ring-like structures and binds to ssDNA (Shinohara *et al.*, 1998).

Interestingly, while phage and eukaryotic SSAPs have been found, they have not been described in bacteria. While it is not known why this is the case, it may be that bacteria have evolved recombination systems that depend on activated enzymes (e.g., RecBCD), hot spots (e.g., Chi), and regulators (e.g., LexA) to keep their recombination systems in check in order to prevent illegitimate recombination events, genome rearrangements, and/or mutations from occurring randomly. For instance, the constitutive expression of the λ Red system in *E. coli* results in a 10-fold increase in the frequency of mutation at the *rpsL* gene (Murphy and Campellone, 2003), perhaps as a result of interference with the mismatch repair system of E. *coli*. Also, if DNA replication intermediates are the true targets of λ and λ-like phage recombinases (see below), they may not be suitable for bacterial systems where a replication fork would be "required" to promote a recombination event. Eukaryotes, perhaps as result of the high levels of regulatory systems, prevent the presence of such recombinases from being problematic. In addition, it is not yet clear yet if such eukaryotic SSAP recombinases, although similar in structure, act in the same way as their phage counterparts.

When the phage recombination systems discussed previously are expressed in a controllable fashion in the absence of any other phage functions, they have the capacity to promote recombination of small linear DNA substrates (i.e., PCR substrate) and ssDNA oligonucleotides into bacterial chromosomes with high efficiency (Datsenko and Wanner, 2000; Murphy, 1998; Yu *et al.*, 2000; Zhang *et al.*, 1998). This process has been termed recombineering (Ellis *et al.*, 2001; Copeland *et al.*, 2001). The collaborative activities of a 5′-3′ dsDNA exonuclease and its cognate SSAP are required for the insertion of dsDNA substrates, while only the SSAP function is necessary for the incorporation of changes encoded by an oligonucleotide (Ellis *et al.*, 2001). Datta *et al.* (2008) have shown that putative SSAPs from nine different phages from both Gram-negative and Gram-positive bacteria have the capacity to promote recombineering of oligonucleotides in *E. coli*, although with varying (≈3000-fold) degrees of efficiency. Five of these SSAP genes have cognate exonuclease functions associated with them in the phage sequence database. When the exogenous gene pairs were expressed in *E. coli*, none of them promoted efficient dsDNA recombineering, even though some of the SSAPs worked as well as Beta for oligonucleotide-mediated recombineering. Also, none of the exogenous SSAPs worked well with λ Exo for dsDNA recombineering in *E. coli*. These results show that SSAPs do not work well with their noncognate partners, in agreement with Muyrers *et al.* (2000), and that species specificity of these functions relies just as much (if not more) on the exonuclease function than the SSAP.

It was proposed that the SSAP–Exo pairs described in the study by Datta *et al.* (2008) would work well within their endogenous hosts for recombineering. To this end, a scan of the mycobacterial phage DNA sequence database by van Kessel and Hatful (2007) revealed two phages (Che9c and Halo) that have functions related to the Rac RecE exonuclease. The Che9c phage also encodes a function similar to RecT. Specifically, the C-terminal domain of Che9c gp60 is 28% identical to a nuclease domain in the RecB family of exonucleases, and the Che9c RecT homologue (gp61) is 39% identical to the Rac RecT protein. Characterization of these Che9c proteins by van Kessel and Hatfull (2007) revealed that gp60 (RecE) encodes for a dsDNA exonuclease and that gp61 (RecT) encodes for a ssDNA-binding protein with a K_D of 163 n*M*. Such a low dissociation constant for gp61 reveals a very high affinity for ssDNA, comparable to that shown by antibodies to their target antigens. Overexpression of Che9c RecET functions in *Mycobacterium smegmatis* and *Mycobacterium tuberculosis* shows that they can promote gene replacement with linear DNA fragments containing drug markers flanked by 500 bp of target DNA (between 30–70% of the transformants are true gene replacements). PCR substrates with 50-bp flanks of targeting homology give rise to resistant clones, but are most often associated with illegitimate recombination events (K. Murphy, unpublished observations). Nonetheless, the system is very useful for making gene knockouts and replacements in mycobacterial species whose genomes have historically been difficult to manipulate.

The Che9 SSAP protein RecT has also been expressed in the absence of the RecE exonuclease to promote oligonucleotide-mediated recombineering of mycobacterial genomes (van Kessel and Hatfull, 2008; van Kessel *et al.*, 2008; K. Murphy, unpublished results). Recombineering with oligonucleotides in *Mycobacteria*, as was seen with *E. coli*, generates >100-fold more recombinants relative to the use of dsDNA substrates. Recombineering frequencies increased when oligonucleotide lengths were varied between 30 to 50 nucleotides; maximal frequencies were obtained with 100 ng of a 60-base oligonucleotide. Oligonucleotides complementary to the chromosomal lagging strand template of the replication fork exhibited much higher frequencies of recombineering relative to oligonucleotides targeting the leading strand template, in some cases reaching 10,000-fold. This bias is quite different from *E. coli*, where the difference in recombination rates between targeting of the lagging and leading strand templates was typically between 3- and 50-fold. This large difference in the bias of targeting observed in *Mycobacteria* relative to *E. coli* likely reflects a disparity in the mechanisms of oligonucleotide-mediated recombineering, or differences in the mechanics of the replication forks in these two bacteria. Use of the SSAP Che9c RecT protein to promote oligonucleotide-mediated recombineering is very useful for the characterization of potential drug targets (van Kessel and Hatfull, 2008) and for the verification of single nucleotide polymorphisms thought to be responsible for generating drug-resistant mutants of *M. tuberculosis* (Murphy and Sassetti, unpublished observations). The usefulness of the Che9c RecT protein in generating mutants of *M. smegmatis* and *M. tuberculosis* should encourage others to find SSAP-like functions in the phage of other (i.e., pathogenic) bacteria.

The *B. subtilis* phage SSP1 also encodes an SSAP function. The gene *35* product (G35P) is a member of the Beta/RecT superfamily of annealing proteins characterized by Iyer *et al.* (2002). G35P shares many of the properties of Beta protein, including the ability to bind ssDNA, promote strand transfer reactions, and initiate strand invasion into AT-rich regions of supercoiled plasmid substrates (Ayora *et al.*, 2002; Rybalchenko and Radding, 2004). Like Beta, it also can form donut-shaped, ring-like structures and left-handed filaments on ssDNA. The SSP1 phage also encodes a neighboring gene, *34.1*, that encodes a dsDNA exonuclease that has 5′-3′ polarity on dsDNA (Martinez-Jimenez *et al.*, 2005; Vellani and Myers, 2003). However, no physical interaction between G35P and this exonuclease has been reported. Instead, G35P interacts with SSP1-encoded replicative DNA helicase G40P and the SSB-like protein G36P (Ayora *et al.*, 2002) as part of an intricate program involved in DNA replication. The ability of the G35P to promote oligonucleotide-mediated recombineering events has not been reported.

IV. RECOMBINEERING

The development of gene replacement technology in *E. coli* took a great stride forward in 1998 when it was shown that a strain expressing the λ Red recombination functions and the λ anti-RecBCD function Gam promoted transformations with linear DNA substrates at rates over 100 times more efficient than the recombinationally-proficient strains in use at that time (Murphy, 1998). The strains used for gene replacement at the time were *recBCD sbcCD* and *recD* strains of *E. coli*. The addition of λ *red* and *gam* to wild-type *E. coli* was an attempt to mimic the principal characteristics of these recombinationally-proficient strains, that is, a knockdown of the endogenous RecBCD recombination pathway (along with its destructive dsDNA exonuclease function) and the substitution of an alternate recombination pathway. For *recBC sbcCD* strains, this alternate pathway was the RecF pathway, a recombination route designed principally for the repair of daughter-strand ssDNA gaps in the chromosome (Kuzminov, 1999). For *E. coli recD* strains, the alternate pathway was provided by the helicase-proficient, exonuclease-deficient RecBC enzyme (Amundsen *et al.*, 1986; Palas and Kushner, 1990), which mimicked, to some extent, the proficiency of a RecBCD enzyme following its encounter with a Chi site (Thaler *et al.*, 1989) (although not entirely; see Anderson *et al.*, 1997). Clearly, however, neither of these alternate recombination pathways are as proficient for successful transformation of *E. coli* with small linear DNA substrates as the λ Red pathway (Murphy, 1998).

The next advance in the use of phage recombinases for gene replacement came with observations by F. Stewart and his group at the University of Dresden. Working with the Rac prophage RecET system, these investigators showed that RecET could promote linear dsDNA transformation with substrates containing only 42 bases of homology to chromosomal target sites (Zhang *et al.*, 1998). This meant that by including homology for a target site in the 5′ ends of primers, PCR products could be used to generate gene replacement substrates. The electroporation of PCR substrates into RecET-expressing cells allowed for one-step gene replacement without the need to clone the gene of interest (or its flanking regions). Interestingly, the recombination observed was independent of *recA*, similar to λ phage crosses when replication was allowed. In addition, this group showed that use of the RecET recombination system, together with site-specific recombination systems such as Cre-*loxP* and Flp-FRT, could be used to make markerless gene deletions (Zhang *et al.*, 1998).

Not surprisingly, the λ Red system could also promote recombination with short homologies and did so at an elevated frequency relative to RecET (Muyrers *et al.*, 2000). An *E. coli* construct that replaced the *recBCD* genes with a P_{lac}–*red* operon allowed efficient gene replacement with linear DNA substrates containing 1 kb of homologies (Murphy, 1998). Recombineering with PCR substrates containing 50-bp flanks occurred at low frequencies with this strain (K. Murphy, unpublished results). However, similar constructs, where *red* is driven by the fivefold stronger P_{tac}, generated high frequencies of gene replacement with PCR substrates containing 50-bp flanks. This result demonstrates that high levels of *red* expression are required for efficient levels of recombination with small regions of homology. However, very high expression levels, such as exhibited by a Ptac–*red* operon on a multicopy plasmid, led to toxic levels of *red* expression and thus low levels of recombineering (unpublished observations).

Datsenko and Wanner (2000) generated a low copy-number plasmid (pKD46) in which the *red* and *gam* functions are driven by the P_{BAD} promoter. The plasmid contained a temperature-sensitive origin of replication that allowed for easy curing of the *red*- and *gam*-expressing vector following the gene replacement event. Don Court and colleagues at NCI-Frederick developed an expression system that took advantage of the high expression levels of *red* and *gam* from the p_L-controlled operon of a defective λ prophage (Yu *et al.*, 2000). The prophage had its lysis, replication, and structural genes removed and contained the *cI857* repressor so that the p_L-controlled operon could be induced by shifting the cells from 30° to 42 °C. This configuration leads to tight control of *red* and *gam* expression, limiting recombination to a "recombinogenic window" following electroporation of heat-induced lysogens with PCR substrates. This is especially important when the targets of recombination are eukaryotic sequences within bacterial artificial chromosomes (BACs) and one has to be careful not to induce unwanted recombination events between repeated sequences within the target plasmids.

Because of its simplicity and high efficiency, recombineering has become the method of choice for genetic engineers. The realization that exogenous DNA sequences from a variety of microorganisms and higher eukaryotes cloned into BACs and cosmids can also serve as templates for recombineering has expanded its usefulness as a genetic tool (Copeland *et al.*, 2001; Muyrers *et al.*, 1999). The value of recombineering is that precise changes can be made in specific regions of BACs containing large inserts or bacterial chromosomes where the use of restriction enzymes would be impractical. Also, the ability to induce and then turn off recombination functions allows one to create a transient hyper-recombinogenic state, preventing unwanted mutations or rearrangements from occurring in nontargeted regions of the chromosome or BAC. What follows are examples of the types of genetic manipulations that have been performed using recombineering. The technical details of how such modifications can be made, and the plasmids and/or strains required, have been described (Savage *et al.*, 2006; Sawitzke *et al.*, 2007; Sharan *et al.*, 2009; Thomason *et al.*, 2007a; Murphy, 2011).

V. GENETIC MANIPULATIONS

A. Chromosomal and BAC modifications

1. Deletions

By far the most common use of recombineering is to create a gene deletion. This is done by amplifying a drug marker by PCR and including 40–50 nucleotides of the target site in the 5′ ends of the primer (see Fig. 1).

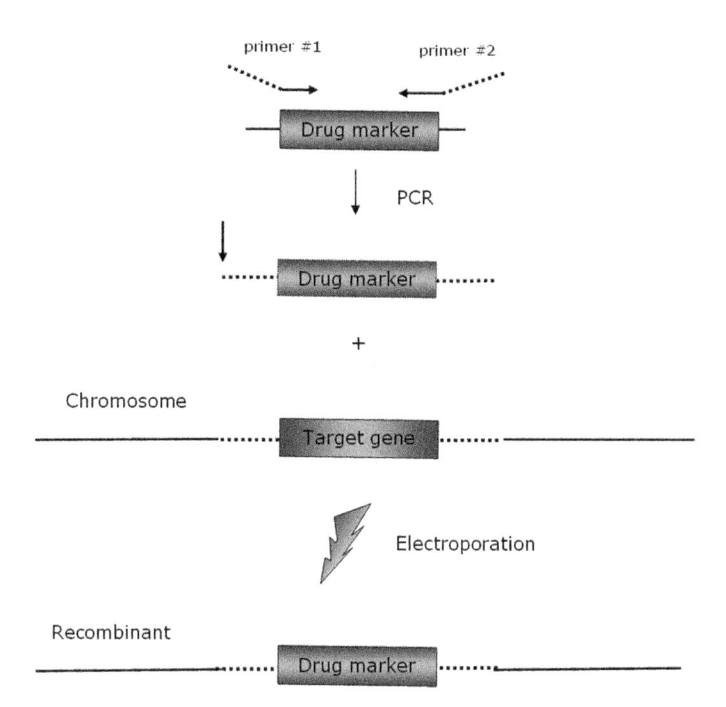

FIGURE 1 Basic diagram of the recombineering method. A drug marker is used as a template for a PCR. The 20 nucleotides at the 3′ end of the primers anneal to the drug marker (solid arrows), and 50 nucleotides at the 5′ ends of the primers are homologous to the upstream and downstream regions of the target gene (dotted lines). Following electroporation of the PCR substrate into electrocompetent recombineering-proficient cells, the culture is grown out and plated on antibiotic selection plates.

Some of the more common antibiotic markers used for this purpose have been genes conferring resistance to chloramphenicol, kanamycin, tetracycline, and streptomycin, although any marker that can be used at single copy can be employed to select the recombinant. Typically, the flanking sequences correspond to regions upstream and downstream of the target gene, although the actual flanking sequences chosen will depend on the investigator's desire to remove additional regulatory sequences, such as promoters and terminators that flank the gene of interest (Fig. 1).

2. Unmarked deletions

Often it is desirable to remove the drug marker. This step is required when one wishes to use the same drug marker at a different locus within the chromosome or when the knockout is within a bacterium that will be tested in animals. Removal of the drug marker is done in two ways: by use of a counterselection marker such as *sacB* (Gay *et al.*, 1983) (the absence of which can be selected for on plates containing 10% sucrose) or by use of site-specific recombination systems (SSRs such as Cre-*loxP*), where two SSR target sites flank the drug marker cassette (Hamilton and Abremski, 1984; Palmeros *et al.*, 2000). In the counterselection scheme, *sacB* (or any other counterselection marker) is contained within a cassette and paired with an antibiotic marker (e.g., *cat-sacB*). The cassette is used as a template for a PCR and target site sequences are included in the 5′ ends of the primer, as described earlier. Once incorporated into the chromosome, a second recombineering event is performed using a substrate containing the desired precise deletion (or any unmarked DNA sequences) containing 40–50 bases that flank the target site. Alternatively, one could use a 70-mer oligonucleotide with 35 nucleotides on each end that corresponds to the end points of the deletion desired. After electroporation of the PCR substrate or an oligonucleotide, cells are spread on plates containing sucrose, which selects for loss of the cassette. Excision of the cassette should be confirmed by loss of the antibiotic marker. This is an important step, as spontaneous sucrose-resistant colonies occur at a moderate frequency ($\approx 10^{-4}$) (Blomfield *et al.*, 1991). One is left with a strain containing a deletion of the target gene, replaced by any sequence desired (e.g., an in-frame deletion consisting of the first few codons of the target gene fused to the last few codons).

Alternatively, one can use as a drug cassette an antibiotic marker that is flanked by SSR target sites, such as *loxP* (Hoess and Abremski, 1984; Zhang *et al.*, 1998; Datsenko and Wanner, 2000). Once recombined into the chromosome in place of the target gene, the strain is then transformed with a plasmid that expresses the SSR recombinase, for example, Cre. Constitutive expression of Cre will promote recombination between direct repeats of the *loxP* sequences and excise the drug marker from the chromosome, leaving behind one *loxP* site (Fig. 2). SSR-expressing plasmids often have counterselection markers or temperature-sensitive

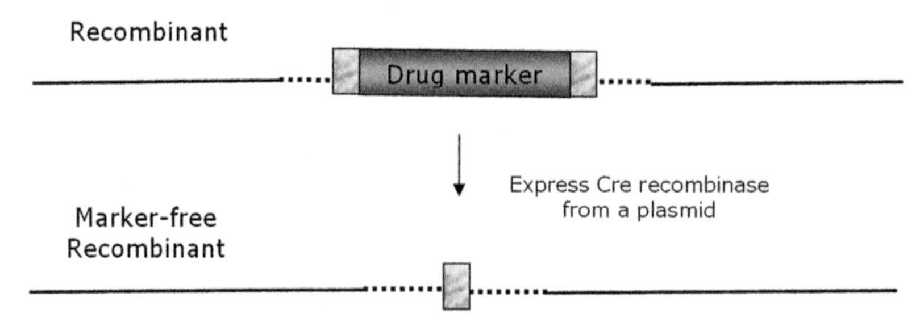

FIGURE 2 Construction of a markerless gene knockout. The same step is performed as described in Figure 1, except that the drug marker has flanking *loxP* sites (small gray rectangles). The recombinant is transformed with a Cre-expressing plasmid, which recombines the *loxP* sequences (which are direct repeats), excising the drug marker and leaving behind one *loxP* site.

origins of replication to allow the unmarked recombinant to be cured of the plasmid. The only drawback with this scheme is that the *loxP* "scar" is left behind. Unwanted Cre-mediated recombination events could occur should this scheme be used in a reiterative fashion.

3. Duplications and inversions

Chromosomal rearrangements such as duplications and inversions are generally ones that genetic engineers would like to avoid. However, if so desired, recombineering offers a direct and easy path for doing so in a controlled manner. One may, for instance, wish to examine if a duplication of a specific region in the *E. coli* chromosome leads to greater rates of amplification or expansion of the genome, as was the case in a study by Poteete (2009). Alternatively, an inversion of a target site within the chromosome may be required to examine if the oligonucleotide-mediated recombineering is dependent on the direction of replication or transcription, as was performed by Ellis *et al.* (2001). For inversions, a drug marker is placed somewhere in the region to be inverted by standard recombineering (as described earlier). One then simply amplifies the region to be inverted by PCR (including the marker), but selects the 50-bp targeting sequences on the 5′ end of the primers so that the region amplified is inverted once incorporated back into a native chromosome. For duplications, a scheme is described (see Fig. 3) that is adapted from a mechanism describing spontaneous chromosomal duplications in *Salmonella* (Galitski and Roth, 1997). To create a duplication by recombineering, an antibiotic marker is amplified, but in this case, the 50-bp targeting sequences of the recombinant PCR substrate are switched relative to a simple deletion substrate (i.e., primers are designed such that the upstream and downstream sequences of the target area are exchanged) (Fig. 3). The first recombination event occurs between one end of the PCR substrate (open arrow in Fig. 3) and a replication fork within the target chromosome (not shown for clarity). This initial event, which proceeds by a mechanism described by the replisome invasion model (see below), creates a dsDNA break in the chromosome. In a second event, this dsDNA end then recombines with a sister chromosome from the original replication fork (at sequences denoted by the black arrows in Fig. 3) or perhaps a second replicating chromosome. Following the second recombination event, duplication of the target area is generated, with the drug marker placed between the sequences that have duplicated. This model assumes that the initial invasion of the recombineering substrate disrupts the replication fork (not shown), allowing the second recombination step (last step in Fig. 3) to proceed via a mechanism that is similar to the repair of spontaneous dsDNA breaks in the *E. coli* chromosome (Kuzminov, 1999), (i.e., with subsequent restoration of the replication fork).

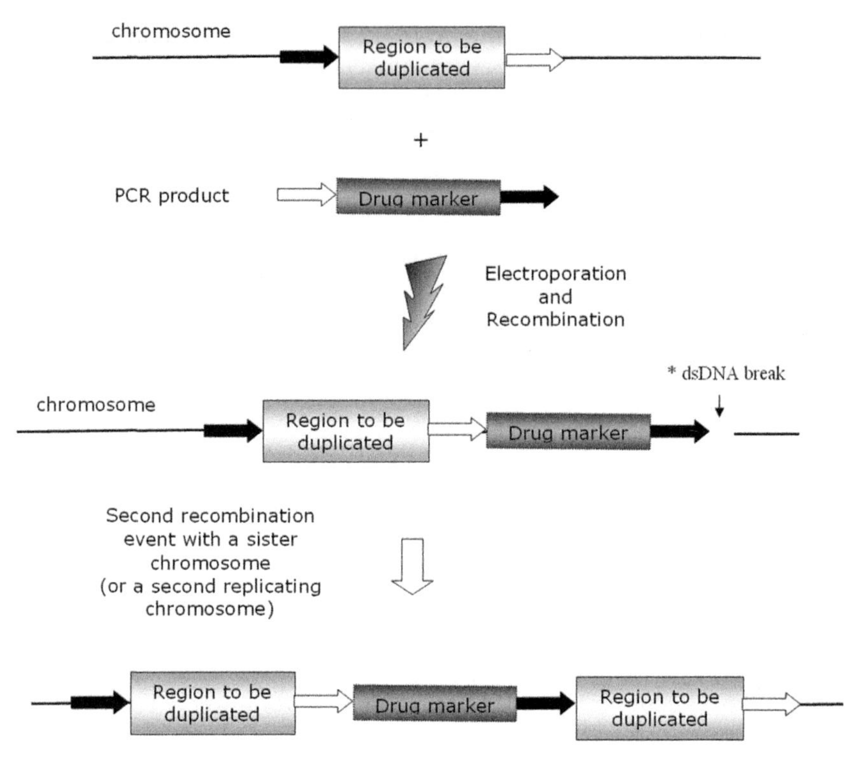

FIGURE 3 Proposed model for Red-promoted duplications. The same step is performed as described in Figure 1, except that the drug marker has its targeted flanking regions switched. The chromosomal region targeted for duplication contains a replication fork (fork not depicted). A recombination event between the chromosome and the PCR substrate (i.e., a crossover at sequences denoted by white arrows) generates a dsDNA break (denoted by asterisk). A second λ Red-promoted recombination event occurs between the dsDNA end generated by the first event and either a sister or another chromosome (which must also be replicating), at sequences denoted by the black arrows. This proposed mechanism is adapted from a model proposed by Galitski and Roth (1997). It is not likely to occur at high frequency, but it is a selectable event.

Duplication can be maintained stably by continued selection for the drug marker, as any subsequent recombination event that loses a copy of the duplicated sequence will also excise the drug marker.

4. Insertions

Recombineering also offers the ability to make precise insertions into the chromosomes of bacteria, or into exogenous DNA contained within a BAC. To do this, 50-bp targeting sequences are chosen that correspond to a 100-bp continuous segment of the targeting site. The insert, often a designed sequence containing a drug marker, is then amplified with primers containing the 50-bp targeting sequences at the 5′ ends. Selection for the drug marker following recombineering generates a precise insertion into the target site. Such insertions allow genetic engineers to construct strains expressing exogenous functions, tagged genes, and promoter replacements though the size of the insertions is generally limited to ~3 kb or lower (Maresca *et al.*, 2010). Very large insertions can be performed by first using recombineering to place a SceI site into a desired location in the chromosome (Kuhlman and Cox, 2009). The SceI meganuclease has an 18-bp recognition sequence not found in the *E. coli* chromosome (Monteilhet *et al.*, 1990). By expressing SceI from a controllable plasmid, a SceI-induced dsDNA break is generated in the chromosome. A predesigned linear dsDNA electroporated into this strain, with appropriate selection of flanking sequences, will recombine at the site of the SceI-induced dsDNA break using the host RecA-promoted recombination pathway. Repair of the break by incorporation of the linear insert occurs at a selectable frequency, as nonrepair is lethal to the cell.

B. Oligonucleotide-mediated mutagenesis

If the role of λ Exo is to generate single-stranded DNA for the Beta protein to bind, then if one starts off with ssDNA, the λ Exo protein might not be required to promote a recombineering event. This supposition was tested by Don Court and colleagues (Ellis *et al.*, 2001), who showed that ssDNA could promote changes in chromosomes and BACs following electroporation of oligonucleotides into cells expressing only the λ Beta protein. By designing a genetic scheme to correct a mutation in the *galK* locus, the authors found that 0.1% of the surviving cells had picked up the single base pair change encoded by the oligonucleotide. As discussed earlier (see Section III), the frequency of oligonucleotide-mediated repair was variable depending on whether the oligonucleotide was complementary to either the leading or the lagging strand template of the replication fork, with annealing to the lagging strand template occurring at a rate 3–50 times higher relative to oligonucleotides targeting the leading strand template. This was the first evidence that the λ Beta protein might target the replication fork as part of the mechanism of recombination promoted by λ Red.

Another key factor determining the rate of oligonucleotide-mediated repair was the type of base pair mismatch generated by the Beta-promoted annealing event. Oligonucleotides designed to create mismatches not recognized by the mismatch repair (MMR) system (e.g., C-C) generated high frequencies of oligonucleotide-mediated repair (Costantino and Court, 2003). Clearly, the mismatch created by annealing the oligonucleotide to ssDNA regions of the replication fork is a substrate for the MMR system. This idea was confirmed by Costantino and Court (2003), who showed that λ Beta production in MMR-deficient cells (*mutS* mutants) produced high levels of oligonucleotide-mediated repair independent of the type of mismatch generated. Frequencies of repair reached as high as 25% of the surviving cells following electroporation, the likely theoretical limit, given that most of the cells at the time of electroporation (which are generated from cells in the logarithmic growth phase) contain four or more chromosomes per cell. Another way of avoiding the MMR system involves creating multiple changes in the vicinity of the base pair change desired. The MMR system will not recognize consecutive mismatches of more than 6 bp. Using denatured PCR fragments, Yang and Sharan (2003) showed that a stretch of up to 20 base mismatches could be inserted into a targeting site of a BAC, using 40 bp of homology on either side. The modified BAC could be identified by the presence of an inserted restriction site and/or PCRs using insertion-specific primers. A second recombineering event changes all the bases back to the original sequence, except for the base pair change desired. More recently, Sawitzke *et al.* (2010) demonstrated alternative ways to evade the MMR system. Increased frequencies of specific gene alteration could be found by including a C-C mismatch 6 bp away from the desired modification, using four or more adjacent mismatches, or by including changes in four consecutive wobble positions, where changes in the DNA are silent with respect to the encoded protein.

C. Gap cloning

The third type of substrate used in recombineering experiments are plasmid backbones that have been generated by PCR. Primers for the PCR contain 20 nucleotides on the 3′ ends that target the plasmid and have 50 nucleotides on the 5′ ends that are homologous to sequences within the *E. coli* chromosome (or BAC) (see Fig. 4). After electroporation into λ Red-expressing cells, the "gap" in the plasmid is filled with sequences from the chromosome or BAC, a process known classically as "gap repair." The repair occurs at high frequency with recombineering strains (Zhang *et al.*, 2000). The generation of empty vectors, due to nonhomologous end joining or a microhomology-directed annealing event, occurs at low frequencies. The process has also been termed "levitation" of genes from the chromosome and offers an alternative to PCR for the amplification of DNA sequences. In effect, PCR errors in amplification of the vector are immaterial, whereas once incorporated into the vector, the target DNA is amplified with high fidelity by the *E. coli* chromosomal DNA replication machinery, along with its accompanying proofreading functions (*dnaQ*) and MMR-editing capabilities.

Zhang *et al.* (2000) demonstrated that the target sequence can also be exogenous DNA that is coelectroporated with PCR-generated vector DNA, in effect generating a system similar to *in vivo* cloning first described

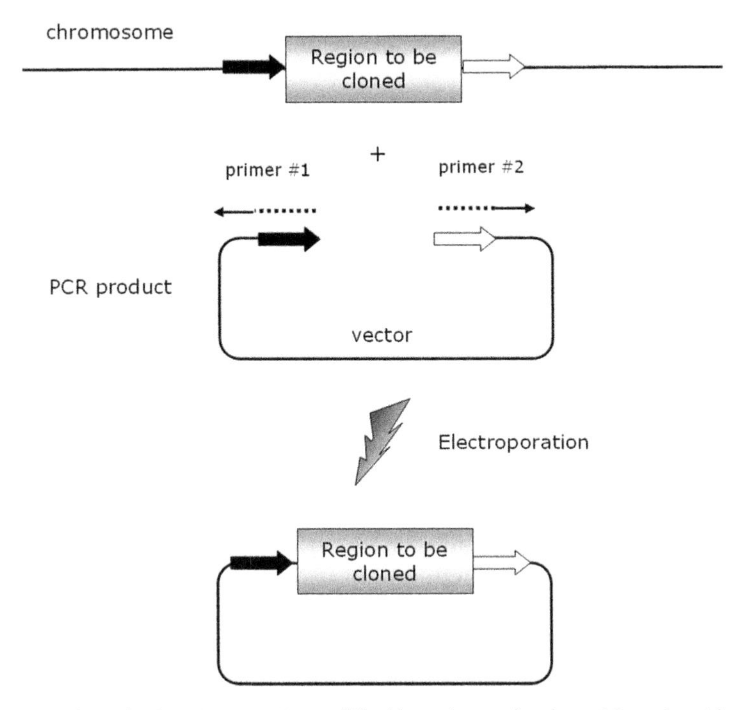

FIGURE 4 Gap cloning. A vector is amplified by primers that have 20 nucleotides on 3′ ends targeting the vector (solid lines) and 50 nucleotides on 5′ ends flanking the region to be cloned (dotted lines). Upon electroporation of the PCR amplicon (the gapped plasmid) into electrocompetent recombineering-proficient cells, the gap is repaired by recombination with the chromosome. The target gene (or region) is "levitated" by recombineering.

by Oliner *et al.* (1993). Note, however, that there are likely mechanistic differences when the targeting sequences are within a replicating *E. coli* chromosome and when target sites are contained in coelectroporated DNA. In the first case, the target site involves a replication fork, an optimal target present in all cells. When the cells are coelectroporated with linear DNA target sequences, the process likely involves a simple ssDNA-annealing event between the λ Red-processed ends of each partner. In this case, the presence of extended nonhomology in the target DNA, and the presence of cells that do not have one or both recombineering partners, may result in lower frequencies of recombination.

D. Plasmid alterations

In addition to chromosomes and BACs, plasmids can also serve as substrates for recombineering. Compared to recombineering of small plasmids (<10 kb), the standard cloning protocol of using restriction enzymes remains a useful tool for the manipulation of plasmids. However, there are times when recombineering offers advantages. These might include cases where the selective drug marker or origin of replication needs to be exchanged; when single base pair alterations are desired; when unnecessary plasmid sequences need to be excised precisely; or when regions of a plasmid needing modification are bereft of restriction sites. Datta *et al.* (2006) generated a valuable set of recombineering plasmids that can be used with ColE1- or pSC101-based target plasmids for the exchange of origins and drug markers. As with chromosomal modifications, PCR amplicons containing selection markers are generally the most convenient substrates to use for targeting plasmids, although oligonucleotides can be used as well.

A study by Thomason *et al.* (2007b) described the benefits and problems associated with recombineering of multicopy plasmids. The study highlighted two issues that should be considered when manipulating plasmids with λ Red. The first is that oligonucleotide-mediated recombineering events in the absence of selection can be performed, but many of the target plasmids in the cell will be left unmodified. To this end, the authors report that by coelectroporation of 10 ng of a target plasmid and 5 pmol of an oligonucleotide in a MMR-deficient host (or by use of a C-C mismatch), unselected mutations could be found at a frequency between 5 and 10%. This result highlights the efficiency of the oligonucleotide-mediated recombineering event. However, it was noted that the initial plasmid preparations are always a mixture of parental and recombinant types, given that a plasmid annealed with a recombinant oligonucleotide will always give rise (in the subsequent replication of the plasmid) to both a parental and a recombinant type. Therefore, retransformation of a minilysate into a *recA* host at low concentrations (to ensure for single-molecule transformation events) is required to isolate a pure recombinant plasmid.

The second consideration is that, for reasons not understood, recombineering of plasmids with either dsDNA substrates or oligonucleotides often generates multimers of the plasmid substrate with mixed populations of both parental and recombinant types (Sawitzke *et al.*, 2007; Thomason *et al.*, 2007b). Digestion of the multimer with a unique restriction enzyme within the parental plasmid, isolation of the linear recombinant plasmid on an agarose gel, and religation/transformation back into *recA* hosts generally allow one to isolate the monomeric recombinant plasmid. Interestingly, the generation of multimers is a consequence of the recombineering event itself and is dependent on the presence of either a dsDNA substrate or an oligonucleotide, not the result of λ Red induction or Gam-induced rolling-circle replication (Thomason *et al.*, 2007b). The mechanism of this recombineering-induced plasmid multimer formation is not known.

E. Bacteriophage engineering

Oppenheim *et al.* (2004) showed that PCR- and oligonucleotide-mediated recombineering of phage λ itself could be performed easily. The phage λ to be modified was mixed with cells containing a defective λ lysogen, which has the p_L-promoted operon under control of the *cI857* repressor. After growth at 42 °C for 15 min to induce the Red functions from the prophage, cells were collected by centrifugation, made electrocompetent, and electroporated with a PCR product or oligonucleotide containing the desired change. Multiple changes in the infecting λ phage were generated, including amber mutants, oligonucleotide-mediated gene deletions, and gene substitutions (Oppenheim *et al.*, 2004). This study also demonstrated that when using ssDNA oligonucleotides to generate mutations, unwanted frameshift mutations occur a fraction of the time. These mutations are not due to Red-promoted mutagenic activity, but to errors in the synthetic oligonucleotides.

VI. PROPOSED MECHANISMS OF λ RED RECOMBINEERING

A. Mechanisms of λ Red-promoted phage recombination

Once a dsDNA end has been processed by λ Red, a 3′ ssDNA tail bound by the λ Beta protein is thought to be the key intermediate that subsequently initiates a recombination event. It is likely the key intermediate in two classical pathways for λ Red recombination of phage chromosomes: the RecA-dependent pathway of recombination and the RecA-independent ssDNA annealing pathway (Stahl *et al.*, 1997). The RecA-dependent pathway (Fig. 5A) is only apparent when phage replication is blocked. The suggestion has been that it may be a "salvage pathway" of λ *in vivo*, which occurs when a Red-processed dsDNA end cannot find an interacting ssDNA region, either in the form of a second Red-processed DNA end or the ssDNA regions of a replication fork (Poteete, 2004). In this scenario, the Beta protein is removed from the ssDNA and replaced with RecA, which then promotes a strand invasion event with a second (unreplicating) phage chromosome. Recombination of λ then proceeds using *E. coli* host recombination functions.

The RecA-independent pathway [i.e., the ssDNA-annealing model of Red recombination (see Fig. 5B)] states that two overlapping regions of the λ chromosome, such as might occur on the dsDNA ends of two rolling-circle replication intermediates whirling in opposite directions, are processed by Red. The Beta bound to ssDNA overhangs, presumably in the form of rings and/or filaments, promotes an annealing event between the complementary ssDNA tails of the rolling circles. Filling in of gaps by DNA polymerase I and subsequent ligation of any nicks complete the

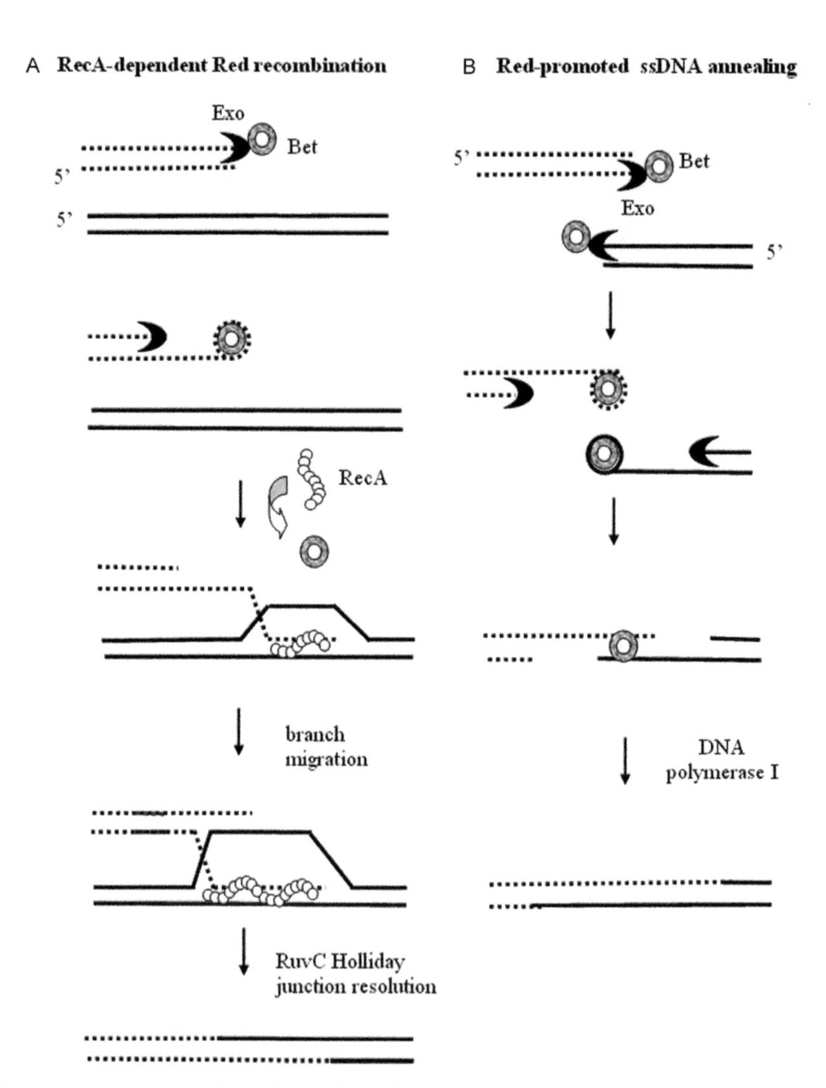

FIGURE 5 Two classical models of λ Red recombination: RecA-dependent Red recombination and Red-promoted ssDNA annealing pathway. (A) In the RecA-dependent pathway, λ Exo generates ssDNA bound by Beta. In the absence of a replicating target substrate, Beta is replaced by the host RecA protein (perhaps with the help of RecF) to promote strand invasion. Host functions carry out branch migration (RecA, RuvAB) and resolution of the Holliday junction (RuvC) to generate the recombinant. (B) In the ssDNA-annealing pathway, ssDNA regions of overlapping segments of DNA are exposed by λ Exo and bound by the Beta protein. Beta promotes annealing of the ssDNA overlaps, polymerase I fills in the ssDNA regions, and nicks are sealed by DNA ligase.

recombination reaction. The role of DNA replication in this model is to provide the DNA ends of overlapping sequences that become the substrates for λ Exo and Beta during infection. In addition, the DNA packaging program, where concatemeric DNA is cut by terminase for insertion into empty phage heads, could also provide a DNA end for processing by λ Red. Use of this pathway as a tool for DNA construction was demonstrated by Narayanan *et al.* (2009) in their description of a λ Red-promoted recombination event between two linear dsDNA substrates *in vivo.*

A third model for Red recombination has been called the replisome invasion/template switch (RITS) model (Poteete, 2008). It draws insight from the oligonucleotide-mediated recombineering mechanisms that have been described (see below) and suggests that a Red-processed phage chromosome end invades a replication fork and promotes annealing of its single-stranded tail to the lagging strand template. It further suggests that the role of Beta is not only to anneal the ssDNA generated by λ Exo, but to redirect the replisome onto the incoming DNA, like a train switching tracks. The template switch, which has precedents in the literature (Branzei and Foiani, 2007; Dutra and Lovett, 2006; Smith *et al.*, 2007), "recombines" the incoming Red-processed DNA to one arm of the replication fork, thus completing the recombination event. This model is consistent with the link between λ replication and λ recombination that has been known for decades.

B. Mechanism of λ Red-promoted oligonucleotide-mediated recombineering

Which of the models discussed earlier for λ Red recombination might explain the mechanistic steps involved in an oligonucleotide-mediated recombineering event? Clearly, the annealing model of λ phage recombination is a scheme that could be adapted easily to describe oligonucleotide-mediated recombineering events. Recombineering with ssDNA oligonucleotides is dependent only on the expression of the λ Beta protein. Given the known annealing capabilities of Beta and the bias of targeting oligonucleotides to the lagging strand template of a replication fork, a model was proposed that oligonucleotides are annealed to ssDNA regions of the replication fork (Ellis *et al.* 2001). The oligonucleotide is then extended by DNA polymerase I as if it were an Okazaki fragment and becomes incorporated into the lagging strand of the fork by the action of DNA ligase (Fig. 6A). Support for this model comes from studies which show that biotinylated oligonucleotides get incorporated into plasmid substrates following electroporation into Red-expressing hosts (Huen *et al.*, 2006).

While the annealing activity of Beta is most likely critical for the high frequency of Beta-promoted oligonucleotide-mediated mutagenesis, an additional important role is protection of the oligonucleotide from host

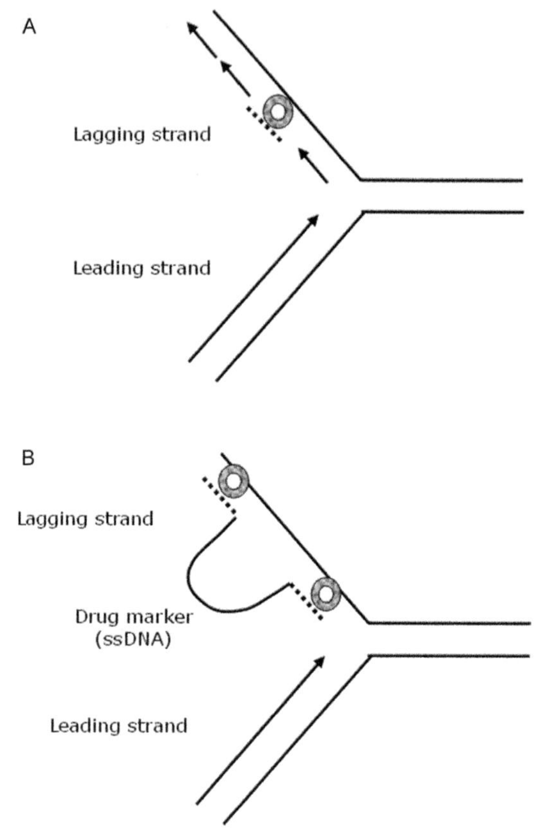

FIGURE 6 (A) Model for oligonucleotide-mediated recombineering. Beta (circle) promotes annealing of an oligonucleotide to ssDNA regions of the lagging strand template of a replication fork. Following elongation by DNA polymerase I and ligation to neighboring Okazaki fragments, the oligonucleotide is incorporated into the chromosome. When the oligonucleotide contains a 1-bp nonhomology relative to the chromosome, the mismatched base pair is a substrate for the mismatch repair system of *E. coli*. Thus, in a MMR-deficient host, Red-promoted oligonucleotide-mediated recombineering is elevated greatly (see Costantino and Court, 2003). (B) Model for dsDNA (PCR)-mediated recombineering. λ Exo interacts with only one end of the PCR substrate *in vivo*, creating an ssDNA intermediate. Beta promotes annealing of the ssDNA generated *in vivo* into the chromosome in the same manner described previously for an oligonucleotide; the nonhomologous drug marker is left unpaired, but is stabilized by the annealing function of Beta (circles). Incorporation into the chromosome occurs following passage of the next replication fork. The model is based on the results of Mosberg *et al.* (2010) and Maresca *et al.* (2010).

exonucleases. In an *E. coli* strain depleted of four of its ssDNA exonucleases, oligonucleotide-mediated recombination could be performed in the absence of any phage-encoded SSAP (Dutra *et al.*, 2007), although not at the frequency seen in cells expressing λ Beta protein. Also, a

Pseudomonas syringae strain, when electroporated with high amounts of oligonucleotide (5 μg), also showed SSAP-independent oligonucleotide-mediated recombination, although again at low frequencies compared to Beta-promoted recombineering (Swingle *et al.*, 2009). The authors suggested that such high amounts of oligonucleotide saturate the endogenous ssDNA exonucleases, allowing low levels of oligonucleotide annealing to occur at the replication fork. These results show that, in addition to promoting annealing to the replication fork, part of the role of λ Beta is to protect oligonucleotides from digestion *in vivo* and is probably the basis for the high rates of λ Beta-promoted oligonucleotide-mediated recombineering in *E. coli*.

C. Mechanism of λ Red-promoted PCR-mediated recombineering

Because recombineering of PCR substrates with 50-bp flanks is largely independent of *recA*, the RecA-dependent pathway of recombination described in Figure 5A cannot be the major route of λ Red-promoted gene replacement. The action of λ Red on dsDNA substrates can be thought of in terms of whether λ Exo acts on one end of the PCR substrate or if it acts on both ends of the substrate at the same time (given that there are no sequence requirements for λ Exo activity, and both ends of a linear DNA are otherwise chemically and structurally identical). This is not an issue for λ phage recombination, as phage biology dictates that only one end is available for each phage chromosome (or multimeric concatemer) during a lytic infection. For instance, there is only one dsDNA end in the structure of a rolling-circle replication intermediate. Also, after generation of a dsDNA break for packaging, the terminase complex stays bound to one end of the dsDNA cut, leaving only one end for processing by λ Red (for review, see Catalano *et al.*, 1995). Application of the RITS model of λ recombination (see earlier discussion) to PCR-mediated recombineering suggests that only one end of the PCR substrate is involved in the invasion step leading to annealing to the lagging strand template. The other dsDNA end has to have undergone limited or no processing by λ Exo. Subsequently, the nonprocessed dsDNA end must participate in a second invasion event with a different replisome, this time necessarily targeting the leading strand template. While the RITS model is consistent with observations from experiments involving plasmid substrates meant to mimic the events of a phage infection (Poteete, 2008), when applied to recombineering, it leaves many mechanistic steps unanswered.

Another report suggests an alternate explanation for how λ Red promotes recombineering of dsDNA PCR substrates. Mosberg *et al.* (2010) generated a 1.2-kb *lacZ::kan* ssDNA substrate and showed that it could be used as a substrate for recombineering, although at only 15% of the rate of a corresponding dsDNA substrate. To demonstrate that the transformants were due to the ssDNA present, and not contaminating dsDNA species, they showed that in hosts where only λ Beta was expressed, the ssDNA recombineering rate remained high, whereas the rate exhibited by a dsDNA substrate fell 200-fold (as expected, as dsDNA recombineering requires both Beta and Exo). In a key experiment, the authors placed mismatches in the ends of the linear PCR dsDNA substrate, where both bases of the mismatch were different from the corresponding bases within the target site on the chromosome. Following recombineering of this substrate, the authors found that most of the recombinants (≈80%) contained sequences corresponding to bases on only one (or the other) of the two strands of the dsDNA substrate. This result suggests a mechanism whereby λ Exo binds to only one end of the dsDNA substrate *in vivo*, converts the dsDNA substrate completely to ssDNA (consistent with its known processivity), followed by Beta-promoted annealing of the ssDNA generated to single-stranded regions of the replication fork, similar to what was described previously for the mechanism of oligonucleotide-mediated recombineering (see Fig. 6B). Thus, given this result, the mechanism of Red-recombineering of PCR substrates seems to proceed through an ssDNA intermediate, at least a majority of the time. The lower rate of recombineering demonstrated for a ssDNA substrate (relative to a corresponding dsDNA substrate) argues for an *in vivo* role for the immediate binding of Beta to ssDNA generated by λ Exo *in vivo*. This role would be to prevent formation of secondary structural features in ssDNA generated by Exo that might otherwise inhibit proper binding of the Beta protein. *In vivo*, one imagines that Beta is bound quickly to ssDNA as soon as it is generated by λ Exo or perhaps is loaded efficiently by λ Exo while the enzyme translocates along dsDNA.

Two other studies lend support to this model. Maresca *et al.* (2010) generated PCR dsDNA substrates containing phosphate groups on one or the other 5′ ends by using one phosphorylated primer in their PCRs. λ Exo has a requirement of a phosphate on the 5′ end of the dsDNA for efficient exonucleolytic processing (Little, 1967; Subramanian *et al.*, 2003). λ Exo acting on these substrates *in vitro* generates long 1.2-kb ssDNA. When the phosphate was positioned on the 5′ end that allows λ Exo to generate a ssDNA complementary to the lagging strand template *in vivo*, 4- to 10-fold higher frequencies of recombineering were obtained relative to a substrate predicted to generate ssDNA complementary to the leading strand template. Thus, ssDNA generated from long dsDNA substrates *in vivo* behaved in a similar way to ssDNA oligonucleotides with regard to targeting the replication fork. This result is in agreement with the results of Mosberg *et al.* (2010) and is consistent with the model depicted in Figure 6B. In another experiment, Lim *et al.* (2008) generated a dsDNA substrate that had homologous target DNA flanking two drug markers, as well as targeting sequences between them. The authors showed that most often, recombinants contained only one of the drug markers, with the double-marker recombinant occurring in only 10% of the transformants. This too is consistent with the model depicted in Figure 6B, where an ssDNA intermediate would make internal homologies easily available for annealing, as opposed to a model suggesting that the two dsDNA ends of the substrate are principally involved in the recombineering event, where one would expect the majority of the transformants to contain both drug markers.

VII. APPLICATIONS

A. *Escherichia coli* genomics

Recombineering has enabled bacterial geneticists to perform manipulations of bacterial chromosomes and BACs beyond the capabilities of procedures that involve the use of restriction enzymes. The technology can target a particular region of a large chromosome-sized DNA substrate based on the extremely efficient recognition specificity of a 100-bp sequence (i.e., the two 50-bp flanking sequences). One of the first major applications of this technology was the construction of a knockout library of every nonessential open reading frame in the *E. coli* chromosome-the Keio collection (Baba *et al.*, 2006; Yamamoto *et al.*, 2009). The mutants contain precisely designed gene deletions, leaving the first two codons and the last seven codons of each gene, interspersed with a *kan* resistance gene and flanked by FRT sites. Out of 4288 open reading frames, mutants were obtained for 3985 of them. Upon further examination (Yamamoto *et al.*, 2009), some of these mutants (at least 25) were found to possess duplications containing both wild-type and deleted versions of the gene. Such chromosomal arrangements can be expected if some of these candidates are essential functions. Out of 25 candidates containing such duplications, 14 have been redefined as essential gene candidates, while the others are still being examined (Yamamoto *et al.*, 2009).

In addition to being a source of genetic mutants for nearly every gene in *E. coli*, the Keio collection has allowed investigators to develop schemes to perform high-throughput studies of genetic interactions in *E. coli*. Two groups (Butland *et al.*, 2008; Typas *et al.*, 2008) have robotically transferred a collection of Cam^R mutants into an array of recipient strains from the Keio collection (marked by Kan^R) by conjugational recombination on solid media. By selecting for exconjugants that show resistance to both drugs and by examining their colony size carefully, investigators describe a methodology that allows them to identify synthetic lethal (or synthetic sickness) interactions, as well as genetic suppressive interactions, where the presence of one mutant suppresses the lethality of another. Studies to examine directly the fitness of interactions between genetically marked mutations have been performed before in *Saccharomyces cerevisiae* (Boone *et al.*, 2007; Tong *et al.*, 2004), but not in *E. coli.* The recombineering-generated Keio collection has also been employed for the study of how individual members of the collection respond to an array of over 300 stress conditions and drug exposures, leading to a rich data set revealing interactions between annotated and unknown genes and insights into the mode of interactions of drugs with their potential targets (Nichols *et al.*, 2011).

Recombineering also played an important role in the construction of multiple gene fusions in the chromosome of *E. coli* (Butland *et al.*, 2005). A total of 857 proteins in *E. coli* had their C termini fused to either a TAP or a SPA affinity tag. The purification of 648 of the proteins 10^6-fold from log-phage cultures and subsequent identification of their interacting partners were performed by peptide mass fingerprinting and/or gel-free shotgun sequencing. An interaction network was generated consisting of 716 nonredundant interactions involving 83 essential genes and 152 nonessential genes. It was determined that 85% of the interactions were new when compared to known interactions from a number of online databases. Such a study would not have been feasible without the ability to efficiently tag multiple chromosomal loci by recombineering. This technology has been applied to more than 1000 orphaned genes, allowing many of the genes to be assigned to functional classes (Hu *et al.*, 2009). Other examples where recombineering has expanded the capabilities of genomic studies in *E. coli* include the development of mutagenic procedures for *E. coli* genes (Kang *et al.*, 2004); the ability to generate strains with reduced genomes (Kolisnychenko *et al.*, 2002; Posfai *et al.*, 2006); a method of programmed and accelerated evolution of *E. coli* (Wang *et al.*, 2009); the ability to observe the position of proteins in living cells (Watt *et al.*, 2007); and genome-wide codon replacement in *E. coli* (Isaacs *et al.*, 2011). Clearly, the manipulation of the genome with recombineering has allowed many novel genomic studies to be performed in *E. coli,* paving the way for a new era in synthetic biology.

B. Mouse genomics

Another area where recombineering has had a great impact is the construction of targeting vectors for manipulation of the mouse genome. The λ Red recombination system has allowed investigators to insert or delete specific regions of eukaryotic DNA cloned into BAC vectors and to modify them by the insertion of promoters, activators, terminators, or other genetic elements with a precision not seen before (Copeland *et al.*, 2001; Lee *et al.*, 2001; Muyrers *et al.*, 1999). In partnership with another phage recombination system, the phage P1 site-specific Cre-*loxP* system (Branda and Dymecki,

2004; Metzger and Chambon, 2001; Nagy, 2000; Sauer and Henderson, 1988, 1989), the combined methodologies have allowed novel constructs in eukaryotic DNA to be constructed in a quicker time period, with precise specificity, and without the need for time-consuming cloning steps involving restriction enzymes and ligation reactions. The methodologies used to take advantage of recombineering- and Cre-based procedures for the construction of gene targeting vectors, and the subsequent planning and assembly of mouse knockouts and transgenic animals, have been described elsewhere (Bouvier and Cheng, 2009; Lee *et al.*, 2009; Parkitna *et al.*, 2009).

C. Metabolic engineering

A third area that has benefited from the use of recombineering technology is the field of metabolic engineering, where the *E. coli* genome is modified for the expressed purpose of producing a particular metabolite for purification, or the knockout of genes encoding undesired metabolites, without creating major changes in the growth rate or a reduction in the fitness of the bacterium. While often these procedures involve deletion of genes within the chromosome that might inhibit efficient production of the metabolite, they also can entail insertion of large biosynthetic operons that have been modified previously by recombineering. Following the paradigm of mouse genome recombineering, foreign DNA is typically inserted into BACs, transferred to λ Red or RecET-producing cells, modified as needed, and transferred back into the original or an exogenous host. As an example of this procedure, a report by Gust *et al.* (2003) demonstrates that a clone containing *Streptomyces coelicolor* DNA sequences was modified by recombineering in *E. coli* to produce a knockout of a gene thought to be responsible for the biosynthesis of the secondary metabolite geosmin, which is known to produce an unfavorable earthy odorant produced by several types of filamentous bacteria (Dionigi *et al.*, 1991). The investigators were able to replace the *cyc2* gene of *S. coelicolor* with an apramycin-resistant cassette that also contained $oriT_{RK2}$ to allow transfer of the clone back to *S. coelicolor* by conjugation. Unable to replicate in *S. coeliocolor*, the transferred DNA recombines into the chromosome to complete the gene replacement. The procedure has been repeated many times to make numerous gene knockouts in this bacterium (Gust *et al.*, 2003).

In another example of recombineering being used for metabolic engineering, a biosynthetic gene cluster from the myxobacterium *Stigmatella aurantiaca* was reconstructed in *E. coli* and modified for overproduction of myxochromide S, a group of cyclic peptides with polyketide side chains (Wenzel *et al.*, 2005). Using recombineering, a thioester domain was added to one of the nonribosomal peptide synthetases, a promoter was inserted to drive expression of the gene cluster, and a *tetR* cassette for selection was added to the construct. In addition, an origin of transfer was included to move the construct to *Pseudomonas putida* by conjugation, along with sequences (*trpE*) to integrate the DNA into the *P. putida* chromosome following transfer. The engineered strain was found to produce a five times greater yield of the cyclic peptides relative to the *S. aurantiaca* host from where the gene cluster originated. Other examples of the use of recombineering in metabolic engineering studies include manipulation of the biosynthetic pathway for geldanamycin (Vetcher *et al.*, 2005), an antibiotic that binds to the heat shock protein HSP90 and alters its function (Whitesell *et al.*, 1994); construction of strains deficient for biofilm formation (Sung *et al.*, 2006); and enhancement of carotenoid production in *E. coli* by promoter replacement (Yuan *et al.*, 2006).

D. Microbial pathogenesis

Attempts have been made to move recombineering technologies from laboratory strains of *E. coli* to bacterial pathogens, where the tools for chromosomal genetic manipulations are often limited to the introduction of nonreplicating plasmid integrants. The establishment of recombineering in these strains has most often involved the plasmid-based expression of the λ *red* and *gam* functions in these exogenous hosts. A list of bacterial hosts where recombineering has been performed is shown in Table II. Not surprisingly, in cases where the pathogen is a close relative of *E. coli* K-12, λ Red recombineering with PCR substrates bearing 50 bp of flanking homology has been demonstrated easily (e.g., *Salmonella*, *Shigella*, and enterohemorrhagic *E. coli*). Most notably, the use of λ Red for gene knockouts has been very useful for the development of vaccine strains for *Shigella flexneri* (Runyen-Janecky *et al.*, 2008; Zurawski *et al.*, 2006). With strains phylogenetically more distant from *E. coli*, the expression of λ Red still allows transformation with linear dsDNA substrates in some cases, but requires an increase in the amount of homology (≈500 bp) to obtain practical frequencies of transformation. Given the results of Datta *et al.* (2008), this result is likely due to the species variability of potential interactions between specific host functions (perhaps components in replication forks) and the λ Beta and Exo proteins.

Alternatively, as in the case of *M. tuberculosis*, the recombineering potential of an endogenous phage recombination system was used to promote recombineering in this pathogenic bacterium (as described in Section III). The advantage of having a large sequence database of mycobacterial phage chromosomes was instrumental in finding the RecET genes from phage Che9c, as this was the only phage that possessed both a synaptase (an annealing protein) and a neighboring exonuclease function. Nonetheless, finding SynExo pairs of recombination functions endogenous to other bacterial pathogens will likely be advantageous for

TABLE II Recombineering in pathogenic and non-*Escherichia coli K-12* hosts

Bacterium	Reference
Enterohemorrhagic *E. coli*	Murphy and Campellone, 2003
Enteropathogenic *E. coli*	Murphy and Campellone, 2003; Nevesinjac and Raivio, 2005
Klebsiella aerogenes	Janes *et al.*, 2001
Mycobacteia abscessus	Medjahed and Reyrat, 2009; Medjahed and Singh, 2010
M. smegmatis	van Kessel and Hatfull, 2007; van Kessel and Hatfull, 2008
M. tuberculosis	van Kessel and Hatfull, 2007; van Kessel and Hatfull, 2008
Pantoea ananatis	Katashkina *et al.*, 2009
Pseudomonas aeruginosa	Lesic and Rahme, 2008
P. syringae	Swingle *et al.*, 2010
Salmonella enterica	Uzzau *et al.*, 2001; Lu *et al.*, 2002; Gerlach *et al.*, 2009
S. typhimurium	Stanley *et al.*, 2000; Price-Carter *et al.*, 2001; Chakravortty *et al.*, 2002; Clegg and Hughes, 2002; Freeman *et al.*, 2002; Havemann *et al.*, 2002; Worlock and Smith, 2002
Serratia marcescens	Rossi *et al.*, 2003
Shigella flexneri	Zurawski *et al.*, 2006; Runyen-Janecky *et al.*, 2008
Uropathogenic *E. coli*	Eto *et al.*, 2007; Wiles *et al.*, 2008
Yesinia pestis	Derbise *et al.*, 2003; Sun *et al.*, 2008

the development of recombineering technology in clinically important strains. The recently described SXT-exonuclease from an integrating conjugative element from *Vibrio cholerae* (Chen *et al.*, 2011) and its corresponding SXT-Beta protein (Datta *et al.*, 2008) may eventually prove useful for genetic manipulation of this important bacterial pathogen.

VIII. PERSPECTIVES

The use of phage recombination systems and SSAPs to precisely alter chromosomal DNA has changed the way investigators approach their research. Modifications that would be too cumbersome to perform with restriction enzymes and standard cloning techniques are now performed with great efficiencies. Libraries of *E. coli* strains have been generated that contain knockouts of every nonessential function, and hundreds of genes can be tagged to allow studies of protein interactions and metabolic networks. Highly complicated constructs can be engineered into mouse or human DNAs cloned into BACs that could not have been done before.

It should not go unnoticed that the most diminutive organisms among us, the bacteriophage, have again given us a great set of tools with which we can manipulate genomes. First it was the P1 phage site-specific recombination system of Cre-*loxP* that eventually was found to work quite well in eukaryotic hosts. It will be curious to know if λ Red, RecET, or perhaps a more endogenous SSAP from the eukaryotic world will allow us to make the same type of genomic modifications in higher eukaryotic cells that can be performed currently in recombineering-proficient strains of *E. coli*. It has been noted that many phage recombinases have more in common structurally and functionally with eukaryotic recombinases (e.g., compare Beta with Rad52) than with recombinases found in bacterial systems. This makes studying the molecular mechanism of phage- and SSAP-promoted recombination events an important area of future research.

REFERENCES

Amundsen, S. K., Taylor, A. F., Chaudhury, A. M., and Smith, G. R. (1986). *recD*: The gene for an essential third subunit of exonuclease V. *Proc. Natl. Acad. Sci. USA* **83:**5558–5562.

Anderson, D. G., Churchill, J. J., and Kowalczykowski, S. C. (1997). Chi-activated RecBCD enzyme possesses $5' \rightarrow 3'$ nucleolytic activity, but RecBC enzyme does not: Evidence suggesting that the alteration induced by Chi is not simply ejection of the RecD subunit. *Genes Cells* **2:**117–128.

Aravind, L., Walker, D. R., and Koonin, E. V. (1999). Conserved domains in DNA repair proteins and evolution of repair systems. *Nucleic Acids Res.* **27:**1223–1242.

Ayora, S., Missich, R., Mesa, P., Lurz, R., Yang, S., Egelman, E. H., and Alonso, J. C. (2002). Homologous-pairing activity of the *Bacillus subtilis* bacteriophage SPP1 replication protein G35P. *J. Biol. Chem.* **277:**35969–35979.

Baba, T., Ara, T., Hasegawa, M., Takai, Y., Okumura, Y., Baba, M., Datsenko, K. A., Tomita, M., Wanner, B. L., and Mori, H. (2006). Construction of *Escherichia coli* K-12 in-frame, single-gene knockout mutants: The Keio collection. *Mol. Syst. Biol.* **2:**2006:0008.

Balasubramanian, N., Bai, P., Buchek, G., Korza, G., and Weller, S. K. (2010). Physical interaction between the herpes simplex virus type 1 exonuclease, UL12, and the DNA double-strand break-sensing MRN complex. *J. Virol.* **84:**12504–12514.

Barbour, S. D., Nagaishi, H., Templin, A., and Clark, A. J. (1970). Biochemical and genetic studies of recombination proficiency in *Escherichia coli*. II. Rec+ revertants caused by indirect suppression of rec- mutations. *Proc. Natl. Acad. Sci. USA* **67:**128–135.

Bi, B., Rybalchenko, N., Golub, E. I., and Radding, C. M. (2004). Human and yeast Rad52 proteins promote DNA strand exchange. *Proc. Natl. Acad. Sci. USA* **101:**9568–9572.

Blomfield, I. C., Vaughn, V., Rest, R. F., and Eisenstein, B. I. (1991). Allelic exchange in *Escherichia coli* using the *Bacillus subtilis sacB* gene and a temperature-sensitive pSC101 replicon. *Mol. Microbiol.* **5:**1447–1457.

Boone, C., Bussey, H., and Andrews, B. J. (2007). Exploring genetic interactions and networks with yeast. *Nat. Rev. Genet.* **8:**437–449.

Botstein, D. (1968). Synthesis and maturation of phage P22 DNA. *J. Mol. Biol.* **34:**621–641.

Botstein, D., and Matz, M. J. (1970). A recombination function essential to the growth of bacteriophage P22. *J. Mol. Biol.* **54:**417–440.

Bouvier, J., and Cheng, J. G. (2009). Recombineering-based procedure for creating Cre/*loxP* conditional knockouts in the mouse. *Curr. Protoc. Mol. Biol.* Chapter 23: Unit 23 13.

Branda, C. S., and Dymecki, S. M. (2004). Talking about a revolution: The impact of site-specific recombinases on genetic analyses in mice. *Dev. Cell.* **6:**7–28.

Branzei, D., and Foiani, M. (2007). Template switching: From replication fork repair to genome rearrangements. *Cell* **131:**1228–1230.

Brooks, K., and Clark, A. J. (1967). Behavior of lambda bacteriophage in a recombination deficienct strain of *Escherichia coli*. *J. Virol.* **1:**283–293.

Butland, G., Peregrin-Alvarez, J. M., Li, J., Yang, W., Yang, X., Canadien, V., Starostine, A., Richards, D., Beattie, B., Krogan, N., Davey, M., Parkinson, J., *et al.* (2005). Interaction network containing conserved and essential protein complexes in *Escherichia coli*. *Nature* **433:**531–537.

Butland, G., Babu, M., Diaz-Mejia, J. J., Bohdana, F., Phanse, S., Gold, B., Yang, W., Li, J., Gagarinova, A. G., Pogoutse, O., Mori, H., Wanner, B. L., *et al.* (2008). eSGA: *E. coli* synthetic genetic array analysis. *Nat. Methods* **5:**789–795.

Carter, D. M., and Radding, C. M. (1971). The role of exonuclease and beta protein of phage lambda in genetic recombination. II. Substrate specificity and the mode of action of lambda exonuclease. *J. Biol. Chem.* **246:**2502–2512.

Cassuto, E., and Radding, C. M. (1971). Mechanism for the action of lambda exonuclease in genetic recombination. *Nat. New. Biol.* **229:**13–16.

Catalano, C. E., Cue, D., and Feiss, M. (1995). Virus DNA packaging: The strategy used by phage lambda. *Mol. Microbiol.* **16:**1075–1086.

Chakravortty, D., Hansen-Wester, I., and Hensel, M. (2002). *Salmonella* pathogenicity island 2 mediates protection of intracellular *Salmonella* from reactive nitrogen intermediates. *J. Exp. Med.* **195:**1155–1166.

Chen, W. Y., Ho, J. W., Huang, J. D., and Watt, R. M. (2011). Functional characterization of an alkaline exonuclease and single strand annealing protein from the SXT genetic element of *Vibrio cholerae*. *BMC Mol. Biol.* **12:**16.

Clark, A. J., Sharma, V., Brenowitz, S., Chu, C. C., Sandler, S., Satin, L., Templin, A., Berger, I., and Cohen, A. (1993). Genetic and molecular analyses of the C-terminal region of the recE gene from the Rac prophage of *Escherichia coli* K-12 reveal the *recT* gene. *J. Bacteriol.* **175:**7673–7682.

Clegg, S., and Hughes, K. T. (2002). FimZ is a molecular link between sticking and swimming in *Salmonella enterica* serovar typhimurium 1213. *J. Bacteriol.* **184:**1209–1213.

Copeland, N. G., Jenkins, N. A., and Court, D. L. (2001). Recombineering: A powerful new tool for mouse functional genomics. *Nat. Rev. Genet.* **2:**769–779.

Costantino, N., and Court, D. L. (2003). Enhanced levels of lambda Red-mediated recombinants in mismatch repair mutants. *Proc. Natl. Acad. Sci USA* **100:**15748–15753.

Court, D. L., Sawitzke, J. A., and Thomason, L. C. (2002). Genetic engineering using homologous recombination. *Annu. Rev. Genet.* **36:**361–388.

Court, R., Cook, N., Saikrishnan, K., and Wigley, D. (2007). The crystal structure of lambda-Gam protein suggests a model for RecBCD inhibition. *J. Mol. Biol.* **371:**25–33.

Datsenko, K. A., and Wanner, B. L. (2000). One-step inactivation of chromosomal genes in *Escherichia coli* K-12 using PCR products. *Proc. Natl. Acad. Sci. USA* **97:**6640–6645.

Datta, S., Costantino, N., and Court, D. L. (2006). A set of recombineering plasmids for gram-negative bacteria. *Gene* **379:**109–115.

Datta, S., Costantino, N., Zhou, X., and Court, D. L. (2008). Identification and analysis of recombineering functions from Gram-negative and Gram-positive bacteria and their phages. *Proc. Natl. Acad. Sci. USA* **105:**1626–1631.

Dawson, P., Hohn, B., Hohn, T., and Skalka, A. (1976). Functional empty capsid precursors produced by lambda mutant defective for late lambda DNA replication. *J. Virol.* **17:**576–583.

Derbise, A., Lesic, B., Dacheux, D., Ghigo, J. M., and Carniel, E. (2003). A rapid and simple method for inactivating chromosomal genes in *Yersinia. FEMS Immunol. Med. Microbiol.* **38:**113–116.

Dillingham, M. S., and Kowalczykowski, S. C. (2008). RecBCD enzyme and the repair of double-stranded DNA breaks. *Microbiol. Mol. Biol. Rev.* **72:**642–671.

Dionigi, C. P., Millie, D. F., and Johnsen, P. B. (1991). Effects of farnesol and the off-flavor derivative geosmin on *Streptomyces tendae. Appl. Environ. Microbiol.* **57:**3429–3432.

Dutra, B. E., and Lovett, S. T. (2006). Cis and trans-acting effects on a mutational hotspot involving a replication template switch. *J. Mol. Biol.* **356:**300–311.

Dutra, B. E., Sutera, V. A., Jr., and Lovett, S. T. (2007). RecA-independent recombination is efficient but limited by exonucleases. *Proc. Natl. Acad. Sci. USA* **104:**216–221.

Echols, H., and Gingery, R. (1968). Mutants of bacteriophage λ defective for vegetative genetic recombination. *J. Mol. Biol.* **34:**239–249.

Ellis, H. M., Yu, D., DiTizio, T., and Court, D. L. (2001). High efficiency mutagenesis, repair, and engineering of chromosomal DNA using single-stranded oligonucleotides. *Proc. Natl. Acad. Sci. USA* **98:**6742–6746.

Enquist, L. W., and Skalka, A. (1973). Replication of bacteriophage λ DNA dependent on the function of host and viral genes. I. Interaction of *red*, *gam* and *rec*. *J. Mol. Biol.* **75:**185–212.

Erler, A., Wegmann, S., Elie-Caille, C., Bradshaw, C. R., Maresca, M., Seidel, R., Habermann, B., Muller, D. J., and Stewart, A. F. (2009). Conformational adaptability of Redbeta during DNA annealing and implications for its structural relationship with Rad52. *J. Mol. Biol.* **391:**586–598.

Eto, D. S., Jones, T. A., Sundsbak, J. L., and Mulvey, M. A. (2007). Integrin-mediated host cell invasion by type 1-piliated uropathogenic *Escherichia coli. PLoS Pathog.* **3:**e100.

Freeman, J. A., Rappl, C. V., Kuhle, V., Hensel, M., and Miller, S. I. (2002). SpiC is required for translocation of *Salmonella* pathogenicity island 2 effectors and secretion of translocon proteins SseB and SseC. *J. Bacteriol.* **184:**4971–4980.

Galitski, T., and Roth, J. R. (1997). Pathways for homologous recombination between chromosomal direct repeats in *Salmonella typhimurium. Genetics* **146:**751–767.

Gay, P., LeCoq, D., Steinmetz, M., Ferrari, E., and Hoch, J. (1983). Cloning structural gene *sacB*, which codes for exoenzyme levansucrase of *Bacillus subtilis*: Expression of the gene in *Escherichia coli. J. Bacteriol.* **153:**1424–1431.

Gerlach, R. G., Jackel, D., Holzer, S. U., and Hensel, M. (2009). Rapid oligonucleotide-based recombineering of the chromosome of *Salmonella enterica. Appl. Environ. Microbiol.* **75:**1575–1580.

Gilbert, W., and Dressler, D. (1968). DNA replication: The rolling circle model. *Cold Spring Harb Symp. Quant. Biol.* **33:**473–484.

Gillen, J. R., and Clark, A. J. (1974). The RecE pathway of bacterial recombination. *In* "Mechanisms of Genetic Recombination" (R. F. Grell, ed.), pp. 123–136. Plenum Press, New York.

Gillen, J. R., Willis, D. K., and Clark, A. J. (1981). Genetic analysis of the RecE pathway of genetic recombination in *Escherichia coli* K-12. *J. Bacteriol.* **145:**521–532.

Gottesman, M. M., Gottesman, M. E., Gottesman, S., and Gellert, M. (1974). Characterization of bacteriophage lambda reverse as an *Escherichia coli* phage carrying a unique set of host-derived recombination functions. *J. Mol. Biol.* **88:**471–487.
Gust, B., Challis, G. L., Fowler, K., Kieser, T., and Chater, K. F. (2003). PCR-targeted *Streptomyces* gene replacement identifies a protein domain needed for biosynthesis of the sesquiterpene soil odor geosmin. *Proc. Natl. Acad. Sci. USA* **100:**1541–1546.
Hall, S. D., and Kolodner, R. D. (1994). Homologous pairing and strand exchange promoted by the *Escherichia coli* RecT protein. *Proc. Natl. Acad. Sci. USA* **91:**3205–3209.
Hall, S. D., Kane, M. F., and Kolodner, R. D. (1993). Identification and characterization of the *Escherichia coli* RecT protein, a protein encoded by the *recE* region that promotes renaturation of homologous single-stranded DNA. *J. Bacteriol.* **175:**277–287.
Hamilton, D. L., and Abremski, K. (1984). Site-specific recombination by the bacteriophage P1 *lox*-Cre system: Cre-mediated synapsis of two *lox* sites. *J. Mol. Biol.* **178:**481–486.
Havemann, G. D., Sampson, E. M., and Bobik, T. A. (2002). PduA is a shell protein of polyhedral organelles involved in coenzyme B12-dependent degradation of 1,2-propanediol in *Salmonella enterica* Serovar Typhimurium LT2. *J. Bacteriol.* **184:**1253–1261.
Hoess, R. H., and Abremski, K. (1984). Interaction of the bacteriophage P1 recombinase Cre with the recombining site *loxP*. *Proc. Natl. Acad. Sci. USA* **81:**1026–1029.
Hu, P., Janga, S. C., Babu, M., Diaz-Mejia, J. J., Butland, G., Yang, W., Pogoutse, O., Guo, X., Phanse, S., Wong, P., Chandran, S., Christopoulos, C., *et al.* (2009). Global functional atlas of *Escherichia coli* encompassing previously uncharacterized proteins. *PLoS Biol.* **7:**e96.
Huen, M. S., Li, X. T., Lu, L. Y., Watt, R. M., Liu, D. P., and Huang, J. D. (2006). The involvement of replication in single stranded oligonucleotide-mediated gene repair. *Nucleic Acids Res.* **34:**6183–6194.
Isaacs, F. J., Carr, P. A., Wang, H. H., Lajoie, M. J., Sterling, B., Kraal, L., Tolonen, A. C., Gianoulis, T. A., Goodman, D. B., Reppas, N. B., Emig, C. J., Bang, D., *et al.* (2011). Precise manipulation of chromosomes in vivo enables genome-wide codon replacement. *Science* **333:**348–353.
Iyer, L. M., Koonin, E. V., and Aravind, L. (2002). Classification and evolutionary history of the single-strand annealing proteins, RecT, Redbeta, ERF and RAD52. *BMC Genomics* **3:**8.
Jacob, F., and Wollman, E. (1954). Etude genetique d'un bacteriophage tempere d'*Echerichia coli*. I. Le systeme genetique du bacteriophage lambda. *Ann. Inst. Pasteur* **87:**653–674.
Janes, B. K., Pomposiello, P. J., Perez-Matos, A., Najarian, D. J., Goss, T. J., and Bender, R. A. (2001). Growth inhibition caused by overexpression of the structural gene for glutamate dehydrogenase (*gdhA*) from *Klebsiella aerogenes*. *J. Bacteriol.* **183:**2709–2714.
Joseph, J. W., and Kolodner, R. (1983a). Exonuclease VIII of *Escherichia coli*. I. Purification and physical properties. *J. Biol. Chem.* **258:**10411–10417.
Joseph, J. W., and Kolodner, R. (1983b). Exonuclease VIII of *Escherichia coli*. II. Mechanism of action. *J. Biol. Chem.* **258:**10418–10424.
Kaiser, A. D. (1955). A genetic study of the temperate coliphage. *Virology* **1:**424–443.
Kaiser, K., and Murray, N. E. (1979). Physical characterisation of the "Rac prophage" in *E. coli* K12 *Mol. Gen. Genet.* **175:**159–174.
Kang, Y., Durfee, T., Glasner, J. D., Qiu, Y., Frisch, D., Winterberg, K. M., and Blattner, F. R. (2004). Systematic mutagenesis of the *Escherichia coli* genome. *J. Bacteriol.* **186:**4921–4930.
Karakousis, G., Ye, N., Li, Z., Chiu, S. K., Reddy, G., and Radding, C. M. (1998). The beta protein of phage lambda binds preferentially to an intermediate in DNA renaturation. *J. Mol. Biol.* **276:**721–731.
Karu, A. E., Sakaki, Y., Echols, H., and Linn, S. (1975). The gamma protein specified by bacteriophage gamma: Structure and inhibitory activity for the *recBC* enzyme of *Escherichia coli*. *J. Biol. Chem.* **250:**7377–7387.

Katashkina, J. I., Hara, Y., Golubeva, L. I., Andreeva, I. G., Kuvaeva, T. M., and Mashko, S. V. (2009). Use of the lambda Red-recombineering method for genetic engineering of *Pantoea ananatis*. *BMC Mol. Biol.* **10**:34.

Kellenberger, G., Zichichi, M. L., and Weigle, J. J. (1961). Exchange of DNA in the recombination of bacteriophage lambda. *Proc. Natl. Acad. Sci. USA* **47**:869–878.

Kmiec, E., and Holloman, W. K. (1981). Beta protein of bacteriophage lambda promotes renaturation of DNA. *J. Biol. Chem.* **256**:12636–12639.

Kolisnychenko, V., Plunkett, G., 3rd, Herring, C. D., Feher, T., Posfai, J., Blattner, F. R., and Posfai, G. (2002). Engineering a reduced *Escherichia coli* genome. *Genome Res.* **12**:640–647.

Kovall, R., and Matthews, B. W. (1997). Toroidal structure of lambda-exonuclease. *Science* **277**:1824–1827.

Kreuzer, K. N., and Brister, J. R. (2010). Initiation of bacteriophage T4 DNA replication and replication fork dynamics: A review in the *Virology Journal* series on bacteriophage T4 and its relatives. *Virol. J.* **7**:358.

Kuhlman, T. E., and Cox, E. C. (2009). Site-specific chromosomal integration of large synthetic constructs. *Nucleic Acids. Res.* **38**:e92.

Kumar, J. K., and Gupta, R. C. (2004). Strand exchange activity of human recombination protein Rad52. *Proc. Natl. Acad. Sci. USA* **101**:9562–9567.

Kuzminov, A. (1999). Recombinational repair of DNA damage in *Escherichia coli* and bacteriophage lambda. *Microbiol. Mol. Biol. Rev.* **63**:751–813.

Lee, E. C., Yu, D., Martinez de Velasco, J., Tessarollo, L., Swing, D. A., Court, D. L., Jenkins, N. A., and Copeland, N. G. (2001). A highly efficient *Escherichia coli*-based chromosome engineering system adapted for recombinogenic targeting and subcloning of BAC DNA. *Genomics* **73**:56–65.

Lee, S. C., Wang, W., and Liu, P. (2009). Construction of gene-targeting vectors by recombineering. *Methods Mol Biol* **530**:15–27.

Lesic, B., and Rahme, L. G. (2008). Use of the lambda Red recombinase system to rapidly generate mutants in *Pseudomonas aeruginosa*. *BMC Mol. Biol.* **9**:20.

Levine, M. (1972). Replication and lysogeny with phage P22 in *Salmonella typhimurium*. *Curr. Top. Microbiol. Immunol.* **58**:135–156.

Li, Z., Karakousis, G., Chiu, S. K., Reddy, G., and Radding, C. M. (1998). The beta protein of phage lambda promotes strand exchange. *J. Mol. Biol.* **276**:733–744.

Lim, S. I., Min, B. E., and Jung, G. Y. (2008). Lagging strand-biased initiation of *red* recombination by linear double-stranded DNAs. *J. Mol. Biol.* **384**:1098–1105.

Little, J. W. (1967). An exonuclease induced by bacteriophage lambda. II. Nature of the enzymatic reaction. *J. Biol. Chem.* **242**:679–686.

Liu, J., and Morrical, S. W. (2010). Assembly and dynamics of the bacteriophage T4 homologous recombination machinery. *Virol. J.* **7**:357.

Lu, S., Killoran, P. B., Fang, F. C., and Riley, L. W. (2002). The global regulator ArcA controls resistance to reactive nitrogen and oxygen intermediates in *Salmonella enterica* serovar enteritidis. *Infect. Immun.* **70**:451–461.

Luisi-DeLuca, C., Lovett, S. T., and Kolodner, R. D. (1989). Genetic and physical analysis of plasmid recombination in *recB recC sbcB* and *recB recC sbcA Escherichia coli* K-12 mutants. *Genetics* **122**:269–278.

Mahdi, A. A., Sharples, G. J., Mandal, T. N., and Lloyd, R. G. (1996). Holliday junction resolvases encoded by homologous *rusA* genes in *Escherichia coli* K-12 and phage 82. *J. Mol. Biol.* **257**:561–573.

Maresca, M., Erler, A., Fu, J., Friedrich, A., Zhang, Y., and Stewart, A. F. (2010). Single-stranded heteroduplex intermediates in lambda Red homologous recombination. *BMC Mol. Biol.* **11**:54.

Marsic, N., Roje, S., Stojiljkovic, I., Salaj-Smic, E., and Trgovcevic, Z. (1993). *In vivo* studies on the interaction of RecBCD enzyme and lambda Gam protein. *J. Bacteriol.* **175**:4738–4743.

Martinez-Jimenez, M. I., Alonso, J. C., and Ayora, S. (2005). *Bacillus subtilis* bacteriophage SPP1-encoded gene 34.1 product is a recombination-dependent DNA replication protein. *J. Mol. Biol.* **351:**1007–1019.
McMilin, K. D., and Russo, V. E. (1972). Maturation and recombination of bacteriophage lambda DNA molecules in the absence of DNA duplication. *J. Mol. Biol.* **68:**49–55.
Medjahed, H., and Reyrat, J. M. (2009). Construction of *Mycobacterium abscessus* defined glycopeptidolipid mutants: Comparison of genetic tools. *Appl. Environ. Microbiol.* **75:**1331–1338.
Medjahed, H., and Singh, A. K. (2010). Genetic manipulation of *Mycobacterium abscessus*. *Curr. Protoc. Microbiol.* **10:**2. Chapter 10.
Meselson, M. (1964). On the mechanism of genetic recombination between DNA molecules. *J. Mol. Biol.* **9:**734–745.
Meselson, M., and Weigle, J. J. (1961). Chromosome breakage accompanying genetic recombination in bacteriophage. *Proc. Natl. Acad. Sci. USA* **47:**857–868.
Metzger, D., and Chambon, P. (2001). Site- and time-specific gene targeting in the mouse. *Methods* **24:**71–80.
Monteilhet, C., Perrin, A., Thierry, A., Colleaux, L., and Dujon, B. (1990). Purification and characterization of the *in vitro* activity of I-Sce I, a novel and highly specific endonuclease encoded by a group I intron. *Nucleic Acids Res.* **18:**1407–1413.
Mosberg, J. A., Lajoie, M. J., and Church, G. M. (2010). Lambda Red recombination in *Escherichia coli* occurs through a fully single-stranded intermediate. *Genetics* **186:**791–799.
Mosig, G. (1998). Recombination and recombination-dependent DNA replication in bacteriophage T4. *Annu. Rev. Genet.* **32:**379–413.
Muniyappa, K., and Radding, C. M. (1986). The homologous recombination system of phage lambda: Pairing activities of beta protein. *J. Biol. Chem.* **261:**7472–7478.
Murphy, K. C. (1991). Lambda Gam protein inhibits the helicase and chi-stimulated recombination activities of *Escherichia coli* RecBCD enzyme. *J. Bacteriol.* **173:**5808–5821.
Murphy, K. C. (1994). Biochemical characterization of P22 phage-modified *Escherichia coli* RecBCD enzyme. *J. Biol. Chem.* **269:**22507–22516.
Murphy, K. C. (1998). Use of bacteriophage lambda recombination functions to promote gene replacement in *Escherichia coli*. *J. Bacteriol.* **180:**2063–2071.
Murphy, K. C. (2000). Bacteriophage P22 Abc2 protein binds to RecC increases the 5′ strand nicking activity of RecBCD and together with lambda Bet, promotes Chi-independent recombination. *J. Mol. Biol.* **296:**385–401.
Murphy, K. C. (2007). The lambda Gam protein inhibits RecBCD binding to dsDNA ends. *J. Mol. Biol.* **371:**19–24.
Murphy, K. C. (2011). Targeted chromosomal gene knockout using PCR fragments. *Methods Mol. Biol.* **765:**27–42.
Murphy, K. C., and Campellone, K. G. (2003). Lambda Red-mediated recombinogenic engineering of enterohemorrhagic and enteropathogenic *E. coli*. *BMC Mol. Biol.* **4:**11.
Murphy, K. C., Casey, L., Yannoutsos, N., Poteete, A. R., and Hendrix, R. W. (1987a). Localization of a DNA-binding determinant in the bacteriophage P22 Erf protein. *J. Mol. Biol.* **194:**105–117.
Murphy, K. C., Fenton, A. C., and Poteete, A. R. (1987b). Sequence of the bacteriophage P22 anti-RecBCD (*abc*) genes and properties of P22 *abc* region deletion mutants. *Virology* **160:**456–464.
Murphy, K. C., and Lewis, L. J. (1993). Properties of *Escherichi coli* expressing bacteriophage P22 Abc (Anti-RecBCD) proteins, including inhibition of Chi activity. *J. Bacteriol.* **175:**1756–1766.

Muyrers, J. P., Zhang, Y., Buchholz, F., and Stewart, A. F. (2000). RecE/RecT and Redalpha/Redbeta initiate double-stranded break repair by specifically interacting with their respective partners. *Genes Dev.* **14:**1971–1982.

Muyrers, J. P., Zhang, Y., Testa, G., and Stewart, A. F. (1999). Rapid modification of bacterial artificial chromosomes by ET-recombination. *Nucleic Acids Res.* **27:**1555–1557.

Mythili, E., and Muniyappa, K. (1993). Formation of linear plasmid multimers promoted by the phage lambda Red-system in *lon* mutants of *Escherichia coli. J. Gen. Microbiol.* **139:**2387–2397.

Nagy, A. (2000). Cre recombinase: The universal reagent for genome tailoring. *Genesis* **26:**99–109.

Narayanan, K., Sim, E. U., Ravin, N. V., and Lee, C. W. (2009). Recombination between linear double-stranded DNA substrates in vivo. *Anal. Biochem.* **387:**139–141.

Nevesinjac, A. Z., and Raivio, T. L. (2005). The Cpx envelope stress response affects expression of the type IV bundle-forming pili of enteropathogenic *Escherichia coli. J. Bacteriol.* **187:**672–686.

Nichols, R. J., Sen, S., Choo, Y. J., Beltrao, P., Zietek, M., Chaba, R., Lee, S., Kazmierczak, K. M., Lee, K. J., Wong, A., Shales, M., Lovett, S., *et al.* (2011). Phenotypic landscape of a bacterial cell. *Cell* **144:**143–156.

Okano, K., Vanarsdall, A. L., and Rohrmann, G. F. (2007). A baculovirus alkaline nuclease knockout construct produces fragmented DNA and aberrant capsids. *Virology* **359:**46–54.

Oliner, J. D., Kinzler, K. W., and Vogelstein, B. (1993). *In vivo* cloning of PCR products in *E. coli. Nucleic Acids. Res.* **21:**5192–5197.

Onrust, R., Finkelstein, J., Turner, J., Naktinis, V., and O'Donnell, M. (1995). Assembly of a chromosomal replication machine: Two DNA polymerases, a clamp loader, and sliding clamps in one holoenzyme particle. III. Interface between two polymerases and the clamp loader. *J. Biol. Chem.* **270:**13366–13377.

Oppenheim, A. B., Rattray, A. J., Bubunenko, M., Thomason, L. C., and Court, D. L. (2004). *In vivo* recombineering of bacteriophage lambda by PCR fragments and single-strand oligonucleotides. *Virology* **319:**3447–3454.

Palmeros, B., Wild, J., Szybalski, W., Le Borgne, S., Hernandez-Chavez, G., Gosset, G., Valle, F., and Bolivar, F. (2000). A family of removable cassettes designed to obtain antibiotic-resistance-free genomic modifications of *Escherichia coli* and other bacteria. *Gene* **247:**255–264.

Parkitna, J. R., Engblom, D., and Schutz, G. (2009). Generation of Cre recombinase-expressing transgenic mice using bacterial artificial chromosomes. *Methods Mol. Biol.* **530:**325–342.

Passy, S. I., Yu, X., Li, Z., Radding, C. M., and Egelman, E. H. (1999). Rings and filaments of beta protein from bacteriophage lambda suggest a superfamily of recombination proteins. *Proc. Natl. Acad. Sci. USA* **96:**4279–4284.

Posfai, G., Plunkett, G., 3rd, Feher, T., Frisch, D., Keil, G. M., Umenhoffer, K., Kolisnychenko, V., Stahl, B., Sharma, S. S., de Arruda, M., Burland, V., *et al.* (2006). Emergent properties of reduced-genome *Escherichia coli. Science* **312:**1044–1046.

Poteete, A. R. (1982). Location and sequence of the *erf* gene of phage P22. *Virology* **119:**422–429.

Poteete, A. R. (1988). Bacteriophage P22. *In* "The Bacteriophages" (R. Calender, ed.), Vol. 2, pp. 647–677. Plenum Press, New York.

Poteete, A. R. (2004). Modulation of DNA repair and recombination by the bacteriophage lambda Orf function in *Escherichia coli* K-12. *J. Bacteriol.* **186:**2699–2707.

Poteete, A. R. (2008). Involvement of DNA replication in phage lambda Red-mediated homologous recombination. *Mol. Microbiol.* **68:**66–74.

Poteete, A. R. (2009). Expansion of a chromosomal repeat in *Escherichia coli*: Roles of replication, repair, and recombination functions. *BMC Mol. Biol.* **10**:14.
Poteete, A. R., and Fenton, A. C. (1983). DNA-binding properties of the Erf protein of bacteriophage P22. *J. Mol. Biol.* **163**:257–275.
Poteete, A. R., and Fenton, A. C. (1984). Lambda *red*-dependent growth and recombination of phage P22. *Virology* **134**:161–167.
Poteete, A. R., Fenton, A. C., and Semerjian, A. V. (1991). Bacteriophage P22 accessory recombination function. *Virology* **182**:316–323.
Poteete, A. R., Sauer, R. T., and Hendrix, R. W. (1983). Domain structure and quaternary organization of the bacteriophage P22 Erf protein. *J. Mol. Biol.* **171**:401–418.
Prell, A., and Wackernagel, W. (1980). Degradation of linear and circular DNA with gaps by the *recBC* enzyme of *Escherichia coli*: Effects of gap length and the presence of cell-free extracts. *Eur. J. Biochem.* **105**:109–116.
Price-Carter, M., Tingey, J., Bobik, T. A., and Roth, J. R. (2001). The alternative electron acceptor tetrathionate supports B12-dependent anaerobic growth of *Salmonella enterica* serovar typhimurium on ethanolamine or 1,2-propanediol. *J. Bacteriol.* **183**:2463–2475.
Radding, C. M. (1970). The role of exonuclease and beta protein of bacteriophage lambda in genetic recombination. *I. Effects of red mutants on protein structure. J. Mol. Biol.* **52**: 491–499.
Radding, C. M. (1971). The purification of β protein and exonuclease made by phage λ. *Methods Enzymol.* **21**:273–280.
Radding, C. M., Rosensweig, J., Richards, F., and Cassuto, E. (1971). Separation and characterization of l exonuclease, β protein and a complex of both. *J. Biol. Chem.* **246**: 2510–2512.
Radding, C. M., and Shreffler, D. C. (1966). Regulation of lambda exonuclease. II. Joint regulation of exonuclease and a new lambda antigen. *J. Mol. Biol.* **18**:251–261.
Reuven, N., Staire, A., Myers, R., and Weller, S. (2003). The herpes simplex virus type 1 alkaline nuclease and single-stranded DNA binding protein mediate strand exchange *in vitro*. *J. Virol.* **77**:7425–7433.
Rossi, M. S., Paquelin, A., Ghigo, J. M., and Wandersman, C. (2003). Haemophore-mediated signal transduction across the bacterial cell envelope in *Serratia marcescens*: The inducer and the transported substrate are different molecules. *Mol. Microbiol.* **48**:1467–1480.
Runyen-Janecky, L., Daugherty, A., Lloyd, B., Wellington, C., Eskandarian, H., and Sagransky, M. (2008). Role and regulation of iron-sulfur cluster biosynthesis genes in *Shigella flexneri* virulence. *Infect. Immun.* **76**:1083–1092.
Rybalchenko, N., Golub, E. I., Bi, B., and Radding, C. M. (2004). Strand invasion promoted by recombination protein beta of coliphage lambda. *Proc. Natl. Acad. Sci. USA* **101**: 17056–17060.
Sakaki, Y., Karu, A. E., Linn, S., and Echols, H. (1973). Purification and properties of the gamma-protein specified by bacteriophage lambda: An inhibitor of the host RecBC recombination enzyme. *Proc. Natl. Acad. Sci. USA* **70**:2215–2219.
Sauer, B., and Henderson, N. (1988). Site-specific DNA recombination in mammalian cells by the Cre recombinase of bacteriophage P1. *Proc. Natl. Acad. Sci. USA* **85**:5166–5170.
Sauer, B., and Henderson, N. (1989). Cre-stimulated recombination at *loxP*-containing DNA sequences placed into the mammalian genome. *Nucleic Acids Res.* **17**:147–161.
Savage, P. J., Leong, J. M., and Murphy, K. C. (2006). Rapid allelic exchange in enterohemorrhagic *Escherichia coli* (EHEC) and other *E. coli* using lambda *red* recombination. *Curr. Protoc. Microbiol.* **5**:2. Chapter 5.
Sawitzke, J. A., Costantino, N., Li, X. T., Thomason, L. C., Bubunenko, M., Court, C., and Court, D. L. (2010). Probing cellular processes with oligo-mediated recombination and using the knowledge gained to optimize recombineering. *J. Mol. Biol.* **407**:45–59.

Sawitzke, J. A., and Stahl, F. W. (1997). Roles for lambda Orf and *Escherichia coli* RecO, RecR and RecF in lambda recombination. *Genetics* **147:**357–369.

Sawitzke, J. A., Thomason, L. C., Costantino, N., Bubunenko, M., Datta, S., and Court, D. L. (2007). Recombineering: *In vivo* genetic engineering in *E. coli, S. enterica,* and beyond. *Methods Enzymol* **421:**171–199.

Semerjian, A. V., Malloy, D. C., and Poteete, A. R. (1989). Genetic structure of the bacteriophage P22 PL operon. *J. Mol. Biol.* **207:**1–13.

Sharan, S. K., Thomason, L. C., Kuznetsov, S. G., and Court, D. L. (2009). Recombineering: A homologous recombination-based method of genetic engineering. *Nat. Protoc.* **4:**206–223.

Shinohara, A., Shinohara, M., Ohta, T., Matsuda, S., and Ogawa, T. (1998). Rad52 forms ring structures and co-operates with RPA in single-strand DNA annealing. *Genes Cells* **3:** 145–156.

Signer, E. R., and Weil, J. (1968). Recombination in bacteriophage lambda. I. Mutants deficient in general recombination. *J. Mol. Biol.* **34:**261–271.

Smith, C. E., Llorente, B., and Symington, L. S. (2007). Template switching during break-induced replication. *Nature* **447:**102–105.

Sriprakash, K. S., Lundh, N., Huh, M.-O., and Radding, C. M. (1975). The specificity of lambda exonuclease: Interactions with single-stranded DNA. *J. Biol. Chem.* **250:**5438–5445.

Stahl, F. W. (1998). Recombination in phage lambda: One geneticist's historical perspective. *Gene* **223:**95–102.

Stahl, F. W., McMilin, K. D., Stahl, M. M., Malone, R. E., Nozu, Y., and Russo, V. E. (1972a). A role for recombination in the production of "free-loader" lambda bacteriophage particles *J. Mol. Biol.* **68:**57–67.

Stahl, F. W., McMilin, K. D., Stahl, M. M., and Nozu, Y. (1972b). An enhancing role for DNA synthesis in formation of bacteriophage lambda recombinants. *Proc. Natl. Acad. Sci. USA* **69:**3598–3601.

Stahl, F. W., and Stahl, M. M. (1971). DNA synthesis associated with recombination, II. Recombination between repressed chromosomes. *In* "The Bacteriophage Lambda" (A. D. Hershey, ed.), pp. 443–453. Cold Spring Harbor Laboratory, Cold Spring Harbor, NY.

Stahl, M. M., Thomason, L., Poteete, A. R., Tarkowski, T., Kuzminov, A., and Stahl, F. W. (1997). Annealing vs. invasion in phage lambda recombination. *Genetics* **147:**961–977.

Stanley, T. L., Ellermeier, C. D., and Slauch, J. M. (2000). Tissue-specific gene expression identifies a gene in the lysogenic phage Gifsy-1 that affects *Salmonella enterica* serovar typhimurium survival in peyer's patches. *J. Bacteriol.* **186:**4406–4413.

Subramanian, K., Rutvisuttinunt, W., Scott, W., and Myers, R. S. (2003). The enzymatic basis of processivity in lambda exonuclease. *Nucleic Acids Res.* **31:**1585–1596.

Sun, W., Wang, S., and Curtiss, R., 3rd (2008). Highly efficient method for introducing successive multiple scarless gene deletions and markerless gene insertions into the *Yersinia pestis* chromosome. *Appl. Environ. Microbiol.* **74:**4241–4245.

Sung, B. H., Lee, C. H., Yu, B. J., Lee, J. H., Lee, J. Y., Kim, M. S., Blattner, F. R., and Kim, S. C. (2006). Development of a biofilm production-deficient *Escherichia coli* strain as a host for biotechnological applications. *Appl. Environ. Microbiol.* **72:**3336–3342.

Susskind, M. M., and Botstein, D. (1978). Molecular genetics of bacteriophage P22. *Microbial. Rev.* **42:**385–413.

Swingle, B., Bao, Z., Markel, E., Chambers, A., and Cartinhour, S. (2010). Recombineering using RecTE from *Pseudomonas syringae. Appl. Environ. Microbiol.* **76:**4960–4968.

Swingle, B., Markel, E., Costantino, N., Bubunenko, M. G., Cartinhour, S., and Court, D. L. (2009). Oligonucleotide recombination in Gram-negative bacteria. *Mol. Microbiol.* **75:** 138–148.

Takano, T. (1966). Behavoir of some episomal elements in a recombination-deficient mutant of *Escherichia coli. Jpn. J. Microbiol.* **10:**201–210.

Tarkowski, T. A., Mooney, D., Thomason, L. C., and Stahl, F. W. (2002). Gene products encoded in the *ninR* region of phage lambda participate in Red-mediated recombination. *Genes Cells* **7**:351–363.

Taylor, A. F., and Smith, G. R. (1985). Substrate specificity of the DNA unwinding activity of the RecBC enzyme of *Escherichia coli*. *J. Mol. Biol.* **185**:431–443.

Templin, A., Kushner, S. R., and Clark, A. J. (1972). Genetic analysis of mutations indirectly suppressing *recB* and *recC* mutations. *Genetics* **72**:105–115.

Thaler, D. S., Sampson, E., Siddiqi, I., Rosenberg, S. M., Thomason, L. C., Stahl, F. W., and Stahl, M. M. (1989). Recombination of bacteriophage lambda in *recD* mutants of *Escherichia coli*. *Genome* **31**:53–67.

Thomason, L., Court, D. L., Bubunenko, M., Costantino, N., Wilson, H., Datta, S., and Oppenheim, A. (2007a). Recombineering: Genetic engineering in bacteria using homologous recombination. *Curr. Protoc. Mol. Biol.* **1**:16. Chapter 1.

Thomason, L. C., Costantino, N., Shaw, D. V., and Court, D. L. (2007b). Multicopy plasmid modification with phage lambda Red recombineering. *Plasmid* **58**:148–158.

Thresher, R. J., Makhov, A. M., Hall, S. D., Kolodner, R., and Griffith, J. D. (1995). Electron microscopic visualization of RecT protein and its complexes with DNA. *J. Mol. Biol.* **254**:364–371.

Tolun, G. (2007). More Than the Sum of Its Parts: Physical and Mechanistic Coupling in the Phage Lambda Red Recombinase. Ph.D. thesis, University of Miami, FL.

Tong, A. H., Lesage, G., Bader, G. D., Ding, H., Xu, H., Xin, X., Young, J., Berriz, G. F., Brost, R. L., Chang, M., Chen, Y., Cheng, X., *et al.* (2004). Global mapping of the yeast genetic interaction network. *Science* **303**:808–813.

Typas, A., Nichols, R. J., Siegele, D. A., Shales, M., Collins, S. R., Lim, B., Braberg, H., Yamamoto, N., Takeuchi, R., Wanner, B. L., Mori, H., Weissman, J. S., *et al.* (2008). High-throughput, quantitative analyses of genetic interactions in *E. coli*. *Nat. Methods* **5**:781–787.

Uzzau, S., Figueroa-Bossi, N., Rubino, S., and Bossi, L. (2001). Epitope tagging of chromosomal genes in *Salmonella*. *Proc. Natl. Acad. Sci. USA* **98**:15264–15269.

van de Putte, P., Zwenk, H., and Rorsch, A. (1966). Properties of four mutants of *Escherichia coli* defective in genetic recombination. *Mutat. Res.* **3**:381–392.

van Kessel, J. C., and Hatfull, G. F. (2007). Recombineering in *Mycobacterium tuberculosis*. *Nat. Methods* **4**:147–152.

van Kessel, J. C., and Hatfull, G. F. (2008). Efficient point mutagenesis in mycobacteria using single-stranded DNA recombineering: Characterization of antimycobacterial drug targets. *Mol. Microbiol.* **67**:1094–1107.

van Kessel, J. C., Marinelli, L. J., and Hatfull, G. F. (2008). Recombineering mycobacteria and their phages. *Nat. Rev. Microbiol.* **6**:851–857.

Vellani, T. S., and Myers, R. S. (2003). Bacteriophage SPP1 Chu is an alkaline exonuclease in the SynExo family of viral two-component recombinases. *J. Bacteriol.* **185**:2465–2474.

Vetcher, L., Tian, Z. Q., McDaniel, R., Rascher, A., Revill, W. P., Hutchinson, C. R., and Hu, Z. (2005). Rapid engineering of the geldanamycin biosynthesis pathway by Red/ET recombination and gene complementation. *Appl. Environ. Microbiol.* **71**:1829–1835.

Wang, H. H., Isaacs, F. J., Carr, P. A., Sun, Z. Z., Xu, G., Forest, C. R., and Church, G. M. (2009). Programming cells by multiplex genome engineering and accelerated evolution. *Nature* **460**:894–898.

Watt, R. M., Wang, J., Leong, M., Kung, H. F., Cheah, K. S., Liu, D., Danchin, A., and Huang, J. D. (2007). Visualizing the proteome of *Escherichia coli*: An efficient and versatile method for labeling chromosomal coding DNA sequences (CDSs) with fluorescent protein genes. *Nucleic Acids Res.* **35**:e37.

Weaver, S., and Levine, M. (1977). Recombinational circularization of *Salmonella* phage P22 DNA. *Virology* **76**:29–38.

Wenzel, S. C., Gross, F., Zhang, Y., Fu, J., Stewart, A. F., and Muller, R. (2005). Heterologous expression of a myxobacterial natural products assembly line in pseudomonads via Red/ ET recombineering. *Chem. Biol.* **12:**349–356.

Whitesell, L., Mimnaugh, E. G., De Costa, B., Myers, C. E., and Neckers, L. M. (1994). Inhibition of heat shock protein HSP90-pp 60v-src heteroprotein complex formation by benzoquinone ansamycins: Essential role for stress proteins in oncogenic transformation. *Proc. Natl. Acad. Sci. USA* **91:**8324–8328.

Wiles, T. J., Dhakal, B. K., Eto, D. S., and Mulvey, M. A. (2008). Inactivation of host Akt/ protein kinase B signaling by bacterial pore-forming toxins. *Mol. Biol. Cell.* **19:**1427–1438.

Willis, D. K., Satin, L. H., and Clark, A. J. (1985). Mutation-dependent suppression of *recB21 recC22* by a region cloned from the Rac prophage of *Escherichia coli* K-12. *J. Bacteriol.* **162:**1166–1172.

Worlock, A. J., and Smith, R. L. (2002). ZntB is a novel Zn2+ transporter in *Salmonella enterica* serovar typhimurium. *J. Bacteriol.* **184:**4369–4373.

Wu, Z., Xing, X., Bohl, C. E., Wisler, J. W., Dalton, J. T., and Bell, C. E. (2006). Domain structure and DNA binding regions of beta protein from bacteriophage lambda. *J. Biol. Chem.* **281:**25205–25214.

Yamamoto, N., Nakahigashi, K., Nakamichi, T., Yoshino, M., Takai, Y., Touda, Y., Furubayashi, A., Kinjyo, S., Dose, H., Hasegawa, M., Datsenko, K. A., Nakayashiki, T., *et al.* (2009). Update on the Keio collection of *Escherichia coli* single-gene deletion mutants. *Mol. Syst. Biol.* **5:**335.

Yang, Y., and Sharan, S. K. (2003). A simple two-step, 'hit and fix' method to generate subtle mutations in BACs using short denatured PCR fragments. *Nucleic Acids Res.* **31:**e80.

Yu, D., Ellis, H. M., Lee, E. C., Jenkins, N. A., Copeland, N. G., and Court, D. L. (2000). An efficient recombination system for chromosome engineering in *Escherichia coli. Proc. Natl. Acad. Sci. USA* **97:**5978–5983.

Yuan, L. Z., Rouviere, P. E., Larossa, R. A., and Suh, W. (2006). Chromosomal promoter replacement of the isoprenoid pathway for enhancing carotenoid production in *E. coli. Metab. Eng.* **8:**79–90.

Zhang, J., Xing, X., Herr, A. B., and Bell, C. E. (2009). Crystal structure of *E. coli* RecE protein reveals a toroidal tetramer for processing double-stranded DNA breaks. *Structure* **17:**690–702.

Zhang, J., McCabe, K. A., and Bell, C. E. (2011). Crystal structures of lambda exonuclease in complex with DNA suggest an electrostatic ratchet mechanism for processivity. *Proc. Natl. Acad. Sci. USA 2011,* **108:**11872–11877.

Zhang, Y., Buchholz, F., Muyrers, J. P., and Stewart, A. F. (1998). A new logic for DNA engineering using recombination in *Escherichia coli. Nat. Genet.* **20:**123–128.

Zhang, Y., Muyrers, J. P., Testa, G., and Stewart, A. F. (2000). DNA cloning by homologous recombination in *Escherichia coli. Nat. Biotechnol.* **18:**1314–1317.

Zinder, N. D., and Lederberg, J. (1952). Genetic exchange in *Salmonella. J. Bacteriol.* **64:**679–699.

Zissler, J., Singer, E., and Schaefer, F. (1971). The role of recombination in the growth of bacteroiophage λ. I. The Gamma gene. *In* "The Bacteriophage Lambda" (A. D. Hershey, ed.), pp. 455–468. Cold Spring Harbor Laboratory, Cold Spring Harbor, NY.

Zurawski, D. V., Mitsuhata, C., Mumy, K. L., McCormick, B. A., and Maurelli, A. T. (2006). OspF and OspC1 are *Shigella flexneri* type III secretion system effectors that are required for postinvasion aspects of virulence. *Infect. Immun.* **74:**5964–5976.

CHAPTER 9

"Push Through One-Way Valve" Mechanism of Viral DNA Packaging

Hui Zhang, Chad Schwartz, Gian Marco De Donatis, and **Peixuan Guo**[1]

Contents

Nanobiotechnology Center, Department of Pharmaceutical Sciences, and Markey Cancer Center, University of Kentucky, Lexington, Kentucky, USA
[1] Corresponding author, E-mail: peixuan.guo@uky.edu

Advances in Virus Research, Volume 83
ISSN 0065-3527, DOI: 10.1016/B978-0-12-394438-2.00009-8

Abstract

Double-stranded (ds)DNA viruses package their genomic DNA into a procapsid using a force-generating nanomotor powered by ATP hydrolysis. Viral DNA packaging motors are mainly composed of the connector channel and two DNA packaging enzymes. In 1998, it was proposed that viral DNA packaging motors exercise a mechanism similar to the action of AAA+ ATPases that assemble into ring-shaped oligomers, often hexamers, with a central channel (Guo *et al.* Molecular Cell, 2:149). This chapter focuses on the most recent findings in the bacteriophage ϕ29 DNA packaging nanomotor to address this intriguing notion. Almost all dsDNA viruses are composed entirely of protein, but in the unique case of ϕ29, packaging RNA (pRNA) plays an intermediate role in the packaging process. Evidence revealed that DNA packaging is accomplished via a "push through one-way valve" mechanism. The ATPase gp16 pushes dsDNA through the connector channel section by section into the procapsid. The dodecameric connector channel functions as a one-way valve that only allows dsDNA to enter but not exit the procapsid during DNA packaging. Although the roles of the ATPase gp16 and the motor connector channel are separate and independent, pRNA bridges these two components to ensure the coordination of an integrated motor. ATP induces a conformational change in gp16, leading to its stronger binding to dsDNA. Furthermore, ATP hydrolysis led to the departure of dsDNA from the ATPase/dsDNA complex, an action used to push dsDNA through the connector channel. It was found unexpectedly that by mutating the basic lysine rings of the connector channel or by changing the pH did not measurably impair DNA translocation or affect the one-way traffic property of the channel, suggesting that the positive charges in the lysine ring are not essential in gearing the dsDNA. The motor channel exercises three discrete, reversible, and controllable steps of gating, with each step altering the channel size by 31% to control the direction of translocation of dsDNA. Many DNA packaging models have been contingent upon the number of base pairs packaged per ATP relative to helical turns for B-type DNA. Both 2 and 2.5 bp per ATP have been used to argue for four, five, or six discrete steps of DNA translocation. The "push through one-way valve" mechanism renews the perception of dsDNA packaging energy calculations and provides insight into the discrepancy between 2 and 2.5 bp per ATP. Application of the DNA packaging motor in nanotechnology and nanomedicine is also addressed. Comparison with nine other DNA packaging models revealed that the "push through one-way valve" is the most agreeable mechanism to interpret most of the findings that led to historical models. The application of viral DNA packaging motors is also discussed.

I. INTRODUCTION

DNA packaging into a preformed protein shell (procapsid) is characteristic of double-stranded (ds)DNA viruses, including bacteriophages, herpesviruses, and adenovirues. Most viral procapsids are a few tens of nanometers in diameter, whereas viral genomes are several micrometers in length. Electron microscopy (EM) images revealed that packaged viral DNA inside the small procapsid can be condensed to 500 mg/ml, comparable to near-liquid crystalline density (Earnshaw and Casjens, 1980; Hohn, 1976). High internal pressure exists inside the procapsid during DNA packaging (Fuller *et al.*, 2007a,b; Rickgauer *et al.*, 2008; Smith *et al.*, 2001); this energetically unfavorable process is accomplished by packaging motors powered by ATP hydrolysis through a DNA and procapsid-dependent ATPase (Guo *et al.*, 1987d).

These natural, nanometer-scale DNA packaging motors are of special interest because of their inherent abilities to recognize specific viral genomes in a pool of DNAs and to package them into such a tiny space against a large internal force. Over the years, studies on the DNA packaging motor have focused primarily on fundamental aspects, including structure, biological/biochemical function, and mechanical or physical behaviors of the viral motor or its components for genome packaging. More recently, powerful biological motors have inspired novel biomimetic designs that have opened up possibilities for building artificial nanomotors operable outside their natural environment for use in nanodevices, nanomedicine, including the sensing of ions, chemicals, or DNA/RNA (Fang *et al.*, 2012; Geng *et al.*, 2011; Hess and Vogel, 2001; Jing *et al.*, 2010a,b; Soong *et al.*, 2000; Wendell *et al.*, 2009), and targeted gene delivery or drug loading (Guo, 2010; Guo *et al.*, 2005a, 2010; Khaled *et al.*, 2005; Liu *et al.*, 2007; Shu *et al.*, 2011b; Zhou *et al.*, 2011). A thorough understanding of how the motor components interact with each other during the packaging process and how the energy from ATP hydrolysis is transferred into physical motion would provide valuable insights into fundamental phenomena and the development and application of nanomotor biomimetics. In addition, possible novel targets for antiviral therapy could be discovered based on studies of the viral DNA packaging mechanism (Bogner, 2002; Trottier *et al.*, 1996; Visalli and van Zeijl , 2003; Zhang *et al.*, 1995). Utilizing motor components in nanotechnology and/or disease treatments has also been pursued actively (Geng *et al.*, 2011; Guo, 2010; Guo *et al.*, 2005a, 2010; Jing *et al.*, 2010a,b; Khaled *et al.*, 2005; Liu *et al.*, 2007; Shu *et al.*, 2011b; Wendell *et al.*, 2009; Zhou *et al.*, 2011).

Various packaging mechanisms have been proposed throughout the years (Black, 1981; Chen and Guo, 1997; Grimes and Anderson, 1997; Guasch *et al.*, 2002; Guo *et al.*, 1998; Hendrix, 1978; Moffitt *et al.*, 2009; Morita *et al.*, 1995; Serwer, 2003, 2010; Sun *et al.*, 2010; Yu *et al.*, 2010). Several thorough reviews on the packaging of dsDNA viruses (Casjens, 2011; Guo and Lee, 2007; Rao and Feiss, 2008; Serwer, 2003, 2010; Sun *et al.*, 2010) and ϕ29 (Grimes *et al.*, 2002; Guo, 2002; Guo and Trottier, 1994; Lee *et al.*, 2009; Yu *et al.*, 2010) have been published. This chapter focuses primarily on recent publications of studies on the structure, function, and mechanism of the ϕ29 DNA packaging motor, but will also compare other phages to this unique system.

II. STRUCTURE OF VIRAL DNA PACKAGING MOTORS

The packaging motor of dsDNA viruses consists of a motor channel for dsDNA translocation (Fig. 1) and two DNA packaging components that are not fixed components in the purified procapsid. These components were classified in 1987 (Guo *et al.*, 1987d) into two categories according to their role in DNA packaging: the first category is the large subunit responsible for binding to the procapsid and contains an ATP-binding consensus sequence for ATP binding and hydrolysis and the second category is the smaller subunit that interacts with DNA (Guo *et al.*, 1987d). In bacteriophage ϕ29, the motor system uniquely involves an RNA ring (Guo *et al.*, 1987b).

A. Motor channel as a one-way valve for unidirectional DNA translocation

The bacteriophage connector is embedded in the procapsid to connect the viral head to the tail. This essential motor component is composed of 12 copies of connector proteins. Detailed three-dimensional structures of

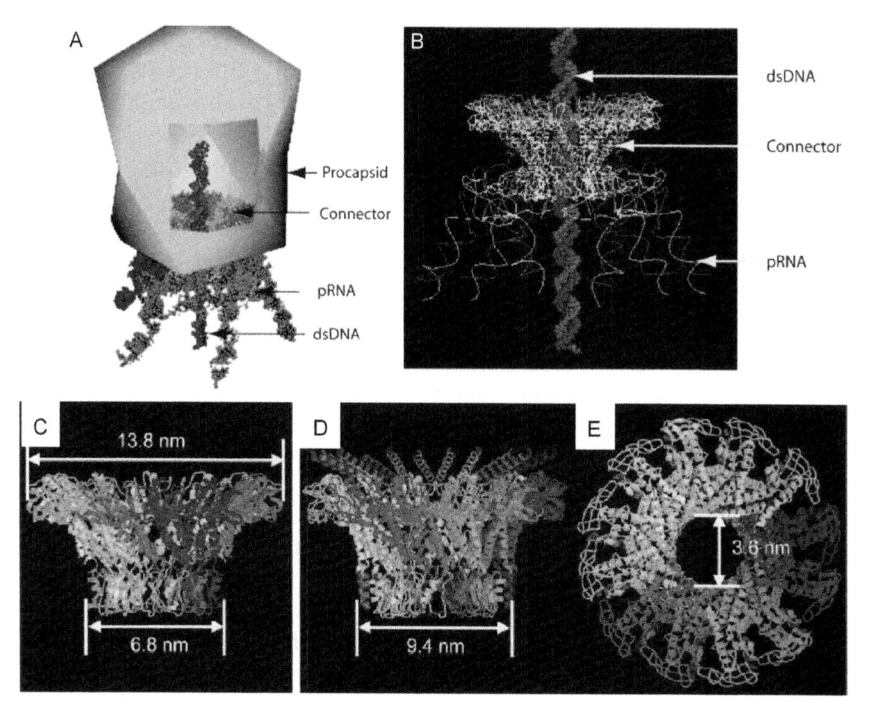

FIGURE 1 The ϕ29 viral DNA packaging motor structure. (A) Illustrated structure of motor *in vivo*. (B) Three-dimensional structure of ϕ29 connector/dsDNA complex showing counterchirality between left-handed connector channel wall and right-handed dsDNA (Jing *et al*., 2010a). (C) Side view of the motor connector without C- and N-terminal fragments (Guasch *et al*., 2002). Side view (D) and bottom view (E) with complete protein sequence (Guo *et al*., 2005b). Adapted with permission from Jing *et al*. (2010a) ©2010 American Chemical Society and from Geng *et al*. (2011) ©2011 with permission from Elsevier. (See Page 9 in Color Section at the back of the book.)

connectors in different phages have been studied and determined at high resolution through both cryo-EM and X-ray crystallography (Agirrezabala *et al.*, 2005; Cingolani *et al.*, 2002; Fokine *et al.*, 2004; Guasch *et al.*, 2002; Jiang *et al.*, 2006; Jimenez *et al.*, 1986; Lander *et al.*, 2006; Lurz *et al.*, 2001; Simpson *et al.*, 2000; Trus *et al.*, 2004). The connectors in various bacteriophages display similar morphology (Agirrezabala *et al.*, 2005; Valpuesta and Carrascosa, 1994), although the sequence alignment showed a lack of homology among them.

The ϕ29 connector is assembled from 12 copies of connector protein gp10 and forms a truncated cone shape (Jimenez *et al.*, 1986). The narrow end, with a diameter of 6.8 nm, extrudes out of the procapsid, while the wider end, with a diameter of 13.8 nm, is buried inside the procapsid (Fig. 1) (Guasch *et al.*, 2002; Simpson *et al.*, 2001). The N-terminal 14 amino acids of gp10 serve as the foothold for packaging RNA (pRNA) of φ29 (Sun *et al.*, 2006; Xiao *et al.*, 2005), as proved by cross-linking (Garver and Guo, 1997). Further point mutation studies of the 14 amino acids revealed that the three basic amino acids, Arg-Lys-Arg, at positions 3–5 of the N terminus of gp10 were responsible for pRNA binding. Mutation of any two of these three amino acids resulted in complete abolishment of pRNA binding to the DNA packaging motor (Atz *et al.*, 2007). In the case of φ29, pRNA most likely serves as the foothold to which the packaging ATPase gp16 binds to exercise its force in pushing the DNA into the procapsid (Lee and Guo, 2006).

Structural analysis showed that the connector contains three layers consisting of two hydrophilic layers separated by a hydrophobic layer. Utilizing this characteristic, the φ29 connector was embedded successfully into a planar lipid membrane, and dsDNA translocation through the channel was studied by electrophysiological measurements (Fang *et al.*, 2012; Jing *et al.*, 2010a,b; Wendell *et al.*, 2009) (Fig. 2). During translocation, DNA physically obstructs the connector channel, and the observed signal indicates a blockage of current. Measurements of dsDNA translocation through connectors in the membrane proved that translocation only occurred in one direction (Figs. 3 and 4). It was also found that the frequency of DNA translocation events changes in agreement with the different orientations

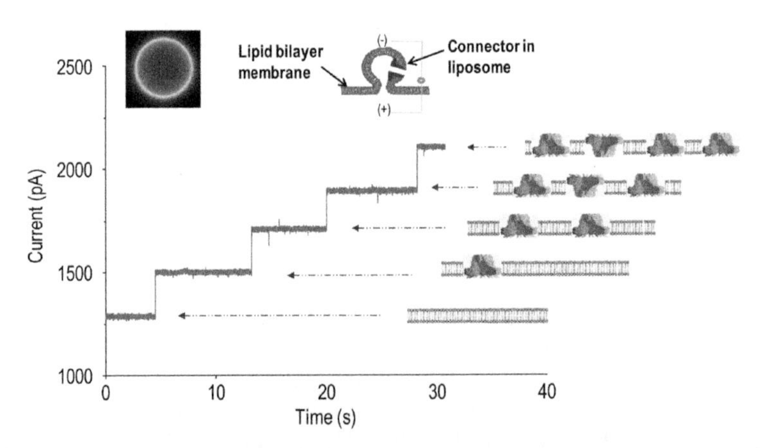

FIGURE 2 Insertion of the re-engineered φ29 connector into lipid membrane and demonstration of robust conductivity property. Red trace: Discrete steps of current representing the insertion of one connector for each step. All these current steps are similar in conductance, indicating the homogeneity of the channel size. (Inserts) Fluorescence image of the fluorescently-labeled connector in the liposome membrane. Connector insertion into the bilayer only occurred when connector-reconstituted proteoliposomes were fused into the bilayer. Adapted with permission from Macmillan Publishers Ltd. (Wendall et al., 2009) ©2009. (See Page 10 in Color Section at the back of the book.)

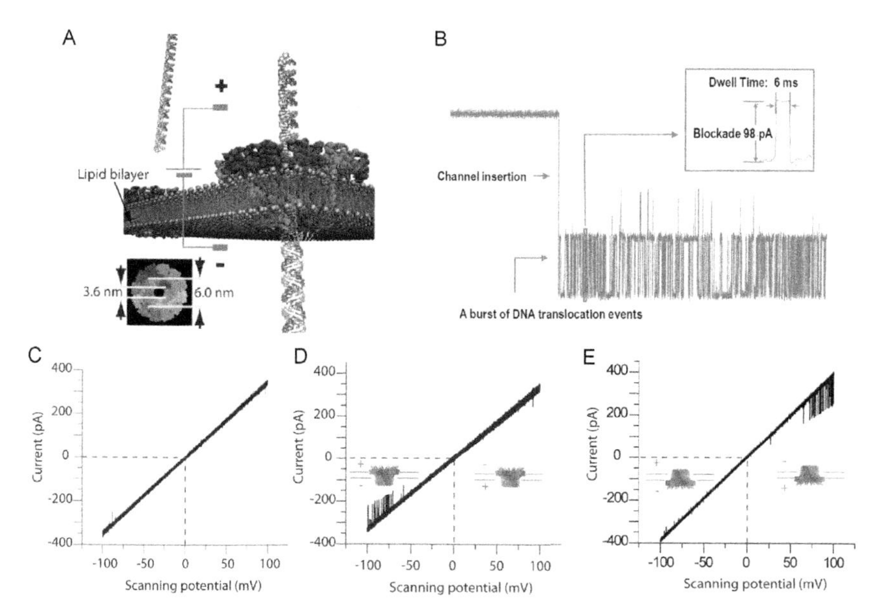

FIGURE 3 The connector acts as a gate allowing DNA passage in one direction. (A) Illustration of DNA translocation through membrane-embedded connector channel. (B) Translocation of linear DNA induced numerous current blockades. (C–E) One-way traffic of DNA translocation through a single connector channel under a ramping potential in conditions without DNA (C) and with DNA in both chambers (D and E). Adapted with permission from Jing *et al.* (2010a) ©2010 American Chemical Society. (See Page 10 in Color Section at the back of the book.)

of the connector when multiple connectors were fused in the membrane. DNA translocation frequency remained the same when additional connectors were inserted with an opposite orientation for DNA traversion or increased if the additional connector was inserted with the same orientation (Fig. 5). Using antibody or gold particles to bind to the tag-conjugated C-terminal, proper orientation of the connector for DNA translocation was probed to be from the N terminus to the C terminus, the same direction that DNA traverses during packaging. In addition, packaged viral dsDNA remained within the procapsid under the strong force of centrifugation (Shu and Guo, 2003b). All the facts strongly support the conclusion that the ϕ29 connector exercises a one-way traffic property for dsDNA packaging into procapsid from the N terminus to the C terminus of the connector (Jing *et al.*, 2010a). Logistically, this brings up an intriguing question of how DNA is expelled from the virus during the infection process; further research revealed that the connector exercises three discrete steps of conformational changes in regulating the direction of DNA translocation (Geng *et al.*, 2011).

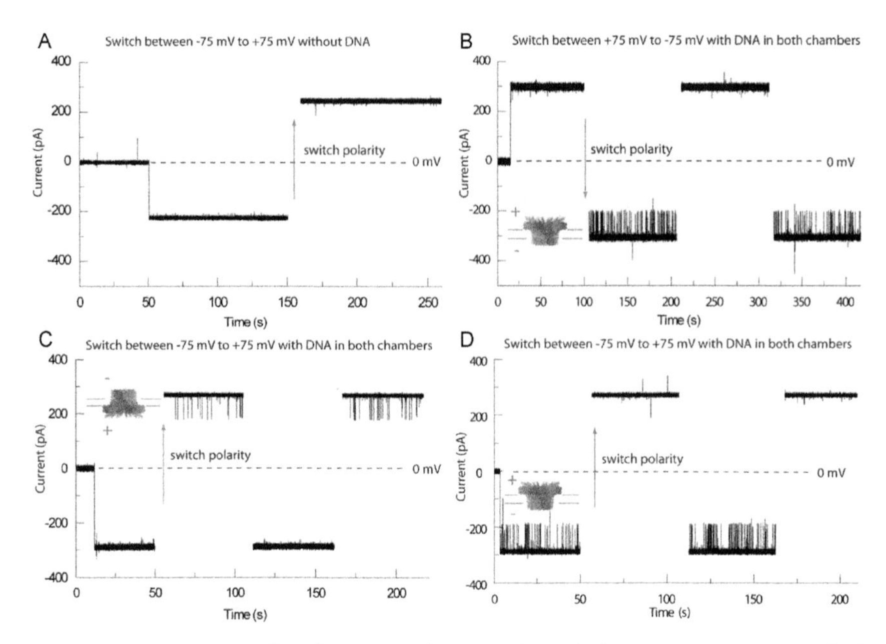

FIGURE 4 One-way traffic of DNA translocation through the connector pore verified by switching polarity. (A) In the absence of DNA. (B–D) In the presence of DNA in both chambers where a single connector was inserted. Adapted with permission from Jing *et al.* (2010a) ©2010 American Chemical Society. (See Page 11 in Color Section at the back of the book.)

B. Effect of pH, channel charge, and lysine ring mutations on DNA translocation

The ϕ29 connector channel was found to be robust and stable under even extreme pH conditions (Jing *et al.*, 2010b). Single pore conductance assays of membrane-embedded connector showed that the connector remains open with uniform channel conductance under extreme pH and that both conductance and membrane insertion orientation of the channel are independent of the pH (Fang *et al.*, 2012). While DNA translocation events were still observed at these extreme conditions, formation of apurinic acid at pH 2 led to shorter current blockade events. Overall, the connector retained its stable channel properties under strong acidic or alkaline conditions, despite the inherent effect on DNA structure.

The structural study of the ϕ29 connector showed that there are 48 positively charged lysine residues in the inner channel (four 12-lysine rings from the 12 gp10 subunits). It was believed that these positively charged rings could play an important role in DNA translocation through the channel in that they may interact with the negatively charged phosphate backbone of dsDNA during DNA translocation (Agirrezabala *et al.*, 2005; Guasch *et al.*, 2002). Different mutations were introduced to the

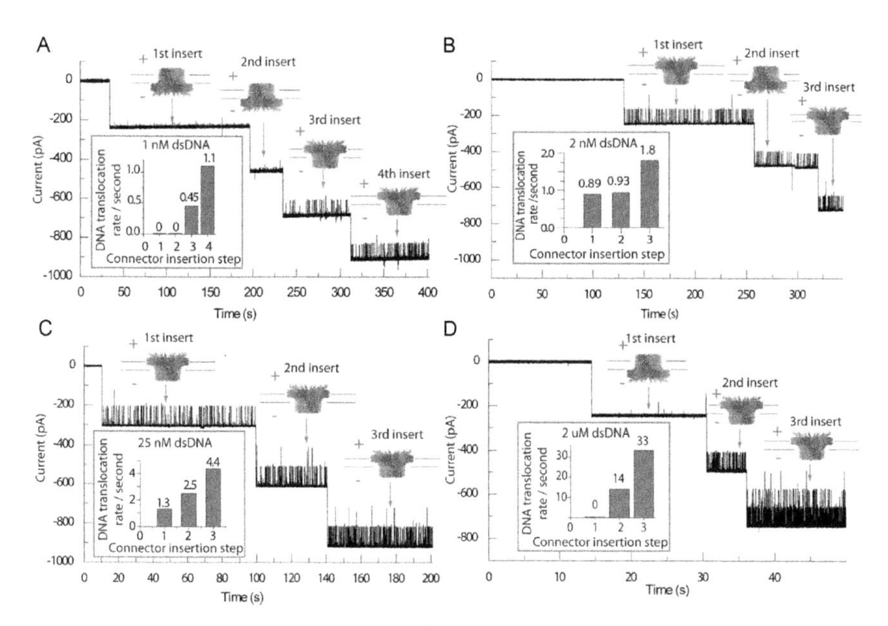

FIGURE 5 Change of DNA translocation frequency in the presence of multiple connectors reflects the one-way traffic pattern of DNA translocation. (Insets a–d) DNA translocation frequency after insertion of each connector channel in the BLM for individual experiments. Adapted with permission from Jing *et al.* (2010a) ©2010 American Chemical Society. (See Page 11 in Color Section at the back of the book.)

lysine residues (Fig. 6) in order to study the effect of lysine rings (Fang *et al.*, 2012). The effect on channel size, procapsid assembly, dsDNA translocation, and connector outer surface charge distribution were assessed through their interaction with the lipid membrane by single pore conductance, direction of dsDNA translocation, efficiency of DNA packaging, and production of infectious virions. The channel size of the lysine mutant connectors can be deduced from the single channel conductance assays. Upon connector insertion into the lipid membrane, discrete current jumps were observed under an applied voltage, representing the open-pore current amplitude. When DNA passes through the channel, the capacity of the electrolyte ion passage is reduced, resulting in transient current blockade events. Because the diameter of dsDNA is 2 nm and the size of the narrowest region of the wild-type connector channel is 3.6 nm, the ratio of the cross-sectional area (represented by the ratio of the open-pore current and the current during the DNA blockade) can be used as a parameter to estimate the channel size at the narrowest point. Furthermore, these assays were performed in both acidic and basic environments to investigate the role of lysine residues in dsDNA translocation and the direction of dsDNA trafficking (Fig. 7). Results indicated that the four lysine rings within the ϕ29 connector channel are not involved

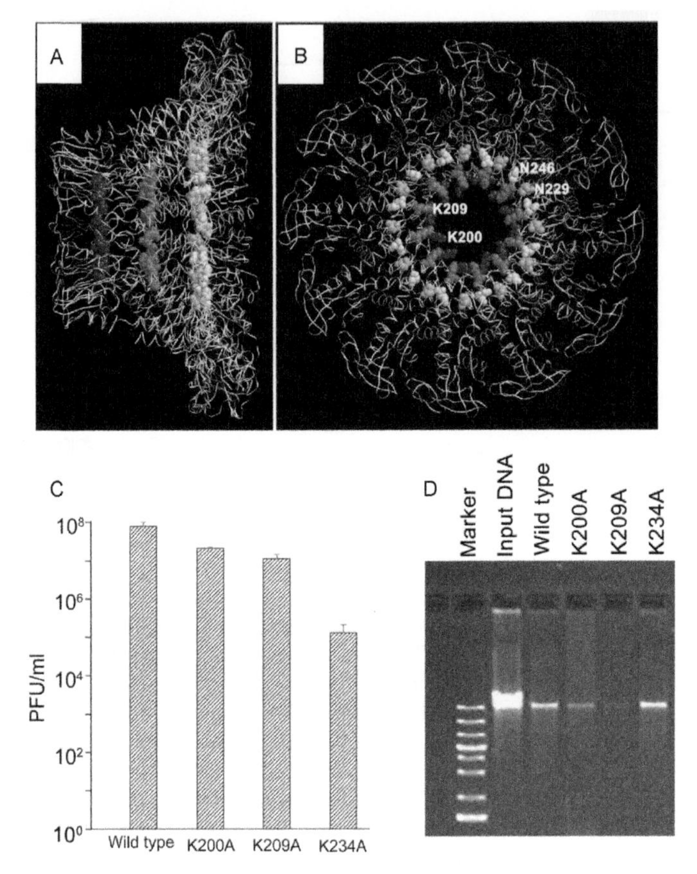

FIGURE 6 Effect of mutations in lysine rings within the connector channel on motor activity. (a and b) Locations of mutations within the connector channel in side view (a) and top view (b). K200 (blue), K209 (red), and the border of tunnel loop N229 (cyan) and N246 (yellow). (c) Virus assembly and (d) DNA packaging activity of procapsids bearing the mutated connectors. Adapted from Fang *et al.* (2012) ©2011 with permission from Elsevier. (See Page 12 in Color Section at the back of the book.)

in the active translocation of dsDNA (Fang *et al.*, 2012) and supported the "push through a one-way valve" model of dsDNA packaging mechanism (Section IV) .

It should be noted that mutation some of the lysine residues can lead to refolding of gp10 and change the charge of the external surface of the connector due to conformational change (Fang *et al.*, 2012). Such change of the external surface charge has hindered the efficiency in serving as a nucleus in procapsid assembly (K200A, K209A, or both) in which the connector acts as a nucleating core (Fu and Prevelige, 2009; Guo *et al.*, 1991; Lee and Guo, 1995b) and in insertion into lipid membrane for

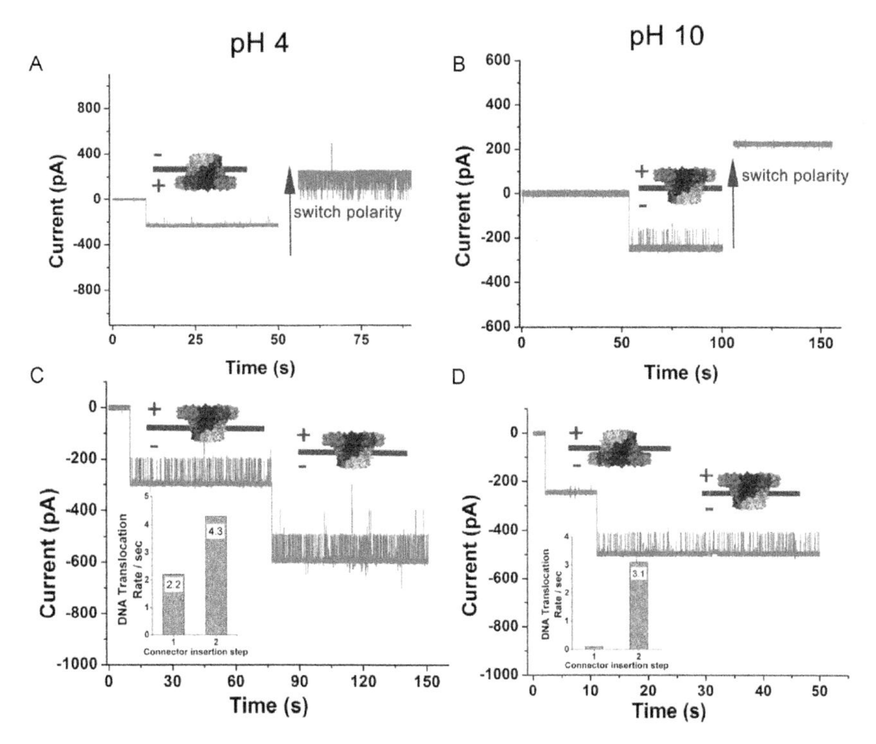

FIGURE 7 One-way traffic property in DNA translocation through connector was not affected by pH. Single pore assay by switching the electrode polarity at pH 4 (a) and pH 10 (b). Comparison of DNA translocation rates at pH 4 (c) and pH 10 (d). Adapted from Fang *et al.* (2012) ©2011 with permission from Elsevier. (See Page 13 in Color Section at the back of the book.)

the measurement of single channel conductance (K200A/K209A)(Fang *et al.*, 2012). Such low efficiency due to the change of external surface charge did not affect the conclusion related to the internal surface of the channel for DNA translocation (Fang *et al.*, 2012).

C. DNA-pushing ATPase enzyme as a member of the AAA+ family

When evidence of the pRNA hexamer was uncovered, it was proposed (Guo *et al.*, 1998), and subsequently supported by other authors (de Haas *et al.*, 1999; Guasch *et al.*, 2002), that viral DNA packaging is similar to the mechanism in DNA replication and RNA transcription and that the mechanism responsible for those important phenomena can be correlated to the mechanism of viral DNA packaging. Moreover, like almost all DNA and RNA packaging motors, it was hypothesized that gp16, the ATPase in the ϕ29 packaging motor, belonged to the superfamily of AAA+ proteins (ATPases Associated with many cellular Activities) (Guo *et al.*, 1998; Lee *et al.*, 2006).

This large family of proteins is extremely diverse in function associated with a multitude of different cellular activities and implied in many others. However, the common characteristic of this family is their ability to convert chemical energy from the hydrolysis of the γ-phosphate bond of ATP into a conformational change inside the protein. This change of conformation generates a loss of affinity for the substrate and a mechanical movement, which is used to make or break contacts between macromolecules, resulting in local or global protein unfolding, assembly or disassembly of complexes, or transport of macromolecules relative to each other. These activities underlie processes critical to DNA replication and recombination, chromosome secretion, membrane sorting, cellular reorganization, and many others (Maurizi and Li, 2001).

Many biochemical and structural aspects of reactions catalyzed by AAA+ proteins have been elucidated, together with interesting allosteric phenomena that occur during ATP hydrolysis. For instance, the crystal structure of sliding clamp loader complex, a system (AAA+ DNA helicase) that helps polymerases overcome the problem of torque generated during the extension of helical dsDNA, has been useful in determining how ATP is required for binding to dsDNA and opening of the clamp (Oyama *et al.*, 2001). The crystal model reveals a spiral structure in the clamp loader with a striking correlation to the grooves of helical dsDNA, suggesting a simple explanation for how the loader/DNA helix interaction triggers ATP hydrolysis and how DNA is released from the sliding clamp. This mechanism may provide hints for understanding the role of the DNA packaging ATPase in many phages.

Notably, structural studies have proven that AAA+ proteins often assemble into homohexameric complexes with a ring structure that acts in a coordinated fashion (Fig. 8) (Maurizi *et al.*, 2001). In past years though, models based on EM reconstruction have suggested that gp16 exists as a pentameric structure (Moffitt *et al.*, 2009; Morais *et al.*, 2008).

D. Packaging RNA as a hinge to gear the motor

Different from other viral systems, an RNA molecule, named pRNA, was discovered to be an essential component in the ϕ29 DNA packaging motor (Guo *et al.*, 1987b). This pRNA was demonstrated to be indispensible in DNA packaging and viral assembly. The detailed structure and

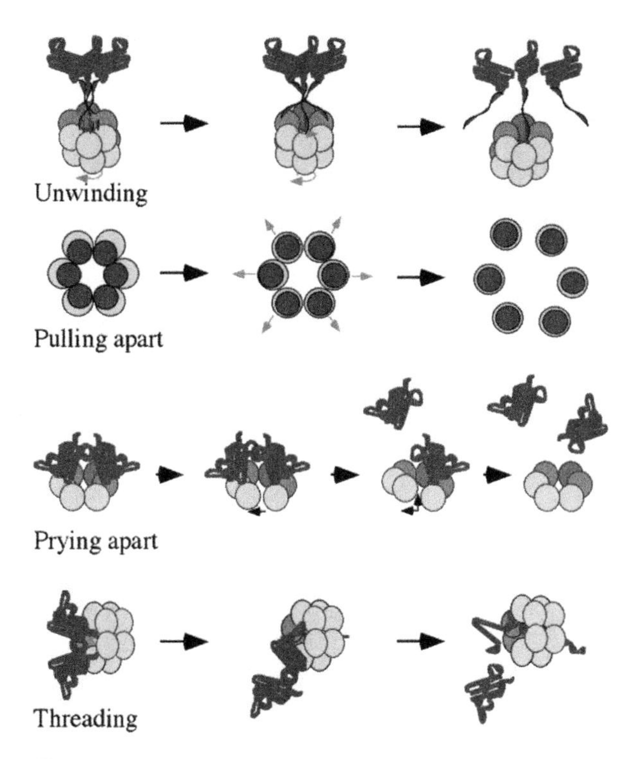

FIGURE 8 Different mechanisms proposed for AAA proteins. Adapted with permission from Macmillan Publishers Ltd. (Maurizi *et al.*, 2001) ©2001. (See Page 13 in Color Section at the back of the book.)

function studies of pRNA has been reviewed thoroughly (Guo, 2002). It has also been found that pRNA mediates binding of the motor ATPase gp16 to procapsid, indicating that the role of pRNA in motor function may be to serve as a hinge to connect motor components and thus gear the motor (Lee *et al.*, 2006).

The pRNA contains two structural domains, which could fold independently (Fig. 9). Its central region contains two interlocking loops responsible for its intermolecular interactions to form a hexameric ring (Figs. 9a and 9c) and binding to the procapsid, while its double-helical 5′/3′ paired region is essential in DNA packaging (Garver *et al.*, 1997; Reid *et al.*, 1994a,b,c; Zhang *et al.*, 1994) and for the binding of the ATPase gp16 to bring it into proximity to the connector (Lee *et al.*, 2006). A three-arms-around-a-hinge model for pRNA function was proposed based on the independent folding and function of the two structural domains (Fang *et al.*, 2005). The two interlocking loop regions and the helical region of the pRNA were regarded as three individual arms connected through a hinge. It is believed that at functional magnesium concentration, the

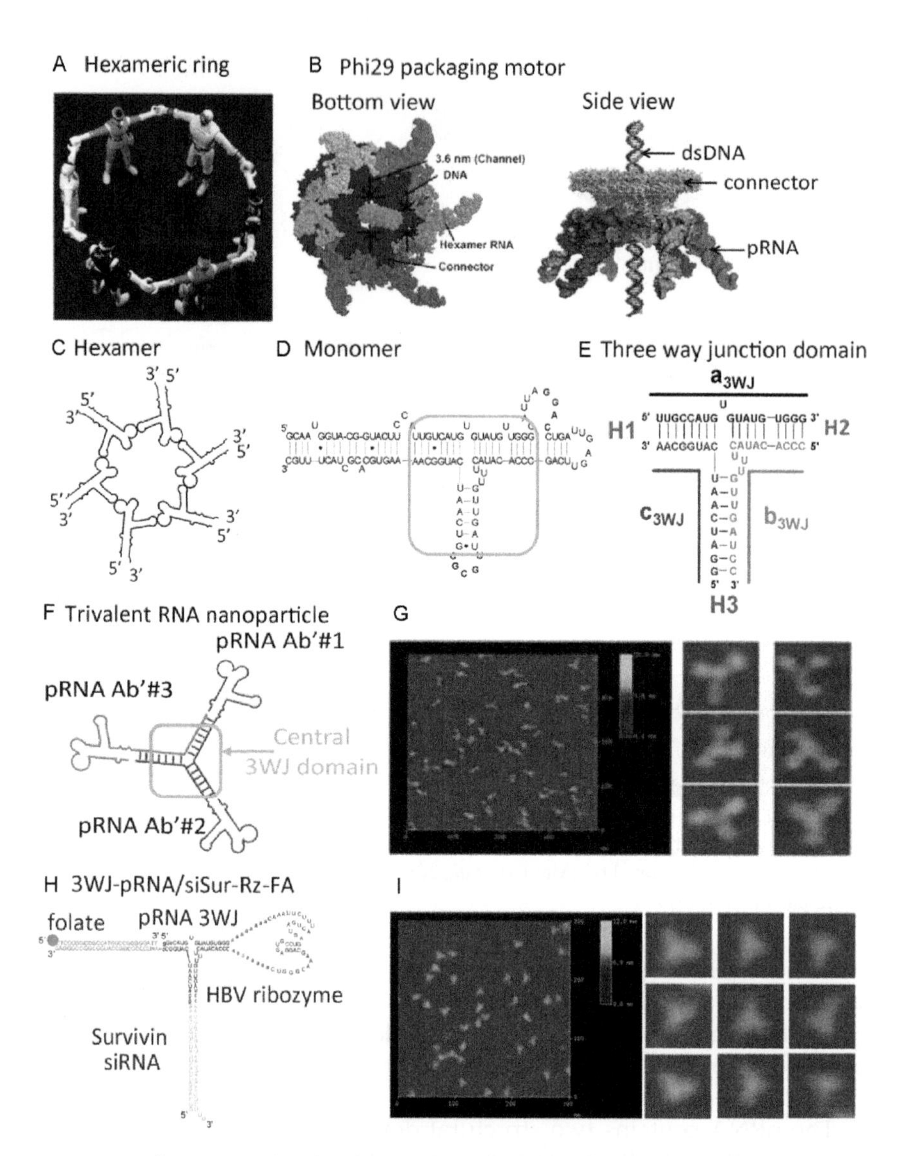

FIGURE 9 Illustration of multivalent pRNA nanoparticle formation. (a) The concept of hand-in-hand interaction in pRNA hexamer formation. Adapted from Guo (2002) ©2002 with permission from Elsevier. (b) Illustration of the ϕ29 packaging motor geared by six pRNAs with bottom view and side view. Adapted from Shu *et al.* (2003c) ©2003 with permission from Elsevier and from Macmillan Publishers Ltd. (Shu *et al.*, 2011a) ©2011. (c) Schematic of pRNA hexamer ring. (d) Secondary structure of pRNA with green box indicating the central 3WJ domain. (e) The 3WJ domain composed of three RNA oligomers. Helical segments are represented as H1, H2, and H3. (f) Schematic of a trivalent RNA nanoparticle using the 3WJ-pRNA core sequence. (g) Atomic force microscopy images of (f). (h) Schematic of 3WJ-pRNA-siSur-Rz-FA nanoparticle using the 3WJ-pRNA core sequence. (i) Atomic force microscopy images of (h) (Chen *et al.*, 1999; Shu *et al.*, 2011a). Adapted with permission from Macmillan Publishers Ltd. (Shu *et al.*, 2011a) ©2011. (See Page 14 in Color Section at the back of the book.)

three arms folded into a conformation in which pRNA would be fully active for motor function. It was also proposed that the addition of ATP would promote relative motion among the three arms of pRNA and thus package DNA (Fang *et al.*, 2005; Shu and Guo, 2003a). This agrees with the hypothesis that the bifurcation polyuridine bulge (U72U73U74) at the helical junction may provide flexibility in orientation of the helices for pRNA to function correctly (Zhang *et al.*, 1997). It has also been revealed that pRNA contains an ATP-binding motif (Shu *et al.*, 2003a). However, pRNA itself does not display ATPase activity. It is expected that the ATPase active center is a complex of the entire motor and that ATP hydrolysis is a collective effort of all motor components. Indeed, in the absence of pRNA, the ATPase activity of the ATPase gp16 is extremely low (Grimes and Anderson, 1990; Guo *et al.*, 1987d; Ibarra *et al.*, 2001; Lee *et al.*, 2008; Shu *et al.*, 2003a).

In addition, the study on the bulged regions of pRNA showed that a C18C19A20 bulge in the double-helix domain is essential for DNA packaging, but dispensable for gp16 binding to the procapsid/pRNA complex (Lee *et al.*, 2006; Reid *et al.*, 1994b; Zhang *et al.*, 1997). Further study on the role of the CCA bulge in DNA packaging revealed that the size and location, rather than the sequence of the bulge, are important for proper orientation of the double-helical domain, which could be critical for the correct contact between gp16 docked on pRNA and genomic DNA. The CCA bulge may also provide flexibility for pRNA to serve as a hinge to drive the motor (Lee *et al.*, 2006; Zhao *et al.*, 2008).

An unexpected finding has been reported regarding an unusual pRNA trifurcation motif (the 3WJ, or three-way junction core) domain. It was found that this pRNA 3WJ domain (Fig. 9d) can be assembled from three pieces of RNA oligonucleotides with remarkable thermodynamically stable properties (Shu *et al.*, 2011a). The deep slope of the T_M melting curve from the three pieces is close to 90°, indicating extremely low free energy and unusually high affinity in 3WJ assembly. The self-assembled RNA nanoparticles with three or six pieces of RNA oligonucleotides guided by the 3WJ core are highly stable, resistant to 8 *M* urea denaturation, and do not dissociate even at extremely low concentrations (Figs. 10a–10c). In comparison with 36 other 3WJ motifs found in different biological RNAs, the 3WJ core of ϕ29 pRNA showed extraordinary stability demonstrated by T_M melting curves and slopes (Fig. 10d). Additionally, metal ions were found to be dispensable for nanoparticle assembly. Because the ϕ29 DNA packaging motor is one of the strongest biomotors discovered to date, the motor module should be stronger than the ordinary counterpart to perform such a special function as a hinge in motor gearing. This 3WJ core domain can serve as a potential platform to construct stable, multivalent nanoparticles for gene therapy purposes (Shu *et al.*, 2011a)(see Section V).

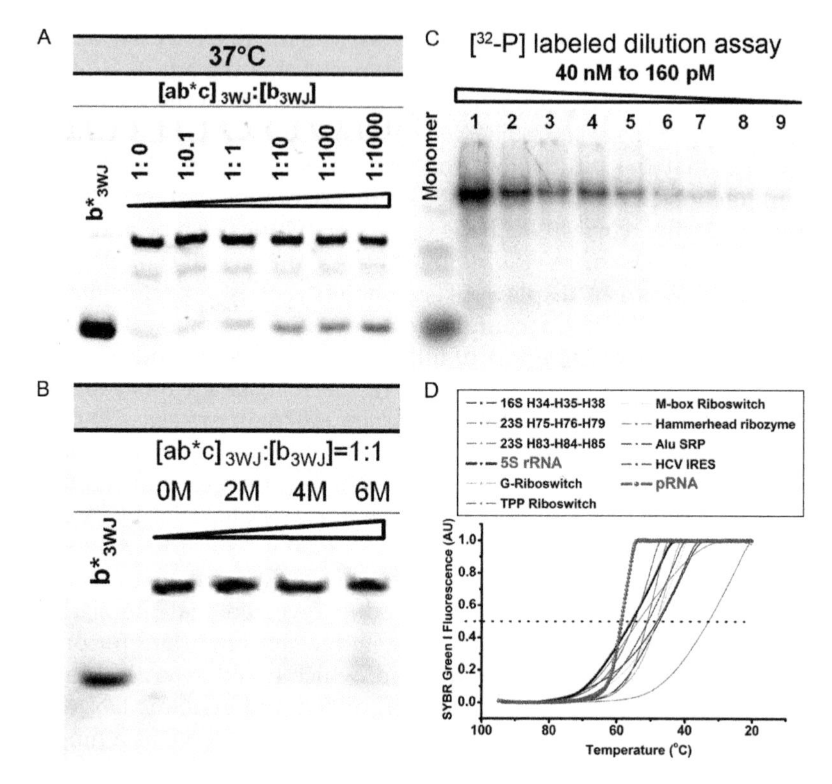

FIGURE 10 The 3WJ-pRNA core motif showed unusual stability. (a) A 16% native gel of the 3WJ-pRNA core at 37°C. Fixed concentration of Cy3-labeled $[ab^*c]_{3WJ}$ was incubated with various concentrations of unlabeled b_{3WJ} at 37°C. (b) Urea-denaturing effects on the stability of $[ab^*c]_{3WJ}$ evaluated by a 16% native gel. A fixed concentration of labeled $[ab^*c]_{3WJ}$ was incubated with unlabeled b_{3WJ} at ratios of 1:1 in the presence of 0–6 *M* urea at 25°C. (c) Dissociation assay for the $[^{32}P]$3WJ-pRNA complex harboring three monomeric pRNA by twofold serial dilution (lanes 1–9). (d) Comparison of melting curves for different RNA 3WJ core motifs. Adapted with permission from Macmillan Publishers Ltd. (Shu *et al.*, 2011a) ©2011. (See Page 15 in Color Section at the back of the book.)

E. Approaches in packaging RNA stoichiometry determinations that lead to hexamer conclusion

1. By concentration-dependent curve

An approach based on the concentration dependence of the dose–response curve was used to estimate the stoichiometry of pRNA (Lee and Guo, 1995a). The principle behind this approach is that the larger the stoichiometry of a component involved in the motor, the more dramatic the influence of the concentration on the motor activity. In comparison with a component of known stoichiometry, the copy number of both pRNA and gp16 can be estimated (Lee *et al.*, 1995a).

2. By binomial distribution using mutant and wild-type pRNA

Mathematical approaches by statistical analysis based on binomial distribution were explored to determine the stoichiometry of pRNA (Chen *et al.*, 1997; Trottier and Guo, 1997). Mutant pRNA with the ability to bind procapsid but not to package DNA were mixed with wild-type pRNA at various ratios in *in vitro* assembly assays. Thus, the pRNA ring on each procapsid would be a mixture of mutant and wild-type pRNA, and the probability could be determined by expansion of a binomial. Viral assembly activity was therefore simulated against different ratios between mutant and wild-type RNA for different stoichiometries of pRNA. By fitting the empirical data with predicted curves, it was determined that the stoichiometry of pRNA is either 5 or 6 (Trottier *et al.*, 1997).

3. By the common factor of 2 and 3

Sequences of the two interlocking loops of pRNA (bases 45–48 of right-hand loop and bases 85–82 of left-hand loop) are complementary and allow pRNA to interact intermolecularly to form homo-oligomers. These sequences can be engineered so that pRNA can form hetero-oligomers. To simplify the description, uppercase and lowercase letters are used to represent the right- and left-hand loop sequences of the pRNA, respectively. The same letter in upper- and lowercases symbolizes a pair of complementary sequences. For example, in pRNA Aa′, right loop A (5′GGAC48) and left loop a′ (3′CCUG82) are complementary, whereas in pRNA Ab′, the four bases of right loop A are not complementary to the sequence of left loop b′ (3′UGCG82). It was found that a mutant pRNA with complementary loop sequences (such as pRNA Aa′) was active in DNA packaging by itself, while a mutant with noncomplementary loops (such as pRNA Ab′) was inactive. However, when pRNA Ab′ is mixed with pRNA Ba′ at a 1:1 molar ratio, DNA packaging activity was restored through interaction of the loops. The stoichiometry of pRNA was therefore predicted to be a multiple of 2 and that dimeric pRNA is the building block of the pRNA ring (Chen *et al.*, 2000; Guo *et al.*, 1998). Similarly, mutant pRNAs J-p′, P-k′, and K-j′ were found to be inactive alone or in a mixture of any two, but regained activity when all three were mixed at a 1:1:1 molar ratio, suggesting that the stoichiometry of pRNA is also a multiple of 3. Combined with studies by mathematical approaches, stoichiometry was therefore concluded to be a common multiple of 2 and 3, and decisively determined to be 6 (Guo *et al.*, 1998).

4. By single molecule photobleaching

A single molecule photobleaching assay in conjunction with statistical analysis was applied to elucidate the copy number of pRNA molecules on active motors (Fig. 11) (Shu *et al.*, 2007). With the pRNA labeled singly

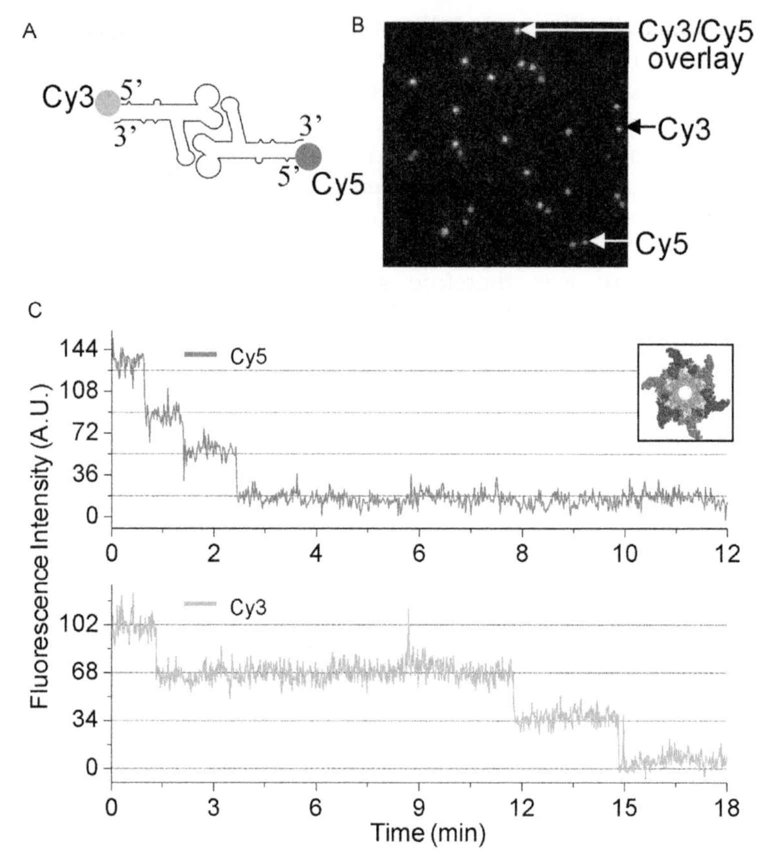

FIGURE 11 Single fluorescence molecule photobleaching assay demonstrating the presence of six pRNA molecules on each procapsid through using a customized dual-channel total internal reflection fluorescence imaging system. (a) Illustration of the dual-labeled dimer design. (b) Dual-color fluorescence image of the pRNA on procapsid. (c) Photobleaching traces of pRNA on a single procapsid. Each discrete step represents one single pRNA molecule (Shu *et al.*, 2007; Xiao *et al.*, 2008; Zhang *et al.*, 2007). Adapted with permission from Macmillan Publishers Ltd. (Shu *et al.*, 2007) ©2007. (See Page 16 in Color Section at the back of the book.)

with Cy3 fluorophore and ϕ29 DNA labeled with Cy5, single active motors or the DNA packaging intermediates were isolated by dual-color fluorescence imaging (Shu *et al.*, 2007). Utilizing the characteristic, quantized drop in intensity for single fluorophores, it was found that there are six photobleaching steps, representing six pRNA molecules on a single motor before or during DNA packaging, confirming the conclusion that stoichiometry of the pRNA on the motor is 6 (Fig. 11). Facilitated by the same single molecule photobleaching approach, a novel mechanism was also discovered in regards to RNA/protein interactions (Xiao *et al.*, 2008).

It was demonstrated that the stable and specific binding of pRNA to the procapsid relies on formation of a closed RNA ring with correct ring size to fit on the connector rather than depending on specific sequences of RNA. Interruption of any one of the interlocking links in the closed RNA ring could impede the formation of the hexameric ring, resulting in abolishment of the pRNA to bind. Extension or reduction of the circumference of the RNA ring by modification of arms in the loop region also abolishes the binding of pRNA to the motor. In addition, an artificial RNA with a designed, proper ring size was found to be active in DNA packaging (Xiao *et al.*, 2008). Formation of the closed pRNA ring from a pure dimer or pure trimer alone strongly supports the argument that the ring is a common multiple of two and three, which must be a hexamer.

5. By gold and ferritin labeling to pRNA

Gold and ferritin conjugation (Fig. 12) was used to observe the number of pRNA on each motor. EM images of motors with single gold- or ferritin-conjugated pRNA revealed that six particles were attached to the vertex of the procapsid (Moll and Guo, 2007). It was found that procapsid particles

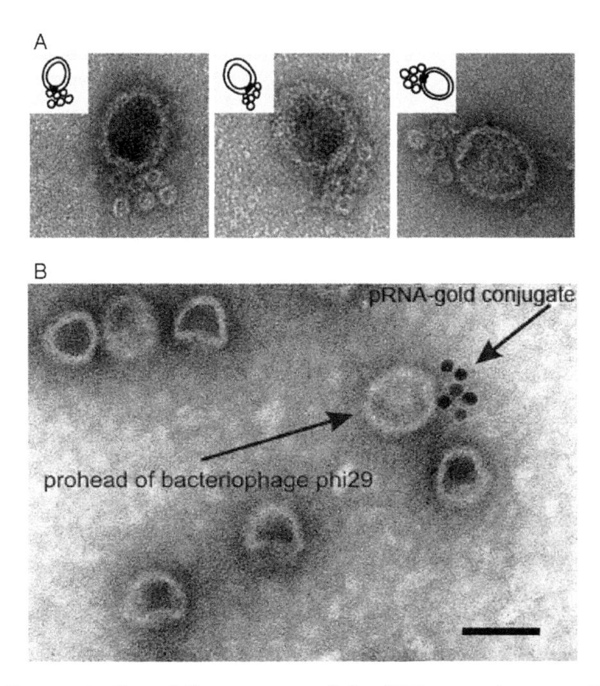

FIGURE 12 Demonstration of the presence of six pRNA on each procapsid by conjugation of one ferritin particle (a) or one gold nanoparticle (b) to each pRNA. Adapted from Xiao *et al.* (2008) by permission of Oxford University Press and Macmillan Publishers Ltd. (Shu *et al.*, 2007) ©2007.

harbored six gold particles or no gold at all (Xiao *et al.*, 2008) (Fig. 12). The result supports the finding that stable and specific binding of pRNA to the procapsid relies on formation of a closed RNA ring with correct ring size to fit on the connector rather than depending on specific sequences of RNA. Interruption of any one of the interlocking links in the closed RNA ring could impede formation of the hexameric ring, resulting in abolishment of the pRNA to bind.

F. Challenge and discrepancy in determination of stoichiometry of motor packaging RNA

To elucidate the role of pRNA in DNA packaging, it is critical to know how many copies of the pRNA are involved in each DNA packaging event. The approaches described earlier strongly demonstrated the pRNA ring as a hexamer. However, a discrepancy in pRNA stoichiometry has been reported. In 1998, the formation of hexameric pRNA ring was demonstrated by two back-to-back papers (Guo *et al.*, 1998; Zhang *et al.*, 1998) and featured in *Cell* by Hendrix (1998). In cryo-EM studies, contradictory results of a sixfold (Ibarra B *et al.*, 2000) or fivefold (Morais *et al.*, 2001, 2008; Simpson *et al.*, 2000) symmetry of the pRNA ring were reported by two separate groups. Traditionally, because cryo-EM was developed to study protein complexes with symmetrical structural arrangement, this technology is very useful to study the viral capsid structure with defined, symmetrical arrangement. The application of cryo-EM to study RNA structure is very intriguing but very challenging for the following reasons. (1) Compared to protein, the chemically unmodified RNA is unstable and very sensitive to RNase, especially the single-stranded region such as the pRNA C18C19C20 bulge and the hand and head loops (Bailey *et al.*, 1990; Chen *et al.*, 1999; Guo *et al.*, 1987a). In cryo-EM image reconstruction, data are averaged from the computation pool of many viral particles. This average will be smaller due to RNA partial degradation or cleavage. (2) Compared to protein, RNA is structurally flexible due to its dynamic nature and kinetic and thermodynamic characteristics (Bothe *et al.*, 2011). Solution-state nuclear magnetic resonance studies revealed that many noncoding RNAs do not fold into a single native conformation but fold into many different conformations along their free-energy landscape. The dynamic motion of RNA spans from the picosecond to second timescale. (3) The RNA structure depends on many ions, including ATP, ADP, and Pi. Alteration of sampling conditions leads to a variety of structures. (4) RNA folding relies on sequences, ligands, and environments. Any alteration in sequences and substrate binding can lead to alterations in RNA structure. (5) It is challenging to use technology for detection of a stationary image to assess a machine in motion undergoing a dynamic process without structural symmetry (Chen *et al.*, 1997; Moffitt *et al.*, 2009). In the meantime, a plethora of studies using biochemical, mathematical, and genetic approaches (Chen *et al.*, 1997; Shu *et al.*, 2007; Trottier *et al.*, 1997; Xiao *et al.*, 2008) were carried out to study the stoichiometry of pRNA on the motor and have supported the hexamer observation (Guo *et al.*, 1998; Shu *et al.*, 2007; Xiao *et al.*, 2008; Zhang *et al.*, 1998). The conclusion of a hexameric ring has also been observed in the T3 system showing that the DNA packaging protein gp19 forms six subunits on the T3 DNA packaging motor (Fujisawa *et al.*, 1991).

The group who asserted the model of a pentameric pRNA ring on the motor argues that pRNA binds to the fivefold procapsid and a pRNA hexamer is formed initially, but one of the pRNA molecules dissociates afterward due to a conformational change (Simpson *et al.*, 2000). However, cross-linking approaches revealed that pRNA did not bind to the procapsid protein but to the connector (Garver *et al.*, 1997). Moreover, another paper (Xiao *et al.*, 2005) provided solid evidence that pRNA binds to the N terminus of the connector protein gp10 via the interaction of the positively charged residues in protein and the negatively charged phosphate backbone of RNA. Further mutation analysis revealed that the three basic amino acids, Arg-Lys-Arg, at the N terminus are responsible for pRNA binding (Atz *et al.*, 2007). Furthermore, it would be difficult to account for how a fivefold pRNA ring docks at the sixfold symmetric connector. In addition, single molecule photobleaching studies of purified DNA packaging intermediates showed six copies of pRNA on active motors filled partially with DNA, disputing the explanation that one pRNA molecule would leave the motor after binding (Shu *et al.*, 2007). Chen and Guo (2000) demonstrated that the pRNA dimer is the building block in hexamer formation. Other literature (Ding *et al.*, 2011; Fang *et al.*, 2005, 2008; Gu and Schroeder, 2011; Robinson *et al.*, 2006) all support the long-standing finding (Chen *et al.*, 2000; Shu *et al.*, 2007; Xiao *et al.*, 2008) that the pRNA dimer is the building block and that the sequential action in hexamer assembly is $2 \rightarrow 4 \rightarrow 6$. A study with RNA alone in the absence of the procapsid, a form of tetramer, was revealed by X-ray crystallography (Ding *et al.*, 2011). However, the authors used the tetramer-based crystal structure to compute a pRNA pentamer on the procapsid by docking the monomeric RNA fragment abstracted from the tetramer by cryo-EM computation. As noted earlier, the RNA structure is flexible, and multiple conformations can result from one RNA sequence, depending on the energy of the landscape, the ligand, and the environment. RNA could change its structure after binding to the procapsid; thus, the pRNA on the procapsid might not be identical to the free RNA in solution that grew into the crystal structure (Ding *et al.*, 2011). In addition, the RNA fragment used in crystallization is not a full pRNA molecule but is a truncated RNA fragment, and several bases were changed in this truncated version (Ding *et al.*, 2011). The truncated and mutated RNA fragment might not display the same conformation as that in the integrated pRNA ring sheathing to the viral motor (Hoeprich and Guo, 2002). Indeed a publication on structural study of the pRNA 3WJ motif using spin labeling established a very different global structure with significant variation from that in the crystal structure (Zhang *et al.*, 2012). Whether such structures and conformations are the real depiction of pRNA on the motor remain to be elucidated.

III. ENERGY SOURCE AND CONVERSION MECHANISM

A. Sequential interaction of motor ATPase with DNA and ATP

Viral DNA packaging has long been recognized to require energy (Hohn, 1976; Kellenberger, 1976; Riemer and Bloomfield, 1978) or ATP (Masker, 1982). In 1986, it was discovered that the larger subunit of the two DNA packaging enzymes of dsDNA bacteriophage is a DNA and procapsid-dependent ATPase that contains both A- and B-type ATP-binding consensus sequences (Guo *et al.*, 1987d). In the past, several studies were aimed at understanding the function of the ϕ29 motor ATPase, gp16, and how it catalyzes ATP hydrolysis in combination with dsDNA. A nonhydrolyzable ATP analog, γ-s-ATP, was used to inhibit ϕ29 DNA packaging (Chemla *et al.*, 2005; Guo *et al.*, 1987c; Llorca *et al.*, 1997; Schwartz *et al.*, 2012; Shu *et al.*, 2003b; Smith *et al.*, 2001). The γ-s-ATP was also used to isolate dsDNA packaging intermediates through sucrose gradient sedimentation (Shu *et al.*, 2003b). These intermediates were then converted into functional phages and infectious viruses by inputting additional ATP and gp16 as demonstrated by phage assembly assay (Shu *et al.*, 2003b). These experiments suggested that the packaging process is strongly dependent on the energy produced by ATP, and blocking the production of energy mediated by gp16 can easily stall packaging of the DNA into the procapsid. The stalling is, however, reversible, as the intermediates can be transformed into functional phages by adding gp16 and ATP, thus restoring the production of energy (Shu *et al.*, 2003b).

The γ-s-ATP was also used to block the departure of the ATPase gp16 from the dsDNA/ATPase complex (Schwartz *et al.*, 2012). Electrophoretic mobility shift assays (EMSA) and fluorescence resonance energy transfer (FRET) studies were used to study the binding affinity of gp16 to dsDNA. The nonhydrolyzable ATP analog, γ-s-ATP, was used to stabilize the binding between gp16 and dsDNA, indicating that gp16 assumes a high-affinity state for DNA when it is complexed with the nucleotide triphosphate. Addition of ATP was capable of releasing the protein from dsDNA (Schwartz *et al.*, 2012). A scheme of reaction, similar to other AAA+ proteins, can be drawn from these experiments in which gp16 alternatively assumes a high- and a low-affinity state for DNA before and after ATP hydrolysis occurs, respectively, and the hydrolysis of ATP induces a power stoke for DNA translocation in which the ATPase pushes the DNA away from itself (Schwartz *et al.*, 2012).

Whether gp16 preferred to bind to dsDNA or ATP first during motor action was also assayed. The EMSA was again employed and dsDNA or γ-s-ATP was applied initially and the missing component subsequently added. The complex was stabilized better when γ-s-ATP was added first, suggesting that the ATPase first binds to ATP, promoting a conformational change to bind tightly to dsDNA (Schwartz *et al.*, 2012).

B. Packaging force and velocity

Many biomotors, found in several different environments (cellular trafficking, ATP-dependent protein degradation, helicases), are powered by breaking down an energy-rich bond of ATP (Dittrich and Schulten, 2006; Grigoriev *et al.*, 2004; Hua *et al.*, 1997; Ishijima *et al.*, 1998; Stock *et al.*, 1999; Walker *et al.*, 1982), similar to combustion engines breaking down molecules in petroleum distillates. Other motors are driven by the flow of protons along a proton gradient, an electrochemical potential across a membrane, just as electrical motors are powered by the flow of electrons along an applied voltage; an example of such can be found in bacterial flagella (Berry, 2001) or in ATP-generating machinery (F_0/F_1 ATP synthase)(Sambongi *et al.*, 1999). Viral packaging motors are powered by chemical energy derived from the hydrolysis of ATP. The active sites of motor ATPases bind ATP and catalyze its degradation to ADP and inorganic phosphate (Pi), thereby releasing a significant quantity of energy and inducing conformational changes in the motor structure that are then converted into mechanical force for the motor to function. Such conformational changes are most likely related to the release of inorganic phosphate to generate the power stroke on the DNA rather than the ATP-binding step. This catalytic process is then restarted by the next ATP molecule, and the motor continues to function in a processive and sequential manner (Chen *et al.*, 1997) until the DNA is completely packaged. Studies have shown that energy converted from the hydrolysis of one ATP molecule corresponds to the packaging of every 2 (Guo *et al.*, 1987d) or 2.5 (Moffitt *et al.*, 2009) bp of DNA in the ϕ29 system, comparable to that of T3 with 1.8 bp per ATP (Morita *et al.*, 1993).

All viral packaging enzymes, including gp16 of ϕ29, possess ATPase activity and contain a conserved Walker-A ATP-binding motif (G/A) X_4GK(S/T), also called the P-loop, and a Walker-B motif (h_4DE) in which the two negatively charged residues, aspartate and glutamate, are involved in γ-phosphate bond hydrolysis (Guo *et al.*, 1987d). Mutations of a single amino acid in the Walker-A or Walker-B regions are capable of abolishing ATP binding or hydrolysis, thereby completely inhibiting

DNA packaging (Huang and Guo, 2003a,b). In ϕ29, gp16 itself was found to be a slow ATPase with low affinity to ATP, while it exhibited much higher activity in ATP hydrolysis when ssDNA, dsDNA, or RNA were bound (Guo *et al.*, 1987d; Shu *et al.*, 2003a). Interestingly, the strength of ATPase stimulation is dependent on the structure and chemistry of the nucleic acids, with an order of pRNA ~ poly d(pyrimidine) > dsDNA > poly d(purine)(Lee *et al.*, 2008). Similar DNA-dependent ATPase activities were also found in other viral systems, including λ (Hwang *et al.*, 1996) and T3 (Morita *et al.*, 1993), and in many AAA+ helicases where ATP hydrolysis is stimulated from 10- to 100-folds.

Procapsids with pRNA also stimulate the ATPase activity of gp16 by 10-fold, compared to the lack of stimulation by procapsids alone, in agreement with the finding that gp16 binds to procapsid through interaction with pRNA on the motor (Lee *et al.*, 2006). ATPase activity was stimulated maximally only when all packaging motor components, including pRNA, procapsid, gp16, and DNA-gp3, were present (Grimes *et al.*, 1990; Guo *et al.*, 1987c,d; Huang *et al.*, 2003a,b; Ibarra *et al.*, 2001; Lee *et al.*, 2006, 2008; Shu *et al.*, 2003a). It was also found in ϕ29 that both the DNA packaging protein gp16 and pRNA possess higher binding affinity for ATP than for ADP or AMP (Shu *et al.*, 2003a), and the stimulation of gp16 to kick away from dsDNA has a stronger effect by ATP than by ADP or AMP (Schwartz *et al.*, 2012).

The coupling of energy production to DNA translocation is a complex task that needs to be fine-tuned by viral packaging motors. Regardless of stoichiometry, packaging motors have the necessities to coordinate the work performed by each single subunit in order to obtain a linear unidirectional translocation of DNA from the cellular environment into the viral capsid. Evidence of such coordination can be found in the negative cooperativity in ATPase activity observed in many members of the viral packaging motors subfamily and the AAA+ proteins in general, as well as the ability of a single inactive subunit in the oligomer of a packaging motor to completely impair the translocation ability of the entire packaging machinery (Chen *et al.*, 1997; Maurizi *et al.*, 2001; Miyagishi *et al.*, 2004; Trottier *et al.*, 1996, 1997). Mechanical and physical explanations of such cooperation are still lacking for the viral packaging motor, albeit some hypotheses have been produced for other subfamilies of AAA+ proteins, such as the Glutamate switch for helicases (Ionel *et al.*, 2011; Zhang and Wigley, 2008) or particular forms of steric hindrance proposed for proteolysis-associated chaperones (Kravats *et al.*, 2011). All the models share the fact that every single subunit of a AAA+ oligomer is able to "sense" the nucleotide state of the others and adjust consequently in order to contribute to the overall directional translocation (Black, 1981; Casjens, 2011; Chen *et al.*, 1997; Grimes *et al.*, 1997; Guasch *et al.*, 2002; Hendrix, 1978; Moffitt *et al.*, 2009; Morita *et al.*, 1995; Serwer, 2003, 2010; Sun *et al.*,

2010; Yu *et al.*, 2010). For gp16 of ϕ29, mechanisms of cooperative function have been deduced from studies demonstrating that DNA translocation is performed in a discrete stepwise manner (10 bp per step), corresponding to the hydrolysis of three to four ATP per step most likely in a coordinated manner (Moffitt *et al.*, 2009; Yu *et al.*, 2010).

To package viral DNA to near-crystalline density inside the procapsid, the packaging motor has to overcome the entropic energy associated with the constrained size and function against the large internal force from the filled procapsid. Single molecule studies with tweezers demonstrated that ϕ29 packaging motors are, to date, the strongest biological motors, generating about 57–110 pN of force acting on beads linked to dsDNA (Casjens, 2011; Rickgauer *et al.*, 2008; Smith *et al.*, 2001). Phage λ (>50 pN) and T4 (>60 pN) package viral DNA with high velocities and high processivity. The initial packaging rate was found to be 165 bp/s for ϕ29 DNA packaging (Rickgauer *et al.*, 2008; Smith *et al.*, 2001), comparable to that of λ with an average velocity of about 600 bp/s (Fuller *et al.*, 2007b). The packaging rate for T4 was found to be strikingly higher than ϕ29 and λ, with the average rate ranging up to about 2000 bp/s, which is in accordance to the much longer length of the T4 genome (Fuller *et al.*, 2007a). The speed of translocation appears to be dependent on the size of the internal capsid cavity. The velocity in packaging varies but tends to decrease after the procapsid is partially filled due to internal pressure. Occasional pauses and slips were observed for all three motors (Fuller *et al.*, 2007a,b; Rickgauer *et al.*, 2008; Smith *et al.*, 2001). Slow speeds are desirable for development of a high-throughput single pore genomic DNA sequencing apparatus, as the current bottleneck for this technology is that the DNA translocation speed is too fast for any available sensing equipment to detect at single base sensitivity. By attaching a bead to the end of DNA and using a magnet, the speed of ϕ29 DNA packaging can be as slow as 20–50 bp per second (Chang *et al.*, 2008)

C. Rationale for the discrepancy between 2 and 2.5 bp of double-stranded DNA per ATP

It has been reported (Guo *et al.*, 1987d) that packaging of 2 bp of DNA requires the hydrolysis of one ATP. This stoichiometry was subsequently supported by the T3 system showing the consumption of one ATP for 1.8 bp (Morita *et al.*, 1993). Subsequently, this rate has been used extensively by biochemists and biophysicists to interpret the mechanism of motor function (Chemla *et al.*, 2005; Mancini *et al.*, 2004; Rao *et al.*, 2008; Roos *et al.*, 2007; Serwer, 2003; Sun *et al.*, 2008). However, it was subsequently reported in ϕ29 that 2.5 bp of DNA are translocated per ATP by single molecule studies using optical tweezer (Moffitt *et al.*, 2009). The newly proposed "push through one-way valve" mechanism for DNA packaging motor (Schwartz *et al.*, 2012) might have reconciled the discrepancy between 2 and 2.5 bp per ATP. Currently, the motion mechanism is interpreted based on structural properties of a connector channel with 12-fold symmetry and dsDNA with 10.5 bp per turn of 360° for the common B-type DNA. The 2 or 2.5 bp per ATP has been used as evidence to argue for four, five, or six discrete steps of DNA translocation. It is logical to believe that an integral number of ATP is required to translocate a definitive number of base pairs of dsDNA if gp16 and connector are an integrated concrete component similar to that in a metal engine. Calculating the ratio of ATP and the number of base pairs packaged might be useful in interpreting the motor mechanism if there is only one functional motor protein or the multiple components act as one module or are synchronized. However, biological machines are more intricate and do not follow simple physical and engineering principles. It has been found that the motor is operated by two distinct unsynchronized components: gp16 for active pushing or driving and the channel as a one-way valve to prevent reverse motion (Fang *et al.*, 2012; Geng *et al.*, 2011; Jing *et al.*, 2010a; Schwartz *et al.*, 2012). Although the pushing or driving force is from ATPase gp16, the connector channel also affects the speed of DNA translocation, as pause or DNA slipping events during translocation through the connector channel have been reported (Aathavan *et al.*, 2009; Chemla *et al.*, 2005; Yu *et al.*, 2010). The two uncoordinated force-generating factors, gp16 and connector, make it impossible to obtain a definite and reproducible number of base pairs per ATP utilized.

IV. MECHANISM FOR DNA PACKAGING MOTOR

A. Historical models

Historically, many DNA packaging models have been proposed for motor action, including (1) five-fold/six-fold mismatch and connector rotating thread (Guasch *et al.*, 2002; Hendrix, 1978; Simpson *et al.*, 2000), (2) electrodipole within central channel (Guasch *et al.*, 2002), (3) force from osmotic pressure (Serwer, 1988), (4) Brownian motion (Astumian, 1997), (5) connector contraction hypothesis (Morita *et al.*, 1995), (6) sequential action of motor components (Chen *et al.*, 1997; Moffitt *et al.*, 2009), (7) ratchet mechanism (Fujisawa and Morita, 1997), (8) supercoiled DNA wrapping(Grimes *et al.*, 1997), and (9) DNA compression and relaxation (Oram *et al.*, 2008; Ray *et al.*, 2010a,b; Sabanayagam *et al.*, 2007). There are cases where some models have been validated in one viral system but disproved in other systems. None of these models have been supported conclusively by experimental data.

Models of the fivefold/sixfold mismatch connector rotation model (Guasch *et al.*, 2002; Hendrix, 1978;Simpson *et al.*, 2000) and the electrodipole within the central channel (Guasch *et al.*, 2002) all portrayed rotation of the connector. However, all of these connector rotation models have been invalidated by single molecule studies using a fluorescently labeled connector (Hugel *et al.*, 2007) and have been argued against through the use of a connector cross-linking assay (Baumann *et al.*, 2006; Maluf and Feiss, 2006).

The negatively charged interior channel wall of the motor is decorated with a total of 48 positively charged lysine residues displayed as four 12-lysine rings from the 12 gp10 subunits that enclose the channel (Badasso *et al.*, 2000; Guasch *et al.*, 1998, 2002; Jimenez *et al.*, 1986; Simpson *et al.*, 2000, 2001). The electrodipole central channel model was also proposed for ϕ29 (Guasch *et al.*, 2002). In this model, positively charged lysine residues inside the highly negative charged connector channel were thought to interact with the negatively charged phosphate backbone of DNA during translocation, and the connector rotates in relation to the dsDNA (Guasch *et al.*, 2002). After contact has been established, the connector rotates by 6°, powered by ATP hydrolysis, dissociating the lysine–phosphate bond, and ultimately leads to packaging of a single base pair. This model fails to explain that an ATP-binding motif has never been discovered on either the connector or DNA, but ATP-binding motifs are clearly present on gp16 (Guo *et al.*, 1987d) and pRNA (Shu *et al.*, 2003a). Furthermore, it was shown that portal rotation is not likely to occur in both T4 and ϕ29 (Baumann *et al.*, 2006; Hugel *et al.*, 2007). To test this hypothesis, a mutation was introduced to change the charge of the lysine ring (Fang *et al.*, 2012). It was demonstrated that the interior channel lysine residues K200, K209, and K234 were not essential for ϕ29 DNA translocation. The conclusion that lysine rings do not play an active role in DNA translocation is also supported by manipulation of the pH. Translocation was retained even at high pH (basic) environments (Fig. 7), where the lysine residues were deprotonated. Furthermore, at a low pH, a state that would reduce the overall net negative charge of the channel, DNA translocation was also not affected (Fig. 7). All data argue against a rotating connector based on charge/charge interaction of dsDNA with the lysine rings of the connector. As mentioned earlier, the connector rotation implied in this model (Guasch *et al.*, 2002) was also invalidated by single molecule studies of a fluorescently labeled connector (Hugel *et al.*, 2007) and testing with a connector cross-linking experiment (Baumann *et al.*, 2006; Maluf *et al.*, 2006).

B. Recently revealed "push through one-way valve" mechanism

A new mechanism of "push through one-way valve" was discovered in the ϕ29 DNA packaging motor (Fang *et al.*, 2012; Geng *et al.*, 2011; Jing *et al.*, 2010a; Schwartz *et al.*, 2012) (Fig. 13). The discovery of this mechanism can

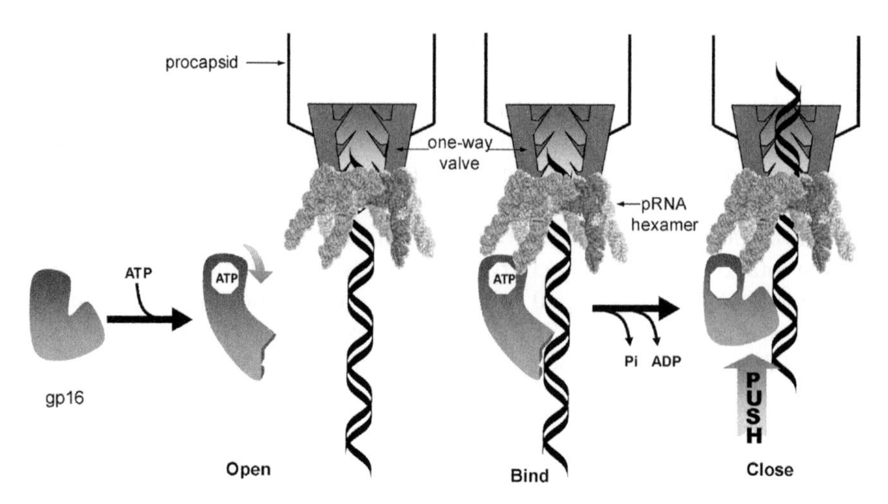

FIGURE 13 Illustration of the "push through a one-way valve" mechanism in ϕ29 DNA packaging. Adapted from Schwartz *et al.* (2012) with permission of Oxford University Press. (See Page 16 in Color Section at the back of the book.)

interpret and explain models based on compression and relaxation (Khan *et al.*, 1995; Oram *et al.*, 2008; Sabanayagam *et al.*, 2007), the ratchet mechanism (Fujisawa *et al.*, 1997), supercoiled DNA wrapping (Grimes *et al.*, 1997), and sequential action of motor components (Chen *et al.*, 1997; Moffitt *et al.*, 2009), and solves many of the puzzles surrounding the investigation of viral DNA packaging throughout the decades (see next section).

The AAA+ family possesses a common adenine nucleotide-binding fold with a ubiquitous characteristic of coupling chemical energy from ATP hydrolysis to mechanical motion. AAA+ ATPases assemble into oligomers, often hexamers, which form ring-shaped structures with a central channel. In 1998, Guo and colleagues proposed that viral DNA packaging motors implement a mechanism similar to the action of the hexameric AAA+ protein acting on dsDNA. A pump-and-valve mechanism was also proposed for the bacteriophage ϕ29 DNA packaging motor (Guo *et al.*, 2007). More and more evidence has emerged to support this mechanism. It has been demonstrated that ϕ29 DNA packaging is accomplished via a "push through one-way valve" mechanism (Fang *et al.*, 2012; Geng *et al.*, 2011; Jing *et al.*, 2010a; Schwartz *et al.*, 2012). The ATPase gp16 forms a hexameric complex geared by the hexameric pRNA ring to push dsDNA through the unrotating connector section by section into the procapsid (Schwartz *et al.*, 2012). The dodecameric connector channel remains unrotating and functions merely as a one-way valve that only allows dsDNA to pass the channel from the narrower N-terminal entrance to the wider C-terminal but not to exit the procapsid during DNA packaging (Fang *et al.*, 2012; Geng *et al.*, 2011; Jing *et al.*, 2010a,b; Wendell *et al.*, 2009).

Although the roles of the ATPase gp16 and the motor connector channel are separate and independent, bridging by the pRNA to link these two components promotes coordination between the two units to become functional modules of an integrated motor. It was found that interaction of ATP with gp16 induced a conformational change of gp16 ATPase that led to the high affinity of binding to dsDNA. ATP hydrolysis led to the departure of dsDNA from the ATPase/dsDNA complex, an action to push dsDNA to pass the connector channel. It was found unexpectedly that neither mutation of the basic residues nor changing of pH to 4 or 10 could impair DNA translocation measurably or affect the one-way traffic property of the channel (Fang *et al.*, 2012), suggesting that positive charges in the lysine ring of the channel are not essential in interacting with the negatively charged DNA phosphate backbone. This also supports the finding that the connector serves only as a valve rather than a rotating machine. The motor channel exercises three discrete, reversible, and controllable steps of gating with each step altering the channel size by 31% to control the direction of motion of dsDNA (Geng *et al.*, 2011). This finding may in fact appear to be quite unique, but is consistent in many facets with data from other phages and is not contradictory to many models proposed previously.

C. Using "push through one-way valve" mechanism to interpret historical DNA packaging models

1. Using "push through one-way valve" to interpret T4 compression and relaxation model (Khan *et al.*, 1995; Oram *et al.*, 2008; Sabanayagam *et al.*, 2007)

Ray and coworkers found that in T4, the portal protein compressed upon DNA entry, resulting in a DNA "crunching" phenomenon (Ray *et al.*, 2010b) (Fig. 14a). Although the authors did not define the role of the channel and interpret that the compression was a result of torsional force from coiling of upstream DNA, the model is not contradictory with the "push through one-way valve" mechanism which suggests that the external force for compression of dsDNA is a result of pushing by the ATPase. The procapsid in their model can also be related to the understanding that the portal acts as a one-way, nonrotating valve through conformational changes associated with the internal loop of the connector protein (Geng *et al.*, 2011; Guasch *et al.*, 2002; Jing *et al.*, 2010a,b; Wendell *et al.*, 2009). Furthermore, the torsion and coiling of dsDNA in this proposal are also in concert with the speculation that ϕ29 DNA packaging protein gp16 is a member of the AAA+ family that tracks along helical dsDNA to generate torsional force (Guo *et al.*, 1998; Lee *et al.*, 2006).

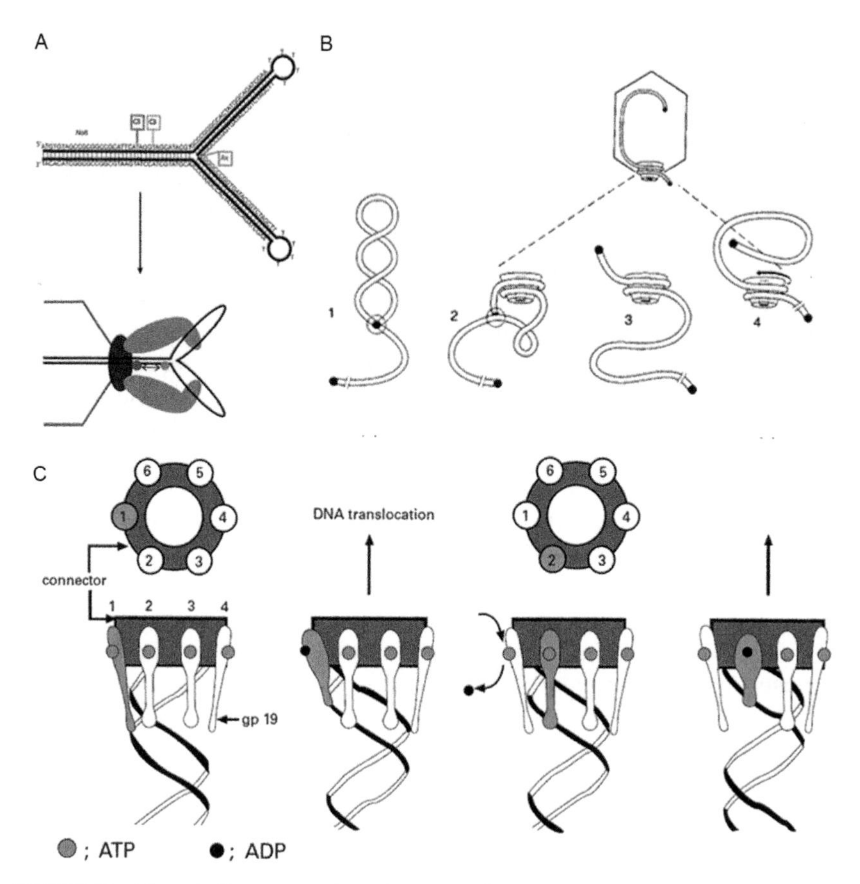

FIGURE 14 Historically proposed mechanisms of viral DNA packaging. (a) Compression of portal stalled Y-shape DNA. (Top) Design of the energy transfer dye pair on the Y-shape DNA. (Bottom) Procapsid portal with stalled Y-shape DNA (Ray *et al.*, 2010b). (b) φ29 DNA packaging motor packages gp3-DNA with gp3 and gp16 supercoiling DNA substrate. Gp3 bound covalently at the left end forms a lariat by interaction with DNA, while gp16 binds at the lariat loop junction (dotted circle), thus producing a supercoiled packaging substrate. The supercoiled DNA wraps around the connector of the prohead and then the left DNA end is freed and enters the capsid (Grimes *et al.*, 1997). (c) Proposed ratchet model for T3 DNA packaging. Blue: gp19 with sugar-phosphate backbone bound (Fujisawa *et al.*, 1997). Adapted from Ray *et al.* (2010b) ©2010; Grimes *et al.* (1997) ©1997 with permission from Elsevier; and Fujisawa *et al.* (1997) ©1997 with permission from John Wiley & Sons Inc. (See Page 17 in Color Section at the back of the book.)

2. Using "push through one-way valve" mechanism to interpret the T3 ratchet model (Fujisawa *et al.*, 1997)

A "ratchet" model (Fig. 14c) has been used to elucidate the packaging of T3 DNA. In this model, it was proposed that one packaging ATPase gp19 is bound to each component of the six domains of the connector. The other domain of one of the gp19 ATPase molecules contacts the sugar-phosphate backbone of the DNA duplex. When ATP is hydrolyzed, a conformational change of gp19 is induced to generate a force to translocate DNA into the procapsid. When a new ATP binds, the ATPase dissociates from the DNA and returns to the original conformation. At the same time, the next gp19 binds to the neighboring sugar-phosphate backbone. When this alternating function processes six total cycles, six molecules of ATP have been hydrolyzed to complete the translocation of one helical turn of dsDNA. Thus, one ATP molecule is utilized to translocate 1.7 bp DNA. Similar to the "push through one-way valve" mechanism, this model describes the control of dsDNA translocation by motor components for the control of the direction of DNA translocation; albeit it fails to elucidate how the motor prevents DNA exit from the high pressure inside the capsid during each hydrolysis transition. The difference between the ratchet model and the "push through one-way valve" mechanism is that, in the ratchet model, the component to control the direction of dsDNA translocation is ATPase, not the connector as in the "push through one-way valve" mechanism (Fang *et al.*, 2012; Geng *et al.*, 2011; Jing *et al.*, 2010a; Schwartz *et al.*, 2012).

3. Using "push through one-way valve" mechanism to interpret supercoiled DNA wrapping model (Grimes and Anderson, 1997)

In 1997, Grimes and Anderson proposed that ϕ29 DNA-gp3 is supercoiled and wrapped around the connector and that translocation is powered by rotation of the connector relative to the viral capsid with the aid of ATP hydrolysis (Fig. 14b). The group observed the formation of DNA–gp3 lariats by EM. Treatment with topoisomerase I shifted fast-sedimenting complexes toward the uncoiled lariat position in sucrose density gradients. They proposed that the packaging proteins gp3 and gp16 supercoil the DNA ends as a prerequisite for efficient interaction with the procapsid. The shared feature of this model with the T4 compression model and the "push through one-way valve" mechanism is that the DNA packaging enzyme twists the dsDNA outside the connector to generate a torsional force for dsDNA translocation. It is very possible that gp16 interacts with gp3 to form a complex for the initiation of dsDNA packaging in ϕ29. However, the proposal that the connector rotates and that dsDNA is wrapped outside the connector channel are contradictory to the "push through one-way valve" mechanism.

4. Using "push through one-way valve" mechanism to interpret sequential action of motor components (Chen *et al.*, 1997; Moffitt *et al.*, 2009)

In 1997, Chen and Guo reported the sequential action of six motor components (Chen *et al.*, 1997; Moffitt *et al.*, 2009). It was proposed that the procapsid contains a sixfold symmetrical connector surrounded by a ring of the packaging component. The relative motion of two rings could provide a driving force for DNA translocation. Analogous to a car engine, the sequential action model was a popular proposal regarding turning of the motor. The finding that a hexameric pRNA complex binds to the connector and that six pRNAs work sequentially, as evidenced from mathematical computations and modeling, lends support to the sequential action model. It was proposed that pRNA contains two domains: one for connector binding and the other 5′/3′ domain, which is free to interact with other components, such as gp16. Furthermore, it was theorized that pRNA is part of the ATPase complex and possesses at least two conformations—a relaxed and a contracted form. Alternating between contraction and relaxation, each member of the hexameric RNA complex powered by ATP hydrolysis helps generate torque to drive the DNA translocation machine. The proposal that pRNA bridges the connector and the ATPase gp16 was proved later (Koti *et al.*, 2008; Lee *et al.*, 2006; Zhao *et al.*, 2008), and the sequential action was also confirmed to regulate motor action by proving coordination between subunits (Chen *et al.*, 1997; Moffitt *et al.*, 2009). The "push through one-way valve" mechanism agrees with the sequential action model in that both describe the action of the ATPase by sequential steps, as many members of the AAA+ family do during dsDNA tracking or translocating (Ammelburg *et al.*, 2006; Frickey and Lupas, 2004; Hanson and Whiteheart, 2005; Iyer *et al.*, 2004a,b; Pyle, 2008; Singleton *et al.*, 2007; Wang, 2004).

Except for the connector rotating model (Simpson *et al.*, 2000, 2001), which has been invalidated, and the osmotic pressure model or the Brownian motion model, which are not supported by the most recent data, all other models, such as sequential action, ratchet mechanism, supercoiled DNA wrapping, and DNA compression, can be interpreted by the "push through one-way valve" mechanism in some aspects. More importantly, most experimental data that led to these historical models can be well interpreted by the "push through one-way valve" mechanism. This comment is reflected in a recent thorough and comprehensive review (Serwer, 2010). Overall, the "push through one-way valve" is the most agreeable mechanism in concert with most models throughout history.

D. Three reversible discrete steps of conformational change and gating of connector channel in controlling direction of DNA translocation

The mechanism for viral motor channel in one-way traffic control raises a question of how dsDNA is ejected during infection if the channel only allows dsDNA to travel in one inward direction. It has been proposed that viral procapsids and motor connectors adopt conformational changes during procapsid maturation and the DNA packaging process. We have proposed that the direction of DNA is controlled by conformational changes of the channel; such conformational changes render the channel to allow dsDNA to travel in one direction. Substantial conformational change upon completion of DNA packaging has been reported for the ϕ29 connector (Gonzalez-Huici *et al.*, 2004; Tang *et al.*, 2008; Tao *et al.*, 1998). A similar significant rearrangement of the connector after DNA packaging has also been reported in other phage systems (Kemp *et al.*, 2004; Lebedev *et al.*, 2007; Lhuillier *et al.*, 2009), indicating that the structural change is a common feature for phages. Such change may reversely favor DNA exit during infection.

A study on the direct observation of the ϕ29 connector with single channel recordings showed that the channel experiences conformational changes induced by molecule binding to the C-terminal located within the capsid, or by a high electrical voltage shift (Geng *et al.*, 2011) (Fig. 15). The conformational change exhibited three discrete steps in conductance, with each step reducing the channel size by 31%. It is therefore possible that the interaction of dsDNA or ϕ29 terminal protein gp3 during DNA packaging can induce a conformational change of the connector that distorts the shape of the channel proteins. This shift may lead to the opening or closing of the channel, which will help control the packaging or release of the viral genome during infection.

The gating properties of the ϕ29 connector under external potentials were also studied (Geng *et al.*, 2011). The significantly different gating

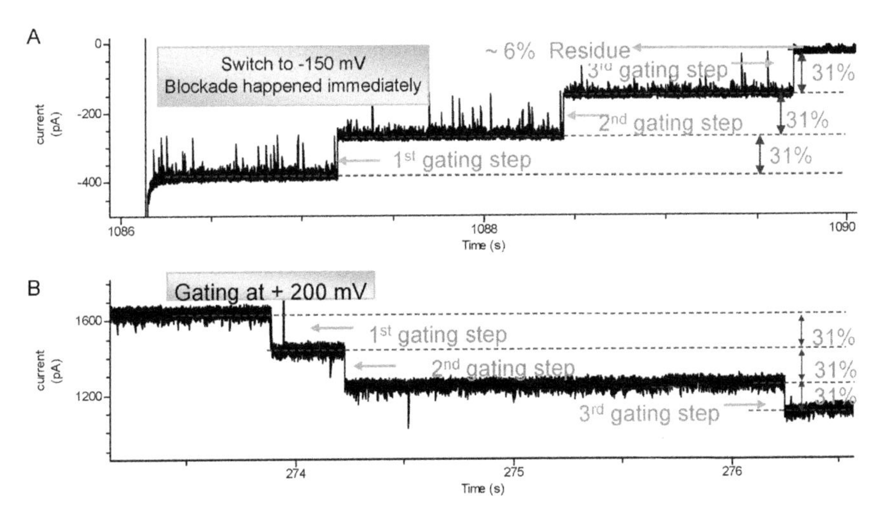

FIGURE 15 Three discrete steps in ϕ29 connector channel gating, with each step representing a reduction of channel size by 31%. Adapted from Geng *et al.* (2011) ©2011 with permission from Elsevier.

behaviors of the internal flexible loop-cleaved connector suggests that that flexible loops may play a key role in voltage gating (Geng *et al.*, 2011). After removal of these loops, both the occurrence and the extent of gating were greatly reduced. The gating mechanism can be explained by conformational changes of α helices in voltage-gated sodium and calcium channels. Based on the crystal structure of the ϕ29 connector, it is therefore possible that the applied potential can induce a conformational change in the protein channel embedded in a membrane. The single chain internal loops possess the flexibility to create a conformational change in response to environmental stimuli. The stepwise conformational change of the connector protein is also reversible and controllable, making it an ideal nanovalve for constructing a nanomachine with potential applications in nanotechnology and nanomedicine.

V. APPLICATION OF VIRAL PACKAGING MOTORS

Inspired by the ingenious configuration of natural biomotors at nanometer precision and better understanding of the assembly and function of these motors, the application of biomotors in nanotechnology and nanomedicine has continued to become a more critical scientific approach (Baneyx and Vogel, 1999; Fang *et al.*, 2012; Geng *et al.*, 2011; Grigoriev *et al.*, 2004; Hess *et al.*, 2001; Jing *et al.*, 2010a,b; Lee *et al.*, 2003; Park *et al.*, 2007; Soong *et al.*, 2000; Wendell *et al.*, 2009). The potential application of biomotors and their modules include integration into nanodevices, such as actuators, molecular sorters, molecular sensors, nanoelectromechanical machines, intricate arrays for diagnostics or computation, electronic or optical devices, and reconstruction as nanocarriers for the delivery of drugs or therapeutic RNA and DNA for the treatment of diseases. One of the most important potential applications is the use of the nanomotor as a high-throughput single pore sequencing apparatus for sequencing of the dsDNA genome of human, plants, and yeast for disease control and health improvement. Due to the strong power of the viral DNA packaging motors, phage designs have, in turn, become stronger and more robust due to natural selection and evolution (Jing *et al.*, 2010b; Shu *et al.*, 2011a). The robust properties of the viral motor and its modules make its application in nanotechnology feasible.

A. Biomimetic machinery

Highly efficient *in vitro* DNA packaging has been accomplished in defined packaging systems for many viral packaging motors (Fujisawa *et al.*, 1991; Guo *et al.*, 1986; Hwang and Feiss, 1997; Oliveira *et al.*, 2005; Rao and Black, 1988). Precise control to turn on and off the motors can also be achieved using ATP analogs, metal ion-chelating reagents, or other regulatory aptamers (Guo *et al.*, 1987c; Ko *et al.*, 2008; Shu *et al.*, 2003b; Smith *et al.*, 2001). It is therefore possible to engineer a controllable packaging motor into nanodevices. To incorporate the motor into nanodevices, techniques generally involve the characterization, manipulation, modification, control, creation, and/or assembly of organized materials on the nanoscale level (Guo, 2005; Niemeyer, 2002; Schmidt and Eberl, 2001). These materials will then be used as building blocks for the construction of larger devices and systems. Nanoporous anodic aluminum oxide (AAO) membranes were fabricated to capture and align the ϕ29 DNA packaging motor (Moon *et al.*, 2009). The pore size of the membranes was controlled precisely by using atomic layer deposition of aluminum oxide. In addition, the AAO membrane was found capable of sorting DNA-filled ϕ29 procapsids from empty procapsids. This nanoporous AAO membrane may be used in the future to integrate the ϕ29 motor with other artificial nanostructures.

B. Single molecule sensing

One emerging field in biology is the development of efficient, inexpensive, and highly sensitive analytical tools that can be used to probe, analyze, interpret, and manipulate single molecules, such as single channel ion transportation pores or nanopore-based DNA-sequencing devices (Deamer and Akeson, 2000). The essential component of the DNA packaging motor, the connector, provides a nanometer-sized channel for DNA's entry and exit of the viral procapsid. It has been demonstrated that the ϕ29 connector can be incorporated into planar lipid membranes and produce a robust single molecule-sensing system (Geng *et al.*, 2011; Jing *et al.*, 2010a,b; Wendell *et al.*, 2009; Haque *et al.*, 2012). Translocation of dsDNA through the membrane-embedded connector pore can be detected by electrophysiological measurements. The structure and length of the dsDNA can be recognized by the characteristic dwell times and current blockades (Wendell *et al.*, 2009). The system is also highly sensitive to ion species and is stable even at extreme pH environments and high salt conditions (Fang *et al.*, 2012; Jing *et al.*, 2010b). As the crystal structure of the connector has been well elucidated, it is possible to reengineer the connector for a wide range of applications, including single molecule sensing, DNA sequencing, gene delivery, drug loading, and bioreactors. It could also be used to provide insights into the mechanism of DNA translocation in viral systems (Jing *et al.*, 2010a). As the connectors in different viral systems share similar morphologies, it is likely that the connector channels of phages other than ϕ29 could be integrated into nanodevices in a similar approach.

C. Bioreactor

Insertion of the motor connector into a lipid membrane has generated new interest in nanobiotechnology (Fang *et al.*, 2012; Geng *et al.*, 2011; Jing *et al.*, 2010a,b; Wendell *et al.*, 2009). Liposomes containing the connector channel are similar to the construction of an active viral DNA packaging motor. This essentially creates an active bioreactor for observing reactions when analytes enter into the reaction chamber gradually through the channel. This will serve as a prototype for other biomimetic DNA packaging motors for future development. It can serve as an active drug or DNA-loading machine against the concentration gradient for drug or gene delivery *in vivo*.

The connector channel-containing liposome can be used for vesicle encapsulation (Okumus *et al.*, 2004; Rhoades *et al.*, 2003) (Fig. 16). The reaction mixture will be encapsulated within this small unilamellar vesicle bioreactor for a variety of reactions such as total internal reflection fluorescence imaging. Vesicles provide an environment close to the native condition of the sample. Vesicles with a diameter of 100–200 nm are smaller than optical resolution and thus lateral motions of the samples within the vesicles will be negligible. α-Hemolysin-containing liposomes have been used to study helicase function and RNA folding by single molecule FRET (Cisse *et al.*, 2007; Okumus *et al.*, 2009). The artificial channel allows the exchange of ions, metabolites, or other small molecules such as ATP, which pass through the channel to reach the encapsulated environment. The ϕ29 connector with a 3.6-nm-wide channel has been inserted successfully into a liposome (Wendell *et al.*, 2009) and its transportation characteristics have been well studied. Similarly, the connector embedded in the liposome can be applied to single molecule observation of a variety of DNA packaging motors, including that of ϕ29 (Fig. 16).

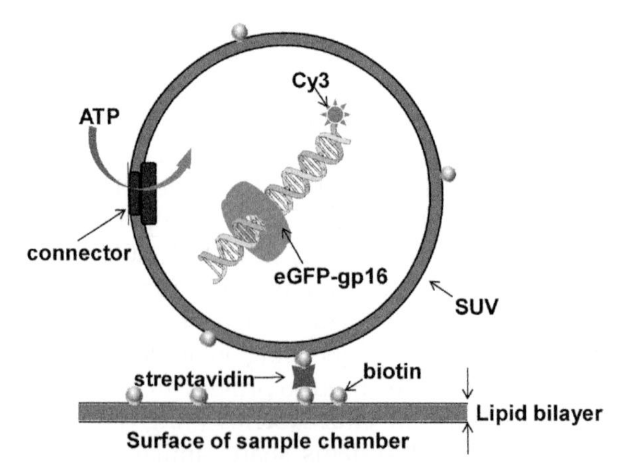

FIGURE 16 Schematic of a porous liposome with ϕ29 connector-embedded single molecule study of motor reaction. (See Page 18 in Color Section at the back of the book.)

The size of the liposomes can be controlled by traditional extrusion through filters of a defined pore size.

For example, to initiate the reaction, buffer containing ATP could be infused into the chamber. Real-time videos would then be recorded by a single molecule fluorescence imaging system (Shu *et al.*, 2007; Zhang *et al.*, 2007, 2010). As the porous liposome allows the exchange of buffers, ATP would diffuse into the liposome through the connector channel. The gp16 ring fixed to the dsDNA chain by γ-s-ATP could then resume its motion by hydrolyzing ATP. Using fluorescently labeled protein and DNA, eGFP-gp16 and Cy3-dsDNA, respectively, changes in FRET efficiency could be observed if the gp16 ring is moving along the chain until it moves out of the FRET range or eventually falls off the chain. Due to the limited volume of the liposome, gp16 may rebind to the DNA chain quickly and start another cycle. Results will provide direct evidence that the ATPase gp16 slides along the dsDNA chain. Other topics concerning motor motion (Balci *et al.*, 2005), interaction of motor components (Wang *et al.*, 2005), or kinetics related to ATP concentration (Lee *et al.*, 2008; Moffitt *et al.*, 2009) can also be examined.

D. Gene therapy in nanomedicine

One possible application of viral packaging motors or its module in nanomedicine is to use the motor to inject therapeutic DNA into targeted cells for disease treatment. The motor, along with its essential components, can be fused into the targeted cell membrane. The membrane-integrated motor may be able to insert functional genes for therapy into the targeted cells, utilizing the energy from ATP hydrolysis, or electrical force, as demonstrated (Fang *et al.*, 2012; Geng *et al.*, 2011; Jing *et al.*, 2010a,b; Wendell *et al.*, 2009).

Additionally, the building blocks of viral packaging motors possess the ability of oligomer formation and thus are ideal candidates in bottom-up assembly to produce functional nanomaterials (Figs. 17 and 18) (Guo, 2005; Khaled *et al.*, 2005; Shu *et al.*, 2004, 2011b; Xiao *et al.*, 2009a,b). The pRNA of the ϕ29 motor has a strong tendency to form dimers, trimers, and hexamers via interlocking loop/loop interactions. A variety of structures and shapes, such as twins, tetramers, rods, triangles, and arrays, several micrometers in size, can be produced using reconstructed ϕ29 pRNA containing a palindrome sequence (Fig. 18). These pRNA nanoparticles were found to be stable and resistant to a wide range of temperatures, salt concentrations, and pH (Shu *et al.*, 2003d, 2004). Utilizing the property of controllable self-assembly of pRNA, RNA nanoparticles can be constructed to carry different therapeutic agents, such as siRNA (Guo *et al.*, 2005a; Khaled *et al.*, 2005), ribozyme (Liu *et al.*, 2007), drugs, and fluorescent or radioactive markers, as well as the targeting ligands of

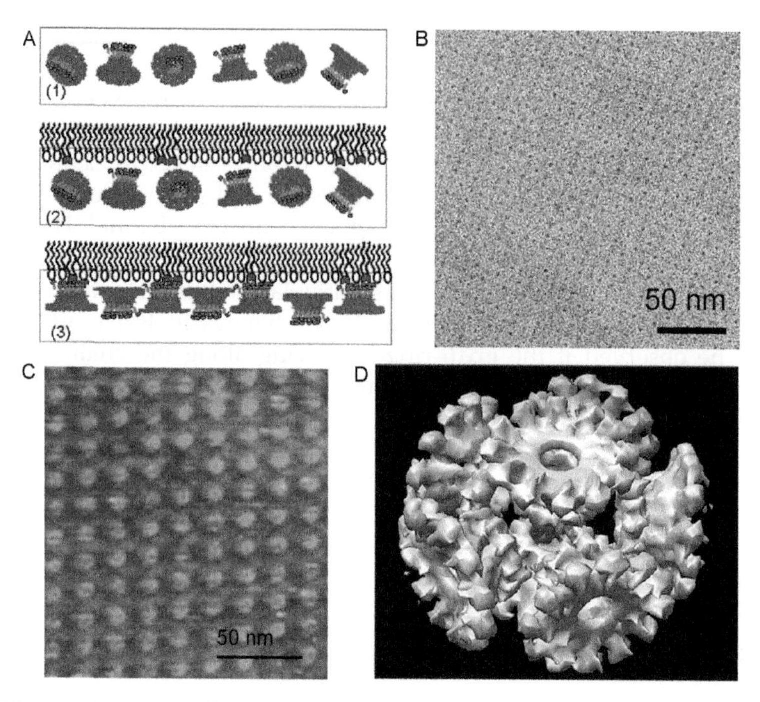

FIGURE 17 Formation of nanostructures by ϕ29 motor connector. (a–c) Formation of massive sheets of connector mediated by lipid bilayers. (a) Scheme of the lipid directed formation of single layer connector array. (b) EM image and (c) Atomic force microscopy image of single layer connector array. (d) Ellipsoid nanoparticles self-assembled from connector. Adapted with permission from Xiao *et al.* (2009b) and Xiao *et al.* (2009a) ©2009 American Chemical Society. (See Page 19 in Color Section at the back of the book.)

folate (Guo *et al.*, 2005a, 2006) or RNA aptamer (Guo *et al.*, 2005a; Khaled *et al.*, 2005; Zhou *et al.*, 2011) for targeted delivery to treat cancer, viral infection, and genetic disease (Fig. 19). Moreover, using the pRNA-based nanoparticle for delivery in nanomedicine has an advantage in that the size of the pRNA particles can be controlled to fall in the range of 10–100 nm, the optimal size to enter the cell while avoiding kidney filtration (Guo, 2010; Guo *et al.*, 2010; Shu *et al.*, 2011b).

RNA holds the advantage over DNA in that it can be manipulated as easily as DNA, but it possesses the versatile structure and catalytic activity similar to that of proteins. RNA is therefore a particularly attractive building block for the bottom-up fabrication of nanostructures (Guo, 2010). As mentioned previously, pRNA of the ϕ29 DNA packaging motor has been used as a carrier for delivery of therapeutics for the treatment of diseases (Guo *et al.*, 2005a, 2006; Hoeprich *et al.*, 2003; Khaled *et al.*, 2005; Zhou *et al.*, 2011). However, *in vivo* dissociation of RNA nanoparticles without covalent modifications or cross-linking has been the main negative

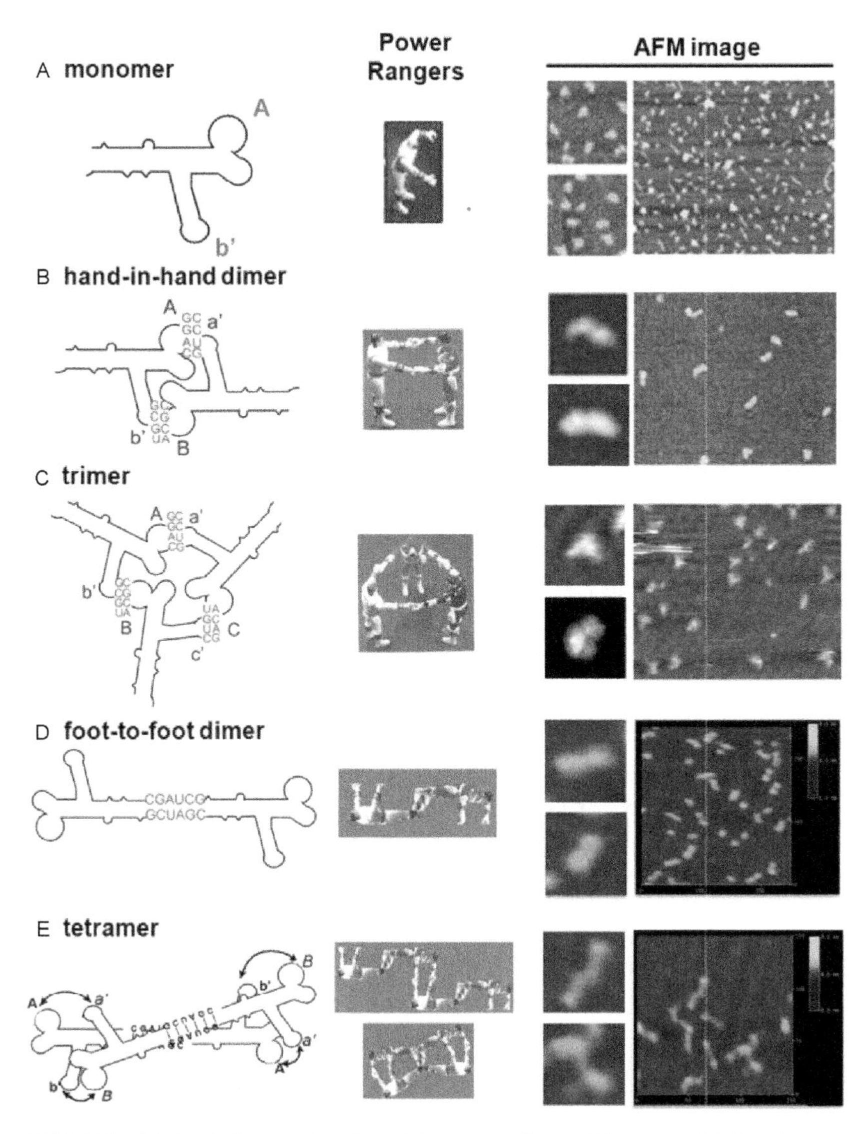

FIGURE 18 Methods for constructing RNA nanoparticles utilizing hand-in-hand and foot-in-foot interactions of ϕ29 motor pRNA systems (Guo *et al.*, 1998; Shu *et al.*, 2011a b). Adapted from Shu *et al.* (2011b) ©2011 with permission from Elsevier. (See Page 20 in Color Section at the back of the book.)

factor that has reduced the delivery efficacy of RNA nanoparticles *in vivo*. The authors discovered the unusual stability of the pRNA three-way junction (3WJ, the pRNA trifurcate domain) of the ϕ29 DNA packaging motor. This pRNA 3WJ (Figs. 9 and 10) has been utilized for fabricating highly stable, multivalent RNA nanoparticles to deliver siRNA to specific

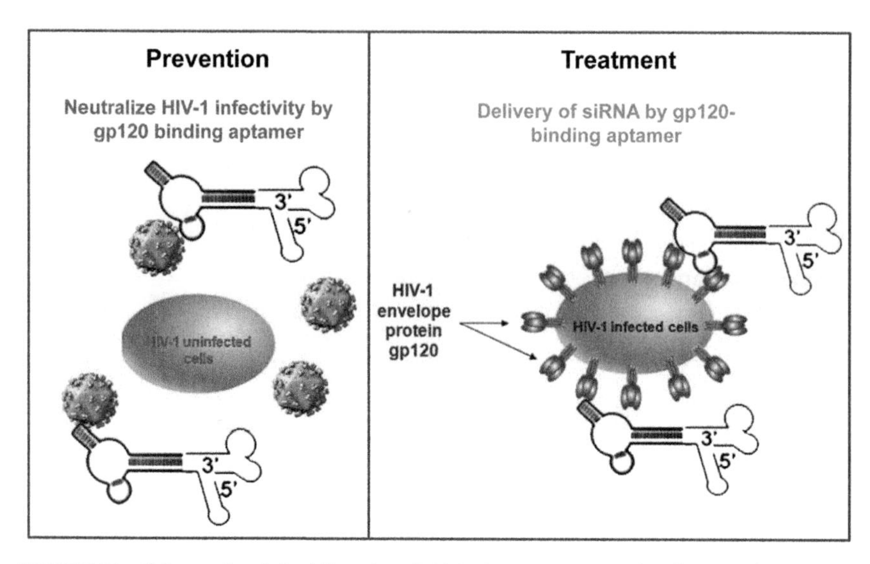

FIGURE 19 Schematic of dual-functional ϕ29 pRNA nanoparticles for HIV therapy. Nanoparticle-carrying p120-binding aptamers can either neutralize HIV-1 infectivity (left) or mediate delivery of siRNA for therapy (right). (Figure courtesy of Dr. Jiehua Zhou and Dr. John Rossi). (See Page 21 in Color Section at the back of the book.)

cancer cells (Shu *et al.*, 2011a). Thermodynamically stable pRNA from three to six pieces of RNA oligomers were assembled without the use of metal salts to form stable multifunctional nanoparticles. Each RNA oligomer contains a receptor-binding ligand, aptamer, siRNA, or ribozyme functional module. When mixed together, they self-assemble into tristar nanoparticles using the 3WJ as a core, as demonstrated by both atomic force microscopy (Fig. 9) and EM imaging (unpublished results). Nanoparticles are resistant to 8 *M* urea denaturation and remain intact at extremely low concentrations *in vitro* or *in vivo*. With a simple chemical modification at the 2′ hydroxyl, these RNA nanoparticles are also resistant to degradation by serum (Shu *et al.*, 2011a). All modules within the nanostructure exhibited independent functionalities for specific cell binding, cell entry, gene silencing, catalytic function, and cancer targeting both *in vitro* and in animal trials (Fig. 20). Modules remain functional *in vitro* and *in vivo*, suggesting that the 3WJ core can be used as a platform for building a variety of multifunctional nanoparticles.

VI. PROSPECTIVE

Studying the mechanism of viral DNA packaging not only promotes understanding of how these strong and powerful motors function, but may also elucidate the mechanism of DNA or RNA translocation through

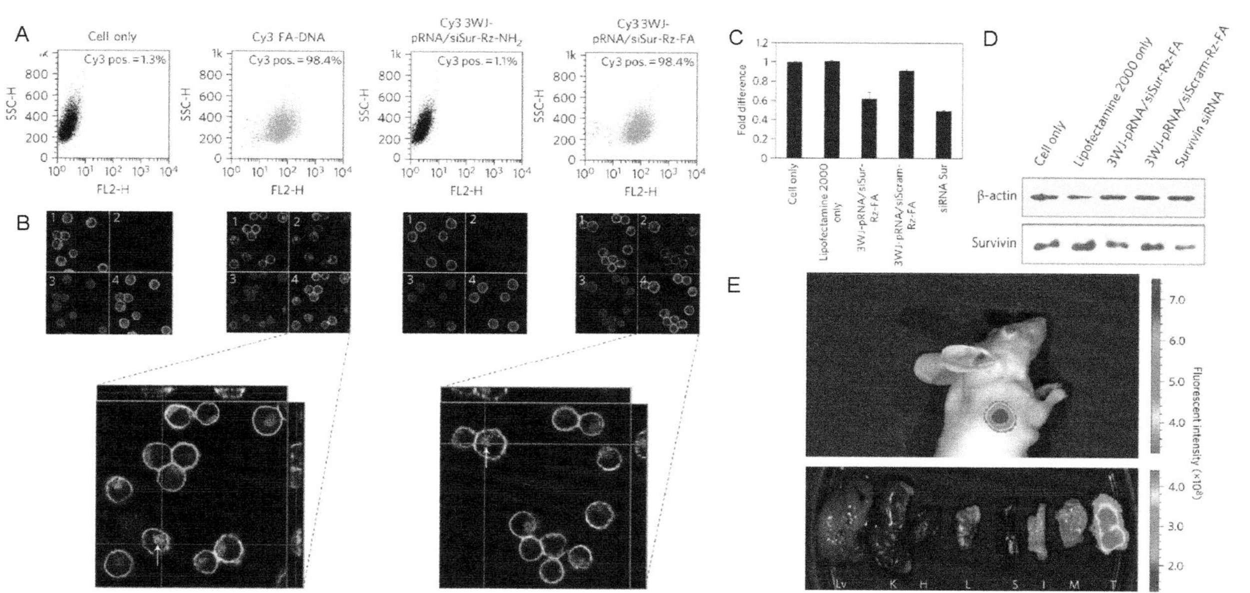

FIGURE 20 *In vitro* and *in vivo* binding and entry of 3WJ-pRNA nanoparticles into targeted cells. Flow cytometry (a) and confocal images (b) revealed binding and specific entry of fluorescent-[3WJ-pRNA-siSur-rZ-FA] nanoparticles into folate receptor-positive (FA^+) cells. Target gene knockdown effects shown by (c) quantitative reverse transcriptase polymerase chain reaction with GADPH as endogenous control and by (d) Western blot assay with β-actin as endogenous control. (e) 3WJ-pRNA nanoparticles target FA^+ tumor xenografts upon systemic administration in nude mice. (Lv, liver; K, kidney; H, heart; L, lung; S, spleen; I, intestine; M, muscle; T, tumor). Adapted with permission from Macmillan Publishers Ltd. (Shu *et al*., 2011a) ©2011. (See Page 22 in Color Section at the back of the book.)

cell membranes, nucleic acid repair, and DNA replication or RNA transcription. Viral motors also possess the potential to be applied to nanotechnology and nanomedicine, including targeted delivery, precise single molecule sensing, high-throughput dsDNA sequencing, or direct incorporation into nanodevices. Further understanding of the mechanism of viral motor function would facilitate the realization of these applications.

ACKNOWLEDGMENTS

Research was supported by NIH R01 EB003730, R01 EB012135, U01 CA151648, R01 GM059944, and NIH Nanomedicine Development Center: ϕ29 DNA Packaging Motor for Nanomedicine, through the NIH Roadmap for Medical Research (PN2 EY 018230) directed by Guo. Guo is a cofounder of Kylin Therapeutics, Inc., and Biomotor and Nucleic Acids Nanotech Development, Ltd.

REFERENCES

Aathavan, K., Politzer, A. T., Kaplan, A., Moffitt, J. R., Chemla, Y. R., Grimes, S., Jardine, P. J., Anderson, D. L., and Bustamante, C. (2009). Substrate interactions and promiscuity in a viral DNA packaging motor. *Nature* **461:**669–673.

Agirrezabala, X., Martin-Benito, J., Valle, M., Gonzalez, J. M., Valencia, A., Valpuesta, J. M., and Carrascosa, J. L. (2005). Structure of the connector of bacteriophage T7 at 8A resolution: Structural homologies of a basic component of a DNA translocating machinery. *J. Mol. Biol.* **347:**895–902.

Ammelburg, M., Frickey, T., and Lupas, A. N. (2006). Classification of AAA+ proteins. *J. Struct. Biol.* **156:**2–11.

Astumian, R. D. (1997). Thermodynamics and kinetics of a Brownian motor. *Science* **276:**917–922.

Atz, R., Ma, S., Gao, J., Anderson, D. L., and Grimes, S. (2007). Alanine scanning and Fe-BABE probing of the bacteriophage ϕ29 prohead RNA-connector interaction. *J. Mol. Biol.* **369:**239–248.

Badasso, M. O., Leiman, P. G., Tao, Y., He, Y., Ohlendorf, D. H., Rossmann, M. G., and Anderson, D. (2000). Purification, crystallization and initial X-ray analysis of the head-tail connector of bacteriophage ϕ29. *Acta Crystallogr. D Biol. Crystallogr.* **56:**1187–1190.

Bailey, S., Wichitwechkarn, J., Johnson, D., Reilly, B., Anderson, D., and Bodley, J. W. (1990). Phylogenetic analysis and secondary structure of the *Bacillus subtilis* bacteriophage RNA required for DNA packaging. *J. Biol. Chem.* **265:**22365–22370.

Balci, H., Ha, T., Sweeney, H. L., and Selvin, P. R. (2005). Interhead distance measurements in myosin VI via SHRImP support a simplified hand-over-hand model. *Biophys. J.* **89:**413–417.

Baneyx, G., and Vogel, V. (1999). Self-assembly of fibronectin into fibrillar networks underneath dipalmitoyl phosphatidylcholine monolayers: Role of lipid matrix and tensile forces. *Proc. Natl. Acad. Sci. USA* **96:**12518–12523.

Baumann, R. G., Mullaney, J., and Black, L. W. (2006). Portal fusion protein constraints on function in DNA packaging of bacteriophage T4. *Mol. Microbiol.* **61:**16–32.

Berry, R. M. (2001). Bacterial flagella: Flagellar motor. *Encyclopedia of Life Sciences.*

Black, L. W. (1981). *In vitro* packaging of bacteriophage T4 DNA. *Virology* **113:**336–344.

Bogner, E. (2002). Human cytomegalovirus terminase as a target for antiviral chemotherapy. *Rev. Med. Virol.* **12:**115–127.

Bothe, J. R., Nikolova, E. N., Eichhorn, C. D., Chugh, J., Hansen, A. L., and Al-Hashimi, H. M. (2011). Characterizing RNA dynamics at atomic resolution using solution-state NMR spectroscopy. *Nat. Methods* **8:**919–931.

Casjens, S. R. (2011). The DNA-packaging nanomotor of tailed bacteriophages. *Nat. Rev. Microbiol.* **9:**647–657.

Chang, C., Zhang, H., Shu, D., Guo, P., and Savran, C. (2008). Bright-field analysis of ϕ29 DNA packaging motor using a magnetomechanical system. *Appl. Phys. Lett.* **93:**153902–153903.

Chemla, Y. R., Aathavan, K., Michaelis, J., Grimes, S., Jardine, P. J., Anderson, D. L., and Bustamante, C. (2005). Mechanism of force generation of a viral DNA packaging motor. *Cell* **122:**683–692.

Chen, C., and Guo, P. (1997). Sequential action of six virus-encoded DNA-packaging RNAs during phage ϕ29 genomic DNA translocation. *J. Virol.* **71:**3864–3871.

Chen, C., Sheng, S., Shao, Z., and Guo, P. (2000). A dimer as a building block in assembling RNA: A hexamer that gears bacterial virus ϕ29 DNA-translocating machinery. *J. Biol. Chem.* **275**(23)**:**17510–17516.

Chen, C., Trottier, M., and Guo, P. (1997). New approaches to stoichiometry determination and mechanism investigation on RNA involved in intermediate reactions. *Nucleic Acids Symp. Ser.* **36:**190–193.

Chen, C., Zhang, C., and Guo, P. (1999). Sequence requirement for hand-in-hand interaction in formation of pRNA dimers and hexamers to gear ϕ29 DNA translocation motor. *RNA* **5:**805–818.

Cingolani, G., Moore, S. D., Prevelige, P. E., Jr., and Johnson, J. E. (2002). Preliminary crystallographic analysis of the bacteriophage P22 portal protein. *J. Struct. Biol.* **139:**46–54.

Cisse, I., Okumus, B., Joo, C., and Ha, T. (2007). Fueling protein DNA interactions inside porous nanocontainers. *Proc. Natl. Acad. Sci. USA* **104:**12646–12650.

Deamer, D. W., and Akeson, M. (2000). Nanopores and nucleic acids: Prospects for ultra-rapid sequencing. *Trends Biotechnol.* **18:**147–151.

de Haas, F., Paatero, A. O., Mindich, L., Bamford, D. H., and Fuller, S. D. (1999). A symmetry mismatch at the site of RNA packaging in the polymerase complex of dsRNA bacteriophage phi6. *J. Mol. Biol.* **294:**357–372.

Ding, F., Lu, C., Zhao, W., Rajashankar, K. R., Anderson, D. L., Jardine, P. J., Grimes, S., and Ke, A. (2011). Structure and assembly of the essential RNA ring component of a viral DNA packaging motor. Proc. Natl. Acad. Sci, USA.

Dittrich, M., and Schulten, K. (2006). PcrA helicase, a prototype ATP-driven molecular motor. *Structure* **14:**1345–1353.

Earnshaw, W. C., and Casjens, S. R. (1980). DNA packaging by the double-stranded DNA bacteriophages. *Cell* **21:**319–331.

Fang, H., Jing, P., Haque, F., and Guo, P. (2012). Role of channel lysines and "push through a one-way valve" mechanism of viral DNA packaging Motor *Biophys. J.* **102:**127–135.

Fang, Y., Cai, Q., and Qin, P. Z. (2005). The procapsid binding domain of ϕ29 packaging RNA has a modular architecture and requires 2′-hydroxyl groups in packaging RNA interaction. *Biochemistry* **44:**9348–9358.

Fang, Y., Shu, D., Xiao, F., Guo, P., and Qin, P. Z. (2008). Modular assembly of chimeric ϕ29 packaging RNAs that support DNA packaging. *Biochem. Biophys. Res. Commun.* **372:**589–594.

Fokine, A., Chipman, P. R., Leiman, P. G., Mesyanzhinov, V. V., Rao, V. B., and Rossmann, M. G. (2004). Molecular architecture of the prolate head of bacteriophage T4. *Proc. Natl. Acad. Sci. USA* **101:**6003–6008.

Frickey, T., and Lupas, A. N. (2004). Phylogenetic analysis of AAA proteins. *J. Struct. Biol.* **146:**2–10.

Fu, C., and Prevelige, P. (2009). *In vitro* incorporation of the phage Phi29 connector complex. *Virology* **394:**149–153.

Fujisawa, H., and Morita, M. (1997). Phage DNA packaging. *Genes Cells* **2:**537–545.

Fujisawa, H., Shibata, H., and Kato, H. (1991). Analysis of interactions among factors involved in the bacteriophage T3 DNA packaging reaction in a defined *in vitro* system. *Virology* **185:**788–794.

Fuller, D. N., Raymer, D. M., Kottadiel, V. I., Rao, V. B., and Smith, D. E. (2007a). Single phage T4 DNA packaging motors exhibit large force generation, high velocity, and dynamic variability. *Proc. Natl. Acad. Sci. USA* **104:**16868–16873.

Fuller, D. N., Raymer, D. M., Rickgauer, J. P., Robertson, R. M., Catalano, C. E., Anderson, D. L., Grimes, S., and Smith, D. E. (2007b). Measurements of single DNA molecule packaging dynamics in bacteriophage lambda reveal high forces, high motor processivity, and capsid transformations. *J. Mol. Biol.* **373:**1113–1122.

Garver, K., and Guo, P. (1997). Boundary of pRNA functional domains and minimum pRNA sequence requirement for specific connector binding and DNA packaging of phage ϕ29. *RNA* **3:**1068–1079.

Geng, J., Fang, H., Haque, F., Zhang, L., and Guo, P. (2011). Three reversible and controllable discrete steps of channel gating of a viral DNA packaging motor. *Biomaterials* **32:**8234–8242.

Gonzalez-Huici, V., Salas, M., and Hermoso, J. M. (2004). The push-pull mechanism of bacteriophage ϕ29 DNA injection. *Mol. Microbiol.* **52:**529–540.

Grigoriev, D. N., Moll, W., Hall, J., and Guo, P. (2004). Bionanomotor. *In* "Encyclopedia of Nanoscience and Nanotechnology" (H. S. Nalwa, ed.), pp. 361–374. American Scientific Publishers.

Grimes, S., and Anderson, D. (1990). RNA dependence of the bateriophage ϕ29 DNA packaging ATPase. *J. Mol. Biol.* **215:**559–566.

Grimes, S., and Anderson, D. (1997). The bacteriophage ϕ29 packaging proteins supercoil the DNA ends. *J. Mol. Biol.* **266:**901–914.

Grimes, S., Jardine, P. J., and Anderson, D. (2002). Bacteriophage phi 29 DNA packaging. *Adv. Virus Res.* **58:**255–294.

Gu, X., and Schroeder, S. J. (2011). Different sequences show similar quaternary interaction stabilities in prohead viral RNA self-assembly. *J. Biol. Chem.* **286:**14419–14426.

Guasch, A., Pous, J., Ibarra, B., Gomis-Ruth, F. X., Valpuesta, J. M., Sousa, N., Carrascosa, J. L., and Coll, M. (2002). Detailed architecture of a DNA translocating machine: The high-resolution structure of the bacteriophage ϕ29 connector particle. *J. Mol. Biol.* **315:**663–676.

Guasch, A., Pous, J., Parraga, A., Valpuesta, J. M., Carrascosa, J. L., and Coll, M. (1998). Crystallographic analysis reveals the 12-fold symmetry of the bacteriophage ϕ29 connector particle. *J. Mol. Biol.* **281:**219–225.

Guo, P. (2002). Structure and function of ϕ29 hexameric RNA that drive viral DNA packaging motor: Review. *Prog. Nucleic Acid Res. Mol. Biol.* **72:**415–472.

Guo, P. (2005). RNA nanotechnology: Engineering, assembly and applications in detection, gene delivery and therapy. *J. Nanosci. Nanotechnol.* **5**(12)**:**1964–1982.

Guo, P. (2010). The emerging field of RNA nanotechnology. *Nat. Nanotechnol.* **5:**833–842.

Guo, P., Bailey, S., Bodley, J. W., and Anderson, D. (1987a). Characterization of the small RNA of the bacteriophage ϕ29 DNA packaging machine. *Nucleic Acids Res.* **15:**7081–7090.

Guo, P., Coban, O., Snead, N. M., Trebley, J., Hoeprich, S., Guo, S., and Shu, Y. (2010). Engineering RNA for targeted siRNA delivery and medical application. *Adv. Drug Deliv. Rev.* **62:**650–666.

Guo, P., Erickson, S., and Anderson, D. (1987b). A small viral RNA is required for *in vitro* packaging of bacteriophage ϕ29 DNA. *Science* **236:**690–694.

Guo, P., Erickson, S., Xu, W., Olson, N., Baker, T. S., and Anderson, D. (1991). Regulation of the phage ϕ29 prohead shape and size by the portal vertex. *Virology* **183:**366–373.

Guo, P., Grimes, S., and Anderson, D. (1986). A defined system for *in vitro* packaging of DNA-gp3 of the *Bacillus subtilis* bacteriophage ϕ29. *Proc. Natl. Acad. Sci. USA* **83:**3505–3509.

Guo, P., Peterson, C., and Anderson, D. (1987c). Initiation events in *in vitro* packaging of bacteriophage ϕ29 DNA-gp3. *J. Mol. Biol.* **197:**219–228.

Guo, P., Peterson, C., and Anderson, D. (1987d). Prohead and DNA-gp3-dependent ATPase activity of the DNA packaging protein gp16 of bacteriophage ϕ29. *J. Mol. Biol.* **197:**229–236.

Guo, P., and Trottier, M. (1994). Biological and biochemical properties of the small viral RNA (pRNA) essential for the packaging of the double-stranded DNA of phage ϕ29. *Semin. Virol.* **5:**27–37.

Guo, P., Zhang, C., Chen, C., Trottier, M., and Garver, K. (1998). Inter-RNA interaction of phage ϕ29 pRNA to form a hexameric complex for viral DNA transportation. *Mol. Cell.* **2:**149–155.

Guo, P. X., and Lee, T. J. (2007). Viral nanomotors for packaging of dsDNA and dsRNA. *Mol. Microbiol.* **64:**886–903.

Guo, S., Huang, F., and Guo, P. (2006). Construction of folate-conjugated pRNA of bacteriophage ϕ29 DNA packaging motor for delivery of chimeric siRNA to nasopharyngeal carcinoma cells. *Gene Ther.* **13:**814–820.

Guo, S., Tschammer, N., Mohammed, S., and Guo, P. (2005a). Specific delivery of therapeutic RNAs to cancer cells via the dimerization mechanism of ϕ29 motor pRNA. *Hum. Gene Ther.* **16:**1097–1109.

Guo, Y., Blocker, F., Xiao, F., and Guo, P. (2005b). Construction and 3-D computer modeling of connector arrays with tetragonal to decagonal transition induced by pRNA of ϕ29 DNA-packaging motor. *J. Nanosci. Nanotechnol.* **5:**856–863.

Hanson, P. I., and Whiteheart, S. W. (2005). AAA+ proteins: Have engine, will work. *Nat. Rev. Mol. Cell Biol.* **6:**519–529.

Haque, F., Lunn, J., Fang, H., Smithrud, D., and Guo, P. (2012). Real-Time Sensing and Discrimination of Single Chemicals Using the Channel of ϕ29 DNA Packaging Nanomotor. *ACS Nano.* In press. DOI: 10.1021/nn3001615.

Hendrix, R. W. (1978). Symmetry mismatch and DNA packaging in large bacteriophages. *Proc. Natl. Acad. Sci. USA* **75:**4779–4783.

Hendrix, R. W. (1998). Bacteriophage DNA packaging: RNA gears in a DNA transport machine (minireview). *Cell* **94:**147–150.

Hess, H., and Vogel, V. (2001). Molecular shuttles based on motor proteins: Active transport in synthetic environments. *J. Biotechnol.* **82:**67–85.

Hoeprich, S., and Guo, P. (2002). Computer modeling of three-dimensional structure of DNA-packaging RNA(pRNA) monomer, dimer, and hexamer of ϕ29 DNA Packaging motor. *J. Biol. Chem.* **277**(23)**:**20794–20803.

Hoeprich, S., Zhou, Q., Guo, S., Qi, G., Wang, Y., and Guo, P. (2003). Bacterial virus ϕ29 pRNA as a hammerhead ribozyme escort to destroy hepatitis B virus. *Gene Ther.* **10:**1258–1267.

Hohn, T. (1976). Packaging of genomes in bacteriophages: A comparison of ssRNA bacteriophages and dsDNA bacteriophages. *Philos. Trans. R Soc. Lond. B* **276:**143–150.

Hua, W., Young, E. C., Fleming, M. L., and Gelles, J. (1997). Coupling of kinesin steps to ATP hydrolysis. *Nature* **388:**390–393.

Huang, L. P., and Guo, P. (2003a). Use of acetone to attain highly active and soluble DNA packaging protein gp16 of ϕ29 for ATPase assay. *Virology* **312**(2)**:**449–457.

Huang, L. P., and Guo, P. (2003b). Use of PEG to acquire highly soluble DNA-packaging enzyme gp16 of bacterial virus ϕ29 for stoichiometry quantification. *J. Virol. Methods* **109:**235–244.

Hugel, T., Michaelis, J., Hetherington, C. L., Jardine, P. J., Grimes, S., Walter, J. M., Faik, W., Anderson, D. L., and Bustamante, C. (2007). Experimental test of connector rotation during DNA packaging into bacteriophage ϕ29 capsids. *Plos Biol.* **5:**558–567.

Hwang, Y., Catalano, C. E., and Feiss, M. (1996). Kinetic and mutational dissection of the two ATPase activities of terminase, the DNA packaging enzyme of bacteriophage lambda. *Biochemistry* **35:**2796–2803.

Hwang, Y., and Feiss, M. (1997). A defined system for *in vitro* lambda DNA packaging. *Virology* **211:**367–376.

Ibarra, B., Caston, J. R., Llorca, O., Valle, M., Valpuesta, J. M., and Carrascosa, J. L. (2000). Topology of the components of the DNA packaging machinery in the phage ϕ29 prohead. *J. Mol. Biol.* **298:**807–815.

Ibarra, B., Valpuesta, J. M., and Carrascosa, J. L. (2001). Purification and functional characterization of p16, the ATPase of the bacteriophage ϕ29 packaging machinary. *Nucleic Acids Res.* **29**(21)**:**4264–4273.

Ionel, A., Velazquez-Muriel, J. A., Luque, D., Cuervo, A., Caston, J. R., Valpuesta, J. M., Martin-Benito, J., and Carrascosa, J. L. (2011). Molecular rearrangements involved in the capsid shell maturation of bacteriophage T7. *J. Biol. Chem.* **286:**234–242.

Ishijima, A., Kojima, H., Funatsu, T., Tokunaga, M., Higuchi, H., Tanaka, H., and Yanagida, T. (1998). Simultaneous observation of individual ATPase and mechanical events by a single myosin molecule during interaction with actin. *Cell* **92:**161–171.

Iyer, L. M., Leipe, D. D., Koonin, E. V., and Aravind, L. (2004a). Evolutionary history and higher order classification of AAA plus ATPases. *J. Struct. Biol.* **146:**11–31.

Iyer, L. M., Makarova, K. S., Koonin, E. V., and Aravind, L. (2004b). Comparative genomics of the FtsK-HerA superfamily of pumping ATPases: Implications for the origins of chromosome segregation, cell division and viral capsid packaging. *Nucleic Acids Res.* **32:**5260–5279.

Jiang, W., Chang, J., Jakana, J., Weigele, P., King, J., and Chiu, W. (2006). Structure of epsilon15 bacteriophage reveals genome organization and DNA packaging/injection apparatus. *Nature* **439:**612–616.

Jimenez, J., Santisteban, A., Carazo, J. M., and Carrascosa, J. L. (1986). Computer graphic display method for visualizing three-dimensional biological structures. *Science* **232:**1113–1115.

Jing, P., Haque, F., Shu, D., Montemagno, C., and Guo, P. (2010a). One-way traffic of a viral motor channel for double-stranded DNA translocation. *Nano Lett.* **10**(9)**:**3620–3627.

Jing, P., Haque, F., Vonderheide, A., Montemagno, C., and Guo, P. (2010b). Robust properties of membrane-embedded connector channel of bacterial virus ϕ29 DNA packaging motor. *Mol. BioSyst.* **6:**1844–1852.

Kellenberger, E. (1976). DNA viruses: Cooperativity and regulation through conformational changes as features of phage assembly. *Philos. Trans. R Soc. Lond. B Biol. Sci.* **276:**3–13.

Kemp, P., Gupta, M., and Molineux, I. J. (2004). Bacteriophage T7 DNA ejection into cells is initiated by an enzyme-like mechanism. *Mol. Microbiol.* **53:**1251–1265.

Khaled, A., Guo, S., Li, F., and Guo, P. (2005). Controllable self-assembly of nanoparticles for specific delivery of multiple therapeutic molecules to cancer cells using RNA nanotechnology. *Nano Lett.* **5:**1797–1808.

Khan, S. A., Hayes, S. J., Wright, E. T., Watson, R. H., and Serwer, P. (1995). Specific single-stranded breaks in mature bacteriophage T7 DNA. *Virology* **211:**329–331.

Ko, S. H., Chen, Y., Shu, D., Guo, P., and Mao, C. (2008). Reversible switching of pRNA activity on the DNA packaging motor of bacteriophage ϕ29. *J. Am. Chem. Soc.* **130:**17684–17687.

Koti, J. S., Morais, M. C., Rajagopal, R., Owen, B. A., McMurray, C. T., and Anderson, D. (2008). DNA packaging motor assembly intermediate of bacteriophage ϕ29. *J. Mol. Biol.* **381:**1114–1132.

Kravats, A., Jayasinghe, M., and Stan, G. (2011). Unfolding and translocation pathway of substrate protein controlled by structure in repetitive allosteric cycles of the ClpY ATPase. *Proc. Natl. Acad. Sci. USA* **108:**2234–2239.

Lander, G. C., Tang, L., Casjens, S. R., Gilcrease, E. B., Prevelige, P., Poliakov, A., Potter, C. S., Carragher, B., and Johnson, J. E. (2006). The structure of an infectious P22 virion shows the signal for headful DNA packaging. *Science* **312:**1791–1795.

Lebedev, A. A., Krause, M. H., Isidro, A. L., Vagin, A. A., Orlova, E. V., Turner, J., Dodson, E. J., Tavares, P., and Antson, A. A. (2007). Structural framework for DNA translocation via the viral portal protein. *EMBO J.* **26:**1984–1994.

Lee, B. S., Lee, S. C., and Holliday, L. S. (2003). Biochemistry of mechanoenzymes: Biological motors for nanotechnology. *Biomed. Microdev.* **5:**269–280.

Lee, C. S., and Guo, P. (1995a). *In vitro* assembly of infectious virions of ds-DNA phage ϕ29 from cloned gene products and synthetic nucleic acids. *J. Virol.* **69:**5018–5023.

Lee, C. S., and Guo, P. (1995b). Sequential interactions of structural proteins in phage ϕ29 procapsid assembly. *J. Virol.* **69:**5024–5032.

Lee, T. J., and Guo, P. (2006). Interaction of gp16 with pRNA and DNA for genome packaging by the motor of bacterial virus ϕ29. *J. Mol. Biol.* **356:**589–599.

Lee, T. J., Schwartz, C., and Guo, P. (2009). Construction of bacteriophage Phi29 DNA packaging motor and its applications in nanotechnology and therapy. *Ann. Biomed. Eng.* **37:**2064–2081.

Lee, T. J., Zhang, H., Liang, D., and Guo, P. (2008). Strand and nucleotide-dependent ATPase activity of gp16 of bacterial virus ϕ29 DNA packaging motor. *Virology* **380:**69–74.

Lhuillier, S., Gallopin, M., Gilquin, B., Brasiles, S., Lancelot, N., Letellier, G., Gilles, M., Dethan, G., Orlova, E. V., Couprie, J., Tavares, P., and Zinn-Justin, S. (2009). Structure of bacteriophage SPP1 head-to-tail connection reveals mechanism for viral DNA gating. *Proc. Natl. Acad. Sci. USA* **106:**8507–8512.

Liu, H., Guo, S., Roll, R., Li, J., Diao, Z., Shao, N., Riley, M. R., Cole, A. M., Robinson, J. P., Snead, N. M., Shen, G., and Guo, P. (2007). Phi29 pRNA vector for efficient escort of hammerhead ribozyme targeting survivin in multiple cancer cells. *Cancer Biol. Ther.* **6:**697–704.

Llorca, O., Marco, S., Carrascosa, J. L., and Valpuesta, J. M. (1997). Conformational changes in the GroEL oligomer during the functional cycle. *J. Struct. Biol.* **118:**31–42.

Lurz, R., Orlova, E. V., Gunther, D., Dube, P., Droge, A., Weise, F., van Heel, M., and Tavares, P. (2001). Structural organisation of the head-to-tail interface of a bacterial virus 1. *J. Mol. Biol.* **310:**1027–1037.

Maluf, N. K., and Feiss, M. (2006). Virus DNA translocation: Progress towards a first ascent of mount pretty difficult. *Mol. Microbiol.* **61:**1–4.

Mancini, E. J., Kainov, D. E., Grimes, J. M., Tuma, R., Bamford, D. H., and Stuart, D. I. (2004). Atomic snapshots of an RNA packaging motor reveal conformational changes linking ATP hydrolysis to RNA translocation. *Cell* **118:**743–755.

Masker, W. E. (1982). *In vitro* packaging of bacteriophage T7 DNA requires ATP. *J. Virol.* **43:**365–367.

Maurizi, M. R., and Li, C. C. (2001). AAA proteins: In search of a common molecular basis. *International Meeting on Cellular Functions of AAA Proteins. EMBO Rep.* **2:**980–985.

Miyagishi, M., Sumimoto, H., Miyoshi, H., Kawakami, Y., and Taira, K. (2004). Optimization of an siRNA-expression system with an improved hairpin and its significant suppressive effects in mammalian cells. *J. Gene Med.* **6:**715–723.

Moffitt, J. R., Chemla, Y. R., Aathavan, K., Grimes, S., Jardine, P. J., Anderson, D. L., and Bustamante, C. (2009). Intersubunit coordination in a homomeric ring ATPase. *Nature* **457:**446–450.

Moll, D., and Guo, P. (2007). Grouping of ferritin and gold nanoparticles conjugated to pRNA of the phage ϕ29 DNA-packaging motor. *J. Nanosci. Nanotech.* **7:**3257–3267.

Moon, J. M., Akin, D., Xuan, Y., Ye, P., Guo, P., and Bashir, R. (2009). Capture and alignment of ϕ29 viral particles in sub-40 nanometer porous alumina membranes. *Biomed. Microdev.* **11:**135–142.

Morais, M. C., Koti, J. S., Bowman, V. D., Reyes-Aldrete, E., Anderson, D., and Rossman, M. G. (2008). Defining molecular and domain boundaries in the bacteriophage ϕ29 DNA packaging motor. *Structure* **16:**1267–1274.

Morais, M. C., Tao, Y., Olsen, N. H., Grimes, S., Jardine, P. J., Anderson, D., Baker, T. S., and Rossmann, M. G. (2001). Cryoelectron-microscopy image reconstruction of symmetry mismatches in bacteriophage ϕ29. *J. Struct. Biol.* **135:**38–46.

Morita, M., Tasaka, M., and Fujisawa, H. (1993). DNA packaging ATPase of bacteriophage T3. *Virology* **193:**748–752.

Morita, M., Tasaka, M., and Fujisawa, H. (1995). Structural and functional domains of the large subunit of the bacteriophage T3 DNA packaging enzyme: Importance of the C-terminal region in prohead binding. *J. Mol. Biol.* **245:**635–644.

Niemeyer, C. M. (2002). The developments of semisynthetic DNA-protein conjugates. *Trends Biotechnol.* **20:**395–401.

Okumus, B., Arslan, S., Fengler, S. M., Myong, S., and Ha, T. (2009). Single molecule nanocontainers made porous using a bacterial toxin. *J. Am. Chem. Soc.* **131:**14844–14849.

Okumus, B., Wilson, T. J., Lilley, D. M., and Ha, T. (2004). Vesicle encapsulation studies reveal that single molecule ribozyme heterogeneities are intrinsic. *Biophys. J.* **87:**2798–2806.

Oliveira, L., Alonso, J. C., and Tavares, P. (2005). A defined *in vitro* system for DNA packaging by the bacteriophage SPP1: Insights into the headful packaging mechanism. *J. Mol. Biol.* **353:**529–539.

Oram, M., Sabanayagam, C., and Black, L. W. (2008). Modulation of the packaging reaction of bacteriophage T4 terminase by DNA structure. *J. Mol. Biol.* **381:**61–72.

Oyama, T., Ishino, Y., Cann, I. K., Ishino, S., and Morikawa, K. (2001). Atomic structure of the clamp loader small subunit from *Pyrococcus furiosus*. *Mol. Cell.* **8:**455–463.

Park, H. H., Jamison, A. C., and Lee, T. R. (2007). Rise of the nanomachine: The evolution of a revolution in medicine. *Nanomed. (Lond.)* **2:**425–439.

Pyle, A. M. (2008). Translocation and unwinding mechanisms of RNA and DNA helicases. *Annu. Rev. Biophys.* **37:**317–336.

Rao, V. B., and Black, L. W. (1988). Cloning, overexpression and purification of the terminase proteins gp16 and gp17 of bacteriophage T4: Construction of a defined *in vitro* DNA packaging system using purified terminase proteins. *J. Mol. Biol.* **200:**475–488.

Rao, V. B., and Feiss, M. (2008). The bacteriophage DNA packaging motor. *Annu. Rev. Genet.* **42:**647–681.

Ray, K., Ma, J., Oram, M., Lakowicz, J. R., and Black, L. W. (2010a). Single-molecule and FRET fluorescence correlation spectroscopy analyses of phage DNA packaging: Colocalization of packaged phage T4 DNA ends within the capsid. *J. Mol. Biol.* **395:**1102–1113.

Ray, K., Sabanayagam, C. R., Lakowicz, J. R., and Black, L. W. (2010b). DNA crunching by a viral packaging motor: Compression of a procapsid-portal stalled Y-DNA substrate. *Virology* **398:**224–232.

Reid, R. J. D., Bodley, J. W., and Anderson, D. (1994a). Characterization of the prohead-pRNA interaction of bacteriophage ϕ29. *J. Biol. Chem.* **269:**5157–5162.

Reid, R. J. D., Bodley, J. W., and Anderson, D. (1994b). Identification of bacteriophage ϕ29 prohead RNA (pRNA) domains necessary for *in vitro* DNA-gp3 packaging. *J. Biol. Chem.* **269:**9084–9089.

Reid, R. J. D., Zhang, F., Benson, S., and Anderson, D. (1994c). Probing the structure of bacteriophage ϕ29 prohead RNA with specific mutations. *J. Biol. Chem.* **269:**18656–18661.

Rhoades, E., Gussakovsky, E., and Haran, G. (2003). Watching proteins fold one molecule at a time. *Proc. Natl. Acad. Sci. USA* **100:**3197–3202.

Rickgauer, J. P., Fuller, D. N., Grimes, S., Jardine, P. J., Anderson, D. L., and Smith, D. E. (2008). Portal motor velocity and internal force resisting viral DNA packaging in bacteriophage phi 29. *Biophys. J.* **94:**159–167.

Riemer, S. C., and Bloomfield, V. A. (1978). Packaging of DNA in bacteriophage heads: Some considerations on energetics. *Biopolymers* **17:**785–794.

Robinson, M. A., Wood, J. P., Capaldi, S. A., Baron, A. J., Gell, C., Smith, D. A., and Stonehouse, N. J. (2006). Affinity of molecular interactions in the bacteriophage ϕ29 DNA packaging motor. *Nucleic Acids Res.* **34:**2698–2709.

Roos, W. H., Ivanovska, I. L., Evilevitch, A., and Wuite, G. J. L. (2007). Viral capsids: Mechanical characteristics, genome packaging and delivery mechanisms. *Cell. Mol. Life Sci.* **64:**1484–1497.

Sabanayagam, C. R., Oram, M., Lakowicz, J. R., and Black, L. W. (2007). Viral DNA packaging studied by fluorescence correlation spectroscopy. *Biophys. J.* **93:**L17–L19.

Sambongi, Y., Iko, Y., Tanabe, M., Omote, H., Iwamoto-Kihara, A., Ueda, I., Yanagida, T., Wada, Y., and Futai, M. (1999). Mechanical rotation of the c subunit oligomer in ATP synthase (F0F1): Direct observation. *Science* **286:**1722–1724.

Schmidt, O. G., and Eberl, K. (2001). Nanotechnology: Thin solid films roll up into nanotubes. *Nature* **410:**168.

Schwartz, C., Fang, H., Huang, L., and Guo, P. (2012). Sequential action of ATPase, ATP, ADP, Pi and dsDNA in procapsid-free system to enlighten mechanism in viral dsDNA packaging. *Nucelic Acids Res.* **40**:2577–2586

Serwer, P. (1988). The source of energy for bacteriophage DNA packaging: An osmotic pump explains the data. *Biopolymers* **27:**165–169.

Serwer, P. (2003). Models of bacteriophage DNA packaging motors. *J. Struct. Biol.* **141:**179–188.

Serwer, P. (2010). A hypothesis for bacteriophage DNA packaging motors. *Viruses-Basel* **2:**1821–1843.

Shu, D., and Guo, P. (2003a). A viral RNA that binds ATP and contains an motif similar to an ATP-binding aptamer from SELEX. *J. Biol. Chem.* **278**(9)**:**7119–7125.

Shu, D., and Guo, P. (2003b). Only one pRNA hexamer but multiple copies of the DNA-packaging protein gp16 are needed for the motor to package bacterial virus ϕ29 genomic DNA. *Virology* **309**(1)**:**108–113.

Shu, D., Huang, L., and Guo, P. (2003c). A simple mathematical formula for stoichiometry quantitation of viral and nanobiological assemblage using slopes of log/log plot curves. *J. Virol. Meth.* **115**(1)**:**19–30.

Shu, D., Huang, L., Hoeprich, S., and Guo, P. (2003d). Construction of ϕ29 DNA-packaging RNA (pRNA) monomers, dimers and trimers with variable sizes and shapes as potential parts for nano-devices. *J. Nanosci. Nanotechnol.* **3:**295–302.

Shu, D., Moll, W. D., Deng, Z., Mao, C., and Guo, P. (2004). Bottom-up assembly of RNA arrays and superstructures as potential parts in nanotechnology. *Nano Lett.* **4:**1717–1723.

Shu, D., Shu, Y., Haque, F., Abdelmawla, S., and Guo, P. (2011a). Thermodynamically stable RNA three-way junctions as platform for constructing multifuntional nanoparticles for delivery of therapeutics. *Nat. Nanotechnol.* **6:**658–667.

Shu, D., Zhang, H., Jin, J., and Guo, P. (2007). Counting of six pRNAs of ϕ29 DNA-packaging motor with customized single molecule dual-view system. *EMBO J.* **26:**527–537.

Shu, Y., Cinier, M., Shu, D., and Guo, P. (2011b). Assembly of multifunctional ϕ29 pRNA nanoparticles for specific delivery of siRNA and other therapeutics to targeted cells. *Methods* **54:**204–214.

Simpson, A. A., Leiman, P. G., Tao, Y., He, Y., Badasso, M. O., Jardine, P. J., Anderson, D. L., and Rossman, M. G. (2001). Structure determination of the head-tail connector of bacteriophage ϕ29. *Acta Cryst.* **D57:**1260–1269.

Simpson, A. A., Tao, Y., Leiman, P. G., Badasso, M. O., He, Y., Jardine, P. J., Olson, N. H., Morais, M. C., Grimes, S., Anderson, D. L., Baker, T. S., and Rossmann, M. G. (2000). Structure of the bacteriophage ϕ29 DNA packaging motor. *Nature* **408:**745–750.

Singleton, M. R., Dillingham, M. S., and Wigley, D. B. (2007). Structure and mechanism of helicases and nucleic acid translocases. *Annu. Rev. Biochem.* **76:**23–50.

Smith, D. E., Tans, S. J., Smith, S. B., Grimes, S., Anderson, D. L., and Bustamante, C. (2001). The bacteriophage ϕ29 portal motor can package DNA against a large internal force. *Nature* **413:**748–752.

Soong, R. K., Bachand, G. D., Neves, H. P., Olkhovets, A. G., Craighead, H. G., and Montemagno, C. D. (2000). Powering an inorganic nanodevice with a biomolecular motor. *Science* **290:**1555–1558.

Stock, D., Leslie, A. G., and Walker, J. E. (1999). Molecular architecture of the rotary motor in ATP synthase. *Science* **286:**1700–1705.

Sun, J., Cai, Y., Moll, W. D., and Guo, P. (2006). Controlling bacteriophage ϕ29 DNA-packaging motor by addition or discharge of a peptide at N-terminus of connector protein that interacts with pRNA. *Nucleic Acids Res.* **34**(19)**:**5482–5490.

Sun, S., Kondabagil, K., Draper, B., Alam, T. I., Bowman, V. D., Zhang, Z., Hegde, S., Fokine, A., Rossmann, M. G., and Rao, V. B. (2008). The structure of the phage T4 DNA packaging motor suggests a mechanism dependent on electrostatic forces. *Cell* **135:**1251–1262.

Sun, S., Rao, V. B., and Rossmann, M. G. (2010). Genome packaging in viruses. *Curr. Opin. Struct. Biol.* **20:**114–120.

Tang, J. H., Olson, N., Jardine, P. J., Girimes, S., Anderson, D. L., and Baker, T. S. (2008). DNA poised for release in bacteriophage ϕ29. *Structure* **16:**935–943.

Tao, Y., Olson, N. H., Xu, W., Anderson, D. L., Rossmann, M. G., and Baker, T. S. (1998). Assembly of a tailed bacterial virus and its genome release studied in three dimensions. *Cell* **95:**431–437.

Trottier, M., and Guo, P. (1997). Approaches to determine stoichiometry of viral assembly components. *J. Virol.* **71:**487–494.

Trottier, M., Zhang, C. L., and Guo, P. (1996). Complete inhibition of virion assembly *in vivo* with mutant pRNA essential for phage ϕ29 DNA packaging. *J. Virol.* **70:**55–61.

Trus, B. L., Cheng, N., Newcomb, W. W., Homa, F. L., Brown, J. C., and Steven, A. C. (2004). Structure and polymorphism of the UL6 portal protein of herpes simplex virus type 1. *J. Virol.* **78:**12668–12671.

Valpuesta, J. M., and Carrascosa, J. (1994). Structure of viral connectors and their funciton in bacteriophage assembly and DNA packaging. *Quart. Rev. Biophys.* **27:**107–155.

Visalli, R. J., and van Zeijl, M. (2003). DNA encapsidation as a target for anti-herpesvirus drug therapy. *Antiviral Res.* **59:**73–87.

Walker, J. E., Saraste, M., Runswick, M. J., and Gay, N. J. (1982). Distantly related sequences in the alpha- and beta-subunits of ATP synthase, myosin, kinases and other ATP-requiring enzymes and a common nucleotide binding fold. *EMBO J.* **1:**945–951.

Wang, J. (2004). Nucleotide-dependent domain motions within rings of the RecA/AAA(+) superfamily. *J. Struct. Biol.* **148:**259–267.

Wang, Y. M., Tegenfeldt, J. O., Reisner, W., Riehn, R., Guan, X. J., Guo, L., Golding, I., Cox, E. C., Sturm, J., and Austin, R. H. (2005). Single-molecule studies of repressor-DNA interactions show long-range interactions. *Proc. Natl. Acad. Sci. USA* **102:**9796–9801.

Wendell, D., Jing, P., Geng, J., Subramaniam, V., Lee, T. J., Montemagno, C., and Guo, P. (2009). Translocation of double-stranded DNA through membrane-adapted ϕ29 motor protein nanopores. *Nat. Nanotechnol.* **4:**765–772.

Xiao, F., Cai, Y., Wang, J. C., Green, D., Cheng, R. H., Demeler, B., and Guo, P. (2009a). Adjustable ellipsoid nanoparticles assembled from re-engineered connectors of the bacteriophage ϕ29 DNA packaging motor. *ACS Nano* **3:**2163–2170.

Xiao, F., Moll, D., Guo, S., and Guo, P. (2005). Binding of pRNA to the N-terminal 14 amino acids of connector protein of bacterial phage ϕ29. *Nucleic Acids Res.* **33:**2640–2649.

Xiao, F., Sun, J., Coban, O., Schoen, P., Wang, J. C., Cheng, R. H., and Guo, P. (2009b). Fabrication of massive sheets of single layer patterned arrays using reengineered Phi29 motor dodecamer. *ACS Nano* **3:**100–107.

Xiao, F., Zhang, H., and Guo, P. (2008). Novel mechanism of hexamer ring assembly in protein/RNA interactions revealed by single molecule imaging. *Nucleic Acids Res.* **36** (20)**:**6620–6632.

Yu, J., Moffitt, J., Hetherington, C. L., Bustamante, C., and Oster, G. (2010). Mechanochemistry of a viral DNA packaging motor. *J. Mol. Biol.* **400:**186–203.

Zhang, C. L., Garver, K., and Guo, P. (1995). Inhibition of phage ϕ29 assembly by antisense oligonucleotides targeting viral pRNA essential for DNA packaging. *Virology* **211:**568–576.

Zhang, C. L., Lee, C.-S., and Guo, P. (1994). The proximate 5′ and 3′ ends of the 120-base viral RNA (pRNA) are crucial for the packaging of bacteriophage ϕ29 DNA. *Virology* **201:**77–85.

Zhang, C. L., Tellinghuisen, T., and Guo, P. (1997). Use of circular permutation to assess six bulges and four loops of DNA-packaging pRNA of bacteriophage ϕ29. *RNA* **3:**315–322.

Zhang, F., Lemieux, S., Wu, X., St.-Arnaud, S., McMurray, C. T., Major, F., and Anderson, D. (1998). Function of hexameric RNA in packaging of bacteriophage ϕ29 DNA *in vitro*. *Mol. Cell* **2:**141–147.

Zhang, H., Shu, D., Huang, F., and Guo, P. (2007). Instrumentation and metrology for single RNA counting in biological complexes or nanoparticles by a single molecule dual-view system. *RNA* **13:**1793–1802.

Zhang, H., Shu, D., Wang, W., and Guo, P. (2010). Design and application of single fluorophore dual-view imaging system containing both the objective- and prism-type TIRF. *Proc. SPIE* **7571:**757107–757108.

Zhang, X., Tung, C. S., Sowa, G. Z., Hatmal, M. M., Haworth, I. S., and Qin, P. Z. (2012). Global structure of a three-way junction in a ϕ29 packaging RNA dimer determined using site-directed spin labeling. *J. Am. Chem. Soc.* **134**(5)**:**2644–2652.

Zhang, X., and Wigley, D. B. (2008). The 'glutamate switch' provides a link between ATPase activity and ligand binding in AAA+ proteins. *Nat. Struct. Mol. Biol.* **15:**1223–1227.

Zhao, W., Morais, M. C., Anderson, D. L., Jardine, P. J., and Grimes, S. (2008). Role of the CCA bulge of prohead RNA of bacteriophage o29 in DNA packaging. *J. Mol. Biol.* **383:**520–528.

Zhou, J., Shu, Y., Guo, P., Smith, D., and Rossi, J. (2011). Dual functional RNA nanoparticles containing ϕ29 motor pRNA and anti-gp120 aptamer for cell-type specific delivery and HIV-1 Inhibition. *Methods* **54:**284–294.

INDEX

Note: Page numbers followed by "*f*" indicate figures, and "*t*" indicate tables.

T

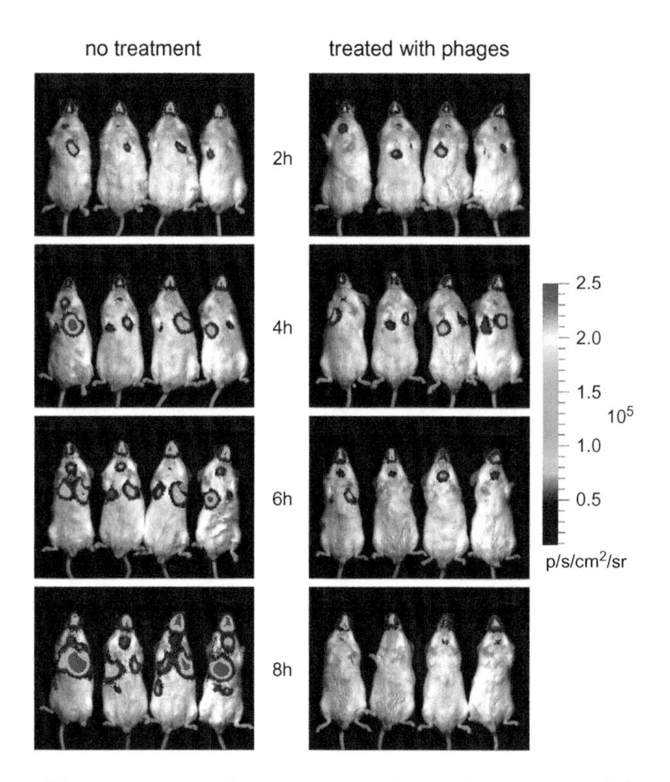

Figure 1, Emilie Saussereau and Laurent Debarbieux (See Page 133 of this Volume)

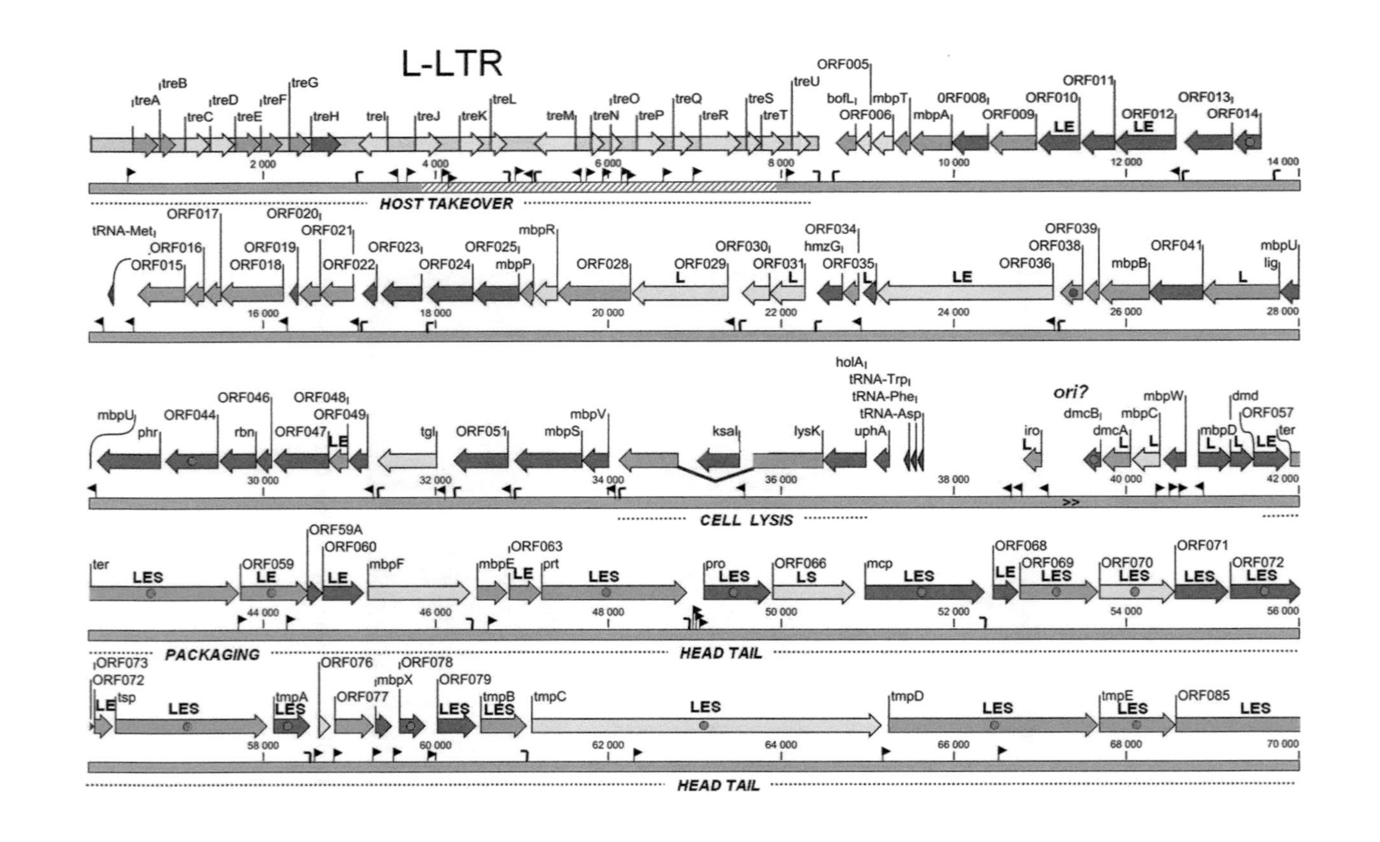

L-LTR
treA
treB
treC
treD
treE
treF
treG
treH
treI
treJ
treK
treL
treM
treN
treO
treP
treQ
treR
treS
treT
treU
ORF005
bofL
ORF006
mbpT
mbpA
ORF008
ORF009
ORF010
LE
ORF011
ORF012
LE
ORF013
ORF014
2 000
4 000
6 000
8 000
10 000
12 000
14 000
HOST TAKEOVER
tRNA-Met
ORF015
ORF016
ORF017
ORF018
ORF019
ORF020
ORF021
ORF022
ORF023
ORF024
ORF025
mbpP
mbpR
ORF028
ORF029
L
ORF030
ORF031
L
ORF034
hmzG
ORF035
L
LE
ORF036
ORF038
ORF039
mbpB
ORF041
L
mbpU
lig
16 000
18 000
20 000
22 000
24 000
26 000
28 000
mbpU
phr
ORF044
rbn
ORF046
ORF047
ORF048
ORF049
LE
tgl
ORF051
mbpS
mbpV
ksal
lysK
holA
tRNA-Trp
tRNA-Phe
tRNA-Asp
uphA
ori?
iro
L
dmcB
dmcA
L
mbpC
L
mbpW
mbpD
L
L
dmd
ORF057
LE
ter
30 000
32 000
34 000
36 000
38 000
40 000
42 000
>>
CELL LYSIS
ter
LES
ORF059
LE
ORF59A
ORF060
LE
mbpF
mbpE
LE
ORF063
prt
LES
pro
LES
ORF066
LS
mcp
LES
ORF068
LE
ORF069
LES
ORF070
LES
ORF071
LES
ORF072
LES
44 000
46 000
48 000
50 000
52 000
54 000
56 000
PACKAGING
HEAD TAIL
ORF073
ORF072
LE
tsp
LES
tmpA
LES
ORF076
ORF077
mbpX
ORF078
ORF079
LES
tmpB
LES
tmpC
LES
tmpD
LES
tmpE
LES
ORF085
LES
58 000
60 000
62 000
64 000
66 000
68 000
70 000
HEAD TAIL

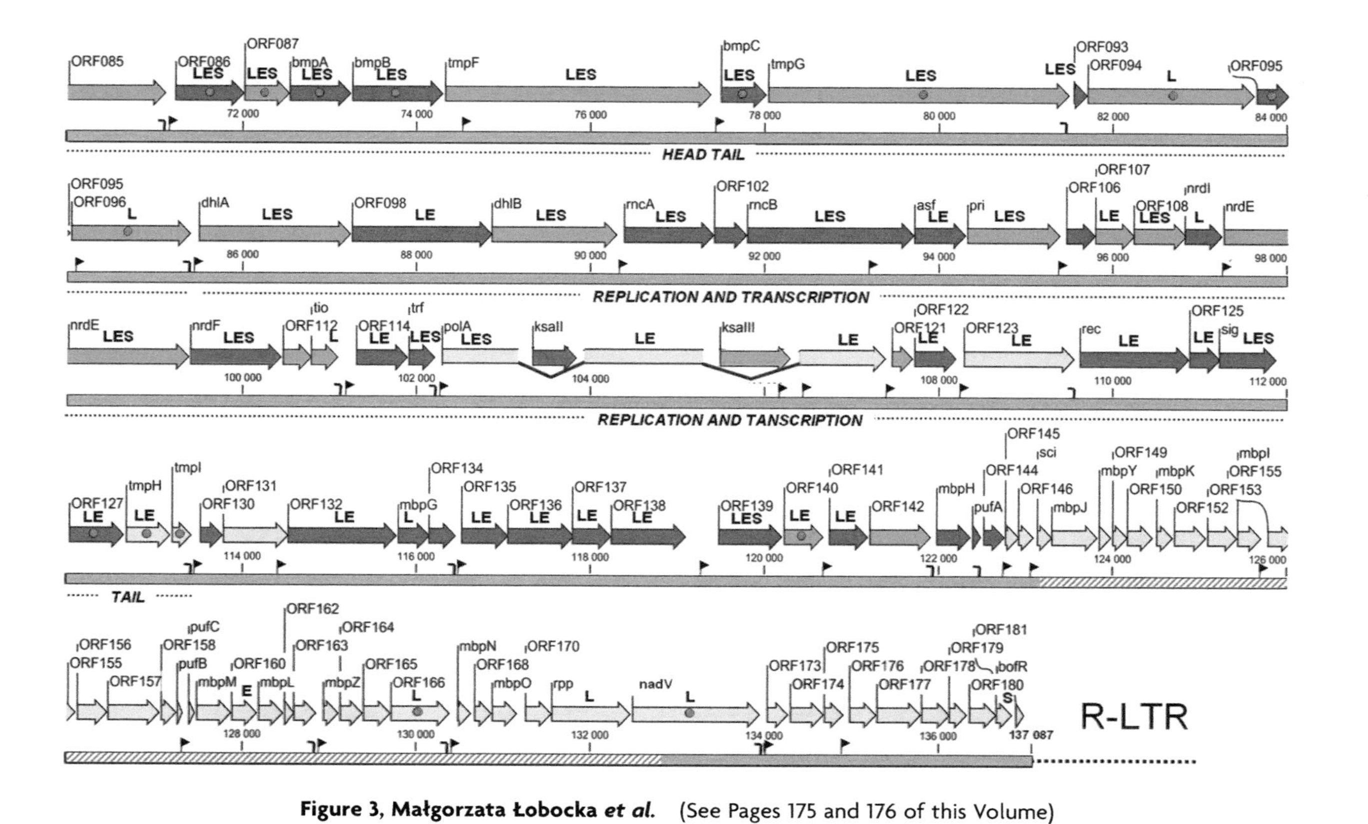

Figure 3, Małgorzata Łobocka *et al.* (See Pages 175 and 176 of this Volume)

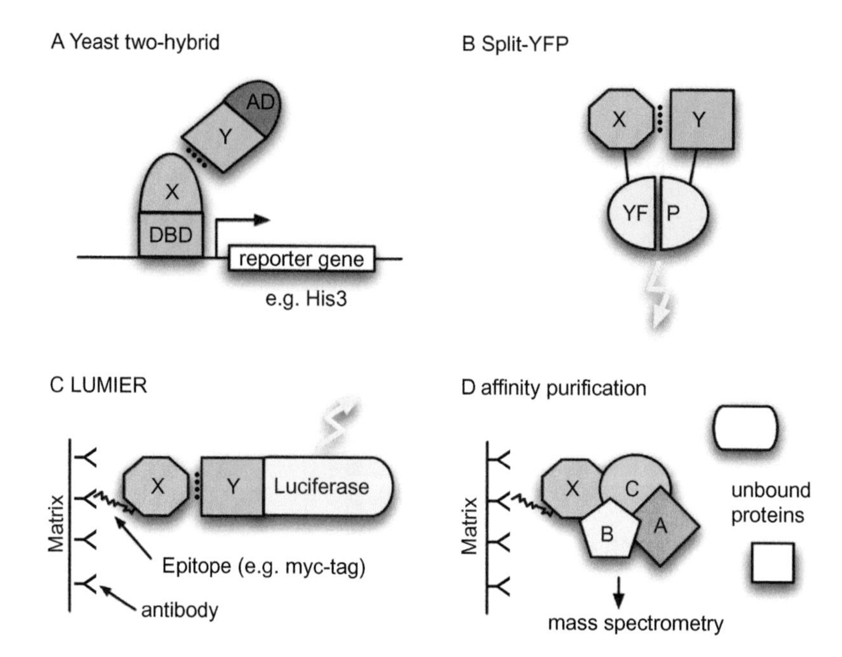

Figure 1, Roman Häuser *et al.* (See Page 223 of this Volume)

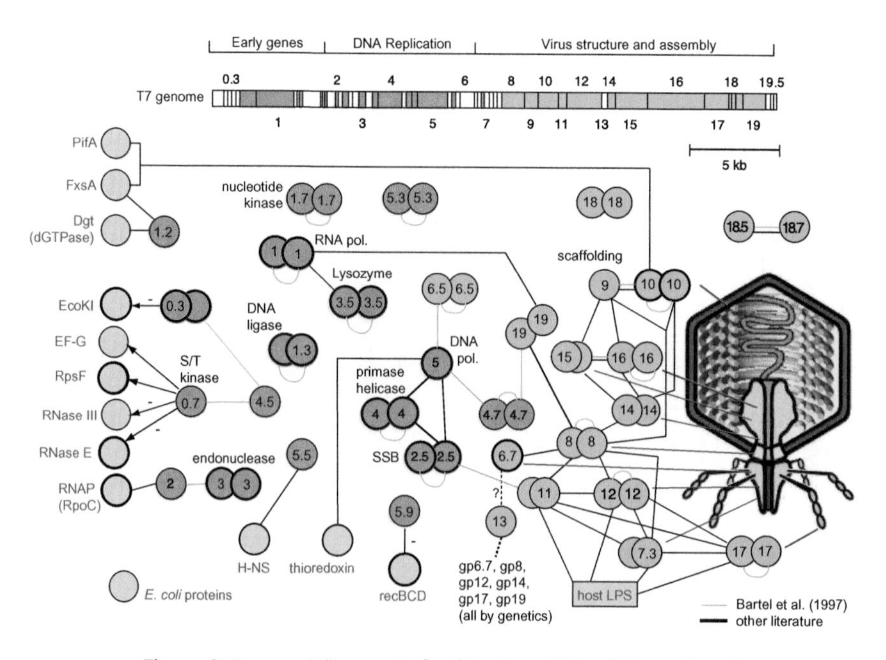

Figure 2, Roman Häuser *et al.* (See Page 228 of this Volume)

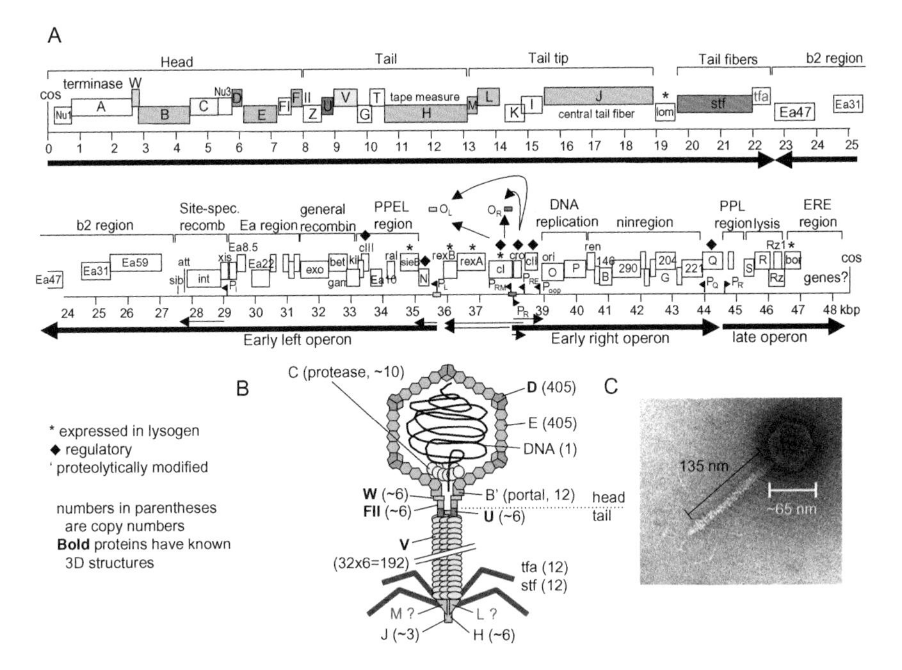

Figure 3, Roman Häuser *et al.* (See Page 237 of this Volume)

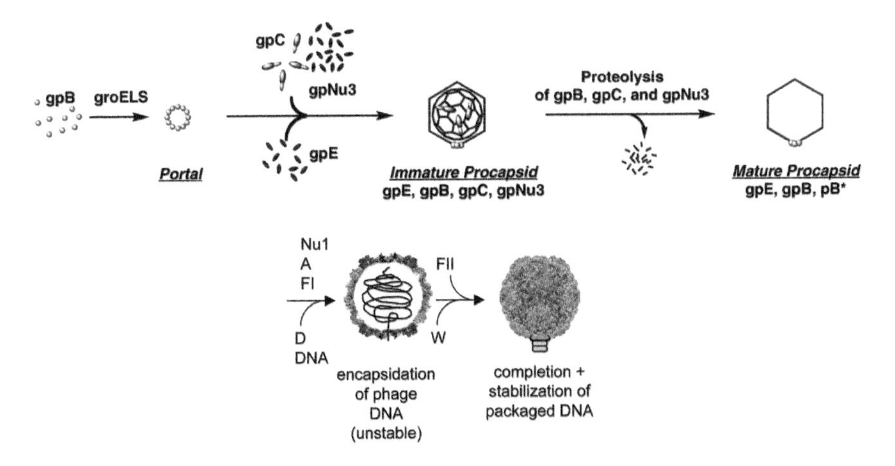

Figure 4, Roman Häuser *et al.* (See Page 240 of this Volume)

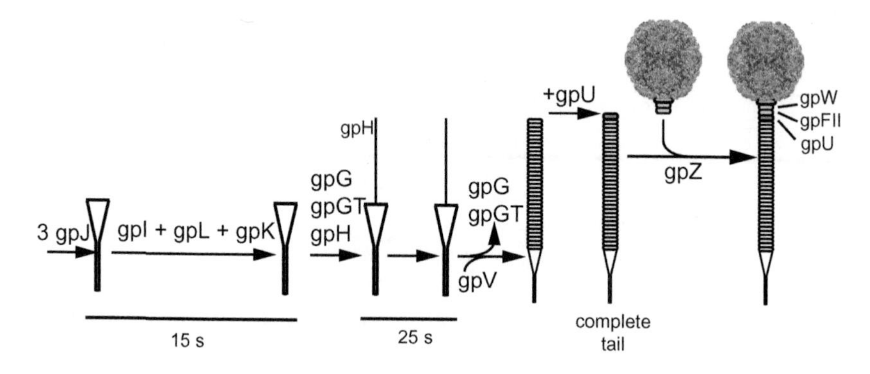

Figure 5, Roman Häuser *et al.* (See Page 241 of this Volume)

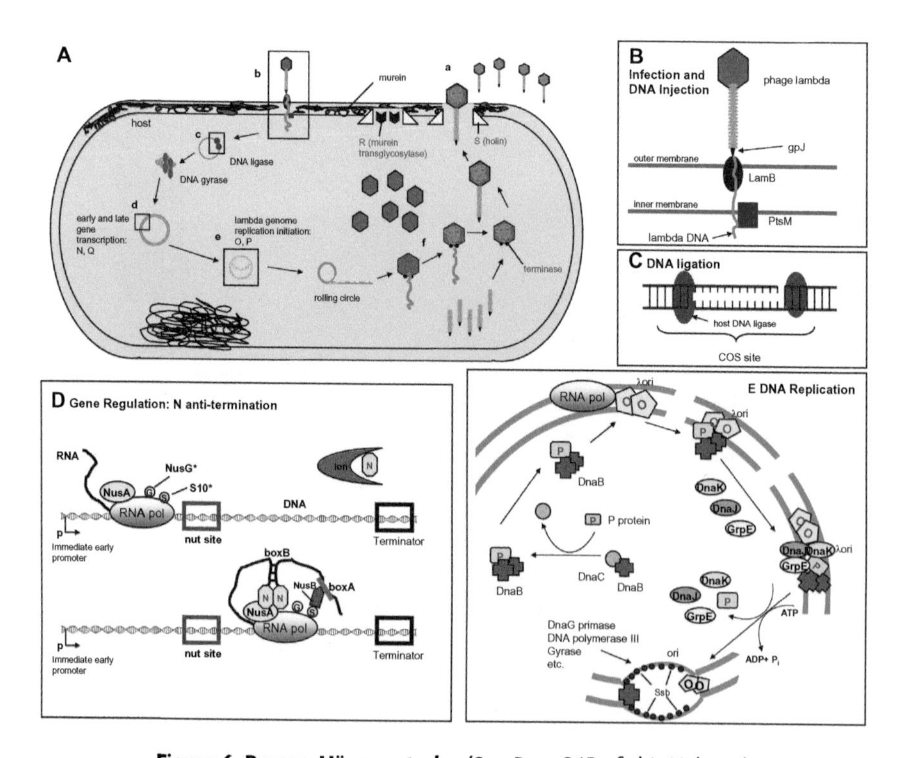

Figure 6, Roman Häuser *et al.* (See Page 245 of this Volume)

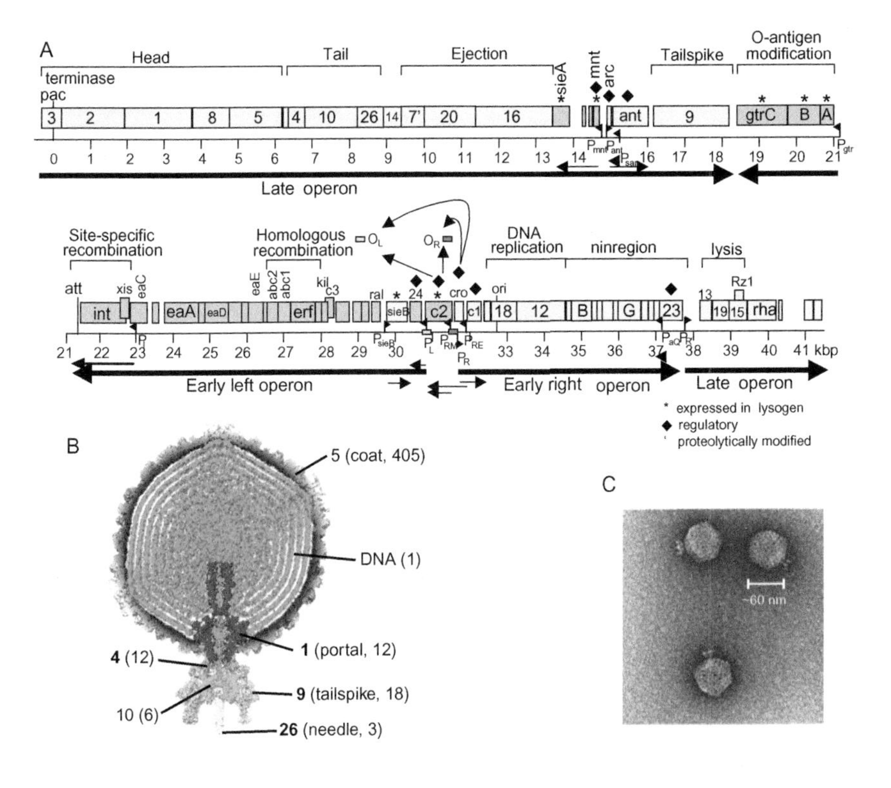

Figure 7, Roman Häuser *et al.* (See Page 248 of this Volume)

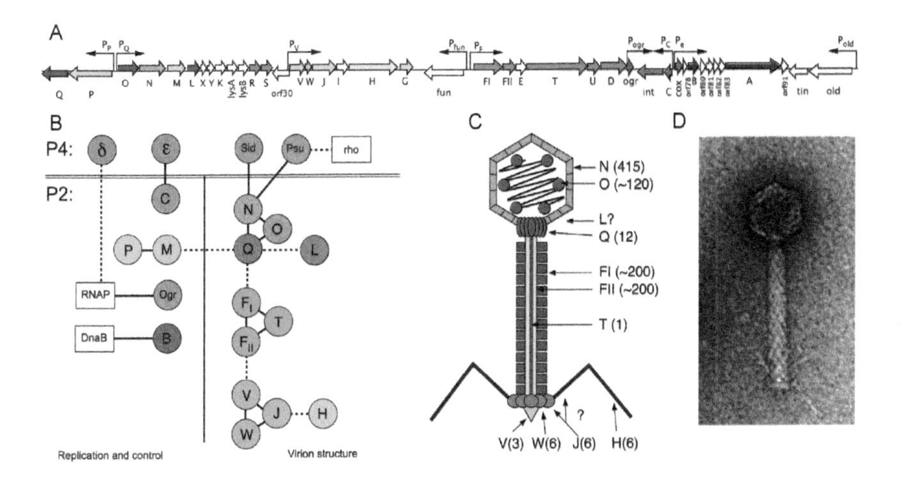

Figure 8, Roman Häuser *et al.* (See Page 257 of this Volume)

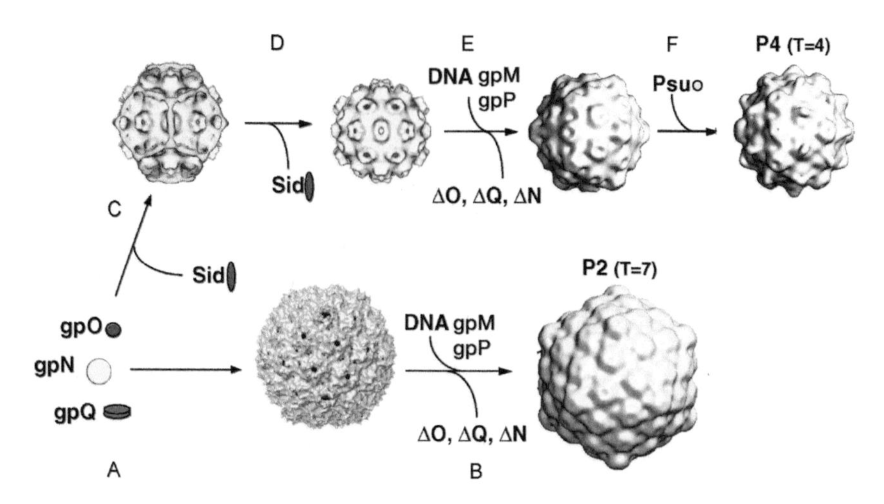

Figure 9, Roman Häuser *et al.* (See Page 261 of this Volume)

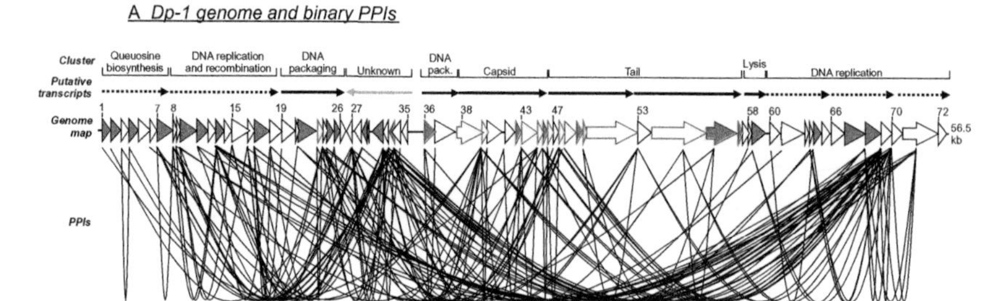

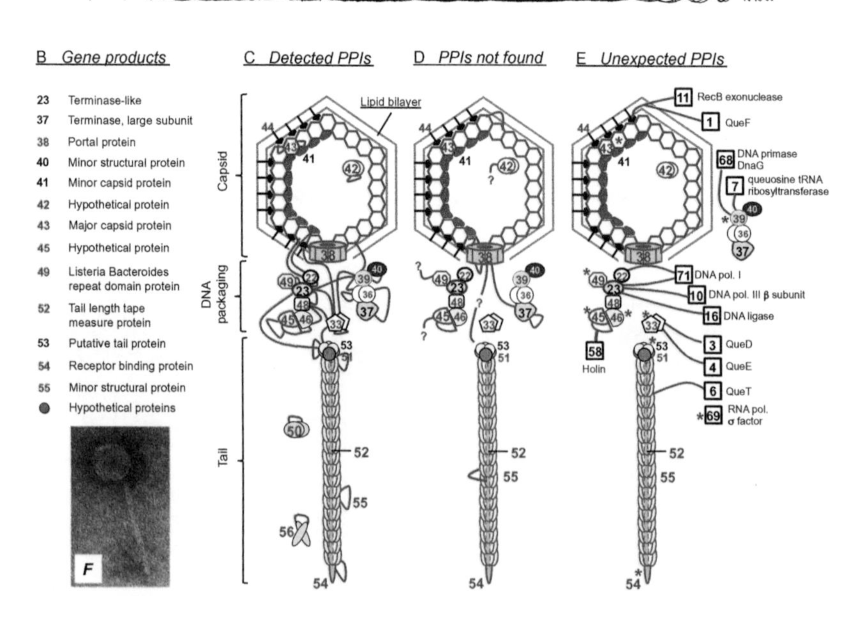

Figure 10, Roman Häuser *et al.* (See Page 264 of this Volume)

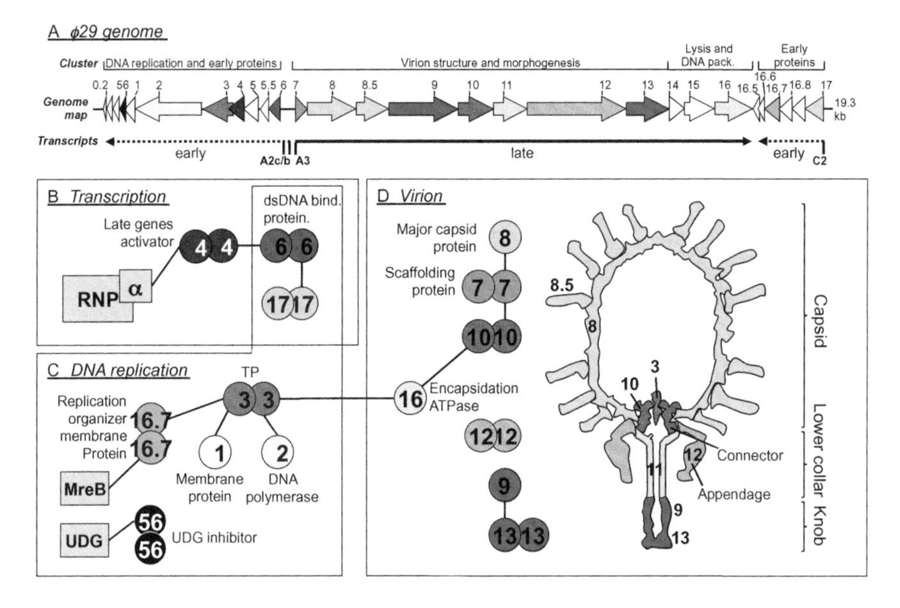

Figure 11, Roman Häuser *et al.* (See Page 267 of this Volume)

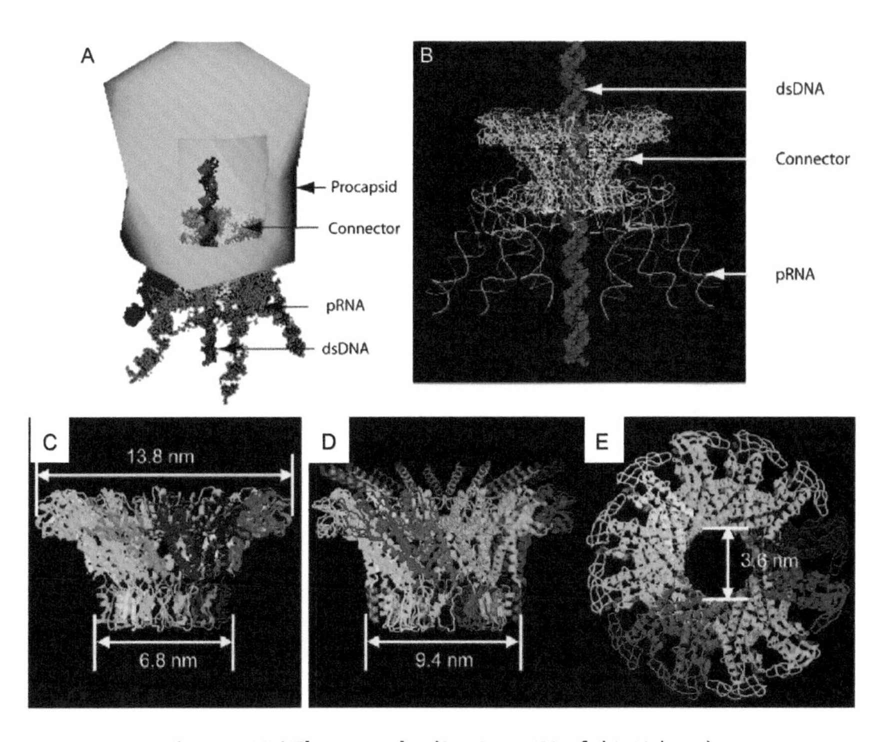

Figure 1, Hui Zhang *et al.* (See Page 419 of this Volume)

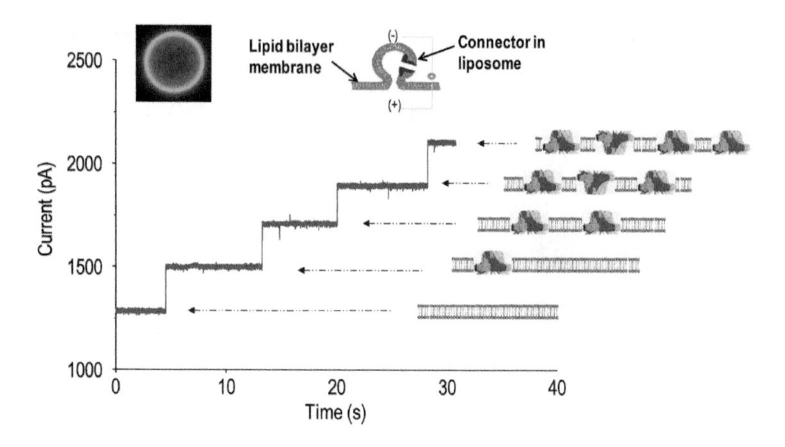

Figure 2, Hui Zhang *et al.* (See Page 420 of this Volume)

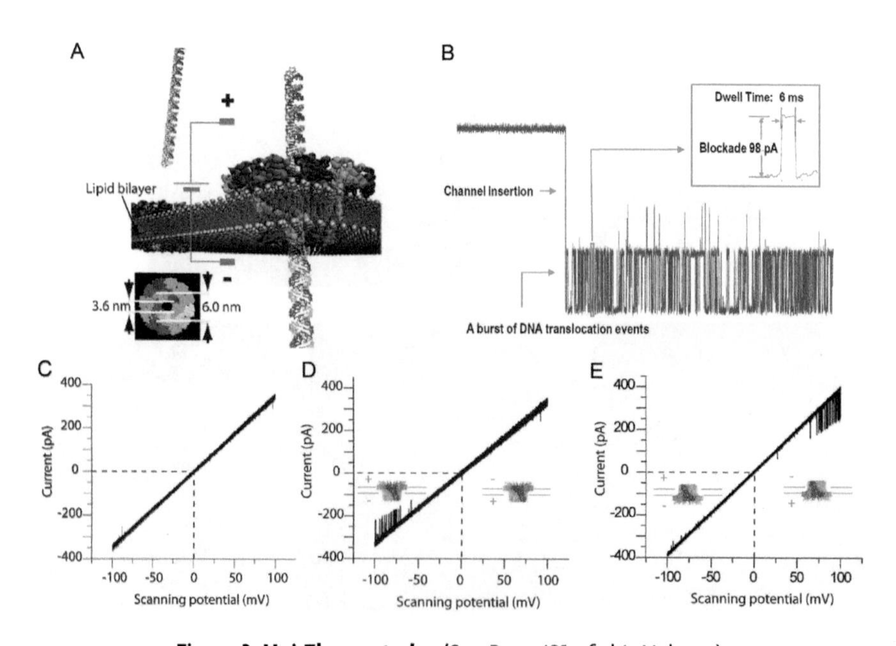

Figure 3, Hui Zhang *et al.* (See Page 421 of this Volume)

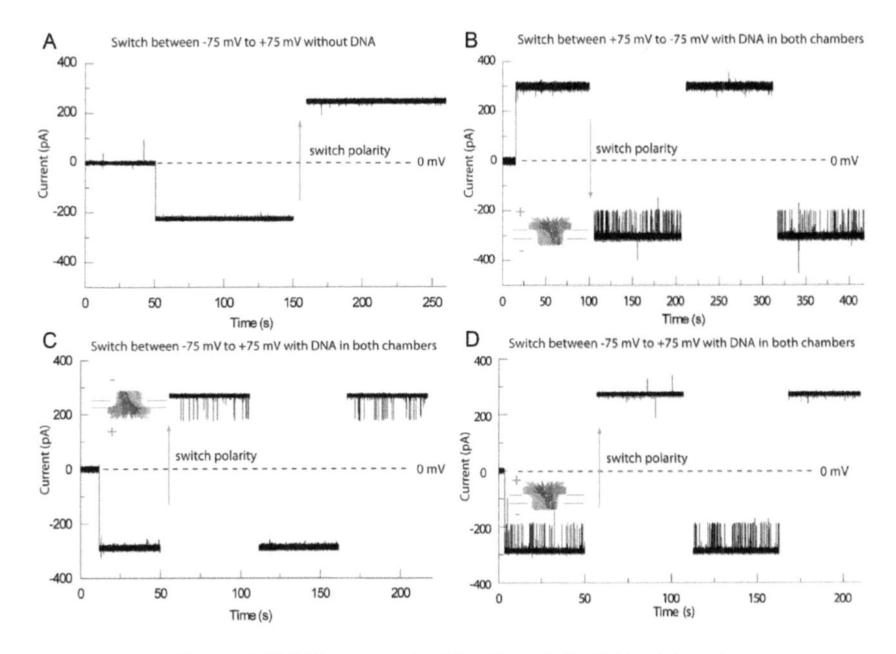

Figure 4, Hui Zhang *et al.* (See Page 422 of this Volume)

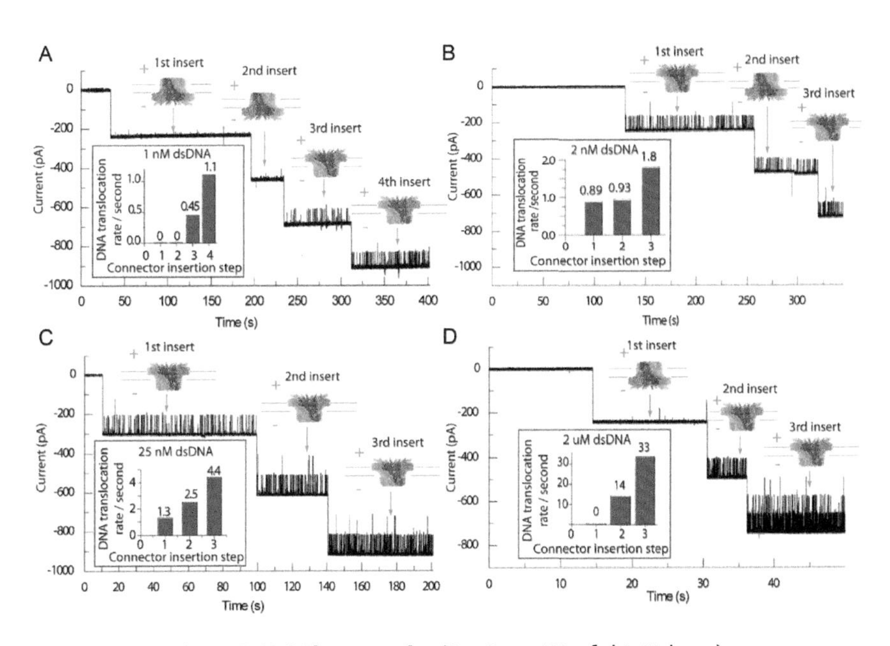

Figure 5, Hui Zhang *et al.* (See Page 423 of this Volume)

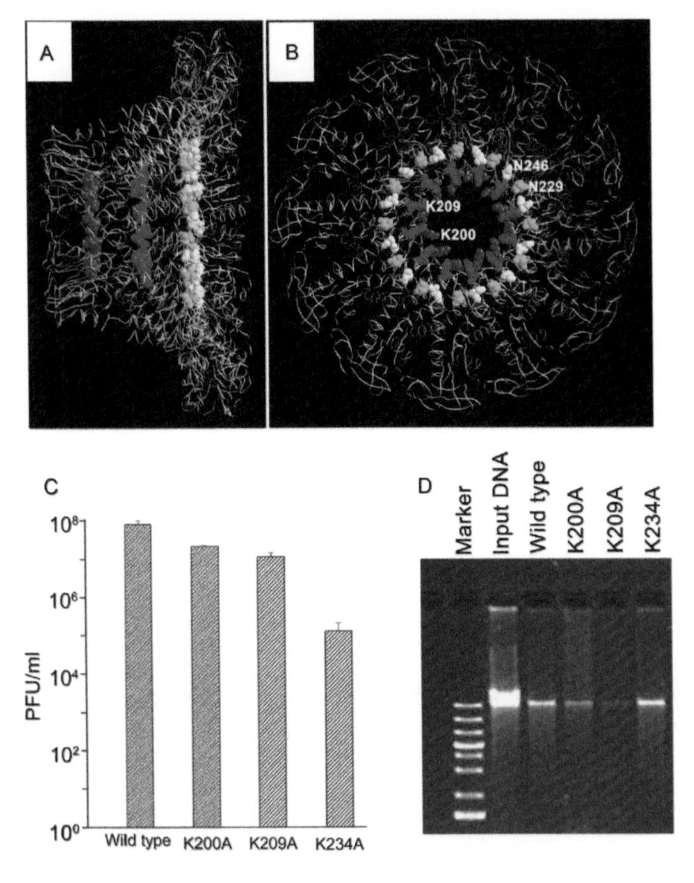

Figure 6, Hui Zhang *et al.* (See Page 424 of this Volume)

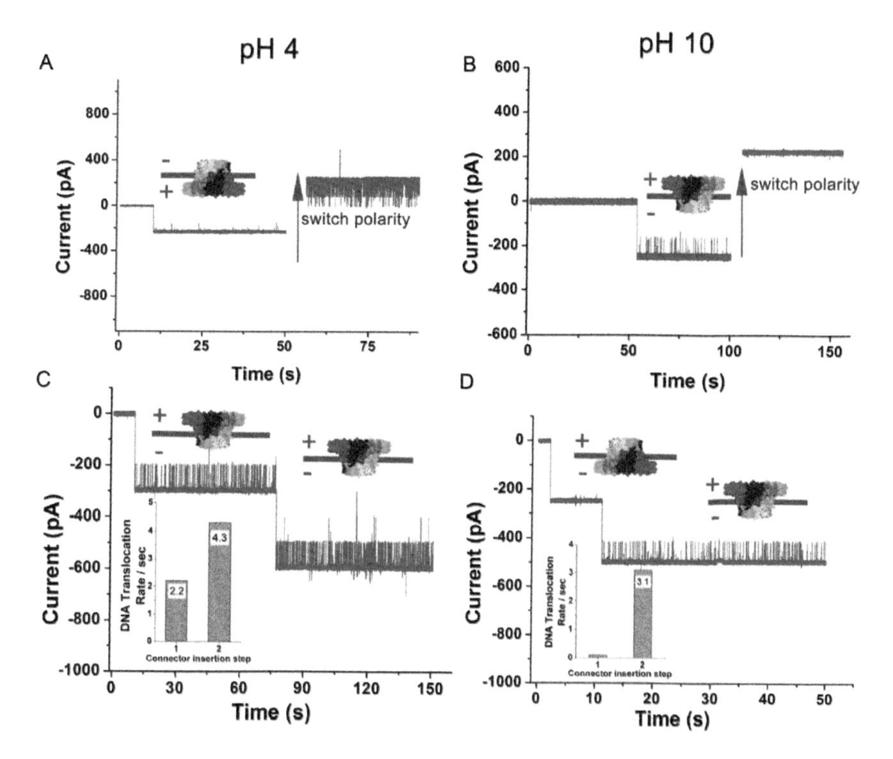

Figure 7, Hui Zhang *et al.* (See Page 425 of this Volume)

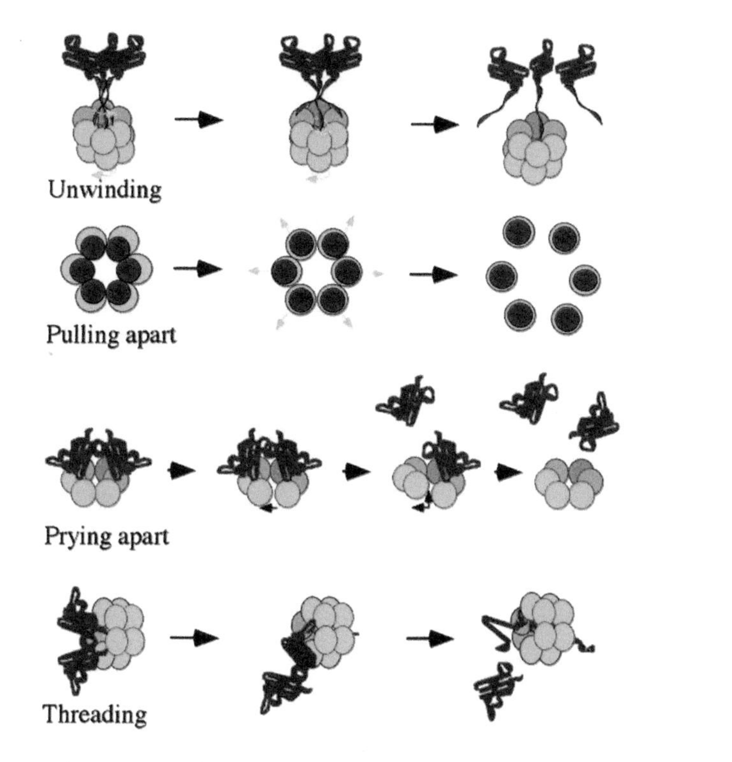

Figure 8, Hui Zhang *et al.* (See Page 427 of this Volume)

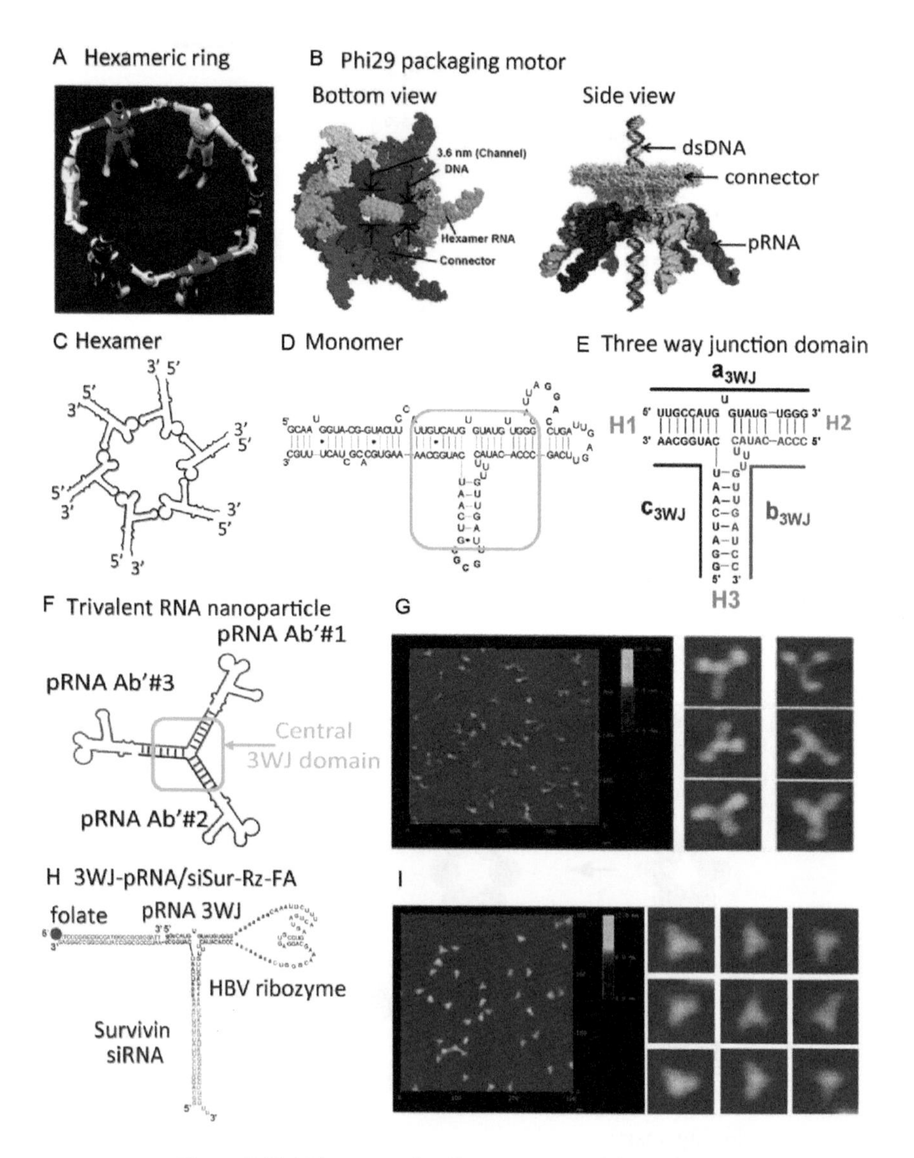

Figure 9, Hui Zhang *et al.* (See Page 428 of this Volume)

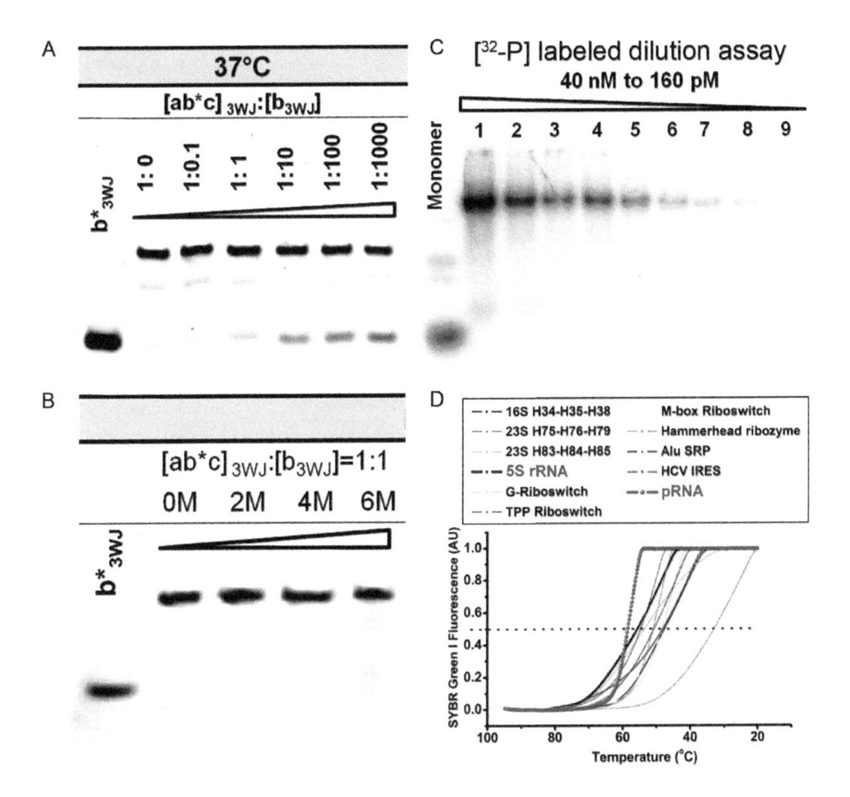

Figure 10, Hui Zhang *et al.* (See Page 430 of this Volume)

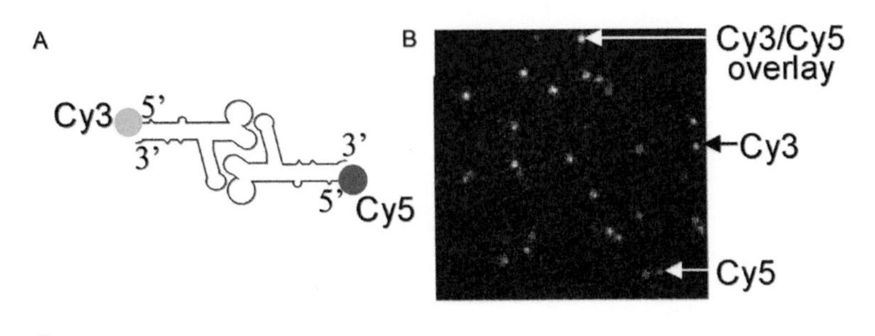

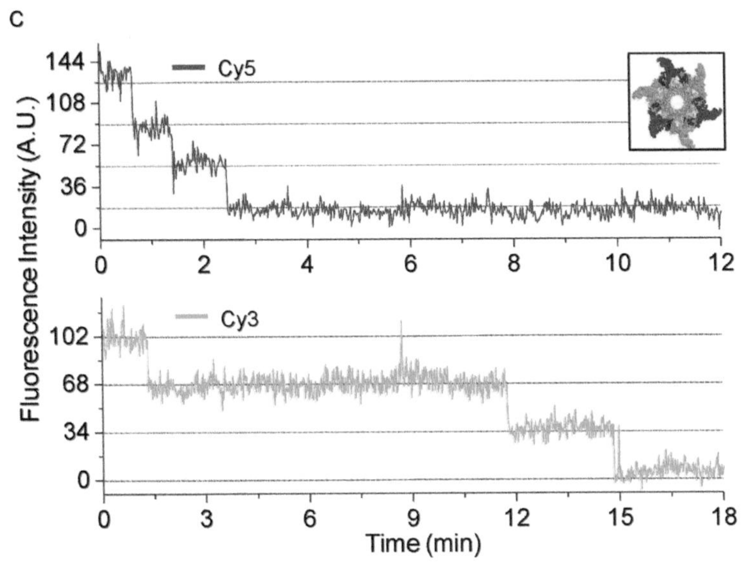

Figure 11, Hui Zhang *et al.* (See Page 432 of this Volume)

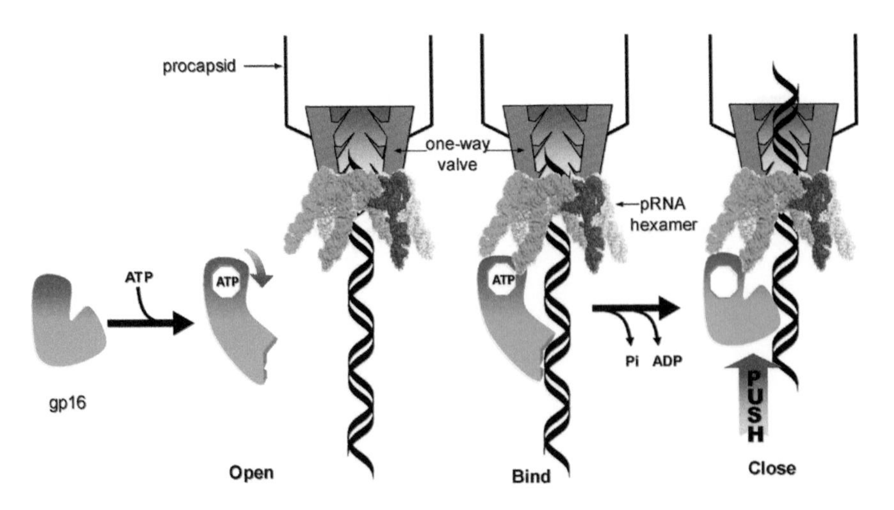

Figure 13, Hui Zhang *et al.* (See Page 442 of this Volume)

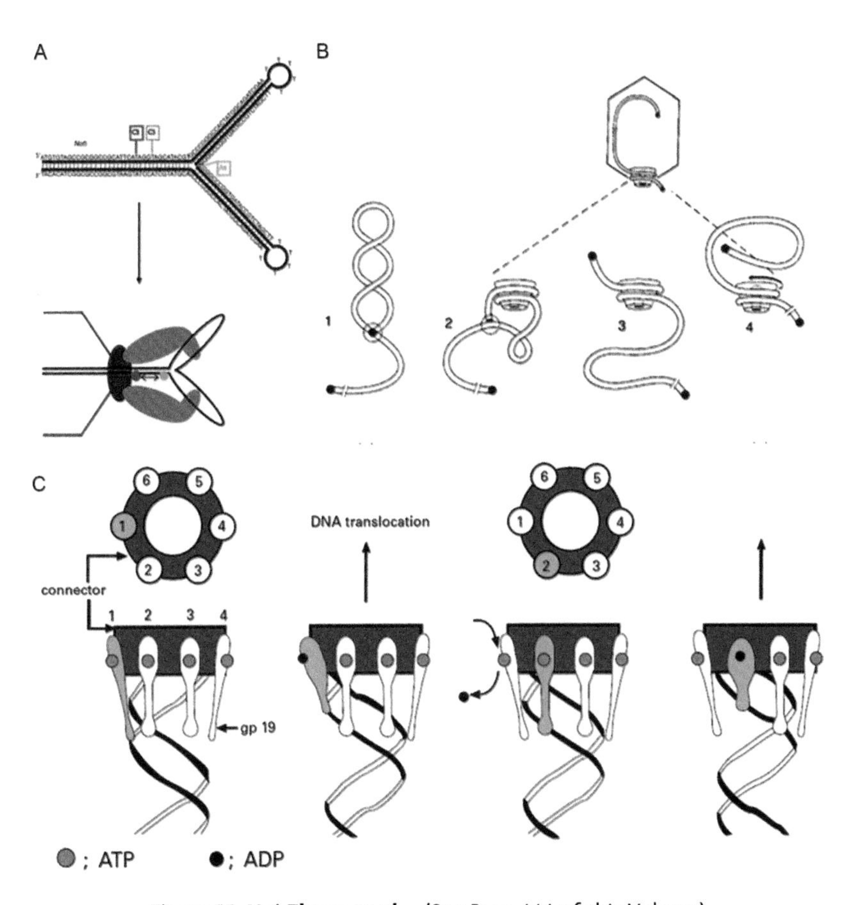

Figure 14, Hui Zhang *et al.* (See Page 444 of this Volume)

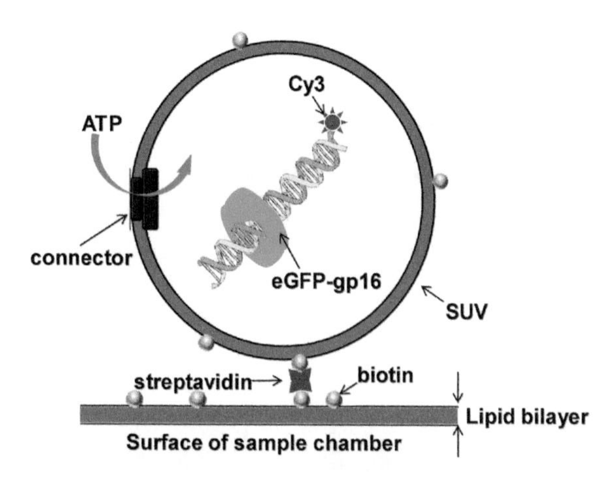

Figure 16, Hui Zhang *et al.* (See Page 450 of this Volume)

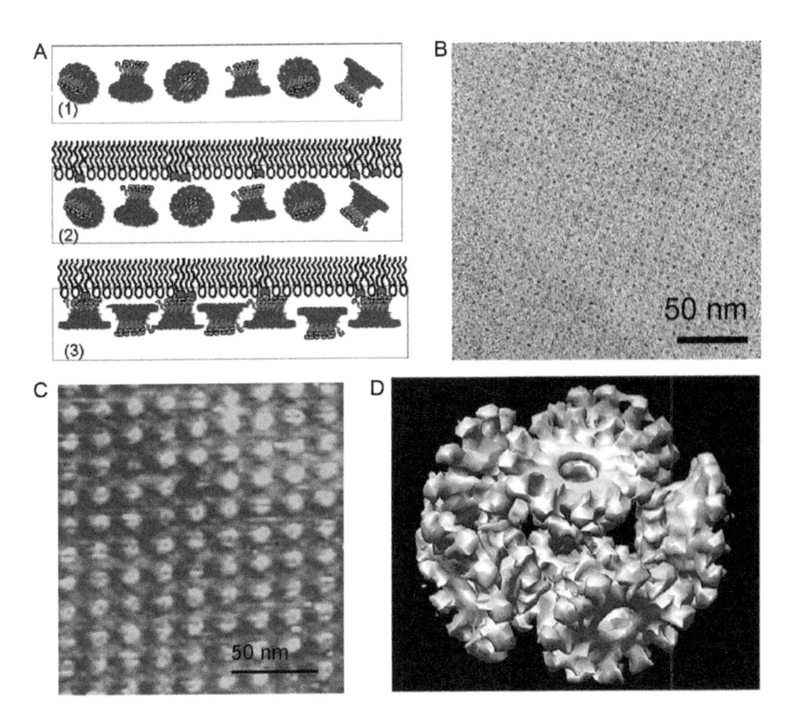

Figure 17, Hui Zhang *et al.* (See Page 452 of this Volume)

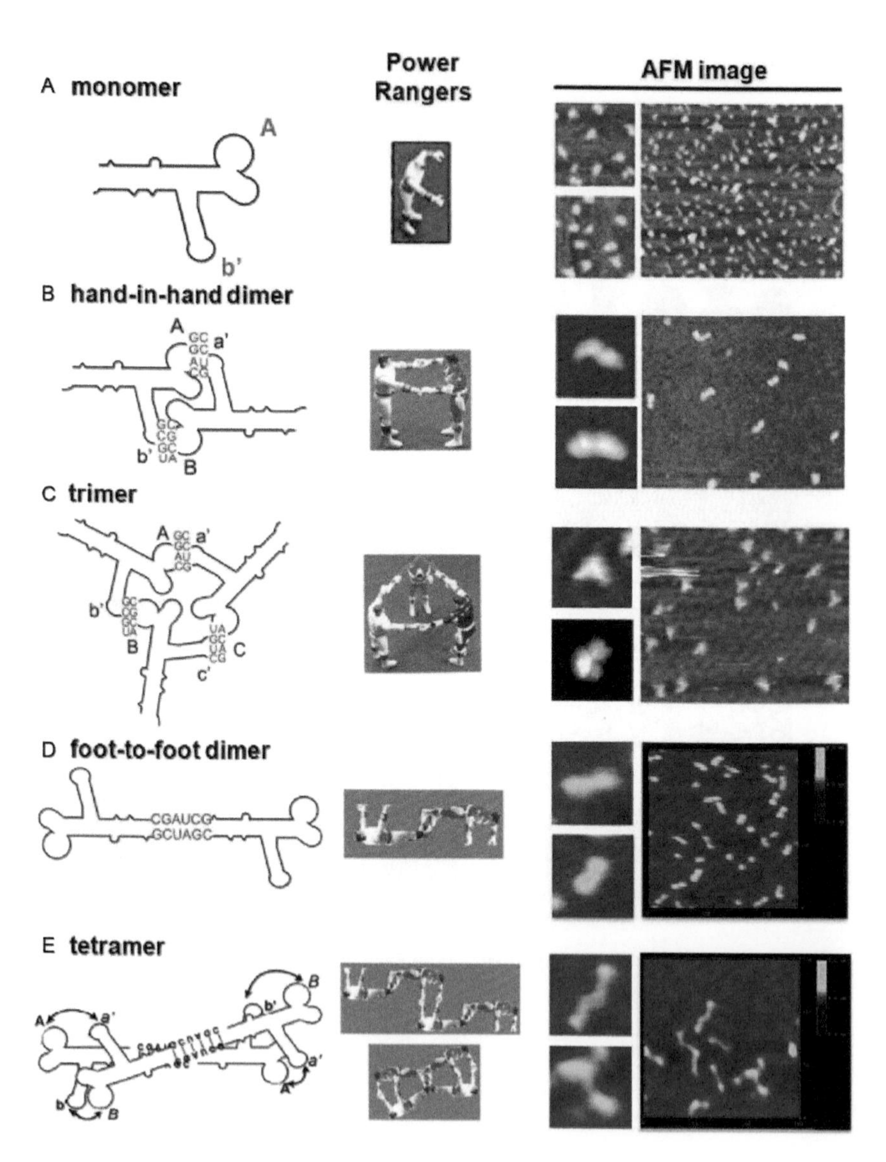

Figure 18, Hui Zhang *et al.* (See Page 453 of this Volume)

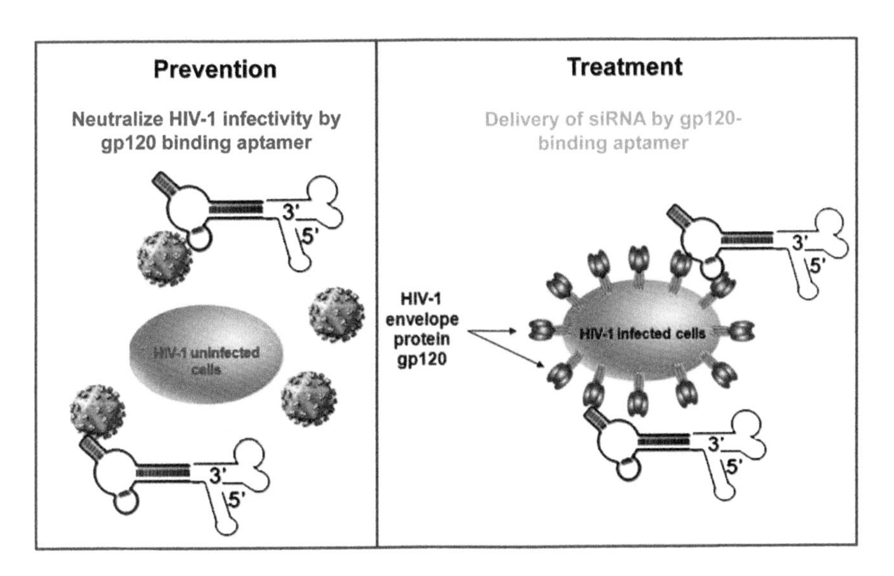

Figure 19, Hui Zhang *et al.* (See Page 454 of this Volume)

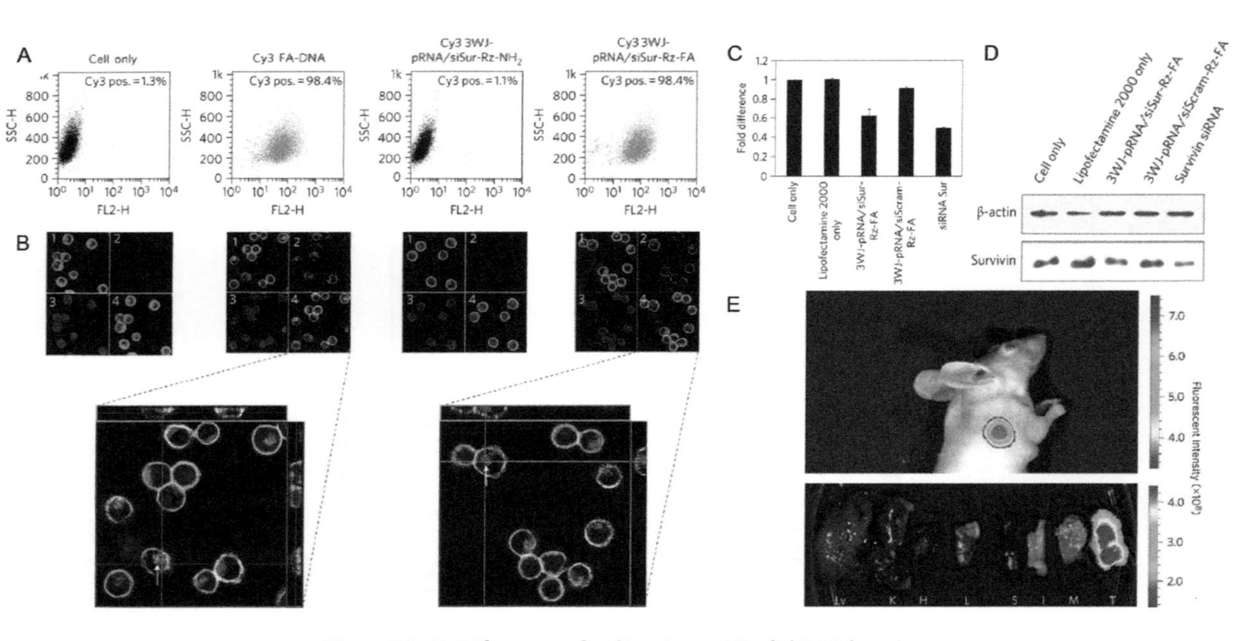

Figure 20, Hui Zhang *et al.* (See Page 455 of this Volume)

Printed and bound by CPI Group (UK) Ltd, Croydon, CR0 4YY
08/05/2025
01864955-0004